# Nanofluids for Thermal Applications

## Synthesis, Stability, and Properties

*Editors:*

**Nikhil S. Mane**
Ajeenkya DY Patil University, Pune, India

**Zafar Said**
Department of Mechanical and Aerospace Engineering
College of Engineering, United Arab Emirates University
United Arab Emirates

**Sayantan Mukherjee**
Gandhi Academy of Technology and Engineering
Berhampur, Odisha, India

**Naser Ali**
Research Scientist
Kuwait Institute for Scientific Research
Kuwait

CRC Press
Taylor & Francis Group
Boca Raton London New York

CRC Press is an imprint of the
Taylor & Francis Group, an **informa** business
A SCIENCE PUBLISHERS BOOK

First edition published 2027
by CRC Press
2385 NW Executive Center Drive, Suite 320, Boca Raton FL 33431

and by CRC Press
4 Park Square, Milton Park, Abingdon, Oxon, OX14 4RN



*CRC Press is an imprint of Taylor & Francis Group, LLC*



*Library of Congress Cataloging-in-Publication Data (applied for)*

ISBN: 978-1-041-11887-9 (hbk)
ISBN: 978-1-041-16710-5 (pbk)
ISBN: 978-1-003-68600-2 (ebk)

DOI: 10.1201/9781003686002

Typeset in Times New Roman
by Prime Publishing Services

# Preface

In recent years, nanofluids have been a focal point of interest among thermal system engineers. This growing attention is driven by rapid developments in advanced thermal systems. The recent trend towards the miniaturization of devices, advancements in battery technology, and an expansion in modern energy systems have increased the importance of nanofluids. Nanofluids are a newly developing class of working fluids that have been shown to have enhanced thermophysical properties. They also exhibit complex heat transfer characteristics. As a result, nanofluids have found applications in thermal systems, tribology, medicine, and energy technologies. This sustained growth in nanofluid research highlights that nanofluid science is a multidimensional field that requires a well-structured and well-written body of literature.

The current edited volume on nanofluid research, Nanofluids for Thermal Applications: Synthesis, Stability, and Properties, attempts to provides a detailed overview on the fundamental principles, synthesis methods, stability issues, and thermophysical properties of nanofluids. This work is intended to be a useful resource to researchers, graduate students, and engineers working in heat transfer and related fields. By bringing together contributions from well-established researchers and scholars in this field, this work attempts to bridge the gap between theoretical research and practical applications.

Collectively, the chapters provide a comprehensive treatment of nanofluid synthesis methods, modeling/prediction of thermophysical properties, and heat transfer characteristics—in all of which aspects system design reliability plays an important role. Emphasis is given to hybrid/multifunctional nanofluids wherein nanofluid performance enhancement strategies using combinations of nanoparticles are discussed. In addition to the fundamental aspects of nanofluid science and technology, application-oriented chapters on nanofluid applications in electronic device cooling, renewable energy technologies, heat exchangers, automotive/aerospace technologies, micro/nanoscale heat transfer, etc., are included. The inclusion of experimental investigations, analytical models, data-driven methods, and multidisciplinary approaches ensures a comprehensive treatment of nanofluid science and technology.

The editors would like to thank all contributing authors and reviewers for their valuable contributions. They would like to thank the CRC Press/Taylor & Francis editorial team for their support throughout the preparation of this volume. It is hoped that, in addition to serving as a useful reference for students, researchers, and engineers, this volume will also stimulate new investigations and innovations in nanofluid science and technology.

# Contents

# List of Contributors

**Aditya Shrivastav**
School of Engineering, Ajeenkya DY Patil University, Charholi Bk. via Lohegaon, Pune-412105, Maharashtra, India.

**Ali Alsayegh**
Department of Design, Manufacturing, and Engineering Management, University of Strathclyde, Glasgow, United Kingdom.

**Alok Kumar Das**
Mechanical Engineering Department, Indian Institute of Technology (Indian School of Mines), Dhanbad, 826004.

**Amit Rai Dixit**
Mechanical Engineering Department, Indian Institute of Technology (Indian School of Mines), Dhanbad, 826004.

**Amrit Shende**
Department of Mechanical Engineering, Chhattisgarh Swami Vivekanand Technical University Bhilai, Chhattisgarh 491107 India.

**Amrita Tripure**
Department of Mechanical Engineering, Chhattisgarh Swami Vivekanand Technical University Bhilai, Chhattisgarh 491107 India.

**Annisa Fatharani**
School of Cancer and Pharmaceutical Sciences, King's College London, London, United Kingdom.

**Arun Kumar Tiwari**
Mechanical Engineering Department, Institute of Engineering & Technology, Dr. A.P.J. Abdul Kalam Technical University, Lucknow, 226021, India.

**Ashish Desai**
Department of Automation and Robotics, Sharad Institute of Technology, College of Engineering, Yadrav, Ichalkaranji

**Avadhut Jadhav**
Department of Mechanical Engineering, Bharati Vidyapeeth's College of Engineering, Kolhapur.

**Avesahemad Husainy**
Department of Automation and Robotics, Sharad Institute of Technology, College of Engineering, Yadrav, Ichalkaranji

**Basant K. Jha**
Department of Mathematics, Ahmadu Bello University, Zaria, Kaduna, Nigeria.

**Bernardo Buonomo**
Dipartimento di Ingegneria, Università degli Studi della Campania "Luigi Vanvitelli", Via Roma 29, 81031 Aversa, Caserta, Italy.

**D.B. Roy**
School of Engineering, Ajeenkya DY Patil University, Charholi, Bk. Via Lohegaon, Pune, Maharashtra-412105, India.

**Gabriel Samaila**
Department of Mathematics, Air Force Institute of Technology, Kaduna, Nigeria.

**Ganesh Moorthy I**
School of Chemical Engineering, Vellore Institute of Technology, Tamilnadu, Vellore 632014, India.

**Hamid-Reza Bahrami**
Assistant Professor, Department of Mechanical Engineering, Qom University of Technology, Qom 37181 46645, Iran. P.O. Box 37195-1519

**Hawraa Alsayegh**
School of Cancer and Pharmaceutical Sciences, King's College London, London, United Kingdom.

**Ihsan Ur Rahman**
Dipartimento di Ingegneria, Università degli Studi della Campania "Luigi Vanvitelli", Via Roma 29, 81031 Aversa, Caserta, Italy.

**Kaajal Anil Rai**
School of Chemical Engineering, Vellore Institute of Technology, Tamilnadu, Vellore 632014, India.

**Manish Kumar**
Department of Mechanical Engineering, Bhilai Institute of Technology, Durg, Chhattisgarh 491001 India.

**Mohamed Aboughaly**
University of Toronto, Canada.

**Oronzio Manca**
Dipartimento di Ingegneria, Università degli Studi della Campania "Luigi Vanvitelli", Via Roma 29, 81031 Aversa, Caserta, Italy.

**Rishi Kant**
School of Natural Sciences, GNA University, Phagwara-144401, Punjab, India.

**Rubania Arshad**
Department of Environmental Sciences, Faculty of Life Sciences, The Women University Multan, Multan, 66000, Pakistan.

**Sahil Sharma**
Department of Medical Laboratory Sciences, School of Allied and Healthcare Sciences, GNA University, Phagwara-144401, Punjab, India

**Saima Nasreen**
Department of Environmental Sciences, Faculty of Life Sciences, The Women University Multan, Multan, 66000, Pakistan.

**Sanatan Ramnath Iyer**
School of Chemical Engineering, Vellore Institute of Technology, Tamilnadu, Vellore 632014, India.

**Sanjay Mane**
Department of Mechanical Engineering, MIT College of Railway Engineering and Research, Barshi-413401, Solapur, Maharashtra, India.

**Santosh Kumar Mishra**
Department of Mechanical Engineering, Bhilai Institute of Technology, Durg, Chhattisgarh 491001 India.

**Santosh Kumar Singh**
Mechanical Engineering Department, Indian Institute of Technology (Indian School of Mines), Dhanbad, 826004.

**Sayli D. Hulwan**
Department of Electronics and Computer Engineering, MIT College of Railway Engineering and Research, Barshi-413401, Solapur, Maharashtra, India.

**Sergio Nardini**
Dipartimento di Ingegneria, Università degli Studi della Campania "Luigi Vanvitelli", Via Roma 29, 81031 Aversa, Caserta, Italy.

**Sonali Biswas**
School of Bio-Sciences and Technology, Vellore Institute of Technology, Tamilnadu, Vellore 632014, India.

**Sujit Kumbhar**

Department of Automation and Robotics, Sharad Institute of Technology, College of Engineering, Yadrav, Ichalkaranji.

**Supriya A. Joshi**

Department of Chemistry, MIT College of Railway Engineering and Research, Barshi-413401, Solapur, Maharashtra, India.

**Susanta Das**

School of Engineering, Ajeenkya DY Patil University, Charholi Bk. via Lohegaon, Pune- 412105, Maharashtra, India. or School of Science, Ajeenkya D Y Patil University, Charholi Bk. via Lohegaon, Pune-412105, Maharashtra, India.

**Vishal Patil**

Department of Mechanical Engineering, Sharad Institute of Technology, College of Engineering, Yadrav, Ichalkaranji.

**Y. Sonwane**

School of Engineering, Ajeenkya DY Patil University, Charholi, Bk. Via Lohegaon, Pune, Maharashtra-412105, India.

# 1 Synthesis Techniques and Surface Functionalization of Nanofluids

Sahil Sharma[1] and Rishi Kant[2*]

The improved thermal properties of engineered nanofluids (created by dispersing nanometric-sized particles into more traditional base fluids) have gained recent attention in thermal management and energy systems, making them a strong contender in these fields. Surface modification and synthesis of nanoparticles are the key factors that influence the performance and stability of nanofluids, and therefore, the two are critical in the process of designing an effective and successful nanofluid. This chapter provides a complete description of the main synthesis methods, including one-step and two-step methods. One-step synthesis synthesizes nanoparticles and optimizes dispersion through a common procedure, such that synthesis and dispersion cannot be separated, and consequently oxidation and agglomeration are constrained. Instead of that, two-step procedures involve preparing the nanoparticles separately and dispersing them into a base fluid that is easier to handle and scale. The significant methods are the ultrasonication, ball milling, and the chemical methods because they are important and effective. The nanoparticles do, however, clump naturally, thus settling and reducing their thermal effectiveness. To this purpose, the significance of surface functionalization is mentioned. Surfactants, coupling agents, and polymer coatings can also be used to enhance nanoparticle dispersion stability and surface properties to fit desired needs. As well, the chapter presents the effect of the synthesis methods and functionalization on the main thermophysical characteristics such as thermal conductivity, viscosity, and long-term thermal stability. Finally, the issues associated with the maintenance of homogeneity, cost-effectiveness, and industrial feasibility are discussed, and we look forward to the future promises of hybrid nanofluids and green methods of synthesis.

## 1. Introduction

To overcome the inherently low thermal conductivity of conventional base fluids such as water, ethylene glycol, and oils, researchers introduced nanofluids—engineered colloidal suspensions of solid nanoparticles (typically 1–100 nm in size)

[1] Department of Medical Laboratory Sciences, School of Allied and Healthcare Sciences, GNA University, Phagwara-144401, Punjab, India.
[2] School of Natural Sciences, GNA University, Phagwara-144401, Punjab, India.
* Corresponding author: rishi.kant@gnauniversity.edu.in

DOI: 10.1201/9781003686002-1

dispersed within these base fluids (Lee *et al.*, 2011). Nanofluids are conceptually different from micro- and millimeter-scale slurries and are intended to provide enhanced thermal conductivity. Nanofluids also provide modified convectional transport (single and two-phase), controllable rheology and stability, and reduced erosion, clogging, and sedimentation of larger size particles (Saidur, Leong, and Mohammed, 2011; Eastman *et al.*, 2001; Yang and Liu, 2010; Keblinski *et al.*, 2002). Following Choi's (1995) findings of an anomalously large enhancement in the effective thermal conductivity of nano-scaffolds, such as copper-ethylene glycol nanofluids, extensive subsequent testing and analysis in this field have been conducted. Since then, improvements have been made to a material-matrix conduction problem, driven a particle-matrix interactions, Brownian-motion-induced micro-mixing, interfacial (Kapitza) resistance, and even potential structuring. Modern nanofluids research is now focused on transport mechanics, metrology, and stability (Yang and Liu, 2010; Keblinski *et al.*, 2002; Mikkola *et al.*, 2018).

In the context of heat-transfer practice, nanofluids are important because they may enhance the compactness of heat exchangers, increase critical heat flux, and reduce thermal bottlenecks in high-power density systems (Wang and Mujumdar, 2007). Research has shown consistent growth in thermal conductivity and, in some cases, improvement in convective heat-transfer coefficients (HTC); despite the improvement, the magnitude varies depending on the particle size, volume, and (quality) of dispersion. Nanofluids can enhance critical heat flux (CHF) at the boiling point by substantially modifying surface characteristics with nanoparticles, but their influence on nucleate boiling heat transfer coefficient (HTC) varies depending on systems (Barber, Brutin, and Tadrist, 2011). This emphasizes the importance of surface science, colloid behavior, and extreme stability control for achieving long-term performance. (Yu and Xie, 2012).

Nanofluids are defined by the convenience of two broad approaches to synthesis and have a close relationship to the conditional ability to stabilize and tune properties. Within the standard two-step method, industrially relevant nano-powders (e.g., oxides, carbons, metals) are dispersed in the base fluid by either ultrasonication, high-shear mixing, or bead milling. Subsequently stabilized by the use of some approaches, such as pre-added surfactant, pH change, and polymer grafts. This pathway can be used and scaled, but is also likely to suffer agglomeration with non-optimal surface chemistry (Das *et al.*, 2007). To minimize oxidation of high activity metals and maximize their long-term dispersion, one-step syntheses prepare nanoparticles in situ during the physical vapor condensation route, chemical reducing route, solvothermal/polyol route-approaches also tend to compromise the ease of the procedure (Chamsa-Ard *et al.*, 2017). All of this proves that synthesis and surface functionalization represent inseparable tools that can be used to regulate nanofluid microstructure and, ultimately, transport performance (Muhammad and Sidik, 2018). It is the surface functionalization of nanofluids that hinges on durable nanofluids. Strong dispersions will generally aim to achieve large values of zeta potential (electrostatic repulsion), steric hindrances through polymer layers or covalent ligand shells, and sometimes combinations of electrosteric approaches. Studies report that an absolute zeta potential of 30-40 mV is a practical colloidal stability criterion in most systems, but the optimum criterion depends on surface chemistry (Ali and Salam, 2020). Other types of generic dispersants and chemically grafted surfaces, such as silane coupling on silica or oxide nanoparticles, can yield months-long stability in the absence of free surfactant and prevent boiling-surface fouling (Yang and Liu, 2010). This type of interfacial engineering directly relates nanoscale surface energy, wettability, and interfacial thermal resistance to the design-relevant macroscale measures (conductivity, HTC, CHF).

Nanofluids are now used in engineering for cooling of electronics (high-density microchannel heat sink, heat pipes), and high-capacity working fluid. Sectoral reviews show performance gains and smaller dimensions, but tradeoffs in viscosity/pumping power, fouling/maintenance, and cost; so, application-ready fluids must be synthesized and surface chemistry-optimized, application-by-application (Bocanegra, Marchitto, and Misale, 2025). The chapter hence focuses on evaluating the various approaches to the choice and verification of synthesis processes (two-step or one-step; chemical or physical process) and functionalization of surfaces (electrostatic/steric stabilization, ligand/silane grafting, pH/ionic control, hybrid shells) to obtain reproducible and scalable nanofluids on the way to more efficient heat-transfer systems.

## 2. Synthesis and Surface Functionalization

Nanofluids are controlled colloidal suspensions of nanoparticles such as metals, oxides, carbides, carbon-based materials, or even mixtures of these, which are dispersed in ordinary heat-transfer liquids. The thermophysical performance, colloidal stability, and long-term applicability of the dispersed phase critically depend on the synthesis route and interfacial chemistry of the dispersed phase in heat-transfer systems. Two paradigms of fabrication, the two-step and one-step methods, have become prominent in the last two decades, and a variety of chemical and physical sub-routes have been explored, involving trade-offs between scalability, particle control, and stability. Simultaneously, surface functionalization such as electrostatic, steric, and covalent grafting has progressed beyond ad-hoc surface addition of surfactants to highly targeted ligand chemistry to control wettability, interfacial thermal resistance, and foul propensity on heated surfaces. This is why such synthesis cum interface approaches can be viewed as the main tool in transferring laboratory improvements in thermal conductivity and heat-transfer coefficients into practice-relevant fluids (Rahman *et al.*, 2024; Kant *et al.*, 2022).

### *2.1 Synthesis Procedures – Theoretical and Practical Implications*

The most popular method is the two-step since it is flexible and economical. This is because oxide powder (carbon nanotubes), metal, and silica nanoparticles can be synthesized by applying traditional chemical and physical processes. They are then dispersed into the base fluid with the help of ultrasonication, high shear mixing, or milling. These can be stabilized by the addition of surfactants, pH balancing, or polymeric dispersants. Although the two-step route can be industrially scaled, re-agglomeration and sedimentation are likely without strongly optimized surface chemistry and processing energy. Other techniques (one-step routes, solvothermal/hydrothermal synthesis, vapor condensation) also produce nanoparticles within the carrier fluid. They are common to drying and storage stages, and can provide better dispersion and oxidative control over reactive metals, although they tend to be more complex methods of reactor control, and are not typically practical for retrofitting existing fluids. Reviews of other studies indicate that one-step approaches tend to provide improved as-prepared stability, whilst two-step approaches are simpler to scale and optimize particle loading (Zainon and Azmi, 2021). Important chemical reduction (of metals, by means of reducing agents or polyols) and sol-gel methods (oxides), solvothermal methods that allow control of particle crystal faces and shape. Variables used as performance metrics in route selection include particle size/distribution, surface chemistry, oxidation state, and energy to break agglomerates on dispersion. All of which directly trade off viscosity, thermal conductivity, and pumping power, in application (Baig, Kammakakam, and Falath, 2021; Dato *et al.*, 2024).

## 2.2 Surface Functionalization – Strategies and Effect

This has been classified as a stabilization technique based on an electrostatic pH or ionic strength to enhance the repellent force between ionized molecules. In more sophisticated approaches, nanoparticle surfaces are grafted to covalent ligands of silane coupling agents to form long-lasting dispersions without surplus free surfactant that is destroyed at elevated temperatures. Silane grafting and polymer-brush coatings are demonstrated to improve the zeta potential, interparticle attraction, and interfacial thermal resistance of many oxides. In practice, guidelines for formulation are usually set at levels to reach electrostatically stabilized systems of $|\zeta| \geq \sim$30-40 mV, whereas sterically stabilized fluids are set at the level of graft density and layer thickness to avoid van der Waals attraction (Mittal and Kant, 2024; Mahtabani *et al.*, 2020).

## 2.3 Comparison between Mono and Hybrid Nanofluids

Mono-nanofluids are one type of nanoparticle (e.g., $Al_2O_3$/water) and are useful in mechanistic experiments and relatively simple optimization. To enhance complementary behaviour, a hybrid nanofluid purposely contains two or more complementary nanoparticles (binary or tertiary mixtures, as in $Al_2O_3$, $SiO_2$, Cu, $Al_2O_3$, carbon-metal hybrids). This enhances their behaviour, either by raising thermal conductivity, increasing the stability, reducing the viscosity, or by a synergy of electron transport processes in a hybrid system. It has been reported that well-designed hybrid nanofluids can outperform simple systems in terms of thermal conduction and convective heat transfer. However, much finer control over the formulation, such as the ratio of individual components, surface compatibilization, and mixing conditions, is required to avoid unexpected aggregation or rheological penalties (Vallejo, Prado and Lugo, 2022; Rashidi *et al.*, 2021; Das *et al.*, 2024).

## 3. Synthesis Techniques of Nanofluids

Nanofluids are created by dispersing nanoparticles within a base fluid, and their effectiveness is crucial in the synthesis process. There are two main methods for nanofluid fabrication, and these are the two-step method and the single-step method, as shown in Table 1. Each offers different benefits and limitations regarding scalability, stability, purity, and control concerning the properties of the particles.

### *3.1 Two-Step Method*

The economically feasible nanofluid production path that is successful and widely utilized is the two-step approach. This technique involves the initial synthesis (via a chemical or physical bonding mechanism) of nanoparticles in the form of dry powders (Figure 2). With the help of such inputs in mechanical energy, these powders are then spread in the base fluid by the mechanisms of ultrasonication, high-shear mixing, homogenization, or ball milling (Urmi *et al.*, 2021).

In particular, ultrasonication relies on acoustic cavitation as a method of dispersing particle agglomerates to provide an efficient technique of size reduction and dispersion at very low concentrations. Alternatively, ball milling uses the highest mechanical impact, top-down, to an aqueous solution or solution in another liquid, such as copper and copper oxide nanoparticles into a water-based translucent nanofluid; these nano-crystallites have been processed using ball milling into an aqueous medium (Angayarkanni and Philip, 2022; Ali *et al.*, 2018). Review studies also show that the two-step method is relatively inexpensive and scalable, but it often faces stability issues because nanoparticles tend to re-aggregate due to increased surface energy. Therefore, stabilization through agents like surfactants, pH adjustments, or polymers is necessary. (Arora and Gupta, 2021).

### *3.2 One-Step Method*

One-step technique incorporates the development of nanoparticles directly into the base fluid, and does not require the baking or drying, storage, transportation, or re-dispersibility of powders. Such methodologies greatly reduce agglomeration and improve as-prepared stability (Jama *et al.*, 2016). In the physical vapour method, it is directly formed by evaporation. Typically, physical vapers produce particles with a vapor phase that condenses to the fluid without any other step. Such processes provide good control over both particle size and dispersion rates, generally demand trap base fluids with low vapor pressure, and may result in

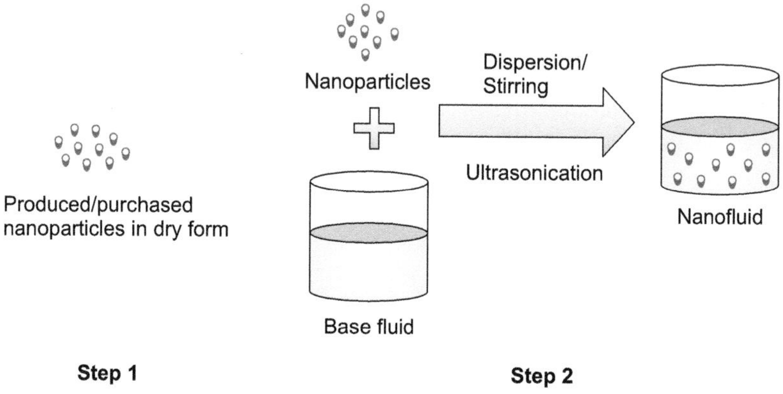


**Fig. 1** Visual description of the two-step method of nanofluids. Adopted from Ref. (Arora and Gupta, 2021) with permission.

impurities or trace materials remaining in the nanofluid. Whereas chemical reduction in the fluid has been used, e.g., to synthesize non-aggregated, stable Cu nanofluids in Ethylene glycol in a single step by chemical reduction (Zhu, Lin and Yin, 2004). Advanced physical one-step methods, such as plasma-arc nanoparticle, magnetron sputtering to condense metal vapor into liquids, better versions of VEROS (Vacuum Evaporation onto a Running Oil Substrate)-all these allow the direct synthesis of Ag, Fe, or Cu nanofluids (Teng, Cheng, and Pai, 2011). One-step plans provide homogeneous dispersion and reduce the number of processing operations, but have limitations generally dependent on the nature of the base liquid, the equipment used, and the exposed chemical residue.

### 3.3 Ultrasonication, Ball Milling, and Chemical Vapor Deposition (CVD)

Even though ultrasonication and ball milling are commonly linked with the two-step pathway, they are crucial steps towards gaining high-quality dispersion. In the production of nano-flowers of PbO, ultrasonication with ball milling resulted in ultra-fine crystallite sizes (~25 nm). The process is directly analogous to the dispersion of nanoparticles into base fluids (Bangi *et al.*, 2017) and ultrasonication, in which aggregates are broken by localized zones of high temperature and pressure. Ultrasonication is a significant technique used in two step method, which facilitates nanoparticles dispersion into the base fluid. This method involves sound waves of high frequencies producing acoustic cavitation, resulting in localised high pressure and temperature, which break up the agglomeration of nanoparticles. The efficiency of ultrasonication is highly affected by factors such as frequency, duration, and power. Usually, lower frequencies in the range of 20–40 kHz create better cavitation and break the large-size agglomerates, whereas higher frequencies of above 40 kHz generate good emulsification but may be less efficient for large cluster molecules. The power also has a significant role as high power increases cavitation capacity with improved dispersion that can also lead to greater particle fragmentation. The time duration of sonication should also be optimized because insufficient or excessive time has an adverse effect on the properties of nanofluids.

Recent studies have given the understandings about the optimized ultrasonication conditions for different nanofluids. For example, Mahbubul et al. have shown that at least 2 hours of ultrasonication produced less viscous and more stable $Al_2O_3$-water nanofluids, thereby increasing their thermal conductivity (Mahbubul *et al.*, 2015). On the contrary, excessive sonication time may lead to higher viscosity and low thermal conductivity, as observed in nanofluids with ultra-long MWCNTs (Multi-walled carbon nanotubes) investigated by Bertrand Jóźwiak and their coworker (Jóźwiak *et al.*, 2021).

Moreover, it has been demonstrated that the use of pulsed sonication or intermittent stirring of the fluid during ultrasonication may promote dispersion homogeneity and fluid overheating, which results in more stable nanofluids. The effectiveness of the synthesis of nanofluids is dependent on optimal parameters of ultrasonication. A combination of frequency, power, and duration, along with improved techniques such as pulsed sonication and controlled cavitation, can be used to achieve a stable dispersion with better thermophysical behavior. As a result, nanofluids can be employed in many applications. Besides the sonication parameters, the amount of dispersant, such as surfactants, polymers, or biopolymers, is also a critical factor that was found to influence the dispersion and ability of nanoparticles and nanofluids. The lack of dispersant coverage will result in aggregation and sedimentation, whereas excessive coverage can produce viscosity, decrease thermal conductivity, or destabilise the suspension due to micelle formation. Research has demonstrated that an optimum PVP concentration of 0.5–1 wt% would give maximum colloidal stability and thermal conductivity in $TiO_2$-water nanofluids, but this concentration would slightly lower the extent of heat transfer in higher concentrations (Jóźwiak *et al.*, 2021). Equally, achieving a balance between rheological properties and steric/electrostatic stabilisation of silver/ copper oxide nanofluids involves moderate concentration levels of dispersants. Thus, to achieve stable and low-viscosity, thermally efficient nanofluids, the company should analyze and optimize dispersant amounts to allow them to be used for practical purposes. This means that ball milling provides mechanical de-agglomeration and size reduction. $TiO_2$-water nanofluids or copper/copper oxide nanofluids are examples wherein extended ball milling has been used to prepare the nanofluids, and ball milling has proved capable of preparing durable and homogeneously dispersed nanofluids.

Chemical Vapor Deposition (CVD) is not a widespread method for nanofluid production, but it can be used as an approach for generating nanoparticles in upstream processes. CVD forms not only nanostructured materials (e.g., ZnO nanocrystals or carbon nanotubes) but also develops them by deposition using vapor-phase precursors (Hachem *et al.*, 2022; Zhang, Wu, Han, and Chen, 2013). Although this type of nanoparticle can then be gathered and spread through the two-step system, CVD in nanofluid one-step synthesis has yet to be explored fully, as it is difficult to make a single method integrate with CVD into liquids.

**Table 1** Summary of synthesis techniques for nanofluids, their advantages, and limitations.

| *Synthesis Technique* | *Description* | *Advantages* | *Limitations* | *References* |
|---|---|---|---|---|
| Two-step method | Nanoparticles are synthesized separately and dispersed in a base fluid (ultrasonication, ball milling, shear mixing) | Scalable, economical, applicable to a wide range of nanoparticle types | Agglomeration and sedimentation require surfactants or stabilizers | (Das et al., 2007), (Urmi et al., 2021), (Arora and Gupta, 2021) |
| One-step method | In-situ synthesis of nanoparticles directly in the fluid (chemical reduction, solvothermal, vapor condensation) | Better dispersion, less oxidation, improved stability | Complex, less scalable, equipment-intensive | (Chamsa-Ard et al., 2017), (Jama et al., 2016), (Hachem et al., 2022) |
| Ultrasonication | Acoustic cavitation disperses aggregates | Effective dispersion, low cost | Limited scalability, may re-agglomerate | (Bangi et al., 2017) |
| Ball milling | Mechanical impact breaks nanoparticle clusters | Homogeneous dispersion, durable suspensions | Time-consuming, risk of contamination | (Angayarkanni and Philip, 2022) |
| Chemical Vapor Deposition (CVD) | Nanoparticles generated from vapor precursors and dispersed | Produces high-purity nanostructures | Limited applicability in direct liquid synthesis | (Hachem et al., 2022), (Zhang, Wu, Han and Chen, 2013) |

## 4. Surface Functionalization Strategies

The dispersion stability of nanofluids is key to sustained dispersion stability for maintaining high thermal transport and operational stability, as shown in Table 2. Since nanoparticles tend to aggregate due to high surface energy from van der Waals interactions, a key strategy in nanofluid design is surface functionalization. This essentially turns volatile colloids into long-life, stable fluids applicable in more advanced heat-transfer means. Surface functionalization is the alteration of the nanoparticle surfaces either by chemical, physical, or biological means to enhance their ability to mix with the underlying fluid as well as to avoid agglomeration. The concept of the strategy is to change the surface energy balance through the introduction of repulsive forces (electrostatic or steric) that oppose attractive van der Waals forces. The functionalization may be based on attaching organic molecules, surfactants, or polymers, or covalently binding a functional group (–OH, –COOH, –NH2, –SH) and customizing the interfacial-energy to provide a better uniformity in dispersion. Indicatively, two common techniques of introducing hydrophilic groups on metal oxide nanoparticles ($Al_2O_3$, $TiO_2$, and $SiO_2$) include hydroxylation and silanization, which increase the affinities of these nanoparticles to polar solvents like water or ethylene glycol. In addition to better drag dispersion, surface functionalization is of significant interest in controlling nanofluid thermo-physical characteristics (e.g., viscosity, electrical conductivity, heat capacity). As an illustration, the carboxylic or amine functional groups, which are attached to carbon-based nanomaterials, provide not only a better solubility, but also a transfer of phonons across interfaces and consequent higher effective thermal conductivity. Differential functionalization of individual nanoparticles will result in even co-dispersion and minimal phase segregation when using hybrid nanofluids, both during shear or temperature gradients. Among the latest developments, the focus on bio-inspired and green functionalization paths, which also use natural extracts, amino acids, or biopolymers as environmentally friendly surface modifiers, is also present. They eliminate the use of toxic solvents, allowing stable dispersions to be made without exploiting thermal or physical performance. Additionally, in-situ functionalization of nanoparticle synthesis process, e.g., sol-gel or hydrothermal pastes of nanoparticles, could produce pre-assembled surface chemistry of nanoparticles, which avoids chemical post-processing and the danger of aggregation. In general, surface functionalization continues to play a vital role in providing both reproducible long-term stability of nanofluids to allow the nanofluids to move beyond laboratory designs and into commercially viable heat transfer fluids.

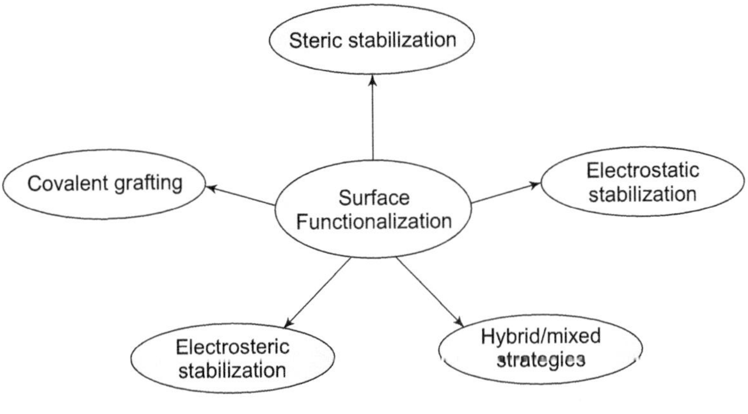


**Fig. 2** Surface Functionalization Strategies.

### 4.1 Need for Functionalization

Untreated nanoparticles tend to aggregate and settle, causing impairment of thermal conduction, distortion of rheology, and blockage of microchannels. Electrostatics (chemical bond) or steric repulsive forces (polymer/surfactant layers) are introduced by functionalization to prevent aggregation. It has been stressed that the stability of nanofluid deteriorates when such interventions are not implemented, regardless of the performances achieved and service life (Yu and Xie, 2012). A review studies of the stability of nanofluids (2023) cites striking variations in retention, such as in $SiO_2$ nanofluids in ethylene glycol-water, where a 36-month stable span was achieved, and all $Al_2O_3$-water nanofluids, stabilized with cetyltrimethylammonium bromide (CTAB), where a 50 percent longer time span existed than in the case of sodium dodecyl sulphate (SDS), showing the importance of the surfactant used (Wang *et al.*, 2023).

Besides these results, Yu and Xie (2012) have emphasized that, in addition to the agglomeration-suppressing functionality of the functionalization process, the process is also used to customize interfacial thermal properties transport. The presentation of functional layers, such as ionic layers, covalent layers, or polymeric layers, forms a potential barrier that acts against the van der Waals attraction and at the same time increases the compatibility between the surfaces of nanoparticles and the base fluid. Wang et al. (2023) added that the balance of electrostatic and steric effects determines the successful performance of functionalization: whereas the electrostatic stabilization (e.g. by using ionic surfactants, like SDS or CTAB) generates repulsive charge on the surface of particles, the steric stabilization (e.g. by using polymeric or non-ionic surfactants, such as PVP or PEG) forms a physical layer over these repulsively charged particles, resulting in some physical separation on the layer interface. Other process factors, including pH, ionic strength, and sonication energy, also have a strong effect on stability other than interface chemistry. Electrostatic repulsion can be dramatically decreased by adjusting the pH near the isoelectric point of the nanoparticles and causing the aggregation of the system, whereas keeping the system far enough below the isoelectric point will allow the system to have higher zeta potential values (> +–30 mV), which is associated with a stable dispersion. The authors proposed the use of hybrid ionic-polymeric-based electrosteric stabilisation by Yu and Xie (2012), suggesting that the hybrid approach provides a higher level of performance, providing both steric and charge barriers. In their most recent results, Wang et al. (2023) were able to build upon this knowledge to achieve longer-term durability, during which this issue produced stable dispersions over the long term when nanoparticles were surface-modified concurrently and kept under optimal thermodynamic conditions. Furthermore, stability is not the only reason for the need for functionalization: surface-modified nanoparticles usually have better interfacial thermal resistance properties and can be more compatible with various base fluids, including ethylene glycol, oils, and ionic solution liquids. This is directly relevant to high-temperature and cooling, which are periodic operations where there is permanence of the thermophysical characteristics of the functionalized nanofluids. Therefore, surface functionalization becomes one of the foundations of nanofluid engineering that makes it possible to ensure that nanoscale advantages of increased heat transfer can be easily reflected in the performance of thermal management systems at larger scales.

## 4.2 *Surfactants: Electrostatic and Steric Stabilization*

Surfactants are amphiphilic molecules that adsorb at the nanoparticle/liquid interface, providing simultaneous nanoparticle-liquid stabilization via two mechanisms. Electrostatic stabilisation arises when a distance potential is created by charged surfactant head groups in which repulsions between particles of similar charge occur. In the second mechanism, surfactant tails or chains can form steric hindrance and physical barriers between different particles that prevent direct particle approach, leading to steric stabilization. An experimental study exploring a range of surfactants, such as CTAB, TSC (trisodium citrate), SDS, and PVP (polyvinylpyrrolidone), was carried out on the $Fe_3O_4$-water, CNT-water, and $Fe_3O_4$-CNT nanofluids. Findings indicated different effects on viscosity and thermal conductivity, which confirmed that the choice of surfactants essentially determines nanofluid stability and performance (Wang *et al.*, 2021). Surfactants such as SDS are being used to improve viscoelasticity and high-temperature resistance in hybrid polymer-nanoparticle systems -e.g., HPAM/SDS/CNT-$NH_2$ nanofluids stay viscous (above 70 m.Pas at 90°C) when using surfactants such as SDS to help disperse nanoparticles (Tang *et al.*, 2021).

It is also important that the effectiveness of surfactant-assisted stabilisation is not only determined by the chemical structure of the surfactant used, but also by the interplay of surfactant molecules and nanoparticle surface. Metal oxide nanofluids are mainly stabilised using anionic surfactants (e.g., SDS, SDBS) to ensure repulsion between ions, and cationic surfactants (e.g., CTAB) may create ion-pair complexes between the surfaces of negatively charged nanoparticles. Instead, steric stabilisation of particles with nonionic surfactants like PVP or Triton X-100 is steric stabilisation because the surfactant adsorbs polymeric chains of the surface but leaves the charge constant. In Wang et al. (2021) and Tang et al. (2021), studies indicated that multi-surfactant systems or mixed stabilisers tend to be more effective than single agents and obtain a better ability in terms of stability and tunable rheological properties that may be applicable during high-shear or high-temperature processes.

Surfactants also affect the interfacial and electrical properties of nanofluids besides enhancing dispersion. As illustrated in recent research by Mane et al. (2022), the conductivity of nanofluids can be substantially altered due to the addition of ionic surfactants by altering the percentage distribution of surface charges and electric charge thickness around nanoparticles. As demonstrated in the study, the presence of surfactants like SDS and CTAB causes the liberation of charge bearers and the creation of micelles in the base fluid to enhance the conduction properties of the electrolyte, as well as to retain its colloidal stability. On the other hand, the surplus concentration of the surfactant may cause aggregation or shielding of the charges of the micelle, which contributes to loss of stability or rise in viscosity. It is thus important to optimise dosage surfactants to achieve a balance between electrical conductivity, viscosity, and performance of thermal transport (Mane, Hemadri, and Tripathi, 2024).

In addition to stabilisation, interfacial heat transfer is also mediated by the surfactants. Surfactant-functionalized nanofluids have higher thermal conductivity and convective heat transfer coefficients than bare nanoparticle suspensions by reducing the interfacial tension and enhancing nanoparticle wettability. In one case, envelope nanofluids made by modifying $Al_2O_3$-water fibres with SDS demonstrated an up to 12 percentage point enhancement in effective thermal conductivity, as well as boosted the dispersion homogeneity at the turbulent flow states. Equally, the CTAB-modified graphene oxide nanofluids exhibited lower viscosity penalties and enhanced stability during several months of storage. However, more recent research has been conducted on green and biodegradable surfactants (including saponins, lecithin, and gum arabic), which give non-synthetic substitute agents but offer equivalent electrostatic stabilisation. The decrease in toxicity of these bio-based surfactants makes them more important, especially since they can be produced economically in large quantities, turning them into eco-friendly nanofluids. In general, electrostatic and steric stabilisation with the aid of surfactants is one of the most diverse approaches towards the development of long-term nanofluids. Nevertheless, its performance relies on the synergistic optimisation of the type of surfactant, surfactant concentration, as well as compatibility of nanoparticles and fluids, which determine the long-term dispersion, rheological, and thermal-electrical properties of the nanofluids in any given application of the respective engineering (Yasmin *et al.*, 2023).

### 4.3 Functional Polymers and Surface Modifiers

Instead of or in addition to surfactants, polymers are used when strong steric layers or grafted interfaces are required. Silver nanofluids are successfully stabilised by making use of PVP to prevent clustering and growth of the particles. Pharmaceutical nanosuspensions have been screened systematically to find SDS, poloxamer188, and HPMC as the most effective stabilizers. In which a combination of surfactant and polymers was used to obtain highly stable sub-250 nm suspensions (Lestari, Müller and Möschwitzer, 2015). In addition to passive surfactants, the use of covalent grafting offers long-term surface adhesion and stable interfaces: Silica nanoparticles (such as GPS or MPPTS) coated with silane produce nanofluids that do not form deposits during boiling and have a very high colloidal stability (Yang and Liu, 2010). By reducing surface energy and enhancing the dispersion thermodynamics, a silane-based modification of metal oxide nanoparticles reduces agglomeration (Ahangaran and Navarchian, 2020).

These advances have led to recent years, biopolymers have been used as alternatives to surface functionalization of nanoparticles in nanofluids by synthetic polymers due to their environmental friendliness. Chitosan, cellulose derivatives, starch, alginate, and gum arabic are biopolymers that may contain functional groups (–OH, –COOH, $-NH_2$) and hydrogen bonding, electrostatic, and van der Waals forces that can bind the nanoparticle surface, thus slowing aggregation and sedimentation. The resulting interactions form a steric and electrostatic barrier around nanoparticles, enhancing long-term colloidal stability at the same time as thermal cycling or long-term storage. Besides the stabilisation of the dispersion, biopolymers may modulate the rheological properties of nanofluids and offer a range of tunable viscosity and flow properties that can be used in a variety of thermal management processes. They are advantageous, especially in the design of sustainable and environmentally friendly nanofluids because of their biodegradability, renewability, and non-toxicity. Recent experiments showed that for these nanoparticles, biopolymer-coated nanoparticles outperform the performance of uncoated systems or those stabilised by surfactants in terms of thermal conductivity and particle settling; hence, they have emerged as promising applications in solar energy systems, biomedical cooling, heat exchangers, and electronic cooling systems. Moreover, biopolymers may be employed separately or together with traditional surfactants or synthetic polymers to create hybrid surfaces of stabilisation with further flexibility in the development of nanofluids with specific characteristics. In general, biopolymers integration can be described as a green and efficient method of getting stable and high-performance nanofluids that can be used in the laboratory sector and industry. It is not only that the dispersion stability is improved by the presence of surfactants, but also that the surfactants have a strong effect on the electrical properties of the nanofluids. When surfactant molecules are adsorbed on nanoparticle surfaces, the distribution of charges at the interface between the nanoparticles and the aqueous solution is altered, and thus the electrical double layer is formed, which consequently changes the conductivity of the entire suspension of nanoparticles. This has been shown by recent studies by Mane *et al* ( 2024) to indicate that not only the type of surfactants but also their concentration can influence conductivity, with optimum concentration improving conductivity, but high amounts could result in being saturated or even reduced conductivity. This underscores the fact that, in addition to thermal and rheological stabilization, the appropriate choice and dosage of surfactants is necessary to regulate the electrical characteristics of nanofluids to use them in sensor applications, cooling of electronic equipment, and energy systems.

Beyond traditional and biopolymer-based methods for functionalization, recent studies have increasingly focused on the significance of stimuli-responsive or smart polymer surface modifications on nanoparticles. These modifications allow for the intelligent adjustment of nanoparticle surface properties in response to environmental changes, such as pH, temperature, or ionic strength. As an example, nanofluids can be obtained by thermo-responsive polymers, including poly(N-isopropylacrylamide) (PNIPAM) grafted onto metal oxide nanoparticles, which can be reversibly measured in terms of viscosity and aggregation behaviour in close ranges to physiological or operational temperature (Ahangaran and Navarchian, 2020; Manimaran *et al.*, 2025). This tunability in specific applications, e.g., biomedical cooling or adaptive heat exchangers, is especially beneficial when conditions of operation vary. Moreover, graft copolymer systems, i.e., PEG-PAA or PEG-PLA, with hydrophilic and hydrophobic segments have higher dispersibility and enhanced thermal conductivity as well, and can be used with various base fluids, i.e., glycol and ionic liquids, and biodegradable oils (Wang, Xu, and Choi, 1999).

A second potential direction is in-situ polymerisation functionalization, where monomers are polymerised on surfaces during synthesis on nanoparticle surfaces. The method offers uniform coating, reduces residues of free surfactants, and has enhanced colloidal stability as opposed to post-synthesis grafting. As an illustration, acrylamide or styrene monomers have been in situ polymerised on $Al_2O_3$ or $SiO_2$ nanoparticles to produce a highly stable suspension with low-sedimentation constants and uniform heat transfer at long times (Rashidi and Zinatloo-Ajabshir, 2024). Equally of interest, zwitterionic polymers have been attractive as dual steric-electrostatic stabilisers, using positive and negative charges on the same chain to prevent aggregation even in high salinity conditions or high temperature places, which makes them applicable to seawater-based cooling systems or extreme industrial conditions (Hermida-Merino *et al.*, 2018). Altogether, functional polymers not only lengthen the nanofluid lifetime but also enable the property tuning at the molecular level, combining many functions in one, including thermal enhancement, anti-corrosion functionality, and electrochemical activity. Researchers can create next-generation nanofluids tailored to sustainability, reuse, and add thermal management devices in renewable energy and in the use of smart electronics by adjusting the polymer chemistry and grafting density.

## 4.4 Mixed and Hybrid Strategies

Coupled systems of surfactants, polymers (or covalent modifiers) -so-called electro-steric or hybrid stabilization-work better by relying on several actions combined. RSC Advances (2025) reported that functionalizing MWCNTs with controlled carboxylic acid group density (-COOH) increased stability and thermal conductivity. Lower -COOH densities yielded zeta potentials of approximately −22 mV and reduced aggregation in hybrid $Al_2O_3$-MWCNT nanofluids by approximately 13.5 to 13.6-fold compared to unfunctionalized CNTs (−16.8 mV) (Manimaran *et al.*, 2025). These more advanced methods let the formulators make nanofluids that can work for a long time at high temperatures. Even when the chemical conditions are bad, electrostatic shear stabilization can be used with polymeric or covalent surface functional group additions (Figure 2).

Recent studies show synergistic benefits of hybrid and mixed stabilisation systems that are not attained in a single-component system. An example of this is surfactant-polymer combinations like SDS-PVP or CTAB-PEG, which are significantly more stable due to their dual-mode mechanism of repulsion by electrostatic force and repulsion by steric hindrance. The surface charge is initially modified by the surfactant molecule, and the polymer chains stick out into the surrounding fluid, creating a shell of defence against coalescence, even in high shear and temperature conditions. The systematic reviews also showed that the use of binary surfactant systems resulted in dramatic improvements in thermal conductivity (up to 18 per cent) and optimisation of viscosity of hybrid $Al_2O_3$-$TiO_2$-water nanofluids, which showed cooperation between ionic and non-ionic surfactants (Ma *et al.*, 2021).

Moreover, it has been demonstrated that the covalent functionalization of the surfaces of nanoparticles coupled with the adsorption of non-covalent surfactants generates strong interfacial architectures. An example is the silane-modified $TiO_2$ nanoparticles pre-treated with polyethene glycol (PEG) that showed higher resistance to the formation of aggregates after repeated heating-cooling intervals than the samples modified with silane alone. This can be attributed to increased stability due to the creation of a dual stabilisation layer, covalent linkages that keep the surface intact, and polymeric covering, which gives it flexibility and steric spacing (Sharma *et al.*, 2025). Equally, carbon nanotubes are functionalized in combination with acid treatment and polymer grafting, with substantial improvement in their solubility and overall compatibility with polar solvents. In 2024, a report in the Journal of Molecular Liquids demonstrated that MWCNT-$Al_2O_3$ hybrid nanofluids modified with the CTAB surfactant and polyaniline significantly improved thermal performance, although technological improvement was achieved only to 35% (Htwe and Mamat, 2024).

New hybrid systems are currently built with biopolymer-surfactant or ionic liquid-polymer systems to achieve more environmentally friendly and tunable stabilisation methods. As an example, in the presence of SDS or ionic liquids, chitosan can be used to form eco-friendly hybrid layers to not only facilitate dispersion but also endow anti-corrosive and antimicrobial properties on the nanofluids, which is useful in the fields of biomedical and solar thermal (Serra *et al.*, 2025). Furthermore, recently, the optimization of hybrid stabilizers using machine learning has been employed to predict the optimal concentrations and ratios of nanoparticles and fluids to enhance nanofluid stability and thermal efficiency. This approach has provided an additional pathway for designing effective and thermally efficient nanofluids. [Mishra *et al.*, 2023].

Generally, hybrid and mixed strategies are a significant development in the field of nanofluid engineering, allowing formulators to manipulate interfacial chemistry and optimize rheological and thermal properties while maintaining multifunctionality even in harsh or variable environments. It is anticipated that the current tendency of incorporating organic-inorganic, hybrid coating schemes, adaptive polymers, and computational modelling will characterise the second generation of nanofluid functionalization technologies that will fill the current gaps between laboratory preparation and applied performance.

**Table 2** Strategies for surface functionalization of nanofluids and their impact on stability and characteristics.

| *Functionalization Strategy* | *Mechanism* | *Typical Agents/ Modifiers* | *Effect on Nanofluid Properties* | *References* |
|---|---|---|---|---|
| Electrostatic stabilization | Charge repulsion between nanoparticles | Ionic surfactants (SDS, CTAB, TSC) | Improves stability, enhances dispersion, and may alter viscosity | (Yu and Xie, 2012), (Ma *et al.*, 2021) |
| Steric stabilization | Polymer/ surfactant layers create physical barriers | PVP, HPMC, poloxamer 188 | Prevents aggregation, long-term stability, and possible viscosity increase | (Zhang, Wu, Han and Chen, 2013), (Wang *et al.*, 2021) |
| Covalent grafting | Strong ligand binding to nanoparticle surfaces | Silane coupling agents (APTES, GPS, MPPTS) | High thermal stability, reduces fouling, improves zeta potential | (Tang *et al.*, 2021), (Mane, Hemadri, and Tripathi, 2024) |
| Electrosteric stabilization | A combination of charge repulsion and steric hindrance | Surfactant + polymer (e.g., SDS + CNT functionalization) | Synergistic stability, higher conductivity, reduced aggregation | (Yasmin *et al.*, 2023) |
| Hybrid/mixed strategies | Multiple stabilizers applied | Silane + polymer, mixed surfactants | Prolonged shelf life, optimized balance of conductivity and viscosity | (Zhang, Wu, Han, and Chen, 2013), (Yasmin *et al.*, 2023) |

## 5. Impact of Synthesis and Functionalization on Properties

Synthesis and further functionalization of nanofluids on surfaces are important processes essential to the characterization of the physicochemical and thermophysical properties of nanofluids. Major characteristics influenced include heat conduction, viscosity, and sustained stability, which define the engineering and medical practice of nanofluids. In addition, the fact that nanofluid can be combined with various base fluids, including water, ethylene glycol, oils, and ionic liquids, is not only directly related to the synthesis pathway but also to the choice of the available options of surface modification. These interrelationships are important in the design of nanofluids to suit definite performance requirements due to the detailed knowledge.

### 5.1 Thermal Conductivity Enhancement

Thermal conductivity of nanofluids has been one of the most widely discussed properties, and it directly governs the efficiency of heat transfer in the nanofluids. Synthesis paths that maximize interaction between nanoparticles and fluids involve processes where both nanoparticle synthesis and dispersion occur through the same method, such as a one-step route. Two-step, in turn, does not presuppose that small-scale production can be influenced by agglomeration of particles until the effective dispersion and functionalization of particles. Interference with surfactants or polymer coating of surfaces overcomes interfacial thermal resistance (Kapitza resistance) and ensures homogeneous dispersion of nanoparticles, which increases heat transport. Using the example of graphene oxide or carbon nanotube nanofluids, they are both more prone to be highly conductive, with attachment of carboxyl or hydroxyl, and thus enable them to address fluids with polar bases more effectively (Murshed and de Castro, 2014).

### 5.2 Influence on Viscosity

The other significant parameter is the viscosity, which determines the power and energy consumption of the pumping process within the cycle of fluid flow. Though the incorporation of nanoparticles is likely to raise the viscosity, the extent of this observation may differ significantly, relative to the synthesis and the surface chemistry. It is inherent to the ultrasonicated nanofluids that they tend to be low-viscosity compared to the nanofluids prepared in the same manner by mechanical stirring, since sonication breaks up aggregates into individually small and well-stabilized suspensions. However, too much functionalization of surfactants or very large concentrations of polymeric surfactants may increase the viscosity and, consequently, may negatively affect the flow properties (Razzaq *et al.*, 2025). A delicate balance between increasing stability and decreasing the viscosity increase is therefore essential to the practical applications. An illustration of this is silica nanoparticles containing short-chain silanes, dispersed as silanes, which have no substantial viscosity penalties when compared to long-chain polymer surfaces [Lau *et al.*, 2013].

### 5.3 Stability Considerations

One of the main issues with nanofluids is stability, because agglomeration and sedimentation cause irregularities in thermophysical properties, leading to ineffective operation due to changes in flow efficiency and particle behavior. There is a direct effect of synthesis route on stability; one-step chemical vapor condensation offers greater stability to nanofluids owing to an exceptionally high level of integration of particles into the fluid frame (Sun *et al.*, 2020). In the meantime, two-step synthesized nanofluids would need added stabilisation measures in the form of surfactant additions, pH corrections, or nanoparticle functional oiling. Electrostatic stabilization obtainable through functionalization with ionic surfactants (e.g., SDS) yields repelling forces to the molecules compared with steric stabilization through non-ionic surfactants and polymer coats (Rahman *et al.*, 2024). It has been demonstrated that hybrid functionalization, as the fusion of electrostatic and steric effects, can prolong the shelf life of nanofluid as much as six months without any noticeable sedimentation.

### 5.4 Compatibility with other Base fluids

The compatibility of nanofluids with different base fluids is also determined by synthesis and surface modification schemes. Such nanofluids as water-based, e.g., take advantage of the hydroxyl and carboxyl surface groups, enabling hydrogen bonding and stability of dispersion (Mebarek-Oudina and Chabani, 2022). Comparatively, nonpolar or oil-based nanofluids tend to need increased hydrophobic nanocomponent surface functionalization using alkyl silane or nonacidic acid groups in place of hydrophilic cleavage phases (Weera, Lee, and Attar, 2020). Ionic liquids that are gaining popularity as heat transfer fluid with a wide liquidus range need the coating of nanoparticles to be selected with careful consideration to ensure neither chemical reactivity nor lack of interactivity (Murshed, 2022). The ability to agglutinate nanoparticle surfaces to specific base liquids broadens the use of nanofluids in atomistic cooling of cars, electronics, and renewable energy systems.

### 5.5 Interdependence of Properties

It is important to note that the effects of thermal conductivity, viscosity, and stability are not isolated; rather, they are interdependent. As an example, the addition of very large concentrations of surfactant can maximise stability at the cost of thermal conductivity by forming layers of thermal resistance (Harikrishnan and Dhass, 2022). Likewise, as nanoparticle concentration increases, so does pumping efficiency, and in addition, the viscosity, which increases conductivity. Therefore, synthesis and functionalization strategies are to be optimized holistically by considering the targeted use. The use of even more advanced techniques, like hybrid nanofluids, which involve more than one type of nanoparticle in a combined application, has the potential to synergistically enhance thermal conductivity while maintaining a reasonably high level of viscosity and stability.

## 6. Challenges and Future Perspectives

The potential of nanofluids, including their high thermal conductivity, high heat transfer coefficient, and multi-functionality, has been overshadowed by ongoing practical issues. By far the most important of these are agglomeration and long-term stability, measurement and standardization, and the barrier of cost-effectiveness and scale-up. These constraints, and other directions open to exploration (green synthesis, hybrid/encapsulated particles, digital optimization), are what will define the near-term research and industrial outlook of nanofluids. The greatest practical barrier is agglomeration and long-lasting stability. Nanoparticles in their pure state have a large surface energy and will form aggregates that settle, block small channels, and negate the original thermal advantages; most extensive reviews indicate that well-prepared nanofluids will degrade over weeks to months unless stabilized by effective surface chemistry or physical entrapment schemes. This issue can be mitigated, but not eradicated, through functionalization, electrosteric stabilization, and in-situ one-step synthesis, and faster ageing and cycling tests are still valid to justify real-world durability (Wang *et al.*, 2023). Close to it are measurement reproducibility and standardization issues. The initial literature was characterized by large and occasionally antagonistic thermal-conductivity gains, subsequently ascribed to methodological artifacts (calibration of instruments, probe effects, sedimentation during measurement). Review and meta-analysis highlight the importance of standardized protocols (e.g., transient hot-wire under ASTM guidance), closely reported sample preparation (sonication energy, surfactants included), and interlaboratory round-robin tests to ensure that results are comparable and reliable. In the absence of this standardization, conversion of the lab claims into engineering design is risky (Souza *et al.*, 2022). Deployment is also limited by viscosity as well as pumping penalties. In most nanofluids, viscosity rises that elevate pumping energy and reduce net system benefits, even though they have thermal benefits. The conductivity gains versus hydrodynamic cost are specific to applications. Techno-economic analysis has demonstrated that even in laminar systems, a rather modest improvement in conductivity gains can be matched by pumping cost, and that in turbulent or throughput-driven systems, the higher pumping cost may only be offset by formulations continuing to be optimized. For example, Mane and Hemadri (2021) investigated hybrid CuO–$Fe_3O_4$ nanofluids and reported that, while thermal conductivity enhancements of up to ~5% can be achieved, the addition of surfactants and stabilizers also increases viscosity, which directly affects pumping power and operational energy costs. Their results highlight that techno-economic evaluation must consider not only the thermal performance improvements but also the associated material, energy, and maintenance costs. Such analyses indicate that optimal nanofluid formulations require balancing thermal gains against hydrodynamic penalties, making it clear that industrial deployment depends on both performance and cost-effectiveness (Mane and Hemadri, 2022). Limited, but increasing, rich lifecycle techno-economic analyses are available (Yu and Xie, 2012). Capturing issues of health, safety, and environmental (HSE) are emerging as significant non-technical barriers. The production, storage, and disposal of nanoparticles require hazard assessment; more environmentally friendly production methods and recovery or recycling strategies (e.g., magnetic separation, immobilized nanoparticles) are being considered to reduce environmental impact and worker exposure. It is thus advisable to integrate lifecycle and environmental impact appraisal into formulation design at an early stage (Parab *et al.*, 202; Cardoso *et al.*, 2025). Pathways to overcome challenges are well defined on the positive side. Agglomeration and fouling can be minimized by hybridization (matching particles with complementary properties), and covalent surface grafting, encapsulation (core shell or polymer shells), and engineered porous/deposit-resistant additives, which do not compromise thermophysical gains. One-step synthesis methods, like vapor condensation and in-situ reduction, make particles that are more spread out and (usually) more stable when they are first made than traditional two-step powder dispersion. However, they need more complicated equipment and process control (Rahman *et al.*, 2024; Bacha *et al.*, 2024).

## 7. Conclusion

Nanofluids have evolved into an interdisciplinary advancement in the science of heat transfer and a research domain encompassing materials chemistry, surface science, fluid mechanics, and applied energy engineering. The chapter has followed the original guiding ideas behind making and using nanofluids, the relationship between how they are made and how they behave thermally, and the trends that show how they go from being lab suspensions to working fluids. The totality of evidence leads to a singular conclusion: the methodology of synthesis and surface functionalization is not merely a discretionary choice, but a critical determinant of nanofluid stability, performance, and scalability.

Active two-step and one-step syntheses are the two main types of synthesis. Each has its own pros and cons. The two-step method is the most popular because it is cost-effective and can be used on a large scale. It can be used to disperse various nanoparticle shapes and concentrations in a wide range of base liquids. The downside is that it has always had the risk of agglomeration and sedimentation, which has made the dispersion very unstable after that.

Much effort has been put into ultrasonication, ball milling, and high-shear mixing to eliminate these problems, although it seldom leaves the system stable indefinitely without surface modification. In contrast, a single-step synthesis, such as chemical reduction or solvothermal reduction and dispersion, combines all essential steps of nanoparticle creation and dispersion into one process, reducing oxidation and promoting better interface compatibility. However, these methods are equipment-dependent, less flexible regarding particle variety, and are not yet optimized for large-scale production. Therefore, the work of the future needs to give emphasis on hybridization or optimization of synthesis methods that integrate the scalability of the two-step method and the stability inherent in one-step fabrication.

Well-dispersed, stable nanoparticles optimize conductive pathways and reduce the Kapitza resistance, giving actual advantages to heat transfer. On the other hand, the viscosity, which is a key factor in pump performance, is usually higher when loads of particles and aggressive, intensive surfactants are greater; thus, the conductivity may be compromised due to hydraulic losses. The third important property that depends on the others is stability, as thermal performance and rheological performance are founded on stability, as sedimentation causes time-dependent performance degradation. Surface modification is also shown to be crucial with multiple base fluids: the presence of hydrophilic functional groups contributes to a higher dispersion in polar fluids, such as water or glycol, whereas hydrophobic layers are required with oils or nonpolar suspensions. Nanofluids are at the intersection of designing materials on a nanoscale and using renewable thermal engineering on a macroscale. The intelligent fusion of synthesis approaches, surface functionalization, and application-specific optimization, not the dispersal of nanoparticles, holds promise. Thermal conductivity, viscosity, and stability are all related to each other. The field can expand beyond the laboratory into industries by considering the cost, reproducibility, and environmental safety. To move forward, we need people from many different fields: materials scientists who can improve the chemistry of nanoparticles, chemical engineers who can make synthesis more efficient on a large scale, thermal scientists who can measure how well nanoparticles move, and systems engineers who can figure out if nanoparticles are worth the money. Nanofluids could well fulfil their initial promise to replace their predecessors as the next generation of heat transfer media that are dense, compact, and carefully tailored to meet the energy and cooling needs of the 21st century.

## References


[1] Ahangaran, F. and Navarchian, A.H., 2020. Recent advances in chemical surface modification of metal oxide nanoparticles with silane coupling agents: A review. *Advances in Colloid and Interface Science*, *286*, p.102298.

[2] Ali, A.R.I. and Salam, B., 2020. A review on nanofluid: preparation, stability, thermophysical properties, heat transfer characteristics and application. *SN Applied Sciences*, *2*(10), p.1636.

[3] Ali, H.M., Babar, H., Shah, T.R., Sajid, M.U., Qasim, M.A. and Javed, S., 2018. Preparation techniques of TiO2 nanofluids and challenges: a review. *Applied Sciences*, *8*(4), p.587.

[4] Angayarkanni, A. and Philip, J., 2022. Synthesis of nanoparticles and nanofluids.

[5] Arora, N. and Gupta, M., 2021. Thermo-hydraulic performance of nanofluids in enhanced tubes-a review. *Heat and Mass Transfer*, *57*(3), pp.377–404.

[6] Bacha, H.B., Ullah, N., Hamid, A. and Shah, N.A., 2024. A comprehensive review on nanofluids: synthesis, cutting-edge applications, and future prospects. *International Journal of Thermofluids*, *22*, p.100595.

[7] Baig, N., Kammakakam, I. and Falath, W., 2021. Nanomaterials: A review of synthesis methods, properties, recent progress, and challenges. *Materials advances*, *2*(6), pp.1821–1871.

[8] Bangi, U.K., Park, H.H., Han, W., Prakshale, V.M. and Deshmukh, L.P., 2017. Ultrasonically assisted synthesis of lead oxide nanoflowers using ball milling. *International Nano Letters*, *7*(2), pp.149–155.

[9] Barber, J., Brutin, D. and Tadrist, L., 2011. A review on boiling heat transfer enhancement with nanofluids. *Nanoscale research letters*, *6*(1), p.280.

[10] Bocanegra, J. A., Marchitto, A. and Misale, M. (2025). Nanofluids in solar collectors: a comprehensive review focused on its sedimentation. *Clean Technologies and Environmental Policy*, *27*(4), 1753–1784.

[11] Cardoso, B.D., Souza, A., Nobrega, G., Afonso, I.S., Neves, L.B., Faria, C., Ribeiro, J. and Lima, R.A., 2025. Progress in nanofluid technology: From conventional to green nanofluids for biomedical, heat transfer, and machining applications. *Nanomaterials, 15*(16), p.1242.

[12] Chamsa-Ard, W., Brundavanam, S., Fung, C.C., Fawcett, D. and Poinern, G., 2017. Nanofluid types, their synthesis, properties and incorporation in direct solar thermal collectors: A review. *Nanomaterials, 7*(6), p.131.

[13] Das, P.K., Santra, A.K., Ganguly, R., Dash, S.K., Muthusamy, S., Sha, M. and Sadasivuni, K.K., 2024. An extensive review of preparation, stabilization, and application of single and hybrid nanofluids. *Journal of Thermal Analysis and Calorimetry, 149*(17), pp.9523–9557.

[14] Das, S.K., Choi, S.U., Yu, W. and Pradeep, T., 2007. *Nanofluids: science and technology*. John Wiley & Sons.

[15] Dato, M.A., Hu, L., McElheny, D. and Cabana, J., 2024. One-Step Solvothermal Synthesis of Perovskite-Type Li3 x La2/3–x□ 1/3–2 x TiO3 Nanomaterials. *Chemistry of Materials, 36*(17), pp.8349-8358.

[16] Eastman, J.A., Choi, S.U.S., Li, S., Yu, W. and Thompson, L.J., 2001. Anomalously increased effective thermal conductivities of ethylene glycol-based nanofluids containing copper nanoparticles. *Applied physics letters, 78*(6), pp.718–720.

[17] Hachem, K., Ansari, M.J., Saleh, R.O., Kzar, H.H., Al-Gazally, M.E., Altimari, U.S., Hussein, S.A., Mohammed, H.T., Hammid, A.T. and Kianfar, E., 2022. Methods of chemical synthesis in the synthesis of nanomaterial and nanoparticles by the chemical deposition method: a review. *BioNanoScience, 12*(3), pp.1032–1057.

[18] Harikrishnan, S. and Dhass, A.D., 2022. *Thermal Transport Characteristics of Phase Change Materials and Nanofluids*. CRC Press.

[19] Hermida-Merino, C., Pérez-Rodríguez, M., Pereiro, A.B., Piñeiro, M.M. and Pastoriza-Gallego, M.J., 2018. Tailoring nanofluid thermophysical profile through graphene nanoplatelets surface functionalization. *ACS omega, 3*(1), pp.744–752.

[20] Htwe, Y.Z.N. and Mamat, H., 2024. Progress and prospects on stability and thermal properties of surfactant assisted graphene, carbon nanotubes based nanofluid and their hybrid for next-generation thermal management. *Journal of Molecular Liquids, 414*, p.126235.

[21] Jama, M., Singh, T., Gamaleldin, S.M., Koc, M., Samara, A., Isaifan, R.J. and Atieh, M.A., 2016. Critical review on nanofluids: preparation, characterization, and applications. *Journal of Nanomaterials, 2016*(1), p.6717624.

[22] Jóźwiak, B., Greer, H.F., Dzido, G., Kolanowska, A., Jędrysiak, R., Dziadosz, J., Dzida, M. and Boncel, S., 2021. Effect of ultrasonication time on microstructure, thermal conductivity, and viscosity of ionanofluids with originally ultra-long multi-walled carbon nanotubes. *Ultrasonics Sonochemistry, 77*, p.105681.

[23] Kant, R., Saini, R., Kaur, M., Karnwal, A., Rather, H.A. and Maji, S., 2022, May. Synthesis and characterization of copper complexes of phenylpiperazine ring-based ligands and evaluation of their antimicrobial, BSA binding, computational, and molecular docking studies. In *Journal of Physics: Conference Series* (Vol. 2267, No. 1, p. 012154). IOP Publishing.

[24] Keblinski, P., Phillpot, S.R., Choi, S.U.S. and Eastman, J.A., 2002. Mechanisms of heat flow in suspensions of nano-sized particles (nanofluids). *International journal of heat and mass transfer, 45*(4), pp.855–863.

[25] Lau, K.Y., Vaughan, A.S., Chen, G., Hosier, I.L. and Holt, A.F., 2013, June. On nanosilica surface functionalization using different aliphatic chain length silane coupling agents. In *2013 IEEE international conference on solid dielectrics (ICSD)* (pp. 896-899). IEEE.

[26] Lee, J.H., Lee, S.H., Choi, C., Jang, S. and Choi, S., 2011. A review of thermal conductivity data, mechanisms and models for nanofluids. *International journal of micro-nano scale transport*.

[27] Lestari, M.L., Müller, R.H. and Möschwitzer, J.P., 2015. Systematic screening of different surface modifiers for the production of physically stable nanosuspensions. *Journal of pharmaceutical sciences, 104*(3), pp.1128–1140.

[28] Ma, M., Zhai, Y., Yao, P., Li, Y. and Wang, H., 2021. Effect of surfactant on the rheological behavior and thermophysical properties of hybrid nanofluids. *Powder Technology*, *379*, pp.373–383.

[29] Mahbubul, I.M., Shahrul, I.M., Khaleduzzaman, S.S., Saidur, R., Amalina, M.A. and Turgut, A.L.P.A.S.L.A.N., 2015. Experimental investigation on effect of ultrasonication duration on colloidal dispersion and thermophysical properties of alumina–water nanofluid. *International Journal of Heat and Mass Transfer*, *88*, pp.73–81.

[30] Mahtabani, A., Rytöluoto, I., Anyszka, R., He, X., Saarimäki, E., Lahti, K., Paajanen, M., Dierkes, W. and Blume, A., 2020. On the silica surface modification and its effect on charge trapping and transport in PP-based dielectric nanocomposites. *ACS Applied Polymer Materials*, *2*(8), pp.3148–3160.

[31] Mane, N.S. and Hemadri, V., 2022. Experimental investigation of stability, properties and thermo-rheological behaviour of water-based hybrid CuO and Fe3O4 nanofluids. *International Journal of Thermophysics*, *43*(1), p.7.

[32] Mane, N.S., Hemadri, V. and Tripathi, S., 2024. Exploring the role of biopolymers and surfactants on the electrical conductivity of water-based CuO, Fe3O4, and hybrid nanofluids. *Journal of Dispersion Science and Technology*, *45*(5), pp.900–908.

[33] Manimaran, M., Norizan, M.N., Kassim, M.H.M., Adam, M.R., Abdullah, N. and Norrahim, M.N.F., 2025. Critical review on the stability and thermal conductivity of water-based hybrid nanofluids for heat transfer applications. *RSC advances*, *15*(18), pp.14088–14125.

[34] Mebarek-Oudina, F. and Chabani, I., 2022. Review on nano-fluids applications and heat transfer enhancement techniques in different enclosures. *Journal of Nanofluids*, *11*(2), pp.155–168.

[35] Mikkola, V., Puupponen, S., Granbohm, H., Saari, K., Ala-Nissila, T. and Seppälä, A., 2018. Influence of particle properties on convective heat transfer of nanofluids. *International Journal of Thermal Sciences*, *124*, pp.187–195.

[36] Mishra, A., Rawat, S.K., Yaseen, M. and Pant, M., 2023. Development of machine learning algorithm for assessment of heat transfer of ternary hybrid nanofluid flow towards three different geometries: case of artificial neural network. *Heliyon*, *9*(11).

[37] Mittal, T. and Kant, R., 2024. Functional Nanohybrid and Nanocomposite Coatings for Sustainable Environmental Remediation. *Lett. Appl. NanoBioScience*, *13*, p.181.

[38] Muhammad, N.M.A. and Sidik, N.A.C., 2018. Applications of nanofluids and various minichannel configurations for heat transfer improvement: A review of numerical study. *Journal of Advanced Research in Fluid Mechanics and Thermal Sciences*, *46*(1), pp.49–61.

[39] Mukherjee, S., Wciślik, S., Mishra, P.C. and Chaudhuri, P., 2024. Nanofluids: Critical issues, economics and sustainability perspectives. *Particuology*, *87*, pp.147–172.

[40] Murshed, S.S. and de Castro, C.N., 2014. Thermophysical properties and heat transfer characteristics of carbon nanotubes dispersed nanofluids. In *Nanofluids: Synthesis, Properties and Applications* (pp. 53–76). Nova Science Publishers, Inc.

[41] Murshed, S.S. ed., 2022. *Fundamentals and transport properties of nanofluids*. Royal Society of Chemistry.

[42] Parab, A.R., Ramlal, A., Gopinath, S.C. and Subramaniam, S., 2025. Forging the future of nanotechnology: embracing greener practices for a resilient today and a sustainable tomorrow. *Frontiers in Nanotechnology*, *6*, p.1506665.

[43] Rahman, M.A., Hasnain, S.M., Pandey, S., Tapalova, A., Akylbekov, N. and Zairov, R., 2024. Review on nanofluids: preparation, properties, stability, and thermal performance augmentation in heat transfer applications. *ACS omega*, *9*(30), pp.32328–32349.

[44] Rahman, M.A., Hasnain, S.M., Pandey, S., Tapalova, A., Akylbekov, N. and Zairov, R., 2024. Review on nanofluids: preparation, properties, stability, and thermal performance augmentation in heat transfer applications. *ACS omega*, *9*(30), pp.32328–32349.

[45] Rashidi, M.M. and Zinatloo-Ajabshir, S. eds., 2024. Nanofluids: preparation, applications and simulation methods.

[46] Rashidi, M.M., Nazari, M.A., Mahariq, I., Assad, M.E.H., Ali, M.E., Almuzaiqer, R., Nuhait, A. and Murshid, N., 2021. Thermophysical properties of hybrid nanofluids and the proposed models: An updated comprehensive study. *Nanomaterials, 11*(11), p.3084.

[47] Razzaq, I., Xinhua, W., Rasool, G., Sun, T., Shflot, A.S., Malik, M.Y., Abbas, K., Ali, S. and Ali, A., 2025. Nanofluids for advanced applications: a comprehensive review on preparation methods, properties, and environmental impact. *ACS omega, 10*(6), pp.5251–5282.

[48] Saidur, R., Leong, K.Y. and Mohammed, H.A., 2011. A review on applications and challenges of nanofluids. *Renewable and sustainable energy reviews, 15*(3), pp.1646–1668.

[49] Serra, J.P., Andonegi, M., Salazar, H., Correia, D.M., de la Caba, K., Guerrero, P., Lanceros-Mendez, S. and Costa, C.M., 2025. Sustainable soft actuators based on chitosan and choline containing ionic liquids with high bending response. *International Journal of Biological Macromolecules*, p.145972.

[50] Sharma, A., Khanal, S., Suvedi, D., Yadav, N., Sharma, S., Verma, R., Kumar, D., Peter, L. and Kalova, M., 2025. The role of nanofluids in enhancing thermal management and biomedical applications: A review. *Thermal Science and Engineering Progress*, p.104437.

[51] Souza, R.R., Faustino, V., Gonçalves, I.M., Moita, A.S., Bañobre-López, M. and Lima, R., 2022. A review of the advances and challenges in measuring the thermal conductivity of nanofluids. *Nanomaterials, 12*(15), p.2526.

[52] Sun, Y., Yang, D., Shi, L., Wu, H., Cao, Y., He, Y. and Xie, T., 2020. Properties of nanofluids and their applications in enhanced oil recovery: a comprehensive review. *Energy & Fuels, 34*(2), pp.1202–1218.

[53] Tang, W., Zou, C., Peng, H., Wang, Y. and Shi, L., 2021. Influence of nanoparticles and surfactants on stability and rheological behavior of polymeric nanofluids and the potential applications in fracturing fluids. *Energy & Fuels, 35*(10), pp.8657–8671.

[54] Teng, T.P., Cheng, C.M. and Pai, F.Y., 2011. Preparation and characterization of carbon nanofluid by a plasma arc nanoparticles synthesis system. *Nanoscale research letters, 6*(1), p.293.

[55] Urmi, W.T., Rahman, M.M., Kadirgama, K., Ramasamy, D. and Maleque, M.A., 2021. An overview on synthesis, stability, opportunities and challenges of nanofluids. *Materials Today: Proceedings, 41*, pp.30–37.

[56] Vallejo, J.P., Prado, J.I. and Lugo, L., 2022. Hybrid or mono nanofluids for convective heat transfer applications. A critical review of experimental research. *Applied Thermal Engineering, 203*, p.117926.

[57] Wang, J., Li, G., Li, T., Zeng, M. and Sundén, B., 2021. Effect of various surfactants on stability and thermophysical properties of nanofluids. *Journal of Thermal Analysis & Calorimetry, 143*(6).

[58] Wang, J., Yang, X., Klemeš, J.J., Tian, K., Ma, T. and Sunden, B., 2023. A review on nanofluid stability: preparation and application. *Renewable and Sustainable Energy Reviews, 188*, p.113854.

[59] Wang, X., Xu, X. and Choi, S.U., 1999. Thermal conductivity of nanoparticle-fluid mixture. *Journal of thermophysics and heat transfer, 13*(4), pp.474–480.

[60] Wang, X.Q. and Mujumdar, A.S., 2007. Heat transfer characteristics of nanofluids: a review. *International journal of thermal sciences, 46*(1), pp.1–19.

[61] Weera, S., Lee, H. and Attar, A., 2020. Utilizing effective material properties to validate the performance of thermoelectric cooler and generator modules. *Energy Conversion and Management, 205*, p.112427.

[62] Yang, X. and Liu, Z.H., 2010. A kind of nanofluid consisting of surface-functionalized nanoparticles. *Nanoscale research letters, 5*(8), p.1324.

[63] Yasmin, H., Giwa, S.O., Noor, S. and Sharifpur, M., 2023. Thermal conductivity enhancement of metal oxide nanofluids: a critical review. *Nanomaterials*, *13*(3), p.597.

[64] Yu, W. and Xie, H., 2012. A review on nanofluids: preparation, stability mechanisms, and applications. *Journal of nanomaterials*, *2012*(1), p.435873.

[65] Zainon, S.N.M. and Azmi, W.H., 2021. Recent progress on stability and thermo-physical properties of mono and hybrid towards green nanofluids. *Micromachines*, *12*(2), p.176.

[66] Zhang, D., Wu, X., Han, N. and Chen, Y., 2013. Chemical vapor deposition preparation of nanostructured ZnO particles and their gas-sensing properties. *Journal of nanoparticle research*, *15*(4), p.1580.

[67] Zhu, H.T., Lin, Y.S. and Yin, Y.S., 2004. A novel one-step chemical method for preparation of copper nanofluids. *Journal of colloid and interface science*, *277*(1), pp.100–103.

# 2 Rheological Behavior and Flow Characteristics of Nanofluids

Amrita Tripure[1], Santosh Kumar Mishra[2*], Amrit Shende[3] and Manish Kumar[4]

Advanced thermal management systems are the new generational technology that focuses on improving efficiency and reducing power consumption. The use of nanofluids makes it more advanced as it is a superfluid with improved rheological and thermophysical properties. These properties include viscosity, particle size and shape, concentration, shear stress responses, and flow behaviour. Some external forces also affect flow behaviour, such as temperature, shear rate, and nanofluid morphology. The changes in flow behavior are transitional from Newtonian to non-Newtonian, which may include shear thinning, shear thickening, yield stress, and thixotropic properties. Additionally, the introduction of a magnetic field also impacts the flow behaviour. By increasing the magnetic field intensity, chain formation increases, and the alignment of particles rearranges. Flow behaviour also shows a significant transition with the tubular structure. The ferrofluids characterize the non-Newtonian behaviour. Among all available models, the Power-law model is best suited for non-Newtonian flow characterization and is widely accepted.

This chapter provides insight into nanofluid flow behaviour for various flow regimes. The characteristics from experimental and computational studies also highlight key insights of Oxide (CuO, $Al_2O_3$ dispersed in water and mineral oils) and carbon-based nanofluids. The hybrid nanofluid systems have also been studied. As rheological properties and thermophysical characteristics are key components for flow behaviour, their impacts have been discussed thoroughly. The impact of the magnetic field introduction on the flow behaviour is also studied. For thermal power analysis and pumping losses, temperature dependant viscosity criterion is observed. Case studies of hybrid nanofluid systems with viscosity variations, base fluids, and nanofluid combinations demonstrate transition from Newtonian to non-Newtonian behaviour, though the base fluid is Newtonian in nature. Finally, the challenges such as variation in numerical and empirical data, availability of universally accepted models have been discussed. The instruments used and limitations, and the technique to overcome non-uniformity caused by

[1,3] Department of Mechanical Engineering, Chhattisgarh Swami Vivekanand Technical University Bhilai, Chhattisgarh 491107 India.

[2,4] Department of Mechanical Engineering, Bhilai Institute of Technology, Durg, Chhattisgarh 491001 India.

* Corresponding author: santosh.mishra@bitdurg.ac.in

DOI: 10.1201/9781003686002-2

agglomeration and segregation have been evaluated, with recommendations provided for future research towards the stability-enhancing techniques, robust predictive framework, and optimizing nanofluid design.

## 1. Introduction

The colloidal solution of fluid with metallic and carbon-based tiny particles has traditionally been used for various applications. The evolution of this tiny particle from mm to nm was done by Choi and Eastman (1995). The particles with a size range between 1 and 100 nm were named as nanoparticles. The oxides of copper, alumina, and titanium, carbon-based, such as CNT and other polymer-based particles, are tested for nanoparticle preparation. The colloidal suspension formed by the process showed improved thermal characteristics. These nanofluids showed better thermal conductivity, surpassing the use of conventionally used mineral oil, for various lubrication as well as thermophysical properties-based applications. Despite being an excellent lubricant, the traditional mineral oil was becoming unstable at higher thermal stress. The introduction of the nanofluid eliminated these problems. These nanoparticles include metallic, non-metallic, oxides, carbon-based, carbides, ceramics, hybrid particles, as well as nanoscale liquid particles. The excellent thermophysical properties, enhanced thermal conductivity, improved thermal stability, and lower flow resistance supported the usefulness of such fluids. Due to their improved properties, these fluids are termed superfluids. For compatibility and stability, the selection of the base fluid and the nanoparticle is equally important. The base fluids currently in use are glycol-based : Ethylene Glycol (EG), propylene glycol, water, and carboxy methyl cellulose (CMC) solution. The rheological properties suitably suggest the use of a type of nanofluid according to the application.

As the nanofluid becomes a colloidal suspension, a transition in flow characteristics is observed from Newtonian to non-Newtonian. These flow characteristics are in better agreement with the power law, as it considers the consistency index as well as the flow behavior index. These are temperature-dependent properties where the flow index increases and the consistency index decreases as the temperature rises, bridging the viscosity and temperature relations (Chen, Ding, and Tan, 2007a). Other models, such as Herschel–Bulkley and Bingham–Plastic formulations, can also be used but are more suitable for shear-thinning fluids (Nik Salimi *et al.*, 2024; Shanmugapriya *et al.*, 2022; Shah *et al.*, 2021; Sharma, Tiwari, and Dixit, 2016).

This study presents an extensive synthesis of experimental and numerical research highlighting the Rheological and thermophysical properties. The different oxide-based (CuO and $Al_2O_3$ in water or mineral oils) and carbon based nanofluid show flow behavioral changes as the rheological properties change. These are studied in this chapter. The insightful effects of magnetic fields and the use of ferrofluids also provide a discussion on the transitional flow characteristics and the different models associated with it. The hybrid nanofluid study is provided to enhance the thermal performance and power utilization. The nanofluid also posses challenge as the stability and uniformity change as the particle is suspended, experiencing different forces. Techniques to achieve stability have also been focused on.

## 2. Fundamentals of Rheology in Nanofluids

Rheology refers to the flow characteristics of any fluid. The rate of shear strain and the variation caused by the shear stress are studied to highlight the effect in terms of rheology. The ratio of shear stress to the rate of shear strain is referred to as viscosity $(\mu)$. This viscosity variation with thermophysical and flow behaviour conditions identifies the type of flow. The flow of a field may show various characteristics that are broadly classified as Newtonian and Non-Newtonian fluids. Viscosity is directly related to flow characteristics in Newtonian fluid; the viscosity remains constant throughout the process, whereas in non-Newtonian fluid flow, the viscosity varies where the shear stress is applied. The complex behavior of viscosity under shear stress also highlights its benefits. The unique flow behavior makes these substances valuable for various industrial uses. Drilling mud, industrial waste, and cement slurry are examples of these fluids (Rahmani, Larachi, and Taghavi, 2024). In daily life, ketchup, toothpaste, blood, milk, etc., are also examples of Non-Newtonian fluids. When Newtonian fluids are characterized by viscosity only, Non-Newtonian fluids are way beyond viscosity. It uses rheometers to examine flow behavior at different shear stress rates. Such rheometers measure pressure during the fluid's flow. At the molecular level, the flow characteristics are determined when chain formation occurs as shear stress is applied. For example, in shear-thinning (pseudoplastic) fluids like blood, fluidity and viscosity increase with shear rate because, in high-tension planes, the molecular chains re-align. (Shende, Niasar, and Babaei, 2021). Alternatively, shear-thickening (dilatant) fluids, such as paints, exhibit enhanced viscosity in their behavior under stress owing to microstructural rearrangement, which amplifies resistance (You, Pan, and Wei, 2024; Wang *et al.*, 2024; Ansari *et al.*, 2020). This inconsistency highlights the need to use certain characterization effectively.

The presence of nanoparticles in nanofluids changes their flow behaviour and thermal characteristics. The characteristics are contingent upon the nanoparticle concentration, fluid temperature, characteristics of the base fluid, flow geometry, and external forces, like magnetic or electric fields (Choi and Eastman, 1995). Nevertheless, Nanofluids do not always act in the same manner. There are those Nanofluids that behave like Newtonian fluids with constant viscosity, whereas others exhibit Non-Newtonian behavior (María José Pastoriza-Gallego *et al.*, 2011; Hojjat *et al.*, 2011; Ahmadi Nadooshan, Eshgarf, and Afrand, 2018). The following are the influential parameters for the rheological properties.

## 2.1 Temperature Effects

Temperature influences Viscosity ($\mu$) and thermal conductivity ($k$) of nanofluids. Nanofluid viscosity also tends to reduce with temperature as fewer intermolecular forces attract each other, causing less resistance to flow (Table 1); this behavior is also described well through Andrade or Vogel Tammann Fulcher (VTF) equations (Mishra *et al.*, 2021). At the same time, the thermal conductivity ($k$) is also more likely to improve, boosting heat transfer functionality (Jama *et al.*, 2016; Patil *et al.*, 2016). Brownian motion in nanofluids is increased by raising the temperature. This also increases the thermal conductivity of fluids. An experimental study highlights the behaviour of thermal and viscous properties. The study also focuses on determining the impacts of agglomeration on these properties with temperature fluctuations (Apmann *et al.*, 2021).

Similar studies suggest that with increasing nanoparticle concentration, thermal conductivity also increases with the transient hot wire (THE). With an increase of 620% in the presence of 15% $Al_2O_3$ at room temperature and 16% at 55°C. The nanofluid showed a 60% increase in viscosity as the concentration increased from 1% to 5% at 25°C. Nevertheless, viscosity reduced exponentially with temperature, and was significantly greater with smaller nanoparticles, suggesting that particle size and temperature have a significant impact on the rheological properties of the $Al_2O_3$ nanofluids (Kwek, Crivoi, and Duan, 2010). With Lubrication, frictional hindrance and wear can be minimized, but mineral-based oils of concern to the environment, as well as biodegradable alternatives, should be used. The lubricants made up of nanofluids, such as hybrid nanofluids (HNFs) of Graphene (flat/platelet/textured) with oxides of Nickel in coconut oil, overcome the limitation of thermal instability as encountered in conventional mineral oils. The findings indicate that viscosity increases as the volume concentration of the nanoparticles rises and falls as the thermal gradient rises. On increasing mass fraction and temperature of silicon oxide ($SiO_2$) nanoparticles, thermal conductivity ($k$) of the nanofluid improved significantly. The empirical correlation formulated regarding the liquefied paraffin-based $SiO_2$ and oleic acid-based nanofluids, which considered the mass fraction of the nanoparticles and the temperature, had a deviation of less than 10% in the relative viscosity, thermal conductivity ($k$), and shape. For aluminum oxide suspensions, the high intrinsic thermal conductivity ($k$) of the alumina traces and their evolution into clusters contributed greatly to the thermophysical and rheological characteristics. The consistency of the obtained thermal diffusivity trends with literature values, together with the distinct representation of thermal propagation through mediate validated the reliability of the proposed discrete measurement methods. Furthermore, the influence of interatomic solid-solid interactive potential considering viscosity of HNFs Ar-Cu highlighted the effect of density variations in the laminar liquid layer at the nanoparticle-basefluid interface (Jabbari, Rajabpour, and Saedodin, 2017; Ghasemi and Karimipour, 2018; Nimdeo and Srivastava, 2018).

**Table 1** Temperature Effects on Nanofluid Flow Behavior. (Jama *et al.*, 2016; Patil *et al.*, 2016; Simpson *et al.*, 2019)

| *Temperature* | *Viscosity Behavior* | *Thermal Conductivity* | *Flow Implications* |
|---|---|---|---|
| Low | High (base fluid level) | Moderate | Higher flow resistance |
| Moderate | Reduced viscosity | Improved conductivity | Stable and improved heat transfer |
| High | Significantly reduced | Enhanced significantly | Reduced resistance |

**Nanoparticle Volume Fraction**

Increasing nanoparticle concentration generally raises both viscosity and thermal conductivity. At moderate concentrations (e.g., 1–4 vol%), the thermal conductivity ($k$) enhancement can be substantial, around 20–32%. In some $Al_2\ O_3$ cases, while viscosity increases, it imposes additional pumping demands (Jama *et al.*, 2016). At low concentrations, limited particle interaction maintains stability, while higher concentrations promote lamellar agglomeration, enhancing viscosity. An artificial neural network (ANN) based theoretical model was proposed to reliably estimate rheological properties of HNFs (Senniangiri, Bensam Raj, and Sunil, 2020). The MgO-MWCNT/ethylene glycol hybrid nanofluid appears Newtonian fluid in nature over the entire span of tested mass fractions and temperatures. Experimental observations showed an increase in dynamic viscosity with rising nanoparticle concentration and a decrease as temperature increased. A maximum relative viscosity enhancement of about 168% was reported as the mass fraction increased from 0.1–1(%), and a noticeable rise in dynamic viscosity was observed. According to experimental reports, a new correlation was developed to estimate the dynamic viscosity ($\mu$) of the HNFs, and the predicted results showed close agreement with the measured values (Soltani and Akbari, 2016). It was confirmed that higher nanoparticle concentrations increase viscosity, while elevated temperatures significantly decrease it. The diameter of carbon nanotubes had a negligible influence on viscosity, while CuO-based nanofluids exhibited a sharp increase in viscosity beyond a concentration of 1.5 wt%, with only minor changes observed below this limit. In addition, empirical correlations derived from molecular dynamics simulations and regression approaches were introduced to determine comparative viscosity as a function of nanoparticle mass fraction and temperature, with prediction errors remaining below 5%. Sensitivity analysis revealed that viscosity is more strongly affected by nanoparticle concentration than by temperature, providing useful insights for industrial applications of CNT/water and CuO nanofluids (Ghasemi and Karimipour, 2018; Jabbari, Saedodin, and Rajabpour, 2019).

Agglomeration of nanoparticles at high concentrations can degrade flow uniformity and stability (D and Surendran, 2016; Giwa *et al.*, 2021; Elfatah *et al.*, 2024). The behaviour is listed in Table 2. The rheological conduct of both the base and nanofluids was effectively described by the power-law model, along with flow behavior indices remaining below unity ($n < 1$). In comparison with the base fluid,

**Table 2** Influence of Particle Volume Fraction.
(D and Surendran, 2016; Giwa *et al.*, 2021; Elfatah *et al.*, 2024)

| *Volume Fraction (%)* | *Viscosity Impact* | *Thermal Conductivity Enhancement* | *Flow Behavior* |
|---|---|---|---|
| 0 (base fluid) | Baseline | Baseline | Newtonian |
| 0.1–1 | Slight increase | Noticeable | Near-Newtonian |
| 1–4 | Moderate increase | Significant (20–30%) | Non-Newtonian possibility |
| >4 | High increase | Plateau or marginal gains | Instabilities risk |

the nanofluids show a marginal drop in the power-law index. Furthermore, the mass fraction increment of the nanoparticle leads to a further decrease in the power-law index across all samples of nanofluid. At a constant nanoparticle concentration, however, the power-law index of both base and nanofluid increases with the temperature rise (Hojjat *et al.*, 2011).

The dynamic viscosity of $Fe_3 O_4$-MWCNT with ethylene glycol HNFs was investigated for mass fractions: 0.1, 0.25, 0.45, 0.8, 1.25, and 1.8. In this formulation, the same amounts of multi-walled CNT and iron oxide were suspended in the base fluid, and the assessments were conducted over a temperature ranging from 25–50°C. At lower concentrations: 0.1–0.8(%), the nanofluid exhibited Newtonian behavior, where viscosity is inversely proportional to temperature but increases with higher particle concentration. In contrast, at higher concentrations of 1.25 and 1.8%, the fluid transitioned to non-Newtonian behavior, which was well captured by the power-law model (Ahmadi Nadooshan, Eshgarf, and Afrand, 2018). Additional investigations showed noticeable variations in mean settling velocity, dispersion time scale, and particle arrangement within the sedimenting suspension as the particle concentration increased. These changes indicate a progressive adaptation from dilute: $\phi \leq 2\%$ to dense: $\phi \geq 10\%$ regime. Within the dense regime, kinematic waves with vertical direction were detected, which appeared as oscillatory patterns in the locally and horizontally averaged particle concentration (Shajahan and Breugem, 2020).

The influence of nanofluids was also examined using a double-pipe counter-current heat exchanger, in which $SiO_2$ nanoparticles were suspended in water at concentrations between 0% and 2%, with particle sizes varying between 25 and 50 nm. Viscosity computations were carried out for different temperatures and particle sizes, while the friction factor and heat transfer coefficient were evaluated under various operating conditions. The results indicated that viscosity increased with higher nanoparticle loading, and both the heat transfer coefficient and friction factor rose as nanoparticle mass concentration increased and particle size reduced. In addition, the effect of nanoparticle agglomeration on thermal conductivity ($k$) was investigated, revealing a significant enhancement in Fe-based nanofluids subjected to high-power sonication. With prolonged duration, a decline in thermal conductivity ($k$) was closely linked to the agglomeration of nanoparticles. For

Fe-based nanofluids, thermal conductivity (*k*) increased in a nonlinear manner with rising nanoparticle volume fraction, resulting in fast particle clustering at increased concentrations. In contrast, a linear relationship was observed only at lower volume fractions. When compared with Cu-based nanofluids, Fe nanofluids demonstrated a more pronounced enhancement in thermal conductivity (*k*) with the rise in nanoparticle volume fraction (Ghasemi and Karimipour, 2018; Hong, Hong, and Yang, 2006).

**Addition of Surfactants**

The performance of various surfactants in stabilizing CNT-water nanofluids and improving their thermo-physical properties is examined. Long-term stability exceeding six months was achieved using Gum Arabic (GA), and Polyvinyl pyrrolidone (PVP) at mixing ratios of 1:0.5 and 1:1. At a CNT concentration of 0.1 wt%, thermal conductivity and specific heat capacity were improved by 1.5 times and 1.36 times, respectively. Experiments were conducted in a shell-and-tube type heat exchanger. The CNT volume fraction is optimized to 0.5% by volume; the heat-transfer rate increases by 65%, accompanied by only a 1.1-fold increase in pressure drop. Additionally, the power required for pumping was approximately one-third of that needed for water, indicating superior thermal performance (Almanassra *et al.*, 2020). An increase in Reynolds number or nanoparticle concentration resulted in higher Nusselt numbers, with all nanofluids outperforming distilled water. Nevertheless, at Reynolds numbers exceeding 30,000, additional pumping power was required to overcome the higher pressure drop associated with nanofluids, even though the relative enhancement in the Nusselt number remained nearly constant across the investigated Reynolds number range (Abbasian Arani and Amani, 2012).

**Nanoparticle Size and Shape**

Particle geometry and scale significantly determine nanofluid thermophysical performance. Smaller particles enhance thermal properties by offering a larger specific surface area, although they also lead to increased viscosity (Jama *et al.*, 2016; Rahman *et al.*, 2024). On increasing particle size from 10 nm to 35 nm, thermal conductivity increases by 60%, followed by a rise to 75% for 150 nm particles. High-aspect-ratio shapes (e.g., rods, needles, wires) outperform spheres in thermal conductivity (*k*) but influence stability and increased viscosity (Razzaq *et al.*, 2025). In certain studies, cubic-shaped nanoparticles exceeded rods and spheres in conductivity; however, viscosity remained relatively unchanged across shapes in low-volume tests. Platelets and bricks yield varied viscosity responses, with platelets showing the highest viscosity (Apmann *et al.*, 2021). Table 3 shows the Viscosity dependence on nanoparticle morphology and dimensions. This study examined how nanoparticle shape, size, and concentration collectively influence rheological properties of nanofluids. Using DLVO theory and laser scattering, it was found that repulsive interaction energy, governed by particle shape, primarily controls suspension stability. Different shapes, such as brick, blade, and platelet, show an increase of nearly 27, 16, and 23%, respectively. The similarity is also observed in stability. Further experiments with $Al_2O_3$ and ZnO nanoparticles (20–100 nm, 1–15 vol%) in engine oil revealed that viscosity depends strongly on both particle size and concentration, exhibiting three regimes: decreasing, constant, and increasing with particle size. Overall, nanoparticle geometry and dispersion stability were identified as key factors influencing the thermophysical behavior of nanofluids (Kim *et al.*, 2015; Hu *et al.*, 2020). The aggregation behavior of hematite nanoparticles with hydrated radii of 12 ± 2 nm, 32 ± 3 nm, and 65 ± 3 nm was investigated under pH and ionic strength conditions relevant to the environment. The findings revealed a clear size-dependent trend, where smaller particles exhibited faster aggregation rates and lower critical coagulation concentrations compared to larger ones. As particle size decreased, rapid aggregation shifted toward lower pH levels, likely due to changes in structure and surface energy affecting the point of zero charge (PZC). Moreover, higher particle concentrations further accelerated aggregation under identical conditions. Although nanoparticles of different sizes responded differently to pH and ionic strength, their overall aggregation behavior could still be qualitatively explained using DLVO theory(He, Wan, and Tokunaga, 2008).

**Table 3** Impact of Nanoparticle Shape and Size.
(Apmann *et al.*, 2021; Li *et al.*, 2024).

| *Particle Shape/Size* | *Thermal Conductivity* | *Viscosity Impact* | *Flow Note* |
|---|---|---|---|
| Small, spherical | Moderate increment | Moderate | Balanced performance |
| Rod/needle/wire (high AR) | High increment | Moderate to high | Great conductivity, higher resistance |
| Cubic | Higher than rods/ spheres | Similar to other shapes | Improved conductivity, stable flow |
| Platelet | Very high | Highest viscosity | Risk of clogging, viscous flow |

### Shear-Rate Dependence

Nanofluid viscosity can vary with shear rate, resulting in shear-thinning or shear-thickening behavior depending on nanoparticle type and concentration (Table 4). Under conditions of low shear rates, fluids exhibit Non-Newtonian characteristics; at higher rates, they often approximate Newtonian behavior(Jiang and Zhou, 2019).

### Aggregation and Interfacial Effects

Aggregation morphology influences both viscosity and heat transfer. Well-dispersed particles typically yield better performance; clusters may improve thermal conductivity *(k)* but at the risk of higher viscosity and reduced stability (Kleinstreuer and Feng, 2011; Razzaq *et al.*, 2025). Interfacial layers around nanoparticles also contribute notably to thermal behavior beyond classical Maxwell models (Li *et al.*, 2022).

**Table 4** Shear Rate Influence on Rheology (Jiang and Zhou, 2019).

| *Shear Rate* | *Viscosity Response* | *Flow Type* | *Implications* |
|---|---|---|---|
| Low | Viscosity may increase or stabilize | Newtonian or shear-thickening | Sensitivity to micro flow |
| Moderate | Viscosity decreases | Shear-thinning | Enhanced pumping and mixing |
| High | Viscosity stabilizes | Newtonian-like | Predictable behavior in turbulent flow |

Nanofluid flow behavior is significantly influenced across a broad spectrum of parameters, including temperature, shear rate, nanoparticle characteristics, and external fields. Understanding these dependencies is crucial for designing efficient microfluidic devices and thermal systems (Jiang and Zhou, 2019; Demirkır and Ertürk, 2021; Miadzvedzeva *et al.*, 2025).

The Effect on flow behavior at different conditions can be understood from Table 5.

**Table 5** Effect on flow behavior at different conditions (Jiang and Zhou, 2019). (Demirkır and Ertürk, 2021)(Miadzvedzeva *et al.*, 2025)

| *S.No.* | *Condition* | *Effect on flow behavior* |
|---|---|---|
| 1 | Temperature Increases | Viscosity decreases, Brownian motion increases, and thermal conductivity increases |
| 2 | Nanoparticle concentration increases | Viscosity increases, Non-Newtonian behavior increases, Thermal conductivity increases |
| 3 | Shear rate increases | Often leads to shear-thinning; sometimes Newtonian behavior |
| 4 | Laminar vs. Turbulent flow | More effective in laminar flow; turbulent flow reduces the benefits |
| 5 | Base fluid change | Affects viscosity, stability, and thermal response |
| 6 | Particle type/shape | Alters thermal and rheological properties |
| 7 | Magnetic/Electric fields | Dynamically control flow in magnetic nanofluids |
| 8 | Confinement (microchannels) | Enhanced effects, slip flow may occur |

## 3. Experimental Methods for Rheological Measurement

The rheological properties of nanofluids can be evaluated using a rotational rheometer (HAAKE RV12) and a programmable Brookfield viscometer (DV-II + Pro). The viscometer, equipped with an S18 spindle, is mounted in a container holding a minimum of 16 ml of nanofluid. Viscosity ($\mu$) and shear stress of water-based alumina and titanium oxide nanofluids recorded at various shear stress rates ($\varphi$) and temperature range (T). The measurement system was interfaced via RS232, with data acquisition and storage managed through RHEOCALC V3.3 software.

Experiments can be performed for volume fraction of 0.1–2.0(%) and temperature ranging between 293–333 K, with a temperature control accuracy of ±0.5 K. With each experiment, viscosity and rate of shear stress were recorded with incremental scale of T = 5K and shear stress and viscosity between 12 and 232 $s^{-1}$ were recorded, corresponding to spindle speeds of 10–190 rpm. The viscometer speed was adjusted in 10 rpm increments and maintained for 30 s at each step. To ensure temperature stability upto ±0.1 K, a viscometer chamber with thermostatic flow throughout the experiments (Das *et al.*, 2022).

Viscosity can be measured using various equipment:

1. Brookfield DV2T: the viscometer is capable of identifying the flow behaviour from Newtonian and Non-Newtonian nature. In these viscometers, other rheological properties like mass fraction and base fluid type significantly influence the flow behavior. This highlights the importance of nano-suspended particles in characterizing the flow field. In this type of viscometer, different cone and plate structures are attached with CPA-40Z, and all these attachments are fitted with a rotational viscometer. The cone specifications are: diameter = 75 mm and cone angle = 1°. The viscosity measurement was facilitated by a 200-Physica rheometer. Each sample should have uniformity in process and combinations to achieve accuracy (Hasan, 2017).
2. Brookfield CAP2000: this can be used to examine the rheological behaviour of HNFs such as MWCNT-MgO/SAE40. Viscosity measurement over the range of temperature 25–50°C and $\varphi = 0.0625 - 1(\%)$ was employed.
3. Anton Paar rheometer (MCR-301): spindle immersed in the NF sample rotates at a controlled speed range of 0.01 to 250 rpm. Calibration was performed using standard fluids provided by Brookfield Engineering Laboratories. All measurements were conducted under steady-state conditions, with 3 experimental sets, and the accuracy achieved was up to 94% (Jamal-abad *et al.*, 2014a).
4. Anton Paar (MCR302, modular compact rheometer: this rheometer has a Peltier temperature-controlled cell (C-PTD200), which records the dynamic viscosity of HNFs. Measurements of rheological parameters such as viscosity, rate of shear stress, and strain rate were performed at spindle speeds ranging from 1 to 100 $s^{-1}$ (Mohammed Zayan *et al.*, 2022). The CC45 DIN spindle works better for an incremental T of 5°C, and the range lies between 25–50°C, in accordance with previously reported studies. At elevated temperatures, convective effects in water-based nanofluids may destabilize rheological behavior; therefore, this temperature range was selected to ensure reliable characterization. The ternary HNF samples were introduced into the rheometer chamber. The spindle is fully immersed and rotating inside. This spindle has a double gap, and when the spindle rotates, the drag force is experienced. The drag force generated by the spindle is used to calculate the deflection amplitude. All thermophysical properties, including shear rate and viscosity, were calculated at room temperature, with the rheometer's data logging system continuously recording the measured

parameters. The instrument demonstrated high measurement precision, with an accuracy of ±1% and spindle speed repeatability within ±0.2%. Notably, the ternary HNFs exhibited counteractive viscous behavior in response to spindle movement and rotation (Zayan *et al.*, 2022).

## 4. Rheological Models Applied to Nanofluids

The clustering of particles, concentration of nanoparticles, base fluid interaction, and use of surfactants often demonstrate Non-Newtonian behavior in Nanofluids. These behaviors are studied using various rheological models depending on the fluid's flow regime, stress response, and time-dependency (Nabwey *et al.*, 2023). Table 6 shows rheological models applied to nanofluids.

**Table 6** Rheological models applied to nanofluids (Nabwey *et al.*, 2023).

| *S.No.* | *Non-Newtonian Model* | *Applicability to nanofluid* | *Practical applications* |
|---|---|---|---|
| 1 | Power Law (Ostwald de Waele) | Shear-thinning/thickening fluids without yield stress | Cooling systems, lubrication, microfluidics |
| 2 | Carreau | Wide shear rate modeling: Newtonian to Non-Newtonian | Biomedical flow, industrial coating |
| 3 | Ellis | Shear-thinning without yield stress | Paints, cosmetics, food nanofluids |
| 4 | Cross Bingham | Yield stress with non-linear shear | Drilling fluids, waste slurry transport |
| 5 | Bingham plastic | Yield stress followed by linear flow | Cement slurries, oil recovery, suspensions |
| 6 | Herschel–Bulkley | Yield stress and shear-thinning | Energy systems, enhanced geothermal fluids |
| 7 | Casson | Smooth shear-thinning; mild yield stress | Biomedical nanofluid flow (arterial, drug delivery) |
| 8 | Maxwell Model | Viscoelastic behavior; relaxation | Medical diagnostics, smart gels |
| 9 | Jeffreys Model | Viscoelastic and rate-dependent | Artificial muscle fluids, oscillatory systems |
| 10 | Upper Convected Maxwell (UCM) Model | Polymeric nanofluids | CFD simulations, soft robotics |
| 11 | Oldroyd-B Model | Bio-nanofluids, structured suspensions | Throat/sinus flows, dynamic nano-drugs |
| 12 | Bautista–Manero Model | Bio-polymeric nanofluids | Biolubricants, tissue engineering gels |
| 13 | Godfrey Model | Surfactant-stabilized nanofluids | Detergents, dynamic coatings, thickeners |

## 5. Flow Characteristics of Nanofluid

Nanofluids show the behaviour of both Newtonian and Non-Newtonian in nature. It is influenced by various factors such as the average size of nanoparticles, concentration (%), the structure or shape of particles, the base fluid used to create the nanofluid, the type of surfactant used, the range of shear rate, and the magnetic field. At near ambient temperature, Brownian diffusion becomes negligible compared to convective forces (Hojjat *et al.*, 2011). Due to this high shear, viscosity becomes independent of temperature. In the theoretical framework, nanofluids are classified depending on the particle loading and structural organization. These nanofluids are termed as semi-dilute, dilute, and semi-concentrated, concentrated nanofluids. The flow characteristics are also classified as:

**Newtonian Fluids**

A fluid that obeys Newton's law of viscosity is termed a Newtonian fluid. According to this law, shear stress in any fluid is directly proportional to the rate of shear strain. Mathematically, it is represented as

$$\tau_f = \mu_f \left( \frac{du}{dy} \right)_f \qquad ...(1)$$

Where $\tau_f$ is shear stress, $\mu_f$ is dynamic viscosity and $\left( \frac{du}{dy} \right)_f$ is velocity gradient.

- Dilute suspensions: $Al_2O_3$–water nanofluids at ≤1 vol% showed Newtonian behavior across wide temperature ranges (Ruvo, Md Shahneoug Shuvo, and Saha, 2024).
- CuO/water nanofluids: Reported nearly constant viscosity under varying shear rates.
- High shear rates: Even fluids with nanoparticles behave Newtonian when shear rate dominates particle interactions(Hojjat *et al.*, 2011).

**Non-Newtonian Fluids**

The fluids that do not obey Newton's law of viscosity are termed Non-Newtonian fluids and can be represented as

$$\tau_f = K \left( \frac{du}{dy} \right)^n + B \qquad ...(2)$$

Where $\tau_f$ is shear stress, $\frac{du}{dy}$ is the velocity gradient. $K$ is the consistency index, and $n$ is the flow behavior index. $B$ is the minimum shear stress. Different flow behaviour has been shown in Table 7.

**Table 7** Various fluid flow characteristics with a power law index.

| *Power-law Index (n)* | *Fluid Behavior* | *Viscosity Trend with Shear Rate* | *Examples/ Observations* | *Remarks/References* |
|---|---|---|---|---|
| **n = 1** | Newtonian fluid | Viscosity remains constant | Water, air, simple liquids | Viscosity independent of shear rate |
| **n < 1** | Pseudoplastic fluid | Viscosity ↓ with shear rate ↑ | Ketchup, paint, blood | Graphene nanofluids: Newtonian behavior below 1 wt% and shear-thinning above 1 wt% ; CMC-based nanofluids consistently exhibit pseudoplastic behavior ($n < 1$); Nanoparticle aggregation induces pseudoplasticity, reversible through ultrasonication |
| **n > 1** | Dilatant fluid | Viscosity ↑ with ↓ shear rate | Cornstarch–water mixture | Exhibits higher shear rates |

- A **Rheopectic fluid:** Non-Newtonian fluid in which viscosity increases with time under continuous stress or shear, a phenomenon referred to as time-thickening. Such fluids become progressively thicker when agitated or stirred. Typical examples include certain printer inks, lubricants, and gypsum pastes, which solidify or stiffen upon prolonged movement.
- A **Thixotropic fluid:** Non-Newtonian fluid in which viscosity decreases over time under constant shear stress (such as stirring or shaking) but gradually regains its original state after removal of stress. This time-dependent thinning behavior is observed in materials, such as paints, ketchup, and mayonnaise, which flow easily when disturbed but regain thickness when left undisturbed. Ethylene glycol/$Fe_2O_3$ nanofluids displayed time-dependent viscosity decay under constant shear (María Jose Pastoriza-Gallego *et al.*, 2011).

Non-Newtonian nanofluids have been further characterized for their properties such as dispersion, transportation, magnetism, optical, and radioactivity. Among these, rheology remains central, as viscosity and shear behavior directly influence flow regimes and energy requirements in thermal systems (Tawalbeh, Shomope, and Al-Othman, 2024). Instruments such as capillary viscometers, rotational rheometers, and piston rheometers are commonly used, each suitable for specific ranges of fluid type and shear rate to measure these properties. With the Experimental results, it is observed that nanoparticles often form aggregates rather than remaining as isolated particles. Suspended solutions behave as highly concentrated non-uniform fluids when hydrodynamic forces cannot disrupt these aggregates. This aggregation leads to shear-thinning responses when nanofluids typically have particle concentrations below 0.15. As the particle concentration and aggregate size increase, the characteristic shear rate decreases, reflecting the reduced contribution of Brownian motion relative to convective effects (Patil *et al.*, 2016; Rudyak *et al.*, 2023).

Hybrid nanofluids (HNFs) are produced by mixing distinct nanoparticles and represent a next-generation class of working fluids with superior thermophysical properties and heat transfer impacts. Their rheological behavior is strongly influenced by shear rate, temperature, and solid volume fraction, with many studies implying their tendency toward Non-Newtonian characteristics. For example, experimental, numerical, and statistical observations on MWCNT (25%)–MgO (75%)/SAE40 nanofluids reported shear-thinning behavior at certain temperatures and volume fractions, where the flow index is n = 0.9233 (Pare and Ghosh, 2021).

Viscosity is the most influential property, as observed by many researchers. This viscosity is also a multilevel property in HNFs. These show that with increasing temperature, the viscosity decreases. The motion of suspended particles is the prime reason for that. On further investigation, increasing the volume concentrations, viscosity also increases (0.02–1%) due to the inter-particle interactions. Shear rate is a key factor in flow regime, as it also categorizes the fluid into Newtonian and non-Newtonian flow characteristics. Various models, like ANN and CNN, are used to capture the trend and development. These models give more accurate and precise results for flow behaviour and characteristics (Hemmat Esfe, Alidoust, *et al.*, 2022; Hemmat Esfe, Toghraie, *et al.*, 2022).

Base fluid is the prime factor in nanofluid preparation. The behaviors of a base fluid: Newtonian and non-Newtonian influences greatly affect the characteristics of the prepared nanofluid. Most non-Newtonian nanofluids use water or polymer solutions like carboxymethyl cellulose (CMC) as base fluids and have a volume concentration of less than 2%. Research reflects that 25% of nanofluid shows a non-Newtonian nature even though the base fluid used is Newtonian. This complex property highlights the interplay between base fluid and nanoparticle (Nabwey *et al.*, 2023). These studies may raise important questions in current research: 1) Which Non-Newtonian models should be used to record the nanofluid flow characteristics? 2) What are the most commonly used base fluids and their flow behaviour? And which numerical methods are most effective for solving Non-Newtonian flow equations?

The synergistic interactions of nanoparticles and their flow behavior in any non-Newtonian nanofluid enhance the heating and cooling efficiency of a heat exchanger, making its design more compact and superior for thermal utilization. An experiment was performed in a square channel by Nasiri et al. (Nasiri *et al.*, 2020) to investigate the cooling capability of such fluids. To form Newtonian nanofluid, EG and water were used as base fluid, and 0.5%. Water and ethylene glycol (EG) were used, while a 0.5% CMC-water solution was used to form a non-Newtonian formulation. $TiO_2$ and $\gamma$-$Al_2O_3$ were used as nanoparticles. By varying the concentration of these particles, the study revealed that non-Newtonian nanofluids consistently achieved better performance irrespective of the type of nanoparticle or base fluid (Hasan, 2017; Nabwey *et al.*, 2023).

With such great properties, nanofluid exhibits great challenges of stability and uniformity. Due to Van-Der-Walls forces, the suspended nanoparticle tends to form a cluster, compromising the stability and altering thermophysical properties (Ali, Teixeira, and Addali, 2018; Dey, Kumar, and Samantaray, 2017). Long-term stability becomes essential for their use in practical applications. To overcome these issues of clustering and segregation, many techniques have been developed, such as a) Chemical and b) Mechanical. In chemical techniques, surfactant, electrostatic, steric, and electrosteric techniques are widely used. The surfactant tends to have a strong dispersion force on nanoparticle surfaces, which makes them resistant to agglomeration (Ghadimi, Saidur, and Metselaar, 2011). But the drawback with surfactant is that it changes the nanoparticle surface, which is often undesirable. The mechanical techniques involve magnetic stirring, sonication, homogenization, ball milling, and vibration. These techniques do not alter the properties of the nanofluid and are hence more preferable in practical systems. Research demonstrated higher stability optimization for water-graphite nanofluid with volume concentration 0.1% with sonication. (Yashawantha and Vinod, 2021).

To study stability, different techniques have been explored. These techniques include Zeta Potential Analysis, Transmission Electron Microscopy (TEM), Dynamic Light Scattering, Spectral Absorbance, Sedimentation, Centrifugation, and Visual Inspection (Yu *et al.*, 2025; Mehta and Subhedar, 2022). Zeta potential shows the most promising results, having higher absolute values depicting a strong repulsive force between particles and enhanced stability (Hojjat *et al.*, 2011). Zeta potential is highly negative in gold, which makes Au-water nanofluids stable for up to a month (Kim *et al.*, 2010).

Colloidal suspensions in polymer-based or complex fluids often show nonlinear shear stress-strain relationships (Wang *et al.*, 2024). Common base fluids include distilled and de-ionized water, EG, hydrogenated oil, engine oil, polymer solutions, CMC solutions, therminol, blood, and other (Tawalbeh, Shomope, and Al-Othman, 2024). Widely used nanoparticles include $Al_2O_3$, $TiO_2$, CuO, Ag, and hybrid combinations. While Newtonian nanofluids typically exhibit constant or temperature-dependent viscosity, Non-Newtonian nanofluids show shear rate- or temperature-dependent viscosity (Hemmat Esfe, Alidoust, *et al.*, 2022).

Beyond stability and viscosity, the improved thermal conductivity ($k$) and enhanced Brownian motion support superior thermal performance capabilities in Non-Newtonian nanofluids in comparison to both base fluids and Newtonian nanofluids (Sheikhpour *et al.*, 2020; Mandal, Anand, and Prabhakar, 2023). These properties are used in solar collectors, cooling devices, heat exchangers, surface coatings, lubricants, and biomedical applications, such as imaging, sensing, separation, and drug delivery. (Shutaywi and Shah, 2021).

Mathematical modeling of Non-Newtonian nanofluids typically relies on mass, momentum, and energy conservation equations, coupled with similarity transformations and boundary conditions. Among numerical techniques, the finite volume method (FVM) is most widely employed for solving these governing equations. Aspects of the fluid flow characteristics have been shown in Table 8.

**Table 8** Aspects of nanofluid flow behaviour.

| *Aspect* | *Newtonian Nanofluids* | *Non-Newtonian Nanofluids* |
|---|---|---|
| Viscosity relation | Shear stress is directly proportional to shear rate; constant viscosity | Non-linear stress–strain relation; viscosity depends on shear rate |
| Typical models | Newton's law of viscosity | Herschel-Bulkley Eq.(2), Bingham-Plastic, and Power-Law models,(Hojjat *et al.*, 2011) |
| Equations used | $\tau = \mu \frac{du}{dy}$, | $\tau = K\left(\frac{du}{dy}\right)^n$ and $\eta = K\left(\frac{du}{dy}\right)^{n-1}$ |
| | | $K$ and $n$ are the Power Law constants, $\eta$ is the apparent viscosity of the fluid. |
| Common base fluids | Water, ethylene glycol, oils | Polymeric solutions (CMC, XG, PVP), gels |
| Practical assumption | Simplifies CFD and thermal design | Requires rheological model fitting, complex analysis |

The base fluid (heat transfer oil) exhibited shear-thickening behavior, whereas the nanofluids showed shear-thinning, indicating that nanoparticle addition altered the rheology. $Al_2O_3$ had the strongest effect, lowering the power-law index by about 5% at 2% volume fraction and increasing the consistency index by ~20% (compared to <14% for CuO and $TiO_2$). The dynamic viscosity ratio (nanofluid/base fluid) was less than one, showing that effective viscosity can decrease at specific concentrations and strain rates. This reduction in viscosity offers practical advantages for lubrication and pumping by minimizing energy losses during operation (Jamal-abad *et al.*, 2014b). The governing factor that significantly defines nanofluid behaviour is mentioned in Table 9

**Table 9** Governing factors of different fluid flow behaviour.

| *Factor* | *Newtonian Response* | *Non-Newtonian Response* |
|---|---|---|
| Concentration ($\phi$) | <1 vol% generally Newtonian (Aladag *et al.*, 2012) | >1–2 vol% → shear-thinning or shear-thickening |
| Temperature (T) | High temperature lowers viscosity, reinforces Newtonian behavior | Low Temperature enhances particle interactions and non-Newtonian transitions |
| Shear Rate ($\dot{\gamma}$) | At high $\dot{\gamma}$, nanofluids approximate Newtonian | At low $\dot{\gamma}$, shear-thinning or viscoelasticity dominates (Rudyak *et al.*, 2023) |

*Contd.*

**Table 9** *Contd.*

| *Factor* | *Newtonian Response* | *Non-Newtonian Response* |
|---|---|---|
| Particle Morphology | Spherical nanoparticles ($Al_2O_3$, CuO) favor Newtonian | Rods, sheets, nanotubes exhibits strong shear-thinning (Rudyak *et al.*, 2023) |
| Base Fluid | Water, EG, oils, Newtonian baseline | Polymeric solutions have inherent pseudoplasticity |
| Stability | Well-dispersed, Newtonian(Ruvo, Md. Shahneoug Shuvo and Saha, 2024) | Aggregated, non-Newtonian, thixotropy (Duan, Kwek and Crivoi, 2011) |

The rheological properties of EG-based nanofluids having α-$Fe_2O_3$ nanoparticles were examined at 303.15 K using a cone–plate rheometer. Nanoparticle concentrations were varied up to 25 wt%. The results showed clear non-Newtonian behavior. The nanofluids exhibited shear-thinning characteristics. Viscosity measured at a constant shear rate changed with time. This confirmed the thixotropic nature of the nanofluids (María Jose Pastoriza-Gallego *et al.*, 2011). Some nanofluids were also found to follow the Herschel–Bulkley model, showing shear thinning behavior. Numerical studies indicated that an increase in the power-law index reduces the average Nusselt number. This reduction becomes stronger at higher Grashof numbers (Kefayati and Bassom, 2021). These rheological properties significantly influence heat transfer performance. Related applications are summarized in Table 10.

**Table 10** Heat Transfer Implications.

| *Aspect* | *Newtonian Nanofluids* | *Non-Newtonian Nanofluids* |
|---|---|---|
| Laminar flow | Predictable enhancement due to higher thermal conductivity (Hojjat *et al.*, 2011) | Shear-thinning increases the velocity gradient near walls, boosting the Nusselt number |
| Turbulent flow | Performance similar to base fluids but with modest thermal gains (Ajeeb *et al.*, 2021) | Higher turbulence and mixing; enhanced heat transfer, but large pressure drops |
| Microchannels | Stable, predictable pressure drop | Strong shear dependence; higher pumping penalties but improved heat transfer (Ajeeb *et al.*, 2021) |
| Applications | Electronic cooling, solar collectors with simple fluids | EV battery cooling, polymeric heat exchangers where shear-thinning helps manage hotspots |

The observation suggests that spherical-shaped nanoparticles exhibit Newtonian behavior, whereas nanotubes containing fluids behave as Non-Newtonian fluids. Higher shear rate predominates the Non-Newtonian behaviour even though low shear behaviour is Newtonian in nature (Kumar, Kumar and Rai, 2016).

**Table 11** Nanofluids and Their Rheological Properties.

| *NP + BF* | *Volume Concentration ($\phi$, %)* | *Size (nm)* | *Shear Rate($s^{-1}$)* | *Flow behaviour* | *Reference* |
|---|---|---|---|---|---|
| $SiO_2$ + $TiO_2$ + Deionized water | 0.468 | 0.16 to 1.73 | 0–500 | $SiO_2$ Newtonian; mixed $SiO_2/TiO_2$-Bingham plastic:with higher plastic viscosity. | (Richmond, Jones, and Fawell, 1998) |
| $TiO_2$ + Pure water | 0.05–0.12 | 7–20 | 10–1000 | Shear thinning; thixotropy appeared above 0.1%. | (Tseng and Lin, 2003) |
| $TiO_2$ + Distilled water | 0.24–1.18 | 20–95 | 0.1–1000 | Shear thinning - Newtonian transition at ≈ 100 $s^{-1}$; viscosity ↑ with $\phi$ and size. | (Michaelides, 2013) |
| $TiO_2$ + EG | 0.1–1.8 | 25 | 0–200 | Newtonian; viscosity ∝ $\phi$, independent of T. | (Chen, Ding and Tan, 2007b) |
| $TiO_2$ + EG | 0.5–8 | 70–100 | 0.5–$10^4$ | Newtonian; viscosity↑ with T↑ and $\phi$↑ | (Chen *et al.*, 2007) |
| $TiO_2$ + water (Distilled) | 0.12–0.6 | 10 | 1–$10^4$ | Shear thinning, intensified with volume fraction; viscosity↓ with T↑. | (Zhang *et al.*, 2015) |
| $TiO_2$ + water(Deionized) | 0.2–3 | 21 | — | $\phi$ ↓- Newtonian<br>$\phi$ ↑- non-Newtonian. | (Turgut *et al.*, 2009) |
| $TiO_2$ + Water | 0.013–0.05 | 5 to 90 | 0–100 | Newtonian at 1–100 $s^{-1}$; shear-thinning above 100 $s^{-1}$. | (Alphonse, Bleta and Soules, 2009) |
| $TiO_2$ + Water | 0.05–0.65 | 20 to 100 | — | Newtonian. | (Penkavova, Tihon and Wein, 2011) |
| $Al_2O_2$ + Water | 0.1–0.5 | 25 | 350–1000 | Shear thinning | (Hojjat *et al.*, 2011) |

*Contd.*

**Table 11** *Contd.*

| **NP + BF* | *Volume Concentration ($\phi$, %)* | *Size (nm)* | *Shear Rate($s^{-1}$)* | *Flow behaviour* | *Reference* |
|---|---|---|---|---|---|
| $TiO_2$ + EG | 1.36–8.83 | 35 to 47 | 0.1–1000 | Shear thinning: viscosity↓ with ϕ↑ | (Ghasemi and Karimipour, 2018) |
| MWCNT + Polycarbonate | 0.5–5 | 10 to 15 | — | >2 wt% → non-Newtonian; ≤1 wt% -Newtonian. | (Pötschke, Fornes and Paul, 2002) |
| MWCNT / PAO6 oil | 0.3–8. | — | 100 | Shear-thinning low & high dispersant; Newtonian at 3 wt%. | (Ueji *et al.*, 2019) |
| MWCNT +EG | 0.5–4 | — | 10–120 | Newtonian; viscosity ↓ with Temperature. | (Meng, Wu, *et al.*, 2012) |
| $SiO_2$ + Ethanol | 1.1–7.0 | 35 to 190 | $0–5\times10^4$ | Newtonian | (Chevalier, Tillement and Ayela, 2007) |
| $SiO_2$ + Ethanol | 0.4–3.1 | 10 to 100 | 10–1000 | Newtonian | (Chevalier, Tillement and Ayela, 2009) |
| $SiO_2$ + Paraffinic oil | 1.0–2.0 | 20 | 1–1000 | Newtonian (30°C); shear thinning (100°C) | (Kanjirakat *et al.*, 2014) |
| CuO +EG + Water (60:40) | 1–6.12 | 29 | 0–8 | Newtonian. | (Namburu *et al.*, 2007) |
| CuO + Xanthan gum solution | 0.1–0.5 wt% | Less than 50 | 10–1200 | Shear thinning | (William *et al.*, 2014) |
| ZnO + EG | 0.01–0.05 | 10 to 20 | 0–100 | <0.02 -Newtonian; >0.03 -shear-thinning. | (Yu *et al.*, 2009) |
| MgO + EG | 0.5–5 | 20 | 10–150 | Newtonian | (Xie, Yu and Chen, 2010) |

*Contd.*

**Table 11** *Contd.*

| *NP + BF | Volume Concentration ($\phi$, %) | Size (nm) | Shear Rate($s^{-1}$) | Flow behaviour | Reference |
|---|---|---|---|---|---|
| Au + Water | 0.01 | 10 to 50 | 200–2000 | Newtonian | (Abdelhalim, Mady and Ghannam, 2011) |
| Carbon black + EG | 2.2–7.8 | 20 | 6–120 | Shear-thinning- increasing with loading | (Meng, Han, *et al.*, 2012) |
| $Y_2O_3$/EG | 1–20 wt% | 31 | 0.01–2000 | Newtonian (low volume fraction), shear-thinning (high volume fraction) | (Cholewa *et al.*, 2013) |

*NP-nanoparticles, BF- base fluid

The rheological conduct of different nanofluids has been analyzed under different conditions, with their specific Newtonian and non-Newtonian characteristics summarized in Table 11.

## 6. Magnetic Field Effects (Ferrofluids)

Magnetic nanofluids, or ferrofluids, are stable colloidal suspensions of ferromagnetic or ferrimagnetic nanoparticles such as $Fe_3O_4$, $Fe_2O_3$, Fe, and $\gamma$-$Fe_2O_3$ in polar or non-polar carrier liquids. Ferrofluids show improved tunable thermal as well as rheological properties when brought to a magnetic field. The magnetic field induced promotion of chain formation in the particles, which enhances heat transfer in the pathways. (Maganti *et al.*, 2016; Prasad, Singh, and Nagar, 2017).

The enhanced behaviour of ferrofluids in terms of viscosity and thermal conductivity ($k$) has been summarized in Table 12. The flow behaviour, as seen previously greatly influenced by concentration. For $Fe_3O_4$-water nanofluids, at lower concentrations, the fluid shows Newtonian flow, whereas at higher concentrations shows non-Newtonian flow behaviour (Hong *et al.*, 2007). The influence of base fluid has been observed in $Fe_2O_3$-ethylene glycol (María Jose Pastoriza-Gallego *et al.*, 2011) and $Fe_2O_3$-glycerol (Abareshi *et al.*, 2011), which shows shear-thinning behaviour, whereas $Fe_2O_3$-oil (transformer) behaves as Newtonian behaviour. Under stronger magnetic fields, Fe–Ni/paraffin-oil and $Fe_3O_4$–$TiO_2$/silicone-oil systems demonstrate transition between Newtonian and Bingham-plastic characteristics. Additionally, it is observed that viscosity increases with magnetic field strength up to a critical limit, after which it decreases despite further field intensification (Wei *et al.*, 2009).

**Table 12** Magnetic Field Influence on Ferrofluid Behavior (Maganti *et al.*, 2016; Prasad, Singh, and Nagar, 2017; Lin and Yang, 2019).

| *Magnetic Field Intensity* | *Particle Alignment* | *Viscosity Change* | *Thermal Conductivity* | *Notes* |
|---|---|---|---|---|
| No magnetic field | Random | Baseline | Standard | General cooling applications |
| Moderate | Partial alignment | Moderate increase | Improved | Useful in microfluidics |
| Strong | Full alignment | Significant increment | Sharp enhancement | Smart cooling, Micro-Electro-Mechanical Systems |

## 7. Conclusion

This chapter focuses on the analytical role of rheological behavior in determining the performance and applicability of nanofluids in advanced thermal management systems. Among available nanoparticles, alumina ($Al_2O_3$) is the most extensively used due to its favorable rheological stability, cost-effectiveness, and ability to enhance heat transfer performance. Nanofluids generally exhibit both Newtonian and non-Newtonian flow behaviors, making them more suitable and commonly used in studies. Among these non-Newtonian flow characteristics, shear thinning, shear thickening, and thixotropic flows are more commonly employed. To analyze the shear stress characteristics of this flow behavior, the Power-law model is the most widely used and applies to most cases. The usefulness of these nanofluids is greatly influenced by their rheological and thermophysical properties, such as the shape and size of nanoparticles, volume concentration, viscosity, etc. These properties help in determining the forces related to the flow resistance, friction factor, and pumping power. Other external parameter such as temperature, shear stress, and ambient conditions, also influence the flow behaviour. The transition of Newtonian to Non-Newtonian flow behaviour can be seen in the grapheme-water nanofluid. This nanofluid, while working under low shear, behaves as Newtonian, while under higher shear, behaves as a non-Newtonian fluid. On introducing a magnetic field, this nanofluid is also influenced. The magnetic field induces the magnetic rheological properties in magnetic as well as tubular structured particles. However, although such modifications improve thermal conductivity *(k)* and convective performance, they also increase viscosity and pressure drop, creating a trade-off between enhanced heat transfer and greater pumping power demand. Despite the progress made through experimental and modeling approaches, significant challenges remain, including variability in reported data, the lack of universally accepted predictive models, and the difficulty of balancing thermal performance with energy demands. Overall, the chapter concludes that the proper selection of nanofluid should be based on specific applications, as the nanofluid acts as a superfluid with highly improved rheological properties. On selecting the nanofluid, the flow behaviour should be coupled with operating conditions for optimum results. The full potential of nanofluid can be realized if stability is also enhanced using effective techniques. While future research can focus on improving stability and reducing agglomeration to promote consistency in the fluid and, hence, the thermophysical properties throughout the life cycle of the nanofluids.

## References


[1] Abareshi, M. et al. (2011) 'Fabrication, characterization, and measurement of viscosity of α-Fe2O3-glycerol nanofluids', *Journal of Molecular Liquids*, 163, pp. 27–32. doi:10.1016/j.molliq.2011.07.007.

[2] Abbasian Arani, A.A. and Amani, J. (2012) 'Experimental study on the effect of $TiO_2$-water nanofluid on heat transfer and pressure drop', *Experimental Thermal and Fluid Science*, 42, pp. 107–115. doi:10.1016/j.expthermflusci.2012.04.017.

[3] Abdelhalim, M.A.K., Mady, M.M. and Ghannam, M.M. (2011) 'Rheological and dielectric properties of different gold nanoparticle sizes', *Lipids in Health and Disease*, 10(1), p. 208. doi:10.1186/1476-511X-10-208.

[4] Ahmadi Nadooshan, A., Eshgarf, H. and Afrand, M. (2018) 'Measuring the viscosity of $Fe_3O_4$-MWCNTs/EG hybrid nanofluid for evaluation of thermal efficiency: Newtonian and

non-Newtonian behavior', *Journal of Molecular Liquids*, 253, pp. 169–177. doi:https://doi.org/10.1016/j.molliq.2018.01.012.

[5] Ajeeb, W. et al. (2021) 'Forced convection heat transfer of non-Newtonian MWCNTs nanofluids in microchannels under laminar flow', *International Communications in Heat and Mass Transfer*, 127, p. 105495. doi:https://doi.org/10.1016/j.icheatmasstransfer.2021.105495.

[6] Aladag, B. et al. (2012) 'Experimental investigations of the viscosity of nanofluids at low temperatures to cite this version : HAL Id : hal-00707410 temperatures'.

[7] Ali, N., Teixeira, J.A. and Addali, A. (2018) 'A Review on Nanofluids: Fabrication, Stability, and Thermophysical Properties', *Journal of Nanomaterials*, 2018(1), p. 6978130. doi:https://doi.org/10.1155/2018/6978130.

[8] Almanassra, I.W. et al. (2020) 'An experimental study on stability and thermal conductivity of water/CNTs nanofluids using different surfactants: A comparison study', *Journal of Molecular Liquids*, 304, p. 111025. doi:10.1016/j.molliq.2019.111025.

[9] Alphonse, P., Bleta, R. and Soules, R. (2009) 'Effect of PEG on rheology and stability of nanocrystalline titania hydrosols.' *Journal of colloid and interface science*, 337(1), pp. 81–87. doi:10.1016/j.jcis.2009.04.087.

[10] Ansari, S. et al. (2020) 'Measurement of the flow behavior index of Newtonian and shear-thinning fluids via analysis of the flow velocity characteristics in a mini-channel', *SN Applied Sciences*, 2(11), p. 1787. doi:10.1007/s42452-020-03561-w.

[11] Apmann, K. et al. (2021) 'Thermal Conductivity and Viscosity: Review and Optimization of Effects of Nanoparticles', *Materials*, 14(5). doi:10.3390/ma14051291.

[12] Chen, H. et al. (2007) 'Rheological behaviour of ethylene glycol based titania nanofluids', *Chemical Physics Letters*, 444, pp. 333–337. doi:10.1016/j.cplett.2007.07.046.

[13] Chen, H., Ding, Y. and Tan, C. (2007a) 'Rheological behaviour of nanofluids', *New Journal of Physics*, 9. doi:10.1088/1367-2630/9/10/367.

[14] Chen, H., Ding, Y. and Tan, C. (2007b) 'Rheological behaviour of nanofluids', *New Journal of Physics*, 9, p. 367. doi:10.1088/1367-2630/9/10/367.

[15] Chevalier, J., Tillement, O. and Ayela, F. (2007) 'Rheological properties of nanofluids flowing through microchannels', *Applied Physics Letters - APPL PHYS LETT*, 91. doi:10.1063/1.2821117.

[16] Chevalier, J., Tillement, O. and Ayela, F. (2009) 'Structure and rheology of SiO2 nanoparticle suspensions under very high shear rates.', Physical review. *E, Statistical, nonlinear, and soft matter physics*, 80(5 Pt 1), p. 51403. doi:10.1103/PhysRevE.80.051403.

[17] Choi, S.U.S. and Eastman, J. (1995) 'Enhancing thermal conductivity of fluids with nanoparticles', in Proceedings of the ASME International Mechanical Engineering Congress and Exposition.

[18] Cholewa, M. et al. (2013) 'Viscosity of yttrium oxide (Y 2 O 3) and magnesium-aluminum spinel (MgAl 2 O 4) nanopowder suspensions in ethyl alcohol', pp. 131–142.

[19] Das, P.K. et al. (2022) 'Effect of particle loading and temperature on the rheological behavior of $Al_2O_3$ and $TiO_2$ nanofluids', *Energy Sources, Part A: Recovery, Utilization, and Environmental Effects*, 44(3), pp. 7062–7079. doi:10.1080/15567036.2022.2103214.

[20] Demirkır, Ç. and Ertürk, H. (2021) 'Convective heat transfer and pressure drop characteristics of graphene-water nanofluids in transitional flow', *International Communications in Heat and Mass Transfer*, 121, p. 105092. doi:https://doi.org/10.1016/j.icheatmasstransfer.2020.105092.

[21] Dey, D., Kumar, P. and Samantaray, S. (2017) 'A review of nanofluid preparation, stability, and thermo-physical properties', *Heat Transfer—Asian Research*, 46(8), pp. 1413–1442. doi:https://doi.org/10.1002/htj.21282.

[22] Duan, F., Kwek, D. and Crivoi, A. (2011) 'Viscosity affected by nanoparticle aggregation in $Al_2O_3$-water nanofluids.', *Nanoscale research letters*, 6(1), p. 248. doi:10.1186/1556-276X-6-248.

[23] Elfatah, M.G.A. et al. (2024) 'State of art on the role of using a nanofluid in enhancing heat transfer in microchannels with different geometries', *Journal of Engineering and Applied Science*, 71(1), p. 218. doi:10.1186/s44147-024-00557-w.

[24] Ghadimi, A., Saidur, R. and Metselaar, H.S.C. (2011) 'A review of nanofluid stability properties and characterization in stationary conditions', *International Journal of Heat and Mass Transfer*, 54(17), pp. 4051–4068. doi:https://doi.org/10.1016/j.ijheatmasstransfer.2011.04.014.

[25] Ghasemi, S. and Karimipour, A. (2018) 'Experimental investigation of the effects of temperature and mass fraction on the dynamic viscosity of CuO-paraffin nanofluid', *Applied Thermal Engineering*, 128, pp. 189–197. doi:10.1016/j.applthermaleng.2017.09.021.

[26] Giwa, S.O. et al. (2021) 'Experimental Investigation on Stability, Viscosity, and Electrical Conductivity of Water-Based Hybrid Nanofluid of MWCNT-$Fe_2O_3$', *Nanomaterials*, 11(1). doi:10.3390/nano11010136.

[27] Hasan, S.M. (2017) 'Huskie Commons Non-Newtonian rheological characteristics of oil-based metal oxide nanofluids'.

[28] He, Y.T., Wan, J. and Tokunaga, T. (2008) 'Kinetic stability of hematite nanoparticles: The effect of particle sizes', *Journal of Nanoparticle Research*, 10(2), pp. 321–332. doi:10.1007/s11051-007-9255-1.

[29] Hemmat Esfe, M., Alidoust, S., et al. (2022) 'Experimental Study of Rheological Behavior of MWCNT-$Al_2O_3$/SAE50 Hybrid Nanofluid to Provide the Best Nano-lubrication Conditions', Nanoscale Research Letters, 17(1). doi:10.1186/s11671-021-03639-3.

[30] Hemmat Esfe, M., Toghraie, D., et al. (2022) 'Investigating the rheological behavior of a hybrid nanofluid (HNF) to present to the industry', Heliyon, 8(12), p. e11561. doi:10.1016/j.heliyon.2022.e11561.

[31] Hojjat, M. et al. (2011) 'Rheological characteristics of non-Newtonian nanofluids: Experimental investigation', *International Communications in Heat and Mass Transfer*, 38(2), pp. 144–148. doi:10.1016/j.icheatmasstransfer.2010.11.019.

[32] Hong, K.S., Hong, T.K. and Yang, H.S. (2006) 'Thermal conductivity of Fe nanofluids depending on the cluster size of nanoparticles', *Applied Physics Letters*, 88(3), pp. 1–3. doi:10.1063/1.2166199.

[33] Hong, R. et al. (2007) 'Magnetic field synthesis of Fe 3O 4 nanoparticles used as a precursor of ferrofluids', *Journal of Magnetism and Magnetic Materials - J MAGN MAGN MATER*, 310, pp. 37–47. doi:10.1016/j.jmmm.2006.07.026.

[34] Hu, X. et al. (2020) 'Experimental investigation and mechanism analysis: Effect of nanoparticle size on viscosity of nanofluids', *Journal of Molecular Liquids*, 314, p. 113604. doi:10.1016/j.molliq.2020.113604.

[35] Jabbari, F., Rajabpour, A. and Saedodin, S. (2017) 'Thermal conductivity and viscosity of nanofluids: A review of recent molecular dynamics studies', *Chemical Engineering Science*, 174, pp. 67–81. doi:10.1016/j.ces.2017.08.034.

[36] Jabbari, F., Saedodin, S. and Rajabpour, A. (2019) 'Experimental Investigation and Molecular Dynamics Simulations of Viscosity of CNT-Water Nanofluid at Different Temperatures and Volume Fractions of Nanoparticles', *Journal of Chemical and Engineering Data*, 64(1), pp. 262–272. doi:10.1021/acs.jced.8b00783.

[37] Jama, M. et al. (2016) 'Critical Review on Nanofluids: Preparation, Characterization, and Applications', *Journal of Nanomaterials*, 2016, pp. 1–22. doi:10.1155/2016/6717624.

[38] Jamal-abad, M.T. et al. (2014a) 'An experimental investigation of rheological characteristics of non-Newtonian nanofluids', *Journal of Heat and Mass Transfer Research*, 00(2013), pp. 17–23.

[39] Jamal-abad, M.T. et al. (2014b) '*Journal of Heat and Mass Transfer Research An experimental investigation of rheological characteristics of non- Newtonian nanofluids*', 00(2013), pp. 17–23.

[40] Jiang, Y. and Zhou, X. (2019) 'Analysis of flow and heat transfer characteristics of nanofluids surface tension driven convection in a rectangular cavity', *International Journal of Mechanical Sciences*, 153. doi:10.1016/j.ijmecsci.2019.01.034.

[41] Kanjirakat, A. et al. (2014) 'Rheology of mineral oil-$SiO_2$ nanofluids at high pressure and high temperatures', *International Journal of Thermal Sciences*, 77, pp. 108–115. doi:10.1016/j.ijthermalsci.2013.10.016.

[42] Kefayati, G. and Bassom, A.P. (2021) 'A lattice Boltzmann method for single- and two-phase models of nanofluids: Newtonian and non-Newtonian nanofluids', Physics of Fluids, 33(10), p. 102008. doi:10.1063/5.0067744.

[43] Kim, H.J. et al. (2015) 'Effect of particle shape on suspension stability and thermal conductivities of water-based bohemite alumina nanofluids', Energy, 90, pp. 1290–1297. doi:10.1016/j.energy.2015.06.084.

[44] Kim, S. et al. (2010) 'Effects of nano-fluid and surfaces with nano structure on the increase of CHF', Experimental Thermal and Fluid Science, 34(4), pp. 487–495. doi:https://doi.org/10.1016/j.expthermflusci.2009.05.006.

[45] Kleinstreuer, C. and Feng, Y. (2011) 'Experimental and theoretical studies of nanofluid thermal conductivity enhancement: a review', *Nanoscale Research Letters*, 6(1), p. 229. doi:10.1186/1556-276X-6-229.

[46] Kumar, Anuj, Kumar, Arun and Rai, A. (2016) 'Rheological behaviour of nano fl uids : A review', 53, pp. 779–791. doi:10.1016/j.rser.2015.09.033.

[47] Kwek, D., Crivoi, A. and Duan, F. (2010) 'Effects of temperature and particle size on the thermal property measurements of $Al_2O_3$ - Water nanofluids', *Journal of Chemical and Engineering Data*, 55(12), pp. 5690–5695. doi:10.1021/je1006407.

[48] Li, C. et al. (2024) 'No Title', *Nanotechnology Reviews*, 13(1), p. 20240058. doi:10.1515/ntrev-2024-0058.

[49] Li, X. et al. (2022) 'Heat Transfer Enhancement of Nanofluids with Non-Spherical Nanoparticles: A Review', Applied Sciences, 12(9). doi:10.3390/app12094767.

[50] Lin, J. and Yang, H. (2019) 'A review on the flow instability of nanofluids', *Applied Mathematics and Mechanics*, 40(9), pp. 1227–1238. doi:10.1007/s10483-019-2521-9.

[51] Maganti, L. et al. (2016) 'Particle and thermohydraulic maldistribution of nanofluids in parallel microchannel systems', *Microfluidics and Nanofluidics*, 20. doi:10.1007/s10404-016-1769-3.

[52] Mandal, D., Anand, K. and Prabhakar, D.P. (2023) 'Biomedical Applications of Nanofluids in Drug Delivery', in, pp. 105–124. doi:10.1007/978-981-99-5281-6_5.

[53] Mehta, B. and Subhedar, D. (2022) 'Review on mechanism and parameters affecting thermal conductivity of nanofluid', *Materials Today: Proceedings*, 56, pp. 2031–2037. doi:https://doi.org/10.1016/j.matpr.2021.11.374.

[54] Meng, Z., Wu, D., et al. (2012) 'Carbon nanotube glycol nanofluids: Photo-thermal properties, thermal conductivities and rheological behavior', Particuology, 10, pp. 614–618. doi:10.1016/j.partic.2012.04.001.

[55] Meng, Z., Han, D., et al. (2012) 'Thermal Conductivities, Rheological Behaviors and Photothermal Properties of Ethylene Glycol-based Nanofluids Containing Carbon Black Nanoparticles', *Procedia Engineering*, 36, pp. 521–527. doi:10.1016/j.proeng.2012.03.076.

[56] Miadzvedzeva, M. et al. (2025) 'Heat Transfer and Flow Dynamics for Natural Convection in $Fe_3O_4$/H2O Nanofluid', *Energies*, 18(11). doi:10.3390/en18112767.

[57] Michaelides, E.E.S. (2013) 'Heat Transfer with Nanofluids', in *Heat and Mass Transfer in Particulate Suspensions. New York, NY*: Springer New York, pp. 121–164. doi:10.1007/978-1-4614-5854-8_4.

[58] Mishra, S.K. et al. (2021) 'CFD study of heat transfer effect on nanofluid of Newtonian and non-Newtonian type under vibration', *Chemical Product and Process Modeling*, 16(4), pp. 251–260. doi:10.1515/cppm-2020-0027.

[59] Mohammed Zayan, J. et al. (2022) 'Investigation on Rheological Properties of Water-Based Novel Ternary Hybrid Nanofluids Using Experimental and Taguchi Method', *Materials*, 15(1). doi:10.3390/ma15010028.

[60] Nabwey, H.A. et al. (2023) 'A Comprehensive Review of Non-Newtonian Nanofluid Heat Transfer', Symmetry, 15(2). doi:10.3390/sym15020362.

[61] Namburu, P. et al. (2007) 'Viscosity of copper oxide nanoparticles dispersed in ethylene glycol and H2O mixture', *Experimental Thermal and Fluid Science*, 32, pp. 397–402. doi:10.1016/j.expthermflusci.2007.05.001.

[62] Nasiri, M. et al. (2020) 'Cooling performance of Newtonian and non-Newtonian nanofluids in a square channel: experimental investigation and ANN modeling', *Journal of Thermal Analysis and Calorimetry*, 142(6), pp. 2189–2202. doi:10.1007/s10973-020-09309-3.

[63] Nik Salimi, N.E.B. et al. (2024) 'Rheological behavior predictions of non-Newtonian nanofluids via correlations and artificial neural network for thermal applications', Digital Chemical Engineering, 12(April), p. 100170. doi:10.1016/j.dche.2024.100170.

[64] Nimdeo, Y.M. and Srivastava, A. (2018) 'Understanding the temperature dependence of thermo-physical properties of nanofluid suspensions using non-intrusive dynamic measurements', Experimental Thermal and Fluid Science, 94, pp. 109–121. doi:10.1016/j.expthermflusci.2018.01.019.

[65] Pare, A. and Ghosh, S.K. (2021) 'A unique thermal conductivity model (ANN) for nanofluid based on experimental study', *Powder Technology*, 377, pp. 429–438. doi:https://doi.org/10.1016/j.powtec.2020.09.011.

[66] Pastoriza-Gallego, María Jose et al. (2011) 'Rheological non-Newtonian behaviour of ethylene glycol-based $Fe_2O_3$ nanofluids', *Nanoscale Research Letters*, 6(1), p. 560. doi:10.1186/1556-276X-6-560.

[67] Pastoriza-Gallego, María José et al. (2011) 'Thermal conductivity and viscosity measurements of ethylene glycol-based Al2O3 nanofluids', Nanoscale Research Letters, 6(1), p. 221. doi:10.1186/1556-276X-6-221.

[68] Patil, M.S. et al. (2016) 'Review on Synthesis, Thermo-Physical Property, and Heat Transfer Mechanism of Nanofluids', Energies, 9(10). doi:10.3390/en9100840.

[69] Penkavova, V., Tihon, J. and Wein, O. (2011) 'Stability and rheology of dilute TiO2-water nanofluids.', Nanoscale research letters, 6(1), p. 273. doi:10.1186/1556-276X-6-273.

[70] Pötschke, P., Fornes, T.D. and Paul, D.R. (2002) 'Rheological behavior of multiwalled carbon nanotube/polycarbonate composites', Polymer, 43, pp. 3247–3255. doi:10.1016/S0032-3861(02)00151-9.

[71] Prasad, A.R., Singh, S. and Nagar, H. (2017) 'A Review on Nanofluids: Properties and Applications', *International Journal Of Advance Research And Innovative Ideas In Education*, 3, pp. 3185–3209.

[72] Raghunath, D. and Surendran, R. (2016) 'Investigation of Rheological Behaviour and Heat Transfer Performance of Alumina Nanofluid', *International Journal of Engineering Research and, V5*. doi:10.17577/IJERTV5IS050181.

[73] Rahman, M.A. et al. (2024) 'Review on Nanofluids: Preparation, Properties, Stability, and Thermal Performance Augmentation in Heat Transfer Applications', *ACS Omega*, 9(30), pp. 32328–32349. doi:10.1021/acsomega.4c03279.

[74] Rahmani, H., Larachi, F. and Taghavi, S.M. (2024) 'Modeling of Shear Flows over Superhydrophobic Surfaces: From Newtonian to Non-Newtonian Fluids', *ACS Engineering Au*, 4(2), pp. 166–192. doi:10.1021/acsengineeringau.3c00048.

[75] Razzaq, I. et al. (2025) 'Nanofluids for Advanced Applications: A Comprehensive Review on Preparation Methods, Properties, and Environmental Impact', *ACS Omega*, 10(6), pp. 5251–5282. doi:10.1021/acsomega.4c10143.

[76] Richmond, W.R., Jones, R.L. and Fawell, P.D. (1998) 'The relationship between particle aggregation and rheology in mixed silica–titania suspensions', *Chemical Engineering Journal*, 71(1), pp. 67–75. doi:https://doi.org/10.1016/S1385-8947(98)00105-3.

[77] Rudyak, V.Y. et al. (2023) 'Comparative characteristics of viscosity and rheology of nanofluids with multi-walled and single-walled carbon nanotubes', *Diamond and Related Materials*, 132, p. 109616. doi:https://doi.org/10.1016/j.diamond.2022.109616.

[78] Ruvo, T.H., Shuvo, Md Shahneoug and Saha, S. (2024) 'Revisiting thermo-physical property models of $Al_{(2)}O_{(3)}$-Water nanofluid for natural convective flow.', *Heliyon*, 10(5), p. e26964. doi:10.1016/j.heliyon.2024.e26964.

[79] Ruvo, T.H., Shuvo, Md. Shahneoug and Saha, S. (2024) 'Revisiting thermo-physical property models of $Al_2O_3$-Water nanofluid for natural convective flow', *Heliyon*, 10(5), p. e26964. doi:https://doi.org/10.1016/j.heliyon.2024.e26964.

[80] Senniangiri, N., Bensam Raj, J. and Sunil, J. (2020) 'Effects of Temperature and Particles Concentration on the Dynamic Viscosity of Graphene-NiO/Coconut Oil Hybrid Nanofluid: Experimental Study', *International Journal of Nanoscience*, 19(3), pp. 1–8. doi:10.1142/S0219581X19500169.

[81] Shah, Z. et al. (2021) 'Impact of nanoparticles shape and radiation on the behavior of nanofluid under the Lorentz forces', *Case Studies in Thermal Engineering*, 26, p. 101161. doi:https://doi.org/10.1016/j.csite.2021.101161.

[82] Shajahan, T. and Breugem, W.P. (2020) 'Influence of Concentration on Sedimentation of a Dense Suspension in a Viscous Fluid', *Flow, Turbulence and Combustion*, 105(2), pp. 537–554. doi:10.1007/s10494-020-00172-8.

[83] Shanmugapriya, M. et al. (2022) 'Impact of nanoparticle shape in enhancing heat transfer of magnetized ternary hybrid nanofluid', *Sustainable Energy Technologies and Assessments*, 53, p. 102700. doi:https://doi.org/10.1016/j.seta.2022.102700.

[84] Sharma, A.K., Tiwari, A.K. and Dixit, A.R. (2016) 'Rheological behaviour of nanofluids: A review', *Renewable and Sustainable Energy Reviews*, 53, pp. 779–791. doi:10.1016/j.rser.2015.09.033.

[85] Sheikhpour, M. et al. (2020) 'Role of Nanofluids in Drug Delivery and Biomedical Technology: Methods and Applications.' *Nanotechnology, science and applications*, 13, pp. 47–59. doi:10.2147/NSA.S260374.

[86] Shende, T., Niasar, V.J. and Babaei, M. (2021) 'An empirical equation for shear viscosity of shear thickening fluids', *Journal of Molecular Liquids*, 325, p. 115220. doi:https://doi.org/10.1016/j.molliq.2020.115220.

[87] Shutaywi, M. and Shah, Z. (2021) 'Mathematical Modeling and numerical simulation for nanofluid flow with entropy optimization', *Case Studies in Thermal Engineering*, 26, p. 101198. doi:https://doi.org/10.1016/j.csite.2021.101198.

[88] Simpson, S. et al. (2019) 'Nanofluid Thermal Conductivity and Effective Parameters', *Applied Sciences*, 9(1). doi:10.3390/app9010087.

[89] Soltani, O. and Akbari, M. (2016) 'Effects of temperature and particles concentration on the dynamic viscosity of MgO-MWCNT/ethylene glycol hybrid nanofluid: Experimental study', *Physica E: Low-Dimensional Systems and Nanostructures*, 84, pp. 564–570. doi:10.1016/j.physe.2016.06.015.

[90] Tawalbeh, M., Shomope, I. and Al-Othman, A. (2024) 'Comprehensive review on non-Newtonian nanofluids, preparation, characterization, and applications', *International Journal of Thermofluids*, 22(May), p. 100705. doi:10.1016/j.ijft.2024.100705.

[91] Tseng, W. and Lin, K.-C. (2003) 'Rheology and colloidal structure of aqueous TiO2 nanoparticle suspensions. Materials Science and Engineering A, 355, 186-192', doi:10.1016/S0921-5093(03)00063-7.

[92] Turgut, A. et al. (2009) 'Thermal Conductivity and Viscosity Measurements of Water-Based $TiO_2$ Nanofluids', *International Journal of Thermophysics*, 30, pp. 1213–1226. doi:10.1007/s10765-009-0594-2.

[93] Ueji, K. et al. (2019) 'Thermophysical properties of a single-wall carbon nanotube thin film on Au electrodes evaluated by a time-domain thermoreflectance method', *Japanese Journal of Applied Physics*, 58. doi:10.7567/1347-4065/ab57b7.

[94] Wang, S. et al. (2024) 'Characteristics of shear thickening fluid and its application in engineering: a state-of-the-art review', *The International Journal of Advanced Manufacturing Technology*, 133(5), pp. 1973–2000. doi:10.1007/s00170-024-13816-0.

[95] Wei, J. et al. (2009) 'Synthesis and magnetorheological effect of $Fe_3O_4$-$TiO_2$ nanocomposite', *Journal of Physics: Conference Series*, 149, p. 12083. doi:10.1088/1742-6596/149/1/012083.

[96] William, J. et al. (2014) 'Effect of CuO and ZnO Nanofluids in Xanthan Gum on Thermal, Electrical and High Pressure Rheology of Water-Based Drilling Fluids', *Journal of Petroleum Science and Engineering*, 117. doi:10.1016/j.petrol.2014.03.005.

[97] Xie, H., Yu, W. and Chen, W. (2010) 'MgO nanofluids: higher thermal conductivity and lower viscosity among ethylene glycol-based nanofluids containing oxide nanoparticles', *Journal of Experimental Nanoscience*, 5(5), pp. 463–472. doi:10.1080/17458081003628949.

[98] Yashawantha, K.M. and Vinod, A.V. (2021) 'ANN modelling and experimental investigation on effective thermal conductivity of ethylene glycol:water nanofluids', *Journal of Thermal Analysis and Calorimetry*, 145(2), pp. 609–630. doi:10.1007/s10973-020-09756-y.

[99] You, A., Pan, Y. and Wei, M. (2024) 'Mechanism of shear-thickening fluid and factors influencing its rheological properties: a review', *Bulletin of Materials Science*, 47(3), p. 212. doi:10.1007/s12034-024-03313-1.

[100] Yu, J. et al. (2025) 'Experimental investigation on nanofluid saturated pool boiling CHF with different chemical additives', *Progress in Nuclear Energy*, 189, p. 105934. doi:https://doi.org/10.1016/j.pnucene.2025.105934.

[101] Yu, W. et al. (2009) 'Investigation of thermal conductivity and viscosity of ethylene glycol based ZnO nanofluid', *Thermochimica Acta - THERMOCHIM ACTA*, 491, pp. 92–96. doi:10.1016/j.tca.2009.03.007.

[102] Zayan, J.M. et al. (2022) 'Investigation on rheological properties of water-based novel ternary hybrid nanofluids using experimental and taguchi method', *Materials*, 15(1). doi:10.3390/ma15010028.

[103] Zhang, J. et al. (2015) 'An experimental study on flow and heat transfer characteristics of nanofluids in multiport minichannel flat tube', *Journal of Engineering Thermophysics*, 36, pp. 1071–1076.

# 3 Thermal and Electrical Conductivity Enhancement Mechanisms in Nanofluids

Sujit Kumbhar[1*], Avesahemad Husainy[2], Ashish Desai[3], Vishal Patil[4] and Avadhut Jadhav[5]

Nanofluids are a revolutionary type of fluid composed of nanoparticles that are suspended and varied in the size range from 1 to 100 nm. Metals such as aluminium, copper (Cu, Al), metal carbides (SiC), oxides (CuO, $Al_2O_3$), and single- or multi-walled carbon nanotubes are some of the most common nanoparticles used. Nanofluids have piqued researchers' interest due to their significant financial benefits in industrial and medical sectors such as enhanced nuclear systems, vehicle cooling, solar water heating, home refrigerators, cancer treatments, and magnetic resonance imaging. Nanofluids have gained a lot of attention as working fluids in a range of heat transfer applications due to their well-known optothermal properties. There are Several factors which influence stability of nanofluids, such as acidity, material selection, base-fluid type, surface deposition, surfactants, concentration, morphological changes, oxidation, aggregation, and cluster formation etc. Such qualities raise severe concerns regarding the long-term viability and utility of nanofluids for practical applications. For example, nanofluids made with sodium dodecyl sulfate (SDS) can be stable for 49 days. Furthermore, the durability of $TiO_2$-based nanofibers produced using various surfactants and surface modification techniques should allow for a significant increase in viscosity and heat conductivity. The essential parameters determining nanofluid performance, such as dynamic motion, optical features, stability, heat transport phase, size, shape, and compositions, fully contribute to these phenomena. In this chapter, experimental and theoretical insights into how nanofluids enhance thermal and electrical conductivity, including the role of particle size, shape, and volume fraction, are discussed.

[1-3] Department of Automation and Robotics, Sharad Institute of Technology, College of Engineering, Yadrav, Ichalkaranji.

[4] Department of Mechanical Engineering, Sharad Institute of Technology, College of Engineering, Yadrav, Ichalkaranji.

[5] Department of Mechanical Engineering, Bharati Vidyapeeth's College of Engineering, Kolhapur.

* Corresponding author: sujit.kumbhar64@gmail.com

DOI: 10.1201/9781003686002-3

## 1. Introduction

Nanofluids are a category of fluids created by dispersing nanoscale particles—typically sized between 1 and 100 nanometers-into conventional base fluids such as water, ethylene glycol, or oils. Since the term Nano fluid was first introduced by Choi and Eastman in 1995, these innovative suspensions have gained significant attention due to their remarkable improvements in thermal, electrical, and transport properties compared to traditional fluids. The suspended nanoparticles, which comprise metals, metal oxides, carbon nanotubes, and other nanostructures, exhibit high thermal conductivity and unique interactions with a base fluid, leading to improved heat transfer capabilities (Sandro Nizetic, 2020). Nanofluids have proven effective in a wide range of applications, from industrial cooling systems and electronics thermal management to energy harvesting and biomedical fields. The preparation of nanofluids requires careful control to ensure stable and uniform dispersion of nanoparticles, as agglomeration adversely affects their thermal performance. Methods such as the single-step and two-step synthesis techniques are commonly employed to achieve optimal stability and homogeneity. Moreover, the choice of base fluid tailored to application needs plays a decisive role in describing the properties of nanofluids (A. Sathishkumar, 2020). Their capability to significantly elevate thermal conductivity of fluid, viscosity, as well as convective heat transfer has positioned nanofluids as promising candidates for overcoming the constraints of conventional heat transfer fluids (Aishwarya S. Patil, 2020). This chapter discusses the fundamentals of nanofluid science, including nanoparticle synthesis, fluid preparation methods, physicochemical properties, heat transfer mechanisms, and broad-spectrum applications driving technological advancements in thermal management and energy systems

Nanofluids are broadly categorized based on type and combination of nanoparticles dispersed within the base fluid, and the nature of the base fluid itself. The primary classification divides nanofluids into mono-nanofluids and hybrid nanofluids. Mono-nanofluids contain a single type of nanoparticle suspended in a traditional fluid like water ($H_2O$), ethylene glycol ($C_2H_6O_2$), or oil, and are the simplest form of nanofluids. These nanoparticles can be metals, metal oxides, carbides, nitrides, or carbon-based materials like graphene and carbon nanotubes. Hybrid nanofluids, on the other hand, comprise two or more distinct nanoparticles simultaneously dispersed in the base fluid to harness synergistic effects that further enhance thermal and electrical properties. (Junhao Li, 2021)

Additionally, Nanofluids may be classified based on the physical and chemical properties of their components. For example, nanoparticles can be (i) metallic, i.e., copper (Cu), silver (Ag), (ii) metal oxides, i.e., alumina ($Al_2O_3$), (iii) titanium oxide ($TiO_2$), (iv) non-metals, i.e., graphite, and (v) composite materials. The base fluids used may also vary, including water, glycol mixtures, oils, bio-fluids, and polymer solutions, chosen based on application-specific requirements such as thermal stability and viscosity. This classification is crucial as it influences the preparation methods, stability, and overall performance of nanofluids in thermal and electrical applications (Avesahemad S. N., Husainy S. K., 2024). Nanofluids are formulated by dispersing several kinds of nanoparticles in the base fluids to improve their thermal conductivity (k) for improved heat transfer applications. The nanoparticles made of metals like copper (Cu) and silver (Ag) offer even higher thermal conductivity enhancements, though their cost and stability can be concerns. Carbon-based nanomaterials such as carbon nanotubes (CNTs) and graphene have shown exceptional improvements in thermal conductivity due to their unique structure and high aspect ratios. (Mohammad Hossein Ahmadi, 2018)

The enhancement techniques to improve thermal conductivity in nanofluids include optimizing nanoparticle concentration, controlling particle size and shape, and stabilizing suspensions using surfactants to prevent agglomeration. Additionally, hybrid nanofluids combining multiple types of nanoparticles leverage synergistic effects for better performance. The methods of preparation, like one-step or two-step synthesis, ensure uniform dispersion, while adding surfactants or applying external fields (e.g., magnetic or electric) further enhances thermal transport. Increasing temperature generally boosts thermal conductivity by enhancing Brownian motion of nanoparticles, leading to micro-convection and improved heat transfer. These strategies collectively contribute to the superior thermal properties of nanofluids, enabling their application in cooling systems, heat exchangers, and electronics thermal management.

Nanofluids used to improve electrical conductivity typically consist of nanoparticles dispersed in base fluids such as water, ethylene glycol, or their mixtures. Common nanoparticles include metal oxides like alumina ($Al_2O_3$), titania ($TiO_2$), silicon carbide (SiC), ferrites, and carbon-based materials such as graphene and carbon nanotubes. (E. Oró a, 2012) These nanoparticles significantly enhance electrical conductivity by increasing charge carrier density and forming conductive pathways within the fluid. (Mohammad Hossein Ahmadi, 2018) Techniques to enhance electrical conductivity of nanofluids involve optimizing nanoparticle concentration, as higher volume fractions generally increase conductivity due to more conductive particle networks. Temperature elevation also boosts conductivity by increasing ion mobility and nanoparticle electrophoretic activity (Marc Martín, 2019). The methods for stabilizing, such as using surfactants, ultrasonication, and pH adjustment, are crucial to maintain uniform dispersion and prevent agglomeration, which otherwise diminishes conductivity. Additionally, hybrid nanofluids combining multiple types of conductive nanoparticles can leverage synergistic effects for better electrical performance. Experimental studies reveal that the formation of an electrical double layer (EDL) at nanoparticle-liquid interfaces plays a key role in conductivity enhancement by facilitating charge transfer mechanisms in the fluid. These advancements enable nanofluids to be effectively employed in electronic cooling, energy storage, and electrochemical applications where improved electrical conduction alongside thermal management is required.

## 2. Thermal and Physical Properties of Nanofluids

The various properties of nanofluids can be broadly categorized into thermal, physical, and chemical properties:

**Thermal Properties:** Nanofluids have better thermal conductivity than their base fluids. This is mostly because nanoparticles like metals, metal oxides, and carbon-based compounds have high thermal conductivity. The thermal conductivity generally increases with nanoparticle concentration and temperature, as Brownian motion increases micro-convection inside the fluid. Specific heat capacity of nanofluids can vary depending on nanoparticle type and concentration.

Thermal properties of nanofluids are significantly enhanced compared with those of their base fluids due to the incorporation of nanoparticles with high thermal conductivity. Nanofluid thermal conductivity changes a lot with the volume percentage of nanoparticles, the temperature, the size of the particles, and how well they are spread out or clumped together. For example, copper oxide (CuO) and alumina ($Al_2O_3$)-based nanofluids exhibit 12–30% thermal conductivity improvements at low volume fractions (around 3–4%). Titania ($TiO_2$) nanofluids also demonstrate enhancements of 14–27%, with variations depending on particle size and preparation methods. The mechanisms liable for this augmentation include the Brownian motion of nanoparticles that causes micro convection, formation of thermally conductive interfacial layers between particles and fluid, and the creation of percolation networks that facilitate heat transfer pathways (Randeep Singh, 2019).

However, results vary across studies due to differences in measurement techniques, nanoparticle properties, and base fluid types. Some debate exists about whether enhancements beyond classical models (like Maxwell's effective medium theory) are genuine or artifacts of experimental conditions. Advanced characterization techniques and theoretical modeling suggest that multiple interacting mechanisms contribute to the overall thermal conductivity enhancement and that particle aggregation state critically affects thermal transport (Mehdi Ghalambaz, 2021). Moreover, recent research on grapheme nanofluids has shown up to 48% enhancement in thermal conductivity linked not only to graphene's intrinsic thermal properties but also to molecular interactions that induce local ordering of solvent molecules—enhancing heat transport further. Such insights highlight the complex interplay of nanoscale physics governing nanofluid thermal behavior, which is crucial for optimizing formulations for applications in electronics cooling, solar thermal systems, and advanced manufacturing (Mohammad Hossein Ahmadi, 2018).

Overall, nanofluid thermal conductivity is influenced by a combination of factors—nanoparticle concentration, size, shape, base fluid characteristics, temperature, and nanoparticle interaction—making continued research vital for reliable and reproducible heat transfer enhancement (Randeep Singh, 2019).

**Physical Properties:** Adding solid nanoparticles to nanofluids makes them denser. Viscosity also rises with nanoparticle concentration, which influences fluid flow and pumping power requirements; however, viscosity usually decreases as temperature rises. Surface tension may change slightly with nanoparticle inclusion, affecting wetting behavior (P. Sivasamy, 2018). Stability is a key physical property, as uniform dispersion of nanoparticles prevents settling and agglomeration, ensuring consistent performance. Particle size, shape, and volume fraction significantly impact these physical properties (Sandro Nizetic, 2020).

The physical properties of nanofluids are crucial in influencing their performance for heat transfer and engineering applications. Key physical properties include density, viscosity, surface tension, specific heat capacity, thermal conductivity, and stability. Nanofluids generally exhibit a higher density than their base fluids due to the addition of solid nanoparticles (Sonali K. Kale, 2021). Viscosity tends to increase with nanoparticle concentration, which can affect the flow of fluid and the need for pumping power; however, viscosity typically decreases with increasing temperature. Surface tension changes may occur depending on particle type and concentration, impacting wetting and spreading behavior. Specific heat capacity of nanofluids can vary based on nanoparticle material and concentration; often, it decreases slightly with nanoparticles because solids generally have lower specific heat than liquids. Thermal conductivity is among the most improved physical qualities, significantly improving heat transfer by leveraging the high conductivity of nanoparticles, Brownian motion, and particle-fluid interactions. Stability is a physical characteristic critical to maintain uniform nanoparticle dispersion—unstable nanofluids with agglomeration or sedimentation will lose enhanced properties (Mohammad Hossein Ahmadi, 2018).

Particle size, shape, and volume fraction are other physical factors that strongly influence nanofluid behavior. Smaller nanoparticles and spherical shapes generally promote better dispersion and stability, leading to improved thermal performance (Shaon Talukdar, 2019). The property changes in nanofluids depend on the base fluid, nanoparticle type, and preparation methods, which together define the overall thermophysical behavior and practical usability of these fluids (Mohammad Hossein Ahmadi, 2018).

**Chemical Properties:** The chemical makeup and interactions amongst nanoparticles and the base fluid determine the nanofluid's stability and reactivity. Surface functionalization of nanoparticles and the use of surfactants help maintain dispersion and prevent chemical reactions that could degrade performance. Together, these properties influence the efficiency of nanofluids in applications like the cooling of electronics appliances, industrial appliances of heat transfer, and guiding the design and formulation of nanofluids for optimal performance. The chemical characteristics of nanofluids are very important for figuring out how stable they are, their performance, and usability in practical applications. Stability is influenced primarily by the chemical interaction amongst nanoparticles and the base fluid, as well as the surface chemistry of the nanoparticles. Nanoparticles can be hydrophilic or hydrophobic, affecting their dispersibility in different base fluids such as water or oils. To maintain long-term stability and prevent agglomeration or sedimentation of nanoparticles, chemical methods like surface functionalization, pH adjustment, and the use of surfactants or dispersants are commonly employed.

Zeta potential is a key indicator of nanofluid stability; high absolute values (typically $> \pm 40$ mV) indicate strong repulsive forces between particles, thus better stability, while low values result in particle agglomeration and sedimentation. Adjusting the pH away from the isoelectric point (IEP) increases zeta potential and improves stability. For example, alumina and titanium nanofluids show better dispersion and higher zeta potential at pH values different from their IEP, ensuring longer stability periods. Chemical stability also relates to resistance against oxidation and chemical reactions that can degrade nanoparticles or alter fluid properties. Metal oxides tend to be chemically stable and corrosion-resistant, while carbon-based materials such as graphene may require protective functionalization to prevent oxidation. Properly enhancing chemical stability in nanofluids ensures consistent thermal and physical properties, essential for reliable heat transfer and other applications.

In summary, chemical properties governing nanoparticle surface characteristics, dispersion chemistry, and fluid interactions are fundamental to nanofluid stability, dictating their effectiveness and longevity (Mohammad Hossein Ahmadi, 2018).

## 3. Thermo-Physical Properties of Nanofluids

| *Nanoparticle* | *Thermal Properties* | *Physical Properties* | *Chemical Properties* |
|---|---|---|---|
| $Al_2O_3$ | Good thermal conductivity (~30 W/m·K) | Density ~3.95 g/cm$^3$, moderate viscosity impact | Chemically stable, resistant to oxidation |
| $TiO_2$ | Moderate thermal conductivity (~8-11 W/m·K) | Density ~4.23 g/cm$^3$, moderate effect on viscosity | Chemically inert, good corrosion resistance |
| Zinc Oxide (ZnO) | Thermal conductivity ~10-15 W/m·K | Density ~5.61 g/cm$^3$, slightly increases viscosity | Chemically stable, photocatalytic activity under UV light |
| Magnesium Oxide (MgO) | High thermal conductivity (~45-60 W/m·K); varies by form | Density ~3.58 g/cm$^3$, moderate viscosity impact | Chemically stable, good thermal stability |
| Iron Oxide ($Fe_3O_4$) | Moderate thermal conductivity (~6-8 W/m·K) | Density ~5.17 g/cm$^3$, magnetic properties | Chemically stable, magnetic, useful for electromagnetic fields |
| Carbon Nanotubes (CNTs) | Extremely high thermal conductivity (~3000 W/m·K) | Low density (~1.3-1.4 g/cm$^3$), minimal viscosity increase | Chemically stable, can be functionalized for dispersion |
| Graphene | Exceptional thermal conductivity (~3000-5000 W/m·K) | Very low density (~2.26 g/cm$^3$), low viscosity effect | Chemically stable, prone to oxidation without protection |

*Contd.*

| *Nanoparticle* | *Thermal Properties* | *Physical Properties* | *Chemical Properties* |
|---|---|---|---|
| Graphene Oxide | Reduced thermal conductivity compared to graphene (10-30 W/m·K) | Moderate density due to oxygen groups, affects viscosity more than graphene | Chemically reactive due to oxygen functional groups |
| Graphene-Gold Hybrid | Enhanced thermal conductivity over graphene alone | Composite properties; density increases due to gold | Chemically stable, gold adds catalytic and electrochemical properties |
| Silver-MWCNTs Hybrid | Enhanced thermal and electrical conductivity | Density depends on silver content; viscosity rises with loading | Silver adds antimicrobial and catalytic chemical traits |

Carbon nanotubes (CNTs) and graphene show the highest thermal conductivities among nanoparticles, often orders of magnitude greater than metal oxides. Metal oxides like magnesium oxide (MgO) and alumina ($Al_2O_3$) provide moderate thermal conductivity improvements but are still effective for many applications. Physically, carbon nanotubes and graphene have low density and cause minimal increases in viscosity, which is beneficial for fluid flow, while metal oxides are denser and typically increase viscosity more significantly (Aishwarya S Patil, 2020). Chemically, metal oxides are generally stable and inert under most conditions, providing good durability and resistance to oxidation, whereas carbon materials require careful handling due to their susceptibility to oxidation. Functionalization of carbon nanoparticles, however, can effectively improve their stability and dispersion in fluids. Hybrid nanoparticles, such as graphene-gold or silver multiwalled carbon nanotubes (MWCNTs), combine the advantageous properties of metals and carbon-based materials, optimizing thermal conductivity, chemical stability, and physical behavior simultaneously for enhanced nanofluid performance (Mohammad Hossein Ahmadi, 2018).

## 4. Thermal Behavior of Nanofluids

The thermal behavior of nanofluids is determined by several interrelated mechanisms and properties, which work in concert to improve the thermal transport characteristics of the fluids. First, the nanoparticles have a high thermal conductivity, which makes the fluids' thermal conductivity better. The improvement in the thermal conductivity of the fluids depends on several key parameters, which include the concentration, size, and temperature of the nanoparticles. The Brownian motion of the nanoparticles also contributes to the enhancement of the thermal transport characteristics of the fluids through micro-convection currents. Moreover, the thermally conductive interfacial nanolayer formed between the nanoparticles and the fluids also contributes to the enhancement of the thermal transport characteristics of the fluids. The percolation network formed by the aggregated nanoparticles also contributes to the enhancement of the thermal transport characteristics of the fluids (Sithara Vinod, 2018).

Temperature plays a critical role; higher temperatures generally boost thermal conductivity via increased Brownian motion and reduced fluid viscosity, although excessive concentration of nanoparticles can lead to agglomeration that deteriorates heat transfer. Particle shape also significantly affects thermal performance—with cylindrical and sheet-like particles often outperforming spherical ones due to larger surface area and better connectivity. The heat capacity of nanofluids is affected by both base fluid and nanoparticles, impacting the system's energy storage and transfer capabilities. Moreover, slip mechanisms such as thermophoretic diffusion, sedimentation, and Brownian diffusion govern nanoparticle movement and thermal behavior, influencing stability and local heat transfer dynamics. Proper dispersion and stabilization of nanoparticles prevent sedimentation and maintain consistent thermal properties during operation (Sithara Vinod, 2018).

In summary, the thermal behavior of nanofluids results from the combined effects of nanoparticle characteristics, their dynamic motion within the fluid, temperature, and interaction between particles and base fluid. These complex phenomena make nanofluids a versatile and highly effective medium for improved heat transfer in various applications, for example, spanning electronics cooling, solar energy systems, and advanced manufacturing (Marc Martín, 2019).

## 5. Thermal Conductivity of Nanofluid and Factors Affecting Thermal Conductivity

It is highly important to research the elements that affect the thermal conductivity of nanofluids since thermal conductivity is a key attribute that determines how well they work in heat transfer applications. The various factors affecting thermal conductivity are shown in Figure 1. Several factors influence this property in detail:

1. Nanoparticle Properties: The size, shape, material, and intrinsic thermal conductivity of nanoparticles are fundamental. Smaller particles with high surface area increase heat transfer, while shapes such as tubes or sheets promote better interaction. Materials like metals and carbon allotropes usually have much higher conductivity than metal oxides.
2. Base Fluid Properties: The type of base fluid (water, oil, ethylene glycol) affects overall conductivity. The base fluid's viscosity and inherent thermal conductivity set the baseline for enhancement (P. Sivasamy, 2018).
3. Concentration and Dispersion: Volume fraction (concentration) of nanoparticles improves conductivity but, beyond a point, causes agglomeration, reducing effectiveness. Stability and uniform dispersion are essential for consistent performance.
4. Temperature Effects: Higher temperatures enhance the Brownian motion of nanoparticles and decrease fluid viscosity, aiding in thermal conductivity improvement (Sithara Vinod, 2018).

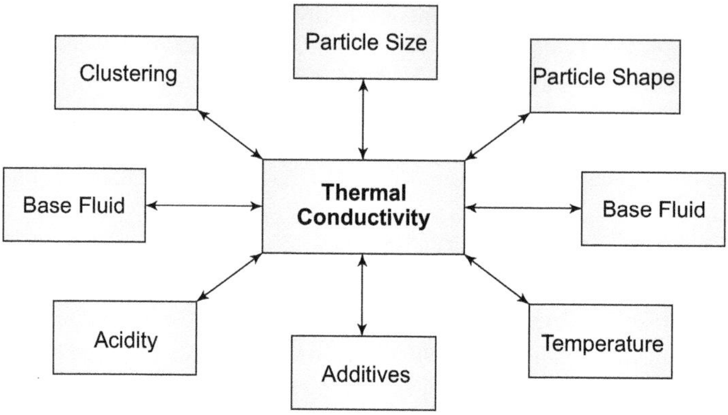


**Fig. 1** Factors influencing thermal conductivity of nanofluids.

5. External Factors: Magnetic or electric fields can influence nanoparticle alignment and motion, potentially modifying heat transfer behavior (E. Oró a, 2012).
6. Nanoparticle Properties: Nanoparticles are the core contributors to the thermal conductivity enhancement in nanofluids, and their properties significantly affect heat transfer (Mohammad Amiri, 2016).
   a. Size: The size of the nanoparticle influences the surface area-to-volume ratio, Brownian motion intensity, and interfacial thermal resistance. The specific surface area of smaller nanoparticles is higher, increasing the heat conduction paths between particles and fluid, thereby enhancing thermal conductivity. However, particles too small may agglomerate due to high surface energy, which adversely affects conductivity. Typical sizes range from 1 to 100 nm (Mohammad Amiri, 2016).
   b. Shape: The shape of nanoparticles influences the effective thermal pathways. Nanoparticles may be spherical, cylindrical (nanotubes), platelets (nanosheets), or irregular shapes. For instance, carbon nanotubes with cylindrical structures provide superior conductive pathways in comparison to spherical nanoparticles because to their greater surface contact area and aspect ratio. Graphene sheets offer high thermal conductivity in-plane due to their two-dimensional lattice (Mohammad Amiri, 2016).
   c. Material: The intrinsic thermal conductivity of the nanoparticle material is fundamental. Metals such as copper, silver, and carbon-containing materials such as graphene, carbon nanotubes, etc. have very high thermal conductivities compared to oxides of metal (e.g., Alumina, Titania). Materials with higher thermal conductivity provide better overall nanofluid performance. However, particle stability and oxidation resistance are also important (Weerapun Duangthongsuk, 15 January 2010).

   d. Thermal Conductivity: The absolute intrinsic thermal conductivity of the nanoparticles sets the maximum enhancement limits. For instance, graphene thermal conductivity is around 3000–5000 W/m·K, significantly higher than base fluids (water ~0.6 W/m·K). Nanoparticles form conductive bridges facilitating faster heat transfer (Sun, 2024).

7. Base Fluid Properties: The base fluid, where nanoparticles are dispersed, plays a critical role in determining the composite nanofluid conductivity. (Shaon Talukdar, 2019)
   a. Type of Base Fluid: Water, ethylene glycol, oils, and synthetic fluids are some of the most common base fluids. Water is extensively used because of its high specific heat and low viscosity. Oils offer higher thermal stability but lower thermal conductivity, affecting the degree of enhancement achievable through nanoparticle addition.
   b. Viscosity: Viscosity impacts nanoparticle motion and fluid flow behavior. High viscosities can suppress Brownian particle movements, diminishing thermal conductivity benefits from micro-convection. Conversely, low viscosity facilitates better nanoparticle movement and improved heat transport, but may affect the mechanical properties of the fluid.
   c. Thermal Conductivity: The baseline thermal conductivity of the base fluid affects the final nanofluid property. Nanoparticles added to fluids with already high thermal conductivity (like water) generally result in modest relative enhancements, whereas those added to low conductivity fluids (such as oils) can yield significant improvements (MdAtiqurRahman, 2024,9 ).
   d. Concentration & Dispersion: Nanoparticle concentration and dispersion quality within the base fluid directly influence nanofluid behavior.
   e. Volume Fraction: Generally, increasing nanoparticle volume fraction leads to higher thermal conductivity due to more conductive pathways. However, beyond a threshold (usually 1–5% by volume), the viscosity dramatically increases, and particles tend to agglomerate, which clogs conductive pathways and reduces heat transfer efficiency.
   f. Stability: Stable dispersions prevent nanoparticle settlement and maintain uniform physical properties over time, ensuring consistent thermal conductivity. Stability is often enhanced by surfactants, pH adjustment, or surface functionalization of nanoparticles. Improved stability maintains nanoparticle separation, preventing the decline in heat transfer.
8. Temperature Effects: Temperature acts in multiple ways to influence nanofluid thermal behavior (Weerapun Duangthongsuk, 15 January 2010).
   a. Fluid Temperature: An increase in fluid temperature generally improves thermal conductivity. Elevated temperatures reduce fluid viscosity, allowing nanoparticles to move more freely, increasing Brownian motion as well as heat transfer due to convection in the fluid (Avesahemad and Husainy, 2020).

   b. Brownian motion of a nanoparticle: The Brownian motion denotes the erratic motion of particles suspended in a fluid. At higher temperatures, nanoparticles move more vigorously, enhancing micro-convection and heat exchange between the nanoparticles and the base fluid. This movement helps overcome limitations caused by thermal interface resistance.

9. External Factors: Several external factors influence nanoparticle behavior and heat transfer efficiency.
   a. Magnetic Fields: Magnetically responsive nanoparticles (such as iron oxide) experience alignment and controlled movement under magnetic fields, which can improve thermal transport pathways and potentially adjust nanofluid properties dynamically for specific applications.
   b. Electric Fields: Electric fields can alter particle distribution and orientation, managing aggregation and improving heat transfer characteristics. This is especially relevant for electrically conductive nanoparticles like metals or carbon-based materials.

The thermal conductivity enhancement of nanofluids is a complex function governed by the interplay between nanoparticle properties, base fluid characteristics, particle concentration and dispersion quality, temperature, and external modulation. Optimizing these parameters allows the development of nanofluids with tailored thermal behavior suited to advanced heat transfer applications such as electronics cooling, solar energy, and industrial processes. However, each factor must be carefully balanced to maximize conductivity gains while maintaining fluidity, stability, and chemical compatibility for long-term, practical use (Weerapun Duangthongsuk, 15 January 2010).

## 6. Electrical Conductivity of Nano Fluid and Factors Affecting Electrical Conductivity

The factors affecting electrical conductivity are shown in Figure 2, which highlights the main contributors to the electrical conductivity of nanofluids. The electrical conductivity depends on the nanofluids and the fluid in which the nanoparticles are suspended, since the two determine the natural charge transport properties. Particle shape and size contribute to the particle surface area and the ability of electrons to move, while particle concentration enhances charge carriers in the fluid. An important aspect of ion transport in the fluid is the electrical double layer that surrounds the nanoparticles. Suspension and stability of the fluid enhance conductivities in the fluid. Temperature enhances ion mobility, hence enhancing conductivities.

1. Type of nanoparticles: The intrinsic electrical conductivity of nanoparticles varies depending on their material. Metallic nanoparticles (e.g., silver, copper, gold) generally provide much higher conductivity compared to metal oxides such as alumina ($Al_2O_3$), Titanium oxides ($TiO_2$), Silicon dioxide ($SiO_2$),

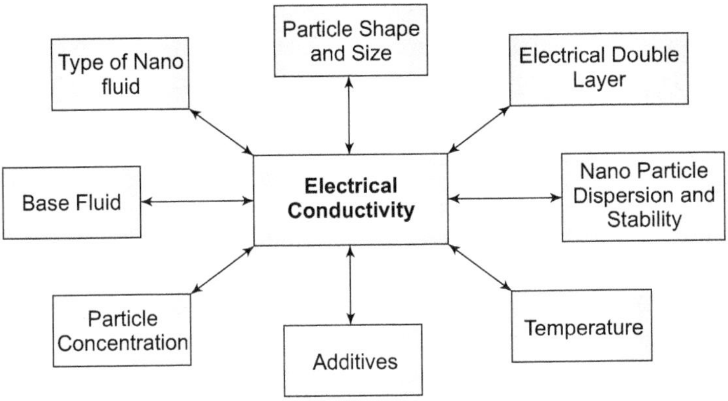


**Fig. 2** Factors affecting the Electrical conductivity of nanofluids.

etc. Carbon nanotubes (CNTs) and graphene are also highly conductive and improve percolation networks in the fluid (P. Sivasamy, 2018).

2. Particle Volume Fraction (Concentration): Electrical conductivity increases with higher nanoparticle concentration because more conductive paths are formed in the fluid. However, at very high concentrations, particles may agglomerate and cause sedimentation, reducing effective conductivity (Habib Ben Bacha, 2024).
3. Particle Size and Shape: Smaller nanoparticles have a larger surface-to-volume ratio, leading to stronger interactions with the base fluid and enhanced electrical conduction. The shape also plays a significant role: rod-like particles and sheet-like structures (e.g., graphene, CNTs) form more effective conduction pathways compared to spherical particles.
4. Base Fluid Properties: The electrical conductivity of the base fluid itself strongly affects the overall conductivity. Polar fluids like water and ethylene glycol allow better ion transport, while non-polar fluids such as oils tend to exhibit lower conductivity.
5. Temperature: As temperature increases, the Brownian motion of nanoparticles becomes more intense, improving dispersion and charge transport in the fluid. However, extremely high temperatures can cause particle agglomeration or chemical instability, reducing long-term performance.
6. Nanoparticle Dispersion and Stability: Stable, well-dispersed nanoparticles ensure continuous conductive pathways throughout the fluid. If particles agglomerate, clusters may block conduction paths and reduce overall conductivity. Surface modification and the use of stabilizers can enhance stability.
7. Acidity of the Fluid: Acidity affects the surface charge and zeta potential of nanoparticles. Adjusting the pH properly can prevent particle aggregation and improve the stability of the nanofluid, leading to higher conductivity.

8. Surface Functionalization and Surfactants: Nanoparticles are often surface-modified to prevent agglomeration. Surfactants can improve dispersion, but their effect on conductivity depends on type: ionic surfactants increase ionic conduction, while non-ionic surfactants may have little impact.
9. Electrical Double Layer (EDL) Effect: When nanoparticles are suspended in polar fluids, they develop an EDL around them. This EDL facilitates additional charge transport and contributes significantly to conductivity, particularly at low nanoparticle concentrations (Mohammad Ghalambaz, 2019).
10. External Fields: Application of external fields, such as electric or magnetic fields, can align nanoparticles (especially magnetic nanoparticles and CNTs). This alignment creates continuous conductive channels, thereby enhancing electrical conductivity.

Thermal Conductivity enhancement techniques:

1. **Stabilizing nanofluids:** When contrasted to a less stable sample of the same nanofluid material, stable nanofluids with optimal amounts of surfactant substance, surfactant weight, and ultrasonication time have a better thermal conductivity. The stable nanofluid offers significantly greater thermal conductivity than unstable nanofluids of the same substance, according to visual observations of the stability of the nanofluid as shown in Figure 3. Because they inhibit agglomeration, surfactants at the ideal concentration can improve the nanofluid's heat conductivity. The stability of nanofluids significantly affects their thermal conductivity. The surfactant material, weight, and ultrasonication time directly affect nanofluid stability and can be considered stability parameters. The stability of a nanofluid affects its thermal conductivity more than its substance. Mane *et al.,* (2021) investigated the impact of preparation parameters on the thermal conductivity of metal oxide nanofluids using the Taguchi method. In comparison to $Fe_3O_4$-water and CuO+ $Fe_3O_4$-water nanofluids, they discovered that CuO-water nanofluids have a greater thermal conductivity. The turbidity of nanofluid samples containing sodium dodecyl sulfate (SDS) and cetyl trimethyl ammonium bromide (CTAB) surfactants is higher than that of samples containing citric acid (CA). The signal-to-noise ratio (SNR) of CA is lower than that of CTAB and SDS. Stable nanofluids with CTAB and SDS surfactants exhibit higher thermal conductivity compared to unstable nanofluids with CA surfactants. ANOVA results indicate that surfactant substance, quantity, and ultrasonication time have a significant impact on thermal conductivity, rather than only nanoparticles.
2. **Interfacial liquid layer:** When nanoparticles are disseminated in a base fluid, a thin liquid layer forms surrounding each one. This layer is more organized and dense than the main liquid because of strong molecular interactions at the solid surface. The structured layer frequently has a greater thermal conductivity than the surrounding fluid. Increasing this layer's thickness or

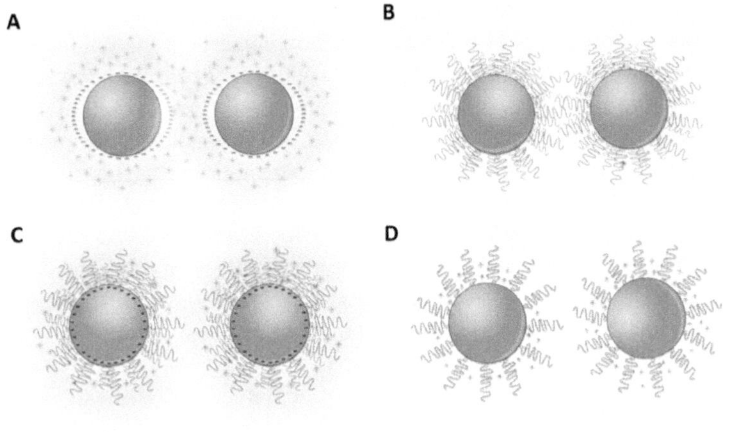


**Fig. 3** Stability of nanofluids: (A) electrostatic stabilization, (B) steric stabilization, (C, D) electrosteric stabilization. Reproduced from Fan Yu et al. (2017) under the Creative Commons CC-BY-NC-ND license.

conductivity improves total heat transmission. Surface modifications, such as functionalization or surfactant coating, can reinforce this layer. Using polar base fluids like water or ethylene glycol improves interfacial stacking. Smaller nanoparticles enhance surface area, which increases total interfacial layer volume. The combined effect results in quicker heat transport routes across solid-liquid interfaces. Effective medium models characterize this as a core-shell structure, with the shell representing the interfacial layer. Thus, manipulating this nanolayer is an excellent technique to improve nanofluid thermal conductivity (Ahmed A. Mohammed, 2023).

3. **Effective Medium Theory (EMT):** The Effective Medium Theory (EMT) estimates the overall thermal conductivity of a composite or nanofluid. It views the nanofluid as a combination of solid nanoparticles and base fluids with varying conductivities. The theory calculates effective thermal conductivity ($k_{eff}$) using particle concentration and characteristics. EMT suggests that adding high-conductivity nanoparticles (such as Cu, $Al_2O_3$, or graphene) increases $k_{eff}$. Smaller particles with a larger surface area make greater contact with the fluid, resulting in better heat transfer. Nanoparticles dispersed uniformly avoid clumping and promote improved heat conduction pathways. Surface modification aids in the stability and correct interaction of particles and fluids. Modified EMT models use interfacial resistance and liquid layer effects to improve forecast accuracy. The optimal mix of particle type, size, and volume fraction increases the nanofluid's $k_{eff}$. Thus, EMT aids in the design of nanofluids with predictably increased thermal conductivity for use in practical applications.
4. **Brownian motion:** Brownian motion causes tiny solid nanoparticles in a nanofluid to move at all times. This random motion arises as a result of

constant collisions between fluid molecules and nanoparticles. The movement of nanoparticles creates minute convection currents in the fluid. These tiny fluid disturbances improve the efficiency with which heat energy is transferred throughout the liquid. Higher temperatures cause particles to move faster, which improves thermal conductivity even more. Smaller nanoparticles travel more quickly, making Brownian motion more noticeable. This dynamic heat transport effect is most effective when nanoparticles are well-dispersed and stable. The effect is stronger in fluids with low viscosity, where particles can flow more freely. Brownian motion is included in models as a feature that contributes to increased thermal energy transmission. Thus, controlled Brownian motion improves the total thermal conductivity of nanofluids.

5. **Agglomeration mechanism:** Nanoparticles in a nanofluid can occasionally gather together to form tiny clusters or agglomerates. These agglomerates form continuous heat conduction pathways between particles. Close contact between particles reduces the fluid's heat resistance. As a result, heat can move more quickly through these coupled particle networks. Moderate agglomeration can thus improve heat conductivity over isolated particles. However, severe agglomeration can lead to sedimentation and reduced fluid stability (Mane *et al.*, 2022). Controlled or partial clustering aids in balancing high conductivity with adequate dispersion. The strength of agglomeration is determined by particle concentration, size, and surface energy. Proper surfactant application or surface treatment can control the degree of aggregation. Thus, a well-controlled agglomeration network can enhance heat transfer characteristics in nanofluids.

**Applications of nanofluids:**

1. **Nanofluid in electronic cooling:** Nanofluids are increasingly employed in electronics cooling because of their superior thermal performance compared to common cooling fluids such as water, air, and ethylene glycol. Adding nanoparticles to a fluid, including metals, oxides of metal, carbides, or carbon-containing compounds, makes the fluid more conductive to heat and allows for better convective heat transfer than a pure fluid. Nanofluids boost the process of heat removal in the case of microprocessors, GPUs, and power electronic devices, as a result of which the devices can run at optimal temperatures. Nanofluids can also be used in the case of heat pipes, which help to enhance the process of boiling heat transfer. Additionally, nanofluids can be used in vapor chambers. For the purpose of cooling data centers, nanofluid cooling solutions can be used to decrease energy consumption by enhancing cooling rates and thereby decreasing the requirement for significant air or cooling infrastructure. For LEDs and high-frequency electronic components, nanofluids can be used to ensure even distribution of temperatures and thereby prevent thermal stresses and variations in performance due to temperature variations in the component. Despite the above benefits, issues concerning stability, elevated viscosity, the possibility of clogging within micro-channels,

and the regulation of conductivity still need caution. In general, it can be concluded that nanofluids provide a promising approach for advanced electronics cooling.

2. **Nanofluid in Solar Systems:** Nanofluids have gained significant attention in the applications of solar energy because of their superior thermal properties. The solar systems, such as flat plate collectors, evacuated tubular collectors, and parabolic trough collectors etc., nanofluids increase the rate of absorption of heat as well as the efficacy of heat transfer, resulting in higher outlet temperatures. The high thermal conductivity of nanofluids helps in quicker dissipation of heat from the surface of the hot plates. Moreover, some nanofluids can work as direct absorption solar collectors. In this case, nanoparticles absorb solar radiation, thereby reducing losses due to the hot plates. Nanofluids can also be utilized in solar photovoltaic and thermal systems to make photovoltaic systems work better by cooling the panels.
3. **Nanofluid in Biomedical applications:** Nanofluids serve a vital function in the biomedical field for targeted drug delivery, medical imaging, and cancer therapy. Their controlled thermal properties are exploited for hyperthermia treatment, where magnetic nanofluids, under an external magnetic field, present localized heat that destroys cancerous cells while sparing healthy tissues. Nanofluids enhance not only biosensing but also diagnostic systems through improved electrical and thermal responsiveness. Regarding medical cooling applications, such as the cooling of medical devices and cryopreservation systems, nanofluids allow efficient temperature control. While impressive, biocompatibility, toxicity, and long-term safety are some of the critical considerations for the use of nanofluids in clinical applications.

There are several important limitations to improving thermal and electrical conductivity in nanoparticles and nanofluids. Among these is agglomeration, when the concentration of nanoparticles is increased to improve conductivity, particles tend to be clustered owing to strong van der Waals forces, which reduces effective surface area and degrades both thermal and electrical performance. Besides, stability issues further restrict long-time use because sedimentation over time leads to properties that are not uniform. Another limitation is the trade-off with viscosity. While higher nanoparticle loading improves conductivity, it also significantly raises fluid viscosity, resulting in higher pumping power, pressure drop, and reduced system efficiency. Effective heat transfer is further restricted by the interfacial thermal resistance between nanoparticles and the base fluid, also known as Kapitza resistance.

Excessive enhancement of electrical conductivity results in unwanted effects, as it can lead to electrical losses, corrosion, or the risk of a short circuit, particularly in applications related to electronic cooling or insulation. Another issue is that of temperature sensitivity: changes in conductivity may behave erratically with temperature. In addition, compatibility with base fluids and additives is limited; some surfactants improve dispersion but reduce conductivity. Finally, high cost, complicated synthesis methods, and scaling further limit industrial applications.

The handling of nanoparticles raises environmental, health, and safety concerns that impose practical limitations on the enhancement of thermal and electrical conductivity.

## 7. Conclusion

This chapter has discussed and analyzed the basic mechanism underpinning the enhancement of thermal and electrical conductivity properties in nanofluids and has explained their importance and applications in the area of technologies for heat and energy. Nanofluids that are composed of nanoparticles mixed with a basic fluid or medium have higher transport properties compared to basic fluids and are used in applications related to heat transfer, cooling technologies, and energy devices. Enhancement of thermal conductivity in nanofluids has been explained through various coupled phenomena, such as Brownian motion among nanoparticles and the base fluid, elevated surface area for optimal heat transfer, micro-convection currents around nanoparticles, and decreased thermal resistance at the solid-liquid interface. The importance of nanoparticle material properties, size, shapes, and concentration has been evidenced in dictating the magnitude of heat transfer rates in nanofluids. At the same time, this chapter highlights that improvement in thermal conductivity is not linear and may be restricted due to various factors such as aggregation of nanoparticles and higher viscosity. Similarly, the underlying physics for electrical conductivity enhancement in nanofluids is bound to charge transport through nanoparticles, the formation of electrical double layers, ion mobility, and particle-fluid interactions. Electrical behavior is substantially influenced by parameters like nanoparticle dispersion quality, base fluid properties, temperature, and the addition of additives. Enhanced electrical conductivity, though desirable in applications pertaining to sensors, electrochemical systems, and smart fluids, may lead to enhanced insulation failure, corrosion-related issues, and electrical safety issues in some systems.

This chapter also indicates that dispersion and long-term stability of nanoparticles are critical, since uniform dispersion is directly related to consistent thermal and electrical performance. Surface functionalization, adding surfactants, and optimization of synthesis methodologies are applied to overcome such dispersion problems, but this adds more complications, such as possible reductions in conductivity, chemical incompatibility, and environmental concerns. On the applications front, the importance of a well-rounded strategy concerning the optimization of the fluids introduced above is brought out, where the improvement in the properties of conductivity, whether thermal or electrical, is attained without affecting the viscosity, cost, stability, or ease of use of the liquids. The absence of widely accepted theoretical predictions on the topic makes the need for experimental correlations all the more apparent.

Nanofluids represent a significant advance in thermal and electrical transport engineering. Although tremendous efforts have been made to ~~identify~~ the fundamentals of the enhancement phenomenon, various issues regarding associated stability, scalability, cost, and application-related restrictions remain to be addressed. Future studies should address the area of hybrid nanofluids, environmentally safe nanoparticles, novel modeling tools, and practical implementation on real-life systems. Overcoming these hurdles would be critical for the commercial success of nanofluids in future thermal and electrical systems.

## References


[1] A. Sathishkumar, M.C. (2020). Influence of thermal transport properties of NEPCM for cool thermal energy storage system. *Journal of Thermal Analysis and Calorimetry.*

[2] Aishwarya, S. and Patil, G.V. (2020). Modern Ethical and Societal Implications of Nanotechnology. *J. Adv. Res. in Manufacturing, Material Science & Metallurgical Engineering,* 1–8.

[3] Avesahemad S.N. and Husainy, S. M. (2024). Heat transfer phenomenon of Nano-enhanced phase change material (NEPCM) incorporated in refrigeration test rig. *ES Energy & Environment.*

[4] Avesahemad, S.N. and Husainy, R. S. (2020). A State of the Art: Graphene Research with their Properties and Applications. *J. Adv. Res. in Manufacturing, Material Science & Metallurgical Engineering,* 1–10.

[5] Bacha, H.B., Ullah, N., Hamid, A. and Shah, N.A. A comprehensive review on nanofluids: Synthesis, cutting-edge applications, and future prospects, *International Journal of Thermofluids,* Volume 22, 2024, 100595, https://doi.org/10.1016/j.ijft.2024.100595.

[6] E. Oró a, A.d. (2012). Review on phase change materials (PCMs) for cold thermal energy storage applications. *Applied Energy,* 513–533.

[7] Husainy, A.S.N. *et al.,* (2024). Investigation of Power Consumption of Refrigeration Model and Its Exploratory Data Analysis (EDA) by Using Machine Learning (ML) Algorithm. In: Garg, D., Rodrigues, J.J.P.C., Gupta, S.K., Cheng, X., Sarao, P., Patel, G.S. (eds) Advanced Computing. IACC 2023. Communications in Computer and Information Science, vol 2053. Springer, Cham. https://doi.org/10.1007/978-3-031-56700-1_12.

[8] Junhao, Li, Xilong Zhang, Bin Xu and Mingyu Yuan. (2021). Nanofluid research and applications: A review. *International Communications in Heat and Mass Transfer, Volume 127,2021,105543, ISSN0735-1933, https://doi.org/10.1016/j.icheatmasstransfer.2021.105543.*

[9] Mane, N.S., Tripathi, S. and Hemadri, V. (2022). Effect of biopolymers on stability and properties of aqueous hybrid metal oxide nanofluids in thermal applications. Colloids and Surfaces A: Physicochemical and Engineering Aspects, 643, 128777.

[10] Marc Martín, A.V. (2019). Development of new nano-enhanced phase change materials (NEPCM) to improve energy efficiency in buildings: Lab-scale characterization. *Energy & Buildings,* 75–83.

[11] Mehdi Ghalambaz, S.M. (2021). Thermal behavior and energy storage of a suspension of nano-encapsulated phase change materials in an enclosure. *Advanced Powder Technology,* 2004–2019.

[12] Mohammad Amiri, S.M. (2016). Thermal conductivity of water and ethylene glycol nanofluids containing new modified surface $SiO_2$ - Cu nanoparticles: experimental and modeling. *Applied Thermal Engineering,* 48–53.

[13] Mohammad Ghalambaz, A.J. (2019). Natural convective flow and heat transfer of Nano-Encapsulated Phase Change Materials (NEPCMs) in a cavity. *International Journal of Heat and Mass Transfer,* 738–749.

[14] Mohammad Hossein Ahmadi, A.M.-N. (2018). A review of thermal conductivity of various nanofluids. *Journal of Molecular Liquids,* 181–188.

[15] Mohammed, A.A., Dawood, H.I. and Onyeaka, H.N. A Review Paper on Properties and Applications of Nanofluids, 3rd International Conference for Civil Engineering Science (ICCES 2023), IOP Publishing, IOP Conf. Series: Earth and Environmental Science 1232 (2023) 012009, doi:10.1088/1755-1315/1232/1/012009 .

[16] Nikhil, M. and Hemadri, V. (2021). Study of the effect of preparation parameters on thermal conductivity of metal oxide nanofluids using Taguchi method. *Journal of Energy Systems*, 5(2), 149–164.

[17] P. Sivasamy, A.D. (2018). Review on Heat Transfer Enhancement of Phase Change Materials. *Materials Today: Proceedings* (pp. 14423–14431). Materials Today: Proceedings.

[18] Rahman, MdAtiqur., Hasnain, S.M. M., Pandey, S., Tapalova, Akylbekov, N. and Zairov, R. Review on Nanofluids: Preparation, Properties, Stability, and Thermal Performance Augmentation in Heat Transfer Applications, *ACS Omega* 2024,9,32328−32349, https://doi.org/10.1021/acsomega.4c03279.

[19] Randeep Singh, S.S. (2019). Thermal Conductivity Enhancement of Phase Change Materials for Low-Temperature Thermal Energy Storage Applications. *Energies, 12, 75*, 1–20.

[20] Sandro Nizetic, M.J. (2020). Nano-enhanced phase change materials and fluids in energy applications:A review. *Renewable and Sustainable Energy Reviews,* 109931.

[21] Shaon Talukdar, H.M. (2019). Heat transfer enhancement of charging and discharging of phase change materials and size optimization of a latent thermal energy storage system for solar cold storage application. *Journal of Energy Storage*, 100797.

[22] Sithara Vinod, J.P. (2018). Experimental evidence for the significant role of initial cluster size and liquid confinement on thermo-physical properties of magnetic nanofluids under applied magnetic field. *Journal of Molecular Liquids*, Pages 1–11.

[23] Sonali K. Kale, G.V. (2021). Emerging Agriculture Applications of Silver Nanoparticles. *ES Food Agrofor.*, 17–22.

[24] Sun, L., Geng, J., Dong, K. and Sun, Q. An Experimental Study on the Effect of Nanofluids on the Thermal Conductivity and Rheological Properties of a Coolant for Liquids. Energies 2024, 17, 1313. https://doi.org/10.3390/ en17061313.

[25] Weerapun Duangthongsuk, S.W. (15 January 2010). An experimental study on the heat transfer performance and pressure drop of $TiO_2$ – water nanofluids flowing under a turbulent flow regime. *International Journal of Heat and Mass Transfer*, 334–344.

[26] Yu, F., Chen, Y., Liang, X., Xu, J., Lee, C., Liang, Q., Tao, P. and Deng, T. (2017). Dispersion stability of thermal nanofluids. *Progress in Natural Science: Materials International*, 27(5), 531–542.

# 4 Wettability and Surface Interaction Phenomena in Nanofluids

Mohamed Aboughaly[1*]

Wettability and surface interactions in nanofluids play a vital role in determining their thermal transport properties, stability, and interfacial behavior. The introduction of nanoparticles alters surface energy and adhesion forces, thereby influencing fluid–solid interactions at the molecular scale through van der Waals and electrostatic forces. Enhanced wettability improves heat transfer efficiency by promoting better spreading, boiling, and convective performance, which are essential for advanced cooling and energy systems. At the nanoscale, the modification of surface phenomena optimizes nanoparticle dispersion and interfacial transport, enabling superior performance in thermal management, lubrication, and biomedical applications. The nanoparticles can alter the solid-liquid interfacial energy, decreasing the nanofluid surface tension and contact angles, while enhancing the effective heat transfer area. Such changes enhance boiling heat transfer, delay vapor film formation, and elevate the critical heat flux (CHF), leading to improved system stability and safety. In porous media, nanoparticle-induced wettability enhances spontaneous imbibition and capillary-driven fluid flow. The addition of oxides, metallic, or carbon-based nanoparticles (e.g., $TiO_2$, $SiO_2$, graphene) promotes hydrophilicity, alters nanolayer formation, and reduces interfacial thermal resistance.

## 1. Introduction

Wettability and surface interactions in nanofluids play an important role in determining fluid stability, thermal properties during transport, and interfacial behaviour. Nanoparticles alter the surface energy and adhesion forces that influence the fluid interactions with various solid substrates (Wang *et al.*, 2023; Kim *et al.*, 2005). These effects are governed by molecular forces such as van der vaal and electrostatic molecular interactions that govern the solid-liquid interface. E nhanced wettability improves heat transfer efficiency and enhances the spreading and boiling characteristics, which are key factors for cooling efficiency and energy systems. The improvement of nanoscale surface phenomena optimizes the dispersion ability and tailors nanofluid interactions in thermal management, lubrication, and

[1] University of Toronto, Canada.
* Corresponding author: abougm1@mcmaster.ca

DOI: 10.1201/9781003686002-4

biomedical systems (Hosseini *et al.*, 2022). The practical modification of surface interactions in nanofluids has implications for its applications, such as pool boiling heat transfer in porous media and enhanced transport in thermal systems (Hosseini *et al.,* 2022; Kasaeian *et al.,* 2017; Nabwey *et al.*, 2023). Wettability is the fluid's ability to adhere to a solid surface that has the molecular balance of cohesive molecular forces and adhesive forces at the nanofluid interface. The suspension of nanofluids, as well as the surface interactions within a nanofluid, govern the nanoparticles' suspension in nanofluids. These nanoparticles are essential for the molecular heat transfer flow rate and nanofluid stability, as well as the cohesive forces at the nanofluid interface. The fluid suspension of nanoparticles plays a vital role in the heat transfer flow rate and stability of the nanofluid, as well as the interfacial transport phenomena Kasaeian *et al.,* 2017; D'Rose *et al.*, 2023; He *et al.*, 2019).

Modification of surface interactions in nanofluids for various applications, including boiling heat transfer and absorption in porous media, enhances fluid transport in thermal systems. The deposition of nanoparticles on a heated surface modifies the wettability and improves the boiling heat transfer performance, which depends on the fluid morphology. In porous media, the nanoparticle-induced wettability improves the spontaneous imbibition of the fluid (Le Ba *et al.*, 2022; Murshed *et al.,* 2011; Kamel *et al.,* 2019). The addition of nanoparticles alters the interaction between the solid and liquid, which improves the nanofluid's spreading over heated surfaces. The enhanced wettability lowers the contact angle, which increases the effective heat transfer surface area during heat exchange, thereby reducing the thermal resistance at the fluid interface (Kuang *et al.,* 2022). The nanoparticle addition in base fluids and its continuous phase improves the wettability and reduces surface tension and modifies the solid-liquid interfacial energy (Roustael *et al.,* 2013; Altgen *et al.*, 2020).

The introduction of nanoparticles to nanofluids improves their wettability and reduces the surface tension, as well as modifies the solid-liquid interfacial energy. Improved wettability reduces the contact angle and promotes efficient spreading and improvement of convective and boiling heat transfer. Such fluid improvement is essential in thermal systems applications (Upot *et al.,* 2022; Qin *et al.*, 2022). The addition of nanoparticles to nanofluids is customized to the design requirements for thermal systems, including solar collectors, heat exchangers, and microelectronic cooling, where interfacial heat transfer controls the fluid's performance. Enhancement of the fluid wettability delays the formation of vapor films during boiling, increases the critical heat flux (CHF), and ensures system safety and stability (Ali *et al.*, 2018). The suspended nanoparticles affect the nature of the base fluid, temperature, and the fluid surface tension and wettability of nanofluids. Nanoparticles are influenced by the surface tension and wettability of dissolved oxides, carbon-based nanofluids such as graphene nanoplatelets, fullerenes, and metallic nanofluids. The factors that affect the nanofluid molecular surface tension are affected by the nanoparticle's concentration and surfactant effects.

## 2. Basics of Wettability and Surface Tension

Wettability plays a vital role in the enhancement of heat transfer processes, such as boiling point, nanofluid condensation, and flow boiling in microchannels. The wetting behaviour is determined by how the fluid interacts with the solid surface and bubble formation that directly influences the thermal performance of nanofluids. For example, $TiO_2$ suspended in nanofluids could modify the surface wettability and nanofluid liquid spreading, which are governed by the fluid pool boiling heat transfer rate (Mukherji *et al.*, 2012). The solid, liquid, or particle fluid interface is controlled by the molecular ordering and nanolayers that control the fluid mobility, viscosity, and its slip behaviour. The nanofluids change their surface wettability by the addition of nanoparticles that decrease the contact angle of various nanofluids through layer formation or particle deposition. For example, silica particles could be added to nanofluids to decrease the contact angle of various fluids, which improves the overall fluid hydrophilicity (Goharzadeh et al., 2022). Wettability also influences nanofluid spreading that affects the nanoparticle-induced structural disjoining pressures. The addition of nanoparticles that form nanolayers alters the fluid mobility, effective viscosity, and fluid slip behaviour. The addition of nanofluids can also strengthen molecular interactions between water and carbon nanotubes, leading to a reduction in interfacial thermal resistance (Castro et al., 2023). Below is a schematic diagram that shows the wettability mechanism in enhancing boiling heat transfer, condensation, or flow boiling as presented below in Figure 1:

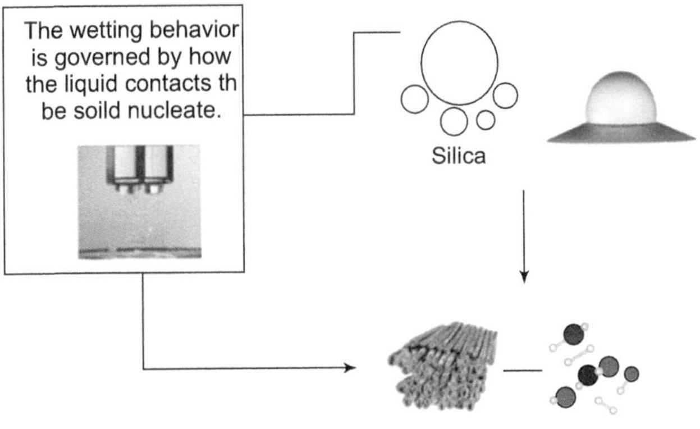


**Fig. 1** Schematic diagram that describes the wettability mechanism in enhancing heat transfer and boiling.

## 3. Current Challenges for Design of Nanofluids

The selection of nanoparticles at different concentrations of base fluid is a major challenge in designing nanofluid engineering systems. The designer of nanofluid systems must consider the surface preparation and compatibility, as well as the solid surface substrate condition and surface roughness that strongly affects the nanofluid interaction (Saleh *et al.*, 2021). Also, smaller effective nanoparticles have stronger spreading driven by structure-disjoining pressure. The substrate condition strongly affects the interaction of nanofluids. To achieve the desired wettability or spreading, the surface preparation and wettability of nanoparticles must be optimized. In various thermal applications such as boiling, cooling, and condensation, surface interactions influence droplet nucleation, bubble flow dynamics, film boiling, and condensation. The reduced contact angle increases the heat-transfer coefficients, thereby improving substrate wettability. The nanoparticle deposition and reduced contact angle lead to an increase in heat transfer coefficients that improves the substrate wettability. The deposition of nanoparticles and the reduction in contact angle increase the heat-transfer coefficients, thereby improving the nanofluid substrate wettability. The nanoparticles deposition also modifies the fluid surface energy and roughness of the heating substrate, which leads to long-term alterations in wettability and thermal performance (Saleh, S., 2021). Many phenomena, including nanolayers, structural forces, and multiscale modeling, operate at molecular and nanoscale levels to capture molecular dynamics and interfacial behavior. Understanding such interactions requires coupling atomistic simulation with continuum-scale models. The molecular dynamics provide insights into layering of fluid molecules near solid surfaces, the interfacial energy, and the effects of nanoparticle fluid interactions on the fluid viscosity and slip length (Chen *et al.,* 2022). The effective interfacial thermal resistance, disjoining pressure, and boundary conditions affect the nanofluid surface interactions. A multiscale modelling approach is used to bridge the gap between macroscopic properties, such as wettability and heat transfer coefficients, for underlying molecular mechanisms, and bridging the gap between atomic level phenomena and engineering scale applications (Zhou *et al.*, 2022; Ali *et al.*, 2023).

One of the challenges in nanoparticle surface interactions is the lack of consistency in experimental methodologies for the measurement of contact angle and interfacial properties in nanofluids. The interplay between fluid surface energy, nanoparticle adsorption, and nanolayer formation is complex, and yet more experimental data is required (Xiao *et al.,* 2019; Yang *et al.,* 2022). Another challenge is the lack of consistency between contact angles and interfacial properties of the nanofluid. The nanoparticle variations in fluid concentration, stability, and surface treatment lead to inconsistency in nanofluid properties. Also, time-dependent surface modifications due to nanoparticle deposition, fouling, and chemical changes pose a major challenge for the reproducibility of nanofluids. These processes typically improve or weaken surface wettability over repeated heating or fluid flow cycles, affecting heat transfer and system reliability. (Joseph *et al.*, 2020; Kilincarsian *et al.*, 2021). Also, time-dependent surface modifications due to nanoparticle depositions, fouling, and chemical changes pose a challenge to the fluid stability and reproducibility. These processes either enhance or degrade the nanofluid surface wettability, flow cycles, and impact the heat transfer and system reliability (Anwar *et al.*, 2024; Wen *et al.*, 2011).

Nanoparticles at different concentrations within the base fluid remain a current challenge in the design and optimization of nanofluid-engineered systems. The nanofluid surface preparation, compatibility, and stability of the nanofluid suspension, as well as the solid substrate and surface roughness, affect the nanofluid solid interaction and overall heat transfer characteristics [see reference: 1–25]. The nanoparticle size, morphology, and concentration determine the nanofluid disjoining pressure and surface energy. The deposition of smaller nanoparticles results in better spreading because of increased surface tension forces and a higher surface area-to-volume ratio, which improves wettability and decreases the contact angle.

**Nanofluid–Surface Interactions**

The interactions between nanofluids and solid surfaces are governed by their thermal and interfacial behaviour. The molecular interactions govern how nanoparticles alter the nanofluid surface energy, molecular adhesion forces, and nanofluid contact angle, as well as the fluid interface heat transfer (Wijesooriya *et al.*, 2023). Also, the structural disjoining pressure from nanoparticle layering and confinement effects could drive spontaneous spreading of the nanoparticle film even on the nanolayer surface that is not inherently wettable. Those interfacial forces that determine the nanofluid film stability, nanoparticle deposition, and dynamic wetting behaviour of nanofluids in solid substrates (Kamel *et al.,* 2020; U.A Muhammad *et al.*, 2021). The microscopic interactions between nanofluids are the foundation of understanding the macroscopic thermal phenomena in nanofluids.

## 3.1 Solid-Liquid Interfaces

The solid-liquid interface affects molecular and nanoparticle interactions that govern the wetting and stability of nanofluids. When nanoparticles are suspended in nanofluids, they adsorb on the solid surface, forming an ordered nanolayer. This nanolayer typically has different viscosity, density, and nanofluid thermal conductivity compared to the bulk fluid, which affects the interfacial transport process (Pavia *et al.*, 2021). The modifications of solid-liquid interfaces using nanoparticles significantly affect the mechanism of heat transfer, such as conduction, convection, and fluid boiling. The added nanoparticles enhance the nanofluid thermal transport and increase the thermal conductivity of modified nanofluids, improve the surface wettability, and modify the interfacial thermal resistance (Kenellopoulos *et al.,* 2012; Noorhism *et al.*, 2021). Molecular dynamics reveal that stronger solid-liquid affinity and a higher degree of ordering in interfacial correlation with improved heat transport across the nanofluid interface (Kenellopoulos *et al.,* 2012; Noorhism *et al.*, 2021; Tarkowski, 2019).

## 3.2 Heat Transfer Implications

The molecular interactions between the nanofluid and the modified solid surface play a vital role in determining the overall heat transfer efficiency in nanofluid thermal systems. The engineered nanofluid colloidal suspension of nanoparticles in conventional-based fluids has improved thermal properties compared to base nanofluids due to their mechanisms that occur both at the molecular and interfacial fluid level. The improvements in heat transfer are due to increased thermal conductivity as well as surface wettability, interfacial thermal resistance, and nanoparticle layering near the solid-liquid interface (Ali and Salam, 2020; Ganvir *et al.*, 2017). Nanofluids are colloidal suspensions that have nanometer-sized particles, typically less than 90 nm in diameter, that are dispersed in conventional-based fluids such as ethylene glycol and water, which enhances the thermal properties compared to base fluids due to various molecular and interfacial level mechanisms. The molecular improvements in heat transfer increase the thermal conductivity and are also linked to changes in surface wettability and interfacial thermal resistance, nanoparticle layering, and micro convective phenomena near the solid-liquid interface (Wen *et al.,* 2022). Wettability plays a vital role in heat transfer applications during convective boiling, condensation, and convective heat transfer. A reduced contact angle improves liquid spreading, enhances capillary rewetting, and increases heat transfer and contact between the heated surface and the nanofluid, which raises the effective heat transfer coefficient. (Zhu *et al.*, 2021). However, excessive nanoparticle fouling or agglomeration in nanofluids deteriorates the thermal performance by increasing the thermal resistance and bubble nucleation in nanofluids. Therefore, it is vital to monitor and control the nanoparticle concentration, size, and morphology deposition since it is essential to sustain consistent heat transfer that enhances the nanofluid's prolonged operations.

## 4. Wettability Modification Using Nanofluids

Surface wettability modification using nanofluids is a phenomenon that impacts the boiling, condensation, and overall heat transfer rate. Wettability is quantified by measuring the contact angle between the solid surface and the liquid droplet, which determines the liquid spreading effectiveness and interaction with the substrate (Kim *et al.*, 2007). Addition of nanoparticles in the base fluid alters the nanofluid surface energy and the microstructure of the substrate, which leads to a measurable change in the nanofluid contact angle (Aksoy *et al.*, 2023). The suspended nanoparticle in the fluid causes deposition onto the heated solid surface during thermal processes, forming a thin coated layer that alters the surface roughness and nanofluid chemistry (Aksoy *et al.*, 2023; Song *et al.*, 2021; Kim *et al.*, 2007).

The deposited nanolayer increases surface hydrophilicity by reducing the contact angle, which enhances fluid spreading and capillary rewetting, benefiting boiling and convective heat transfer [see references1–20]. The mechanistic effects caused by nanofluids alter the surface wettability, which involves both chemical and physical effects. The added nanoparticles introduce oxygen-containing functional groups or surface hydroxyl groups through chemical addition, which increases the nanofluid's surface affinity for polar fluids like water. The formation of nano-rough textures that are formed physically due to the particle deposition improves the surface contact angle area between the fluid surface and the solid phase, which amplifies the wetting effects that are described by Wenzel and Cassie Baxter models (Parvate *et al.*, 2020). These effects improve the adhesion of the liquid phase and enhance the stable film during different heat transfer phases, such as boiling, and delay the heat flux limit (CHF). The improved surface wetting characteristics are beneficial in heat exchangers design, nuclear cooling systems design, and microelectronic thermal management, where surface dryout causes a major operational limitation. The degree of wettability modification is determined by several parameters, including nanoparticle concentration and surface functionalization, as well as nanoparticle size and concentration. For example, the addition of wettability modification is determined by several process parameters, including nanoparticle concentration, size, and surface functionalization. The addition of metal oxide nanoparticles such as $TiO_2$, $SiO_2$, and $Al_2O_3$ is used in the synthesis of hydrophilic coatings compared to carbon-based nanofluids (Yu *et al.*, 2019).

The addition of nanoparticles and increasing their concentration tends to lower the nanofluid contact angle and causes excessive loading and agglomeration in nanofluids, as well as increasing surface fouling and reducing long-term fluid performance (Chen *et al.*, 2013). Moreover, the repeated heating and cooling cycles cause structural rearrangements in the deposited nanoparticle layer that alter the surface morphology of the nanofluid and lead to hysteresis in nanofluid wettability behaviour (Coccia *et al.,* 2021). The increase in nanoparticle concentrations lowers the contact angle, which reduces long-term performance (Z Chen *et al.*, 2021). The heating and cooling cycles cause structural rearrangements in the deposited nanoparticle layer that alter the fluid surface morphology and lead to hysteresis in wettability behaviour (Z.A. Chen *et al.*, 2022; Guo *et al.*, 2022). The repeated heating and cooling cycles cause structural rearrangements in the deposited nanoparticle layer that alter the surface morphology of nanofluids. The nanofluid formulations and operational parameters are used in optimization that are essential to achieve stable wettability enhancement without detrimental fouling effects. The van der Waal forces, as well as the electrostatic and steric forces in nanoparticles, are modified through surface functionalization, PH control, and surfactant adsorption, which leads to enhanced dispersion stability and reduction in the nanoparticle agglomeration (Arab *et al.,* 2017). Also, the interfacial thermal resistance at the fluid interface and nanoparticles is reduced by optimizing the surface chemistry and improving the phonon coupling, and therefore increasing the overall nanofluid thermal conductivity (Eastman *et al.*, 2001). The nanoparticles' surface functionalization through surfactant adsorption enhances the surface compatibility between the base fluid and its nanoparticles. The improvement in wettability decreases the contact angle, which allows more uniform nanolayer film formation and better heat spreading (Hansjosten *et al.*, 2022; Menichetti *et al.*, 2023). Experimental investigations have shown that the nanoparticle concentration and surface coating are directly influenced by the dynamic contact angle and wetting kinetics of heated substrates. The incorporation of oxide nanoparticles, such as $Al_2O_3$ and $TiO_2$, transforms the hydrophobic surfaces into hydrophilic nanoparticles that improve the boiling performance and condensation rates. The modification of wettability using nanofluids has also gained interest in antifouling coatings and self-cleaning surfaces. The $TiO_2$ and $SiO_2$ nanofluids are used in superhydrophilic and superhydrophobic surfaces through controlled nanoparticle deposition and surface activation

The tunable wettability of nanofluids allows for the creation of multifunctional fluid interfaces with customized adhesion, fluid friction, and capillary characteristics that enhance performance in both industrial and environmental systems. The linking at the molecular level of interfacial interactions, such as nanolayer formation and electrostatic forces, with macroscopic wettability behaviour, enables predictive control over nanolayer surfaces modified using nanofluids (Wang *et al.*, 2016). Also, the addition of nanoparticles (NP), such as hydrophilic silica nanoparticles, adsorbs oil at the oil-water interface, which are surface-activated and function as emulsifiers. Various approaches, including in-situ and ex-situ techniques, are employed to modify the hydrophilicity of nanoparticles, with in-situ methods being particularly common in oil field applications. Different surfactants influence the hydrophilic behavior of silica nanoparticles when used to emulsify simple oils such as decane, dodecane, 1-octadecene, and n-hexane in water (Novoselov *et al.*, 1979; Pei *et al.*, 2021). The electrostatic forces between nanoparticles and surface molecules contribute to the stabilization of oil-in-water emulsions.

Interfacial nanofluids exhibiting tailored wettability are applied in a wide range of fields, including self-cleaning textiles, oil–water separation, anti-icing and antifogging coatings, atmospheric water harvesting, and chemical protection (Liu *et al.*, 2017; Di *et al.*, 2021). Wettability is a critical surface property that is strongly influenced by interfacial phenomena. Additionally, nanofluids play a significant role in processes such as membrane-based separation, adsorption, biological fouling prevention, and corrosion control.

## 5. Practical Applications in Boiling and Condensation

Modification of nanofluids with properties such as surface wettability significantly improves industrial fluids, particularly in condensation and boiling systems. The fluids used in boilers require enhanced surface wetting and nucleation sites, which is achieved using nanoparticles and by altering nanofluids' surface energy. The modification of the nanofluid contact angle results in the promotion of bubble nucleation, increases the bubble departure frequency, and improves the replenishment at the heated surface of the liquid (Z Chen *et al.*, 2020). For example, the usage of nanoparticles such as $Al_2O_3$ and $TiO_2$ in nanofluids enhances and improves wettability by 200% and formation of stable nanoparticle coatings that facilitate the efficient vapor-liquid interaction (Nguyen et al., 2021).

Nanofluids could create thermal adaptive surfaces through the deposition of nanoparticles, which results in evolving micro and nanotexture that sustains enhanced capillary and delays dryout. This is a desired behaviour in heat transfer systems such as microchannel heat sinks, cooling of nuclear reactor heat exchange systems, and electronics thermal management that maintains stable nuclear boiling. Improved wettability leads to better heat transfer, and excessive nanoparticle accumulation is to be avoided due to the formation of insulating layers that degrade over time. In condensation processes, nanofluids influence fluid dynamics and the formation of nanolayer films through altering the surface energy that promotes heterogeneous nucleation. The addition of hydrophilic nanoparticle coatings improves the film-wise condensation by improving the liquid spreading and increasing the overall heat transfer rates on vertical and inclined surfaces. The engineering hydrophobic and mixed wettability coatings using nanostructured materials promote dropwise condensation that boosts the heat transfer coefficients through facilitating rapid drop removal. $TiO_2$ and $SiO_2$-based nanofluids are used to create tunable surfaces (PW Wang *et al.*, 2016; Zhang *et al.*, 2017).

The modification of fluid wettability using nanofluids in thermal applications has gained interest in antifouling coatings, self-cleaning surfaces, and oil-water separation technologies. Nanofluid particles such as $TiO_2$ and $SiO_2$ are used to synthesize hydrophilic or hydrophobic surfaces through controlled nanoparticle deposition and surface activation processes. The tunable wettability is achieved using nanofluids, which enables the development of multifunctional interfaces with adjusted adhesion, friction, and capillary properties that enhance the performance in both industrial systems. In engineering systems, controlled usage of nanofluids allows engineered wettability tuning, such as boiling surfaces, microchannel coolers, and condensation heat exchangers. The addition of water-based CuO and $Al_2O_3$ nanofluids is used to synthesize hydrophilic coatings that increase the critical heat flux by more than 100%. The manipulation of nanoparticle concentration, surface chemistry, and operating conditions of nanofluids enables the design of thermally efficient interfaces tailored for advanced energy and thermal cooling technologies (Cieślinński and Krygier, 2014). The current challenges in controlling wettability at the nanoscale include the interdependence of nanoparticle size, surface energy, and concentration over the fluid-solid interaction. There is a demand for the development of machine learning approaches and molecular dynamics simulations to predict interfacial properties of nanofluids. The development of biocompatible and eco-friendly nanofluids requires further surface modifications that enhance wettability without compromising environmental concerns (Babanezhad *et al.*, 2012; Sharma *et al.*, 2022).

Nanofluids are also used in refrigeration and air-cooling systems to improve the thermal performance of the evaporator and condenser efficiency through increasing surface wettability and heat flux density. In power generation systems, nanoparticles enhance boiling and improve the safety and reliability of nuclear fuel cooling systems, while in solar thermal collectors, nanofluid coatings reduce thermal resistance and enhance condensation oil recovery. The current challenges remain related to nanofluid stability, corrosion, and long-term coating during cyclic thermal loads (Baharin *et al.*, 224; Mukherjee *et al.*, 2023). In refrigeration and air conditioning systems, nanofluids are used to improve evaporator and condenser efficiency through increasing surface wettability and heat flux density. The nanoparticles enhance boiling and enhance the safety and reliability of nuclear fuel cooling systems, while in solar thermal collectors, nanofluid coatings are used to reduce thermal resistance and enhance condensation recovery. The current challenges remain related to nanofluid thermal stability, corrosion, and coating durability under thermal loads. The nanofluid enhanced boiling and condensation is identified through multiscale modelling, optimized nanoparticle selection, and surface engineering design.

## 6. Measurement Techniques for Nanofluids

There are various measurement techniques essential for nanofluids to assess thermal conductivity, particle size distribution, and nanofluid viscosity. To measure the particle size distribution for nanofluids, dynamic light scattering (DLS), as well as various analytical equipment such as transmission electron or scanning electron microscopy (TEM or SEM), is used. These techniques determine the nanoparticle size, morphology, and dispersion uniformity [see references 1–25]. The thermal conductivity of nanofluids is measured using laser flash analysis and the transient hot wire method. The transient hot wire (THW) method is preferred for higher precision and detection of variations due to nanoparticle addition. The viscosity is measured using capillary viscometers and vibrational viscometers, depending on the nanoparticle concentration, shape, and temperature, which influence the flow and heat transfer performance. [see references 1–30]. Also, the fluid thermal diffusivity is measured using the laser flash method and photothermal radiometers to measure heat propagation in nanofluids. The optical and spectral properties of nanoparticle dispersions in nanofluids are measured using optical absorption, which confirms the fluid stability and concentration [see references 1–40].

Various measurement techniques are employed to characterize nanofluids, determining particle size distribution, thermal conductivity, viscosity, and other physicochemical properties that affect the nanofluid's thermal performance in heat transfer applications. The particle size distribution is measured using dynamic light scattering (DLS) due to its ability to estimate the hydrodynamic diameter and nanoparticle size distribution suspended in fluids. The usage of Transmission Electron Microscopy (TEM) and Scanning Electron Microscopy (SEM) provides insights about particle morphology, surface roughness, and nanofluid agglomeration behaviour [see references 1–30]. This analytical equipment is essential to assess the nanofluid dispersion stability and nanoparticles uniformity to ensure thermal performance and avoid sedimentation in flow systems [see references 30–40].

The thermal diffusivity and nanofluid heat transfer properties, such as specific heat capacity, are measured using laser flash calorimetry and photothermal radiometry, which enables accurate evaluation of heat propagation characteristics and transient energy storage within nanofluids [see references 30–40]. The usage of optical and spectroscopic techniques, such as Fourier Transform Infrared (FTIR) spectroscopy and UV-visible spectroscopy, is used to identify the chemical bonding and nanofluid dispersion stability, which confirms the particle fluid interactions and nanofluid concentration stability over time. The optical absorption spectra are considered an indirect indicator of nanoparticle dispersion and nanofluid agglomeration tendencies.

## 7. Emerging Trends and Future Research in Nanofluids

Future research focuses on nanofluids with tailored nanoparticles, such as hybrid and doped structures, to optimize thermal conductivity, stability, and rheological performance. Hybrid nanofluids such as $Al_2O_3$–Cu and graphene-based $TiO_2$ that exhibit synergistic effects that enhance the heat transport and stability. Also, the addition of functional groups using surface functionalization with surfactants and polymer coatings that mitigate agglomeration and improve long-term dispersion. E merging fabrication methods, such as green synthesis and laser ablation, can provide sustainable and reproducible nanoparticles in nanofluids. Smart and adaptive nanofluids, such as magneto and electrorheological nanofluids, that adjust the flow and viscosity dynamically based on external fields. These nanofluids are used in intelligent cooling and energy harvesting systems in electronics and biomedical devices.

Innovative nanofluids used in renewable energy systems, such as solar collectors, phase change materials (PCM), and thermoelectric devices, are employed to improve solar thermal efficiency by utilizing optical properties and absorption. Also, the usage of nano PCM composites to boost latent heat storage with higher thermal conductivity. The integration with heat pipes and photovoltaic thermal systems enhances energy conversion rates. In addition, using computational and AI-assisted modelling, such as molecular dynamics (MD), provides insights into nanoscale interfacial phenomena of nanofluids and machine learning algorithms that predict the thermophysical properties and optimize nanofluid formulations, and accelerate experimental design. The nanofluid thermal conductivity is measured using the transient hot wire (THW) and laser flash analysis (LFA) methods. The THW is recommended for its high precision due to its transient nature that allows *in-situ* thermal conductivity evaluation under different concentrations and operating temperatures [see references 40–60]. The viscosity of nanofluids influences pumping power and energy efficiency, which is measured using rotational and vibrational viscometers that depend on the viscosity range and flow regime [24–28]. These instruments assess the shear thinning and Newtonian properties of nanofluids, influenced by nanoparticle concentration, temperature, and particle aspect ratio. The high nanoparticle loading results in enhanced viscosity that leads to increased flow resistance and a reduction in overall heat transfer efficiency.

## References


[1] Wang, K.Y. *et al.*, "Bioinspired Framework Catalysts: From Enzyme Immobilization to Biomimetic Catalysis," 2023. doi: 10.1021/acs.chemrev.2c00879.

[2] Kim, S.H., Na, S.W., Lee, N.E., Nam, Y.W. and Y.H. Kim, "Effect of surface roughness on the adhesion properties of Cu/Cr films on polyimide substrate treated by inductively coupled oxygen plasma," *Surf Coat Technol*, vol. 200, no. 7, 2005, doi: 10.1016/j.surfcoat.2005.05.021.

[3] Hosseini, M.S., Khazaei, M., Misaghi, M. and Koosheshi, M.H. "Improving the stability of nanofluids via surface-modified titanium dioxide nanoparticles for wettability alteration of oil-wet carbonate reservoirs," *Mater Res Express*, vol. 9, no. 3, 2022, doi: 10.1088/2053-1591/ac4fdf.

[4] Kasaeian, A. *et al.*, "Nanofluid flow and heat transfer in porous media: A review of the latest developments," 2017. doi: 10.1016/j.ijheatmasstransfer.2016.11.074.

[5] Nabwey, H.A., Armaghani, T., Azizimehr, B., Rashad, A.M. and Chamkha, A.J. "A Comprehensive Review of Nanofluid Heat Transfer in Porous Media," 2023. doi: 10.3390/nano13050937.

[6] D'Rose, R., Willemsz, M. and Smeulders, D. "Transient Convective Heat Transfer in Porous Media," *Mathematics*, vol. 11, no. 21, 2023, doi: 10.3390/math11214415.

[7] He, Y.L., Liu, Q., Li, Q. and Tao, W.Q. "Lattice Boltzmann methods for single-phase and solid-liquid phase-change heat transfer in porous media: A review," 2019. doi: 10.1016/j.ijheatmasstransfer.2018.08.135.

[8] T. Le Ba *et al.*, "Experimental Study of Halloysite Nanofluids in Pool Boiling Heat Transfer," *Molecules*, vol. 27, no. 3, 2022, doi: 10.3390/molecules27030729.

[9] Murshed, S.M.S., Nieto De Castro, C., Loureno, A.M.J.V., Lopes, M.L.M. and Santos, F.J.V. "A review of boiling and convective heat transfer with nanofluids," 2011. doi: 10.1016/j.rser.2011.02.016.

[10] Kamel M.S. and Lezsovits, F. "Boiling heat transfer of nanofluids a review of recent studies," 2019. doi: 10.2298/TSCI170419216K.

[11] Kuang H. *et al.*, "Wettability and thermal contact resistance of thermal interface material composited by gallium-based liquid metal on copper foam," *Int J Heat Mass Transf*, vol. 199, 2022, doi: 10.1016/j.ijheatmasstransfer.2022.123444.

[12] Roustaei, A., Saffarzadeh, S. and Mohammadi, M. "An evaluation of modified silica nanoparticles' efficiency in enhancing oil recovery of light and intermediate oil reservoirs," *Egyptian Journal of Petroleum*, vol. 22, no. 3, 2013, doi: 10.1016/j.ejpe.2013.06.010.

[13] Altgen, D., Altgen, M., Kyyrö, S., Rautkari, L. and C. Mai, "Time-dependent wettability changes on plasma-treated surfaces of unmodified and thermally modified European beech wood," *European Journal of Wood and Wood Products*, vol. 78, no. 2, 2020, doi: 10.1007/s00107-020-01505-0.

[14] Upot, N.V., Fazle Rabbi, K., Khodakarami, S., Ho, J.Y., Kohler Mendizabal, J. and Miljkovic, N. "Advances in micro and nanoengineered surfaces for enhancing boiling and condensation heat transfer: a review," 2022. doi: 10.1039/d2na00669c.

[15] Qin, S., Ji, R., Miao, C., Jin, L., Yang, C. and Meng, X. "Review of enhancing boiling and condensation heat transfer: Surface modification," *Renewable and Sustainable Energy Reviews*, vol. 189, 2024, doi: 10.1016/j.rser.2023.113882.

[16] Ali, J.A., Kolo, K., Manshad, A.K. and Mohammadi, A.H.,"Recent advances in application of nanotechnology in chemical enhanced oil recovery: Effects of nanoparticles on wettability alteration, interfacial tension reduction, and flooding," 2018. doi: 10.1016/j.ejpe.2018.09.006.

[17] Mukherjee, S., Mishra, P.C. and Chaudhuri, P. "Pool boiling performance of aqueous $Al_2O_3$ and $TiO_2$ nanofluids on a horizontally placed flat polished surface: an experimental investigation," *J Therm Anal Calorim*, vol. 146, no. 1, 2021, doi: 10.1007/s10973-020-09995-z.

[18] Goharzadeh A. and Fatt, Y. Y. "Combined effect of silica nanofluid and wettability on enhanced oil recovery process," *J Pet Sci Eng*, vol. 215, 2022, doi: 10.1016/j.petrol.2022.110663.
[19] Casto, A. *et al.*, "Water filling in carbon nanotubes with different wettability and implications on nanotube/water heat transfer via atomistic simulations," *Int J Heat Mass Transf*, vol. 205, 2023, doi: 10.1016/j.ijheatmasstransfer.2023.123868.
[20] Saleh, S., Neubauer, E., Borovina, A., Hincapie, R.E., Clemens, T. and Ness, D. "Wettability changes due to nanomaterials and alkali—a proposed formulation for eor," *Nanomaterials*, vol. 11, no. 9, 2021, doi: 10.3390/nano11092351.
[21] Chen, S.H., Bell, D.R. and Luan, B. "Understanding interactions between biomolecules and two-dimensional nanomaterials using in silico microscopes," 2022. doi: 10.1016/j.addr.2022.114336.
[22] Zhou, L., Zhu, J., Zhao, Y. and Ma, H. "A molecular dynamics study on thermal conductivity enhancement mechanism of nanofluids – Effect of nanoparticle aggregation," *Int J Heat Mass Transf*, vol. 183, 2022, doi: 10.1016/j.ijheatmasstransfer.2021.122124.
[23] Ali, I. *et al.*, "Preparation of graphene based nanofluids: Rheology determination and theoretical analysis of the molecular interactions of graphene nanoparticles," *J Mol Liq*, vol. 390, 2023, doi: 10.1016/j.molliq.2023.122954.
[24] Xiao, C., Shi, P., Yan, W., Chen, L., Qian, L. and Kim, S.H. "Thickness and structure of adsorbed water layer and effects on adhesion and friction at nanoasperity contact," 2019. doi: 10.3390/colloids3030055.
[25] Yang, D. *et al.*, "Probing the Interfacial Forces and Surface Interaction Mechanisms in Petroleum Production Processes," 2022. doi: 10.1016/j.eng.2022.06.012.
[26] Joseph, O.O., Afolalu, S.A., Okeniyi, J. and Emetere, M.E. "Dataset on corrosion degradation behaviour of mild steel in developed nano-cutting coolant from agro waste," *Data Brief*, vol. 33, 2020, doi: 10.1016/j.dib.2020.106435.
[27] Kilincarslan, E., Kabave Kilincarslan, S. and Cetin, M.H. "Evaluation of the clean nano-cutting fluid by considering the tribological performance and cost parameters," *Tribol Int*, vol. 157, 2021, doi: 10.1016/j.triboint.2021.106916.
[28] Anwar, Asifa, T., Kumam, P., El-Zahar, E.R., Muhammad, S. and Seddek, L.F. "Thermal analysis of mineral oil-based hybrid nanofluid subject to time-dependent energy and flow conditions and multishaped nanoparticles," *J Therm Anal Calorim*, vol. 149, no. 13, 2024, doi: 10.1007/s10973-023-12622-2.
[29] Wen, D., Corr, M., Hu, X. and Lin, G. "Boiling heat transfer of nanofluids: The effect of heating surface modification," *International Journal of Thermal Sciences*, vol. 50, no. 4, 2011, doi: 10.1016/j.ijthermalsci.2010.10.017.
[30] Wijesooriya, N., Brambilla, A. and Markauskaite, L. "Biophilic design frameworks: A review of structure, development techniques and their compatibility with LEED sustainable design criteria," 2023. doi: 10.1016/j.clpl.2023.100033.
[31] Kamel, M.S., Al-agha, M.S., Lezsovits, F. and Mahian, O. "Simulation of pool boiling of nanofluids by using Eulerian multiphase model," *J Therm Anal Calorim*, vol. 142, no. 1, 2020, doi: 10.1007/s10973-019-09180-x.
[32] Muhammad, U.A., Bhattacharyya, D., Endrino, J.L. and Fereres, S. "Preparation of binary nanofluid with heat transfer additives by particle surface functionalisation," *Emergent Mater*, vol. 4, no. 6, 2021, doi: 10.1007/s42247-021-00260-z.
[33] Pavía, M., Alajami, K., Estellé, K., Desforges, A. and Vigolo, B. "A critical review on thermal conductivity enhancement of graphene-based nanofluids," 2021. doi: 10.1016/j.cis.2021.102452.
[34] Kanellopoulos, Nick, and Kanellopoulos, N. *Nanoporous Materials: Advanced Techniques for Characterization, Modeling, and Processing*. 2012.
[35] Noorhisham, N.A. *et al.*, "Characterisation techniques for analysis of imidazolium-based ionic liquids and application in polymer preparation: A review," 2021. doi: 10.1016/j.molliq.2021.115340.

[36] Tarkowski, R. "Underground hydrogen storage: Characteristics and prospects," 2019. doi: 10.1016/j.rser.2019.01.051.

[37] Ali, A.R.I. and Salam, B. "A review on nanofluid: preparation, stability, thermophysical properties, heat transfer characteristics and application," 2020. doi: 10.1007/s42452-020-03427-1.

[38] Ganvir, R.B., Walke, P.V. and Kriplani, V.M. "Heat transfer characteristics in nanofluid—A review," 2017. doi: 10.1016/j.rser.2016.11.010.

[39] Wen, T., Luo, J., Jia, K. and Lu, L. "Pool boiling heat transfer enhancement of aqueous solution with quaternary ammonium cationic surfactants on copper surface," *Int J Heat Mass Transf*, vol. 190, 2022, doi: 10.1016/j.ijheatmasstransfer.2022.122761.

[40] Zhu, M., Liu, Y., Chen, M., Xu, Z., Li, L. and Zhou, Y. "Metal mesh-based special wettability materials for oil-water separation: A review of the recent development," 2021. doi: 10.1016/j.petrol.2021.108889.

[41] Kim, S.J., Bang, I.C., Buongiorno, J. and Hu, L.W. "Surface wettability change during pool boiling of nanofluids and its effect on critical heat flux," *Int J Heat Mass Transf*, vol. 50, no. 19–20, 2007, doi: 10.1016/j.ijheatmasstransfer.2007.02.002.

[42] Aksoy, Y. T., Liu, L., Abboud, M., Vetrano, M.R. and Koos, E. "Role of Nanoparticles in Nanofluid Droplet Impact on Solid Surfaces," 2023. doi: 10.1021/acs.langmuir.2c02578.

[43] Song, Y., Ma, X., Wang, Y., Yao, S. and Vafai, K. "Amelioration of boiling heat transfer by 3D deposition structure of graphene-silver hybrid nanoparticle," *Energy Conversion and Management: X*, vol. 12, 2021, doi: 10.1016/j.ecmx.2021.100109.

[44] Kim, S.J., Bang, I.C., Buongiorno, J. and Hu, L.W. "Study of pool boiling and critical heat flux enhancement in nanofluids," *Bulletin of the Polish Academy of Sciences: Technical Sciences*, vol. 55, no. 2, 2007.

[45] Parvate, S., Dixit, P. and Chattopadhyay, S. "Superhydrophobic Surfaces: Insights from Theory and Experiment," 2020. doi: 10.1021/acs.jpcb.9b08567.

[46] Yu, L., Bian, Y., Liu, Y. and Xu, X. "Experimental investigation on rheological properties of water based nanofluids with low MWCNT concentrations," *Int J Heat Mass Transf*, vol. 135, 2019, doi: 10.1016/j.ijheatmasstransfer.2019.01.120.

[47] Chen, Y.J., Wang, P.Y. and Liu, Z.H. "Application of water-based $SiO_2$ functionalized nanofluid in a loop thermosyphon," *Int J Heat Mass Transf*, vol. 56, no. 1–2, 2013, doi: 10.1016/j.ijheatmasstransfer.2012.09.048.

[48] Coccia, G., Tomassetti, S. and G. Di Nicola, "Thermal conductivity of nanofluids: A review of the existing correlations and a scaled semi-empirical equation," 2021. doi: 10.1016/j.rser.2021.111573.

[49] Chen, Z., Chen, Z., Zhou, D., Xia, T. and Pan, E. "Reliability evaluation for multi-state manufacturing systems with quality-reliability dependency," *Comput Ind Eng*, vol. 154, 2021, doi: 10.1016/j.cie.2021.107166.

[50] Z. (Ashley) Chen, F. Huang, P.A. Tsai and A. Komrakova, "Numerical study of microfluidic emulsion dynamics under the influence of heterogeneous surface wettability," *International Journal of Multiphase Flow*, vol. 147, 2022, doi: 10.1016/j.ijmultiphaseflow.2021.103863.

[51] Guo, R. *et al.*, "Role of heterogeneous surface wettability on dynamic immiscible displacement, capillary pressure, and relative permeability in a CO2-water-rock system," *Adv Water Resour*, vol. 165, 2022, doi: 10.1016/j.advwatres.2022.104226.

[52] Arab, D., Kantzas, A. and Bryant, S.L. "Nanoparticle stabilized oil in water emulsions: A critical review," 2018. doi: 10.1016/j.petrol.2017.12.091.

[53] Eastman, J.A., Choi, S.U.S., Li, S., Yu, W. and Thompson, L.J. "Anomalously increased effective thermal conductivities of ethylene glycol-based nanofluids containing copper nanoparticles," *Appl Phys Lett*, vol. 78, no. 6, 2001, doi: 10.1063/1.1341218.

[54] Hansjosten, I. *et al.*, "Surface functionalisation-dependent adverse effects of metal nanoparticles and nanoplastics in zebrafish embryos," *Environ Sci Nano*, vol. 9, no. 1, 2022, doi: 10.1039/d1en00299f.

[55] Menichetti, A., Mavridi-Printezi, A., Mordini, D. and Montalti, M. "Effect of Size, Shape and Surface Functionalization on the Antibacterial Activity of Silver Nanoparticles," 2023. doi: 10.3390/jfb14050244.

[56] Wang, Z., Elimelech, M. and Lin, S. "Environmental Applications of Interfacial Materials with Special Wettability," 2016. doi: 10.1021/acs.est.5b04351.

[57] Novoselov, K.S. *et al.*, "Electric Field Effect in Atomically Thin Carbon Films Supplementary," *Science (1979)*, vol. 5, no. 1, 2004, doi: 10.1126/science.aab1343.

[58] Pei, H., Shan, J., Cao, X., Zhang, G. and Jiang, P. "Research Progresses on Nanoparticle-stabilized Emulsions for Enhanced Oil Recovery," 2021. doi: 10.11896/cldb.20050023.

[59] Liu, M., Hou, Y., Li, J. and Guo, Z. "Stable Superwetting Meshes for On-Demand Separation of Immiscible Oil/Water Mixtures and Emulsions," *Langmuir*, vol. 33, no. 15, 2017, doi: 10.1021/acs.langmuir.7b00658.

[60] Di, J., Li, L., Wang, Q. and Yu, J. "Porous membranes with special wettabilities: Designed fabrication and emerging application," 2021. doi: 10.31635/ccschem.020.202000457.

[61] Chen, Z., Zheng, D., Wang, J., Chen, L. and Sundén, B. "Experimental investigation on heat transfer characteristics of various nanofluids in an indoor electric heater," *Renew Energy*, vol. 147, 2020, doi: 10.1016/j.renene.2019.09.036.

[62] Nguyen, D.H. and Ahn, H.S. "A comprehensive review on micro/nanoscale surface modification techniques for heat transfer enhancement in heat exchanger," 2021. doi: 10.1016/j.ijheatmasstransfer.2021.121601.

[63] Wang, P.W., Liu, M.J. and Jiang, L. "Bioinspired multiscale interfacial materials with Superwettability," 2016. doi: 10.7498/aps.65.186801.

[64] Zhang, P., Lin, L., Zang, D., Guo, X. and Liu, M. "Designing Bioinspired Anti-Biofouling Surfaces based on a Superwettability Strategy," *Small*, vol. 13, no. 4, 2017, doi: 10.1002/smll.201503334.

[65] Cieśliński, J.T. and Krygier, K. "Augmentation of the critical heat flux in water-Al2O3, water-$TiO_2$ and water-Cu nanofluids," in *MATEC Web of Conferences*, 2014. doi: 10.1051/matecconf/20141801012.

[66] Babanezhad, M., Behroyan, I., Marjani, A. and Shirazian, S. "Velocity prediction of nanofluid in a heated porous pipe: DEFIS learning of CFD results," *Sci Rep*, vol. 11, no. 1, 2021, doi: 10.1038/s41598-020-79913-8.

[67] Sharma, P. *et al.*, "Recent Advances in Machine Learning Research for Nanofluid-Based Heat Transfer in Renewable Energy System," 2022. doi: 10.1021/acs.energyfuels.2c01006.

[68] Baharin, K.W. *et al.*, "Assessing material selection and thermophysical considerations for sustainable nanofluids: A comprehensive review," 2024. doi: 10.1016/j.nanoso.2024.101090.

[69] Mukherjee, S., Wciślik, S., Mishra, P. Chandra and Chaudhuri, P. "Nanofluids: Critical issues, economics and sustainability perspectives," *Particuology*, vol. 87, 2024, doi: 10.1016/j.partic.2023.06.021.

# 5 Advanced Characterization Methods for Nanofluids

Supriya A. Joshi[1*], Sayli D. Hulwan[2] and Sanjay Mane[3]

Nanofluids have attracted considerable interest in thermal engineering due to their superior heat transfer characteristics compared to conventional base fluids. Accurate experimental optimization and performance evaluation depend on detailed characterization using advanced analytical techniques capable of probing the complex interactions between nanoparticles and the host fluid. Such characterization techniques are essential for elucidating the structural, thermal, rheological, and stability-related properties of nanofluids. This chapter focuses on advanced characterization methods employed for the systematic evaluation of nanofluids used in thermal applications. It discusses state-of-the-art techniques for analyzing nanoparticle morphology, size distribution, and dispersion behaviour using microscopic and spectroscopic tools, including transmission electron microscopy (TEM), scanning electron microscopy (SEM), X-ray diffraction (XRD), and dynamic light scattering (DLS). In addition, stability assessment methods such as zeta potential analysis, Thermal Conductivity analysis (THW Method), Fourier-transform infrared (FTIR) spectroscopy, and UV–visible spectroscopy are addressed to ensure the long-term performance and reliability of nanofluids. This chapter provides an application-oriented framework for developing high-performance nanofluids through advanced characterizations.

## 1. Introduction

Effective characterization is essential to understand nanofluid properties like particle size, stability, thermal conductivity, viscosity, and interfacial interactions, which directly affect performance in practical applications (Angayarkanni and Philip, 2015; Li *et al.*, 2009; Sedaghat and Yousefi, 2019). Advanced techniques combining experimental, analytical, and computational methods help optimize formulations for heat transfer, energy storage, and multifunctional applications. Nanoparticles such as metals (Cu, Al), metal oxides ($TiO_2$, $Al_2O_3$), carbon-based

1 Department of Chemistry, MIT College of Railway Engineering and Research, Barshi-413401, Solapur, Maharashtra, India.
2 Department of Electronics and Computer Engineering, MIT College of Railway Engineering and Research, Barshi-413401, Solapur, Maharashtra, India.
3 Department of Mechanical Engineering, MIT College of Railway Engineering and Research, Barshi-413401, Solapur, Maharashtra, India.
* Corresponding author: supriyajoshijjj@gmail.com

DOI: 10.1201/9781003686002-5

materials (CNTs, graphene), or hybrid nanomaterials enhance the thermophysical and functional properties of base fluids (Das *et al.*, 2024; Razzaq *et al.*, 2025). Stability is assessed using zeta potential and DLS, while morphology is analyzed via SEM, TEM, and XRD (Nigam *et al.*, 2018; Sedaghat and Yousefi, 2019). Thermophysical properties like thermal conductivity, viscosity, density, and specific heat are measured with transient hot-wire methods, rheometers, and calorimetry. Spectroscopic methods such as UV–Vis, photoluminescence, FTIR, and Raman provide data on particle concentration, band gaps, functional groups, and nanoparticle–fluid interactions (Li *et al.*, 2009; Yu *et al.*, 2011). Figure 1 illustrates the use of these techniques for analyzing nanoparticle morphology, stability, and dispersion behavior.

Characterizing nanofluids is challenging due to the complex interplay of solid nanoparticles in liquid at the nanoscale. Particle size, surface chemistry, dispersion, and nanoparticle–fluid interactions strongly influence thermal and transport behavior. Variations in synthesis, dispersion, and measurement procedures often lead to inconsistent results, highlighting the need for standardized, reproducible methods (Wong & De Leon, 2010; Yu *et al.*, 2011). Standardized characterization links synthesis to performance, enabling rational design of nanofluid systems. Integrated spectroscopic, microscopic, and thermophysical analyses are crucial for controlling design, stability, and performance, supporting energy-efficient and sustainable applications.

## 2. Importance of Nanofluids Characterization

Nanofluids are used in electronics, medicine, and energy, but their performance depends on particle size, stability, surface chemistry, and concentration (Angayarkanni & Philip, 2015; M *et al.*, 2024). Characterization links structure to properties, ensuring stable dispersions, consistent thermal performance, and reliable heat transfer. Techniques like DLS, zeta potential, and UV–Vis assess stability, while SEM, TEM, and XRD reveal morphology and crystallinity (Figure 1). Thermophysical testing predicts real-world performance, and reproducibility, scalability, and reliability studies support standard protocols and industrial applications (Devendiran & Amirtham, 2016; Sengwa *et al.*, 2022). Despite advances, characterization is often fragmented; integrated, standardized methods are essential for accurate, reproducible, and meaningful nanofluid analysis.

## 3. Microscopic Techniques

Microscopic techniques play an essential role in characterizing the nanofluids, as they give details and confirm the particle size, morphology, and dispersion. Scanning Electron Microscopy (SEM), Transmission Electron Microscopy (TEM), Dynamic Light Scattering (DLS), and atomic force microscopy (AFM) are frequently employed to visualize and analyse nanoparticles within the nanofluid. All these microscopic techniques help to determine the stability and properties of

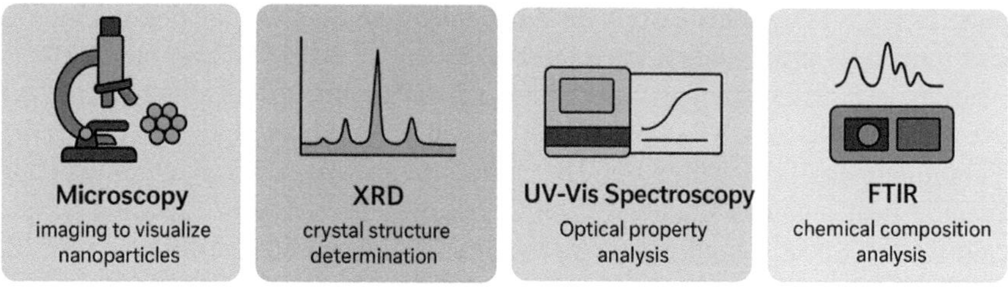


**Fig. 1** Stability and morphological analysis of nanofluids using microscopy, XRD, spectroscopy, and FTIR.

nanofluids, which are essential for their application in various fields and are briefly described as follows:

### *3.1 Scanning Electron Microscopy (SEM)*

SEM analyzes nanofluid morphology, structure, and composition using a focused electron beam (Manimaran *et al.*, 2013). It reveals particle size, shape, surface features, and dispersion, linking microstructure to stability and thermal properties (Habibzadeh *et al.*, 2010; Sengwa *et al.*, 2022). SEM shows metal oxide nanoparticles ($Al_2O_3$, $TiO_2$, ZnO) as 20–50 nm spheres, carbon-based nanofluids as CNT networks or graphene layers, and hybrid nanofluids as well-dispersed multi-phase systems (Rubaiee *et al.*, 2023; Sati *et al.*, 2018; Abu-Oqail *et al.*, 2023). Proper sample preparation (dilution, ultrasonication, drop-casting, drying) is essential for accurate imaging of size, morphology, and agglomeration, which affect dispersion, stability, and thermophysical performance (Chopkar *et al.*, 2006; Paul *et al.*, 2010; Kumar and Subudhi, 2019).

Thuan *et al.* (*Thuan et al., 2024*) employed SEM to investigate the surface morphology and dispersion of palladium on carbon (Pd/C) nanoparticles, as illustrated in Figure 2. SEM analysis was carried out using a field-emission SEM under native conditions, enabling high-resolution visualization of palladium nanoparticles deposited on carbon substrates. The images clearly reveal the surface topography of the carbon support and the successful formation of finely distributed Pd nanoparticles, allowing qualitative assessment of particle size, shape, and spatial distribution. As shown in Figure 2, SEM provides direct morphological evidence of nanoparticle deposition and uniformity, which is essential for understanding structure–property relationships and validating synthesis routes in nanoparticle-based materials. With this, we have also studied the characteristics parameter and related observations and summarized them in Table 1.

**Limitations**

Even though SEM is indispensable, a certain limitation must be considered in nanofluid analysis.

1. SEM analysis needed dried samples, which may not show the actual dispersion state of nanoparticles in fluid. The agglomeration observed in the same analysis may be partly due to solvent evaporation during sample preparation.

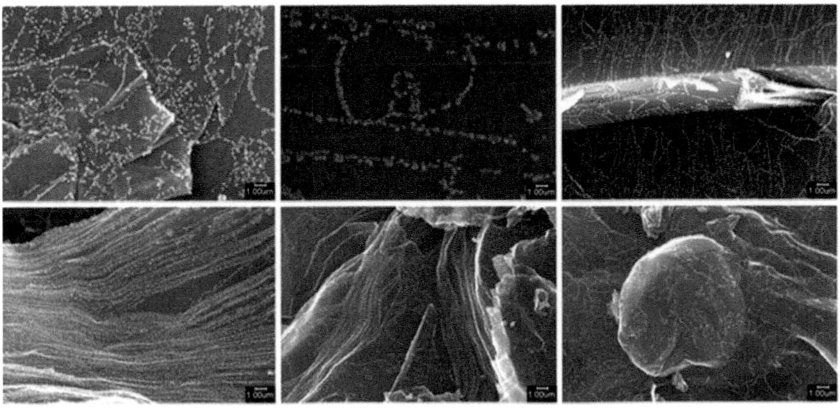

**Fig. 2** Representative SEM images of Pd/C nanoparticles acquired under field-emission conditions, showing the surface morphology of the carbon support and the dispersion of palladium nanoparticles at different locations on the sample surface. Adapted from (Thuan *et al.*, 2024), RSC Advances, licensed under CC BY-NC 3.0.

2. Non-conductive nanoparticles need to be coated with conductive coatings for accurate analysis.
3. The SEM analysis alone does not provide quantitative data on particle size distribution or zeta potential; it needs other techniques.

**Table 1** Critical evaluation of SEM for nanofluid characterization.

| *S. No.* | *Characterization Parameter (SEM)* | *Observation* |
|---|---|---|
| 1. | Information obtained | Particle size, morphology, surface texture |
| 2. | Technique sensitivity | High (nanoscale surface resolution) |
| 3. | Key limitations | Dry-state analysis, possibility of agglomeration during sample preparation |
| 4. | Application suitability | Structural and morphological validation of nanoparticles |
| 5. | Correlation with thermal conductivity | Weak (indirect structural influence) |
| 6. | Correlation with viscosity | Weak |
| 7. | Correlation with long-term stability | Weak–Moderate |
| 8. | Overall data correlation strength | Weak |

### Field Emission Scanning Electron Microscopy (FESEM)

FESEM offers higher resolution and better surface detail than SEM, using a field emission gun instead of a thermionic source. Equipped with BSED, EDS, and EBSD, it images nanoparticles, agglomerates, and nanostructures at low voltages with minimal charging, providing excellent depth of field (Mehrali *et al.*, 2014; Yadav *et al.*, 2022). Like SEM, FESEM determines particle shape, size distribution, clustering, surface modifications, coatings, and nanoparticle interactions in nanofluids (Lo *et al.*, 2005).

Saidi *et al.* (Mohd Saidi *et al.*, 2022) chemically modified multi-walled carbon nanotubes (MWCNTs) using three different acid oxidation methods and investigated their morphological, structural, and thermal characteristics for nanofluid applications. FESEM analysis (Figure 3) shows that MWCNTs treated using methods A and B exhibit partial debundling and disentanglement of nanotubes, with relatively preserved tubular structures but noticeable surface roughness and defect formation due to the introduction of oxygen-containing functional groups.

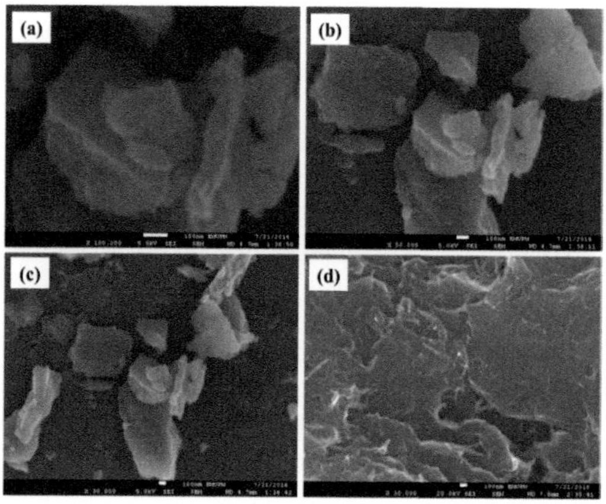


**Fig. 3** FESEM micrographs of (a) commercial MWCNTs, (b) MWCNT-MA, (c) MWCNT-MB, and (d) MWCNT-MC, captured at different magnifications along with their corresponding nanotube diameter distribution histograms. Adapted from (Mohd Saidi *et al.*, 2022), MDPI, licensed under CC BY 4.0.

In contrast, the method C sample displays severe wall etching, dense packing, and collapse of the nanotube structure caused by prolonged oxidation and ultrasonication, resulting in a highly disordered morphology where individual tube diameters are difficult to distinguish. These observations highlight that while moderate oxidation (Methods A and B) improves surface functionality with acceptable structural integrity, excessive treatment (Method C) leads to significant morphological degradation, which can adversely influence nanofluid stability and heat transfer performance.

**Limitations:** Despite several advantages, FESEM needs dried samples, which means the structure may not reflect the exact real dispersion state in liquid form. In case of non-conductive coatings, they generally require the application of thin conductive coatings (such as gold or platinum), which may sometimes obscure very fine details of the nanoparticle surfaces.

### 3.2 Transmission Electron Microscopy (TEM)

Transmission Electron Microscopy (TEM) provides ultra-high-resolution imaging of nanoparticles at atomic to nanometre scales, enabling analysis of morphology, particle size, crystallinity, dispersion, lattice planes, and phase identification through Selected Area Electron Diffraction (SAED) (Cham Sa-Ard *et al.*, 2021; Paul *et al.*, 2011). TEM reveals surface functionalization and characteristic structures such as agglomerated $TiO_2$, spherical $Al_2O_3$, layered graphene, and tubular CNTs, which explain differences in dispersion stability and thermal performance of metal oxide, carbon-based, and hybrid nanofluids (Flower *et al.*, 2016; Moosavi *et al.*, 2010; Qamar *et al.*, 2020; Kanti *et al.*, 2022; Balaji *et al.*, 2021). TEM is commonly combined with XRD, DLS, and zeta potential analysis for comprehensive characterization; however, results may be affected by sample preparation and drying artifacts, making careful preparation essential (Jama *et al.*, 2016). A critical evaluation of TEM for nanofluid characterization is summarized in Table 2.

**Table 2** Critical evaluation of TEM for nanofluid characterization.

| *S.No.* | *Characterization Parameter* | *Observation* |
|---|---|---|
| 1. | Information obtained | Size, shape of particle, image of internal structure, information on crystallinity |
| 2. | Sensitivity | Very high |
| 3. | Limitations | Complex sample preparation, small sampling volume, dry-state analysis |
| 4. | Application suitability | For the identification of particle size and crystallographic analysis |
| 5. | Correlation with thermal conductivity | Moderate (size and shape effects) |
| 6. | Correlation with viscosity | Weak |
| 7. | Correlation with long-term stability | Weak |
| 8. | Overall data correlation strength | Weak–Moderate |

Augustyniak *et al.* (Augustyniak *et al.*, 2022) dispersed 0.01 g of silica nanoparticles in water using ultrasonication and reported a mean hydrodynamic diameter of ~186 nm with excellent stability indicated by a zeta potential of −60 mV. Figure 4(a) shows the particle size distribution with minimal change after 24 hours, confirming low agglomeration, while TEM images verified spherical $SiO_2$ particles below 100 nm (Augustyniak *et al.*, 2022).

**Applications:** Transmission Electron Microscopy (TEM) is widely used to observe changes in nanoparticle dispersion and agglomeration over time. In multicomponent systems, it helps identify and distinguish the ratio, size, and distribution of different nanoparticles. TEM also confirms important morphological

(a)

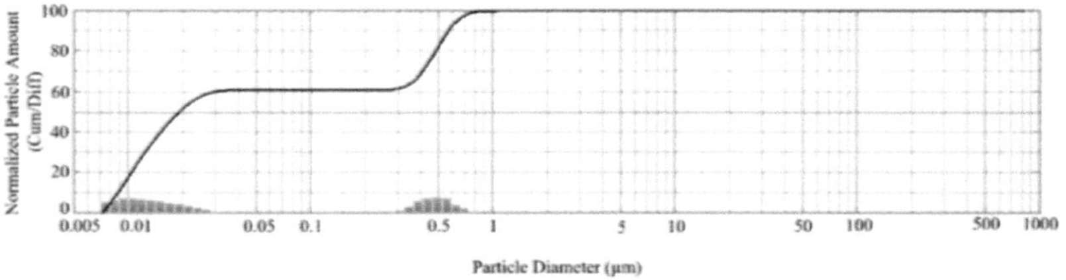


| | Cum% Q(%) | Diam x(μm) | | Cum% Q(%) | Diam x(μm) | | Cum% Q(%) | Diam x(μm) | | Cum% Q(%) | Diam x(μm) |
|---|---|---|---|---|---|---|---|---|---|---|---|
| 1 | 98.0 | 0.709 | 14 | 72.0 | 0.413 | 27 | 46.0 | 0.017 | 40 | 20.0 | 0.010 |
| 2 | 96.0 | 0.657 | 15 | 70.0 | 0.398 | 28 | 44.0 | 0.016 | 41 | 18.0 | 0.010 |
| 3 | 94.0 | 0.609 | 16 | 68.0 | 0.383 | 29 | 42.0 | 0.015 | 42 | 16.0 | 0.010 |
| 4 | 92.0 | 0.576 | 17 | 66.0 | 0.369 | 30 | 40.0 | 0.015 | 43 | 14.0 | 0.009 |
| 5 | 90.0 | 0.558 | 18 | 64.0 | 0.334 | 31 | 38.0 | 0.014 | 44 | 12.0 | 0.009 |
| 6 | 88.0 | 0.541 | 19 | 62.0 | 0.297 | 32 | 36.0 | 0.014 | 45 | 10.0 | 0.009 |
| 7 | 86.0 | 0.524 | 20 | 60.0 | 0.028 | 33 | 34.0 | 0.013 | 46 | 8.0 | 0.008 |
| 8 | 84.0 | 0.507 | 21 | 58.0 | 0.025 | 34 | 32.0 | 0.013 | 47 | 6.0 | 0.008 |
| 9 | 82.0 | 0.492 | 22 | 56.0 | 0.022 | 35 | 30.0 | 0.012 | 48 | 4.0 | 0.008 |
| 10 | 80.0 | 0.476 | 23 | 54.0 | 0.021 | 36 | 28.0 | 0.012 | 49 | 2.0 | 0.007 |
| 11 | 78.0 | 0.461 | 24 | 52.0 | 0.020 | 37 | 26.0 | 0.011 | | | |
| 12 | 76.0 | 0.444 | 25 | 50.0 | 0.019 | 38 | 24.0 | 0.011 | | | |
| 13 | 74.0 | 0.428 | 26 | 48.0 | 0.018 | 39 | 22.0 | 0.011 | | | |

(b)

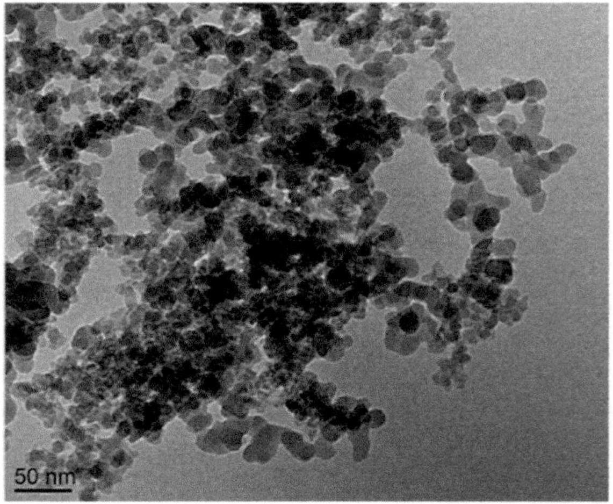


**Fig. 4** Microscopic analysis of (a) Silica particle size distribution for the sample water + $SiO_2$ immediately after the preparation of the liquid, (b) TEM image of raw $SiO_2$ powder. Adapted from Ref (Augustyniak *et al.*, 2022), Scientific Reports, licensed under CC BY 4.0.

and structural features such as core–shell structures, surface coatings, and encapsulated particles. Additionally, it provides valuable insight into the interface between nanoparticles and surrounding surfactant or polymer layers in functionalized nanofluids. (Yu *et al.*, 2011)

**Limitations:** Although TEM has many applications, it also has certain limitations.

1. TEM requires ultrathin samples for analysis.
2. In the case of nanofluids, sample preparation usually involves drop-casting onto grids followed by drying, which may not accurately represent the actual dispersion state.
3. Prolonged exposure to high-energy electron beams can damage organic coatings, polymers, or even cause melting or structural changes in nanoparticles.
4. TEM images capture only a very small area of the sample, which may not fully represent the overall characteristics of the entire nanofluid system (Rubaiee *et al.*, 2023).

### 3.3 Dynamic Light Scattering (DLS)

Dynamic Light Scattering (DLS) widely used technique for determining the hydrodynamic particle size, size distribution, and dispersion quality of nanoparticles in nanofluids (Shanbedi *et al.*, 2015). DLS works by measuring fluctuations in the intensity of scattered laser light caused by the Brownian motion of particles suspended in a fluid. These fluctuations are analysed to calculate the diffusion coefficient, which is then related to particle size using the Stokes–Einstein equation (Hunter, 1981) (Gupta *et al.*, 2017). Unlike SEM or TEM, DLS provides in-situ measurements in liquid media, making it highly suitable for nanofluid characterization. A typical DLS setup consists of a laser source, sample cell, detector, and correlator. The laser beam passes through the nanofluid sample, and scattered light at a fixed angle is collected by the detector (Khullar *et al.*, 2018). The correlator converts intensity fluctuations into an autocorrelation function, from which the average hydrodynamic diameter and polydispersity index (PDI) are obtained.

Sahamifar *et al.* (Sahamifar *et al.*, 2024) employed DLS to characterize the nanoparticle size distribution in an $Al_2O_3$–water nanofluid and to support the analysis of nanofluid stability. The DLS technique measures the hydrodynamic diameter of nanoparticles by analyzing fluctuations in scattered laser light arising from Brownian motion. As shown in Figure 4, the particle size distribution is presented both in terms of number percentage and volume percentage, revealing that although most particles are nanosized by number, a small fraction of larger agglomerates contributes significantly by volume. These larger particles play a dominant role in the gravitational settling and instability behavior of the nanofluid. Thus, DLS analysis provides essential quantitative insight into particle dispersion, agglomeration, and stability characteristics, making it a vital diagnostic tool in nanofluid research. The Critical evaluation of DLS for nanofluid characterization is summarised in Table 3.

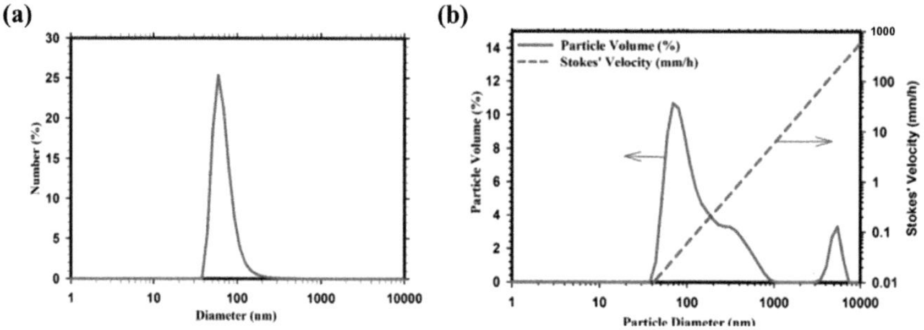


**Fig. 5** Size distribution measurement of nanoparticles in $Al_2O_3$-water nanofluid using DLS, categorized by: (a) number of particles (%) and (b) particle volume (%). Adapted from Ref (Sahamifar *et al.*, 2024), Elsevier, licensed under CC BY.

## Applications

Dynamic Light Scattering (DLS) is widely used to study the stability of metal oxide, carbon-based, and hybrid nanofluids. It helps evaluate the effects of surfactants and pH on dispersion and is useful for monitoring particle growth or agglomeration during storage. DLS is also commonly used to assess the effectiveness of dispersion methods such as ultrasonication and surface functionalization.

## Limitations

Despite its advantages, DLS has certain limitations: (1) it assumes spherical particles, which may lead to errors for anisotropic shapes like CNTs or graphene; (2) results are intensity-weighted, making them sensitive to even small amounts of agglomerates; (3) highly concentrated or polydisperse samples can reduce accuracy; and (4) DLS provides no direct information on particle shape or internal structure (Safiei *et al.*, 2020).

**Table 3** Critical evaluation of DLS for nanofluid characterization.

| *S. No.* | *Characterization Parameter* | *Observation* |
|---|---|---|
| 1. | Information obtained | Aggregation behavior, Hydrodynamic particle size |
| 2. | Sensitivity | High |
| 3. | Limitations | Assumes spherical particles; sensitive to polydispersity |
| 4. | Application suitability | Dispersion quality and aggregation assessment |
| 5. | Correlation with thermal conductivity | Moderate |
| 6. | Correlation with viscosity | Moderate |
| 7. | Correlation with long-term stability | Strong |
| 8. | Overall data correlation strength | Moderate–Strong |

**Summary of Microscopic Techniques**

SEM/FESEM show surface morphology, TEM reveals atomic-scale structure, and DLS provides in-situ particle size and stability. No single method is sufficient; a combined SEM–TEM–DLS approach is recommended, with proper sample preparation, dilution, ultrasonication, and support from zeta potential data. At least one imaging and one dynamic technique should be used, with clear reporting of conditions.

## 3.4 *Spectroscopic and Analytical Methods*

In addition to microscopic techniques, spectroscopic and analytical methods play an important role in evaluating the physical and chemical properties of nanofluids, including nanoparticle dispersion, surface functionalization, stability, and concentration. These properties directly affect the thermal, electrical, and rheological performance of nanofluids. Spectroscopic techniques provide detailed information about particle size, surface chemistry, morphology, and stability. They help detect agglomeration, degradation, and interactions between nanoparticles and surfactants in suspension. Accurate and standardized spectroscopic characterization is therefore essential to ensure consistent formulation, establish clear relationships between concentration and thermophysical properties, and minimize inconsistencies in nanofluid research. This ultimately supports reliable performance evaluation and practical thermal management applications.

**FTIR**

Fourier Transform Infrared Spectroscopy (FTIR) is used to identify surface functional groups and chemical bonding in nanofluids by analyzing infrared absorption of molecular vibrations. It provides a molecular fingerprint that confirms nanoparticle functionalization, surfactant coating, and interactions with the base fluid through characteristic peaks such as O–H, C–H, C=O, and Si–O–Si. Changes in peak position or intensity indicate bonding interactions that affect dispersion stability, thermal conductivity, and degradation. Time-dependent FTIR can also monitor surfactant breakdown and chemical changes during heating, helping to explain performance variations and improve consistency in nanofluid studies. (Berthomieu and Hienerwadel, 2009; Kumar *et al.*, 2009). A critical evaluation of FTIR for nanofluid characterization is summarized in Table 4.

**Working Principle and Sampling Techniques**

FTIR spectroscopy operates on the absorption of infrared radiation by molecular vibrations, producing a characteristic spectrum that serves as a chemical fingerprint of the material. This enables identification of surface functional groups such as hydroxyl, carboxyl, and carbonyl moieties introduced during nanoparticle functionalization (Tkachenko and Niedzielski, 2022). FTIR systems comprise an IR source, interferometer, sample compartment, and detector, with absorption peaks corresponding to specific bond vibrations. The relationship between peak position, bond strength, and atomic mass allows FTIR to effectively identify chemical bonding and surface modifications in nanofluids (Bacsik *et al.*, 2004).

Das *et al.* (Das *et al.*, 2020) employed Fourier Transform Infrared (FTIR) spectroscopy to examine the chemical stability and dispersion of $Ti_3C_2$ (MXene) in ionic liquid (IL)–water-based ionanofluids, as illustrated in Figure 6. The FTIR spectra show a broad peak at 620 cm$^{-1}$ corresponding to the Ti–O bond, confirming effective dispersion of MXene in the IL aqueous medium. Additional characteristic peaks at 3337 cm$^{-1}$ (O–H), 1643 cm$^{-1}$ (C=N), 1579 cm$^{-1}$ (C=C), 1176 cm$^{-1}$ (C–O), 1037 cm$^{-1}$ (C–N), and 816 cm$^{-1}$ (C–H) are attributed to the functional groups of imidazolium-based ionic liquids, in agreement with reported literature. These observations indicate that MXene dispersion occurs through stable physical interactions without inducing new chemical reactions, supporting the suitability of the formulated ionanofluid for heat transfer applications.

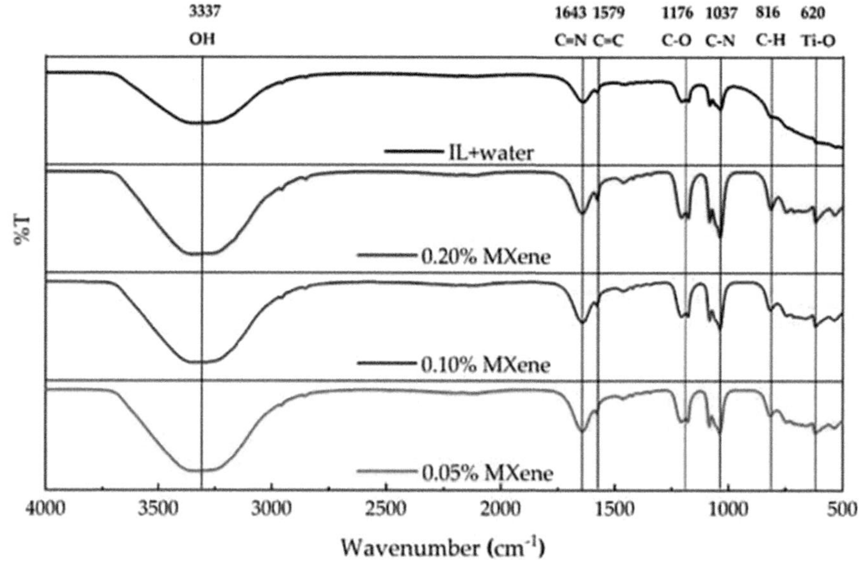


**Fig. 6** FTIR spectra of nanofluids at different concentrations. Adapted from (Das *et al.*, 2020), MDPI, licensed under CC BY 4.0.

Several sampling methods are employed for FTIR-based nanofluid analysis. Attenuated Total Reflectance FTIR (ATR-FTIR) is widely used for both liquid and solid samples due to minimal sample preparation and the ability to analyze nanofluids directly. Transmission FTIR spectroscopy (FTIR-TS) involves drying the nanofluid and preparing potassium bromide (KBr) pellets; however, this method is more susceptible to moisture interference and preparation-induced artifacts. Advanced techniques such as nano-FTIR, which integrates FTIR with Atomic Force Microscopy (AFM), provide nanoscale spatial resolution and enable chemical mapping of individual nanoparticles and nanocomposites.

**Fourier Transform Infrared Nano-Spectroscopy (Nano-FTIR)**

Conventional FTIR is limited by the Abbe diffraction limit, restricting spatial resolution to the micron scale and preventing direct nanoscale analysis. Nano-FTIR overcomes this limitation by combining FTIR with scattering-type near-field optical microscopy (s-SNOM), using a metal-coated AFM tip to confine IR radiation to ~20 nm. The back-scattered signal provides localized chemical information with nanometre resolution, enabling nanoscale chemical identification and mapping, supported by tip–sample interaction models such as the point-dipole and finite-dipole models (Kotov *et al.*, 2025).

**Applications in Nanofluid Analysis**

FTIR is widely used to verify nanoparticle surface chemistry by identifying functional groups associated with capping agents or surfactants and to analyze base fluid–nanoparticle interactions affecting dispersion stability. In hybrid nanofluids, FTIR helps distinguish different nanoparticles and evaluate their interactions. Time-dependent spectral changes enable monitoring of stability, degradation, and agglomeration, while thermal analysis reveals chemical property evolution under heating. Advanced FTIR methods also allow differentiation between surface and subsurface layers in multilayer systems (Mohammadi *et al.*, 2021; Berthomieu and Hienerwadel, 2009). With this, we have also studied the characteristics parameter and related observations and summarized them in Table 4.

**Advantages and Limitations**

FTIR and nano-FTIR offer non-destructive analysis, minimal sample preparation, and detailed chemical information, but face challenges such as overlapping peaks, water interference in aqueous nanofluids, and low sensitivity at low concentrations. Nano-FTIR further involves technical limitations, including long acquisition times and difficulty in liquid-phase measurements. Despite these constraints, FTIR, particularly when combined with nano-FTIR, remains a powerful tool for understanding nanofluid chemistry, stability, and performance (Berthomieu and Hienerwadel, 2009).

**Table 4** Critical evaluation of FTIR for nanofluid characterization.

| *S.No.* | *Characterization Parameter* | *Observation* |
|---|---|---|
| 1. | Obtained Information | Surface functional groups, bonding |
| 2. | Sensitivity | Moderate |
| 3. | Limitations | Limited quantitative information on dispersion |
| 4. | Application suitability | Surface chemistry and functionalization analysis |
| 5. | Correlation with thermal conductivity | Indirect |
| 6. | Correlation with viscosity | Weak |
| 7. | Correlation with long-term stability | Moderate |
| 8. | Overall data correlation strength | Weak–Moderate |

### 3.5 UV-Vis Spectroscopy

UV–Visible (UV–Vis) spectroscopy is used to study the optical absorption, dispersion quality, concentration, and stability of nanofluids by measuring light absorption in the 190–1100 nm range. A typical UV–Vis spectrophotometer includes a light source, monochromator, cuvette holder, detector, and digital display system. Absorbance is directly proportional to nanoparticle concentration, allowing quantitative analysis. UV–Vis is especially useful for evaluating dispersion stability, as changes in absorbance over time indicate particle aggregation or sedimentation, which can significantly affect thermophysical performance (Sadeghi *et al.*, 2014; Mukesh Kumar *et al.*, 2020). A critical evaluation of UV–Vis for nanofluid characterization is summarized in Table 5.

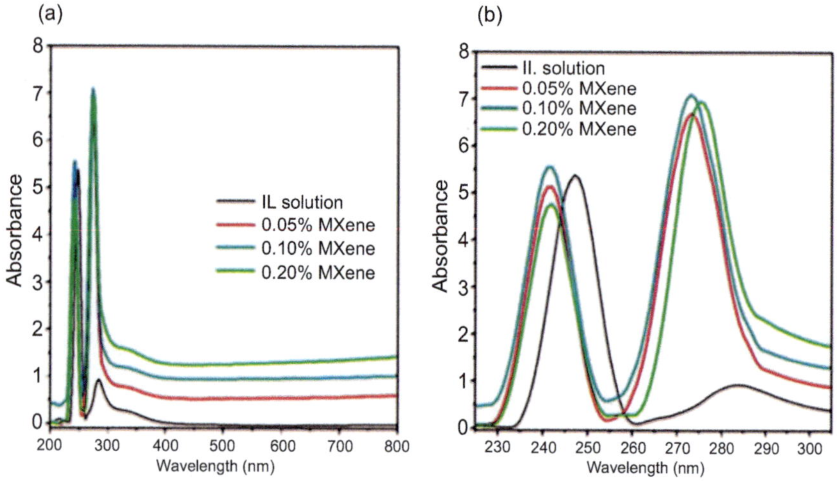


**Fig. 7** (a) UV–vis spectra of pure IL solution and nanofluids, (b) magnified view from 225 nm to 305 nm wavelength. Adapted from (Das *et al.*, 2020), MDPI, licensed under CC BY 4.0.

Das *et al.* (Das *et al.*, 2020) analysed the optical characteristics of the ionic liquid (IL) solution and $Ti_3$ $C_2$ (MXene)–based ionanofluids using UV–Visible spectroscopy, as depicted in Figure 7(a,b). The UV–vis spectra reveal a prominent absorbance peak at 291 nm, which increases significantly with the addition of $Ti_3$ $C_2$ nanosheets, indicating enhanced optical absorption of the ionanofluids. This trend, consistent with earlier studies on carbon-based two-dimensional nanomaterials, shows a linear relationship between nanoparticle concentration and absorbance, satisfying Beer's law. The improved absorption behavior is attributed to the inherently high light absorption capability of MXene, highlighting the suitability of the formulated ionanofluids for solar energy storage and related thermal applications.

**Principle of UV-Vis Spectroscopy**

UV-Vis spectroscopy works on the principle of absorption of electromagnetic radiation by matter. When monochromatic light passes through a nanofluid sample, some of the radiation is absorbed, some is transmitted, and some is scattered. The absorption comes from electronic transitions in atoms or molecules: $\pi \rightarrow \pi^*$ transitions occur in conjugated systems and are often seen in organic ligands or stabilizers. $n \rightarrow \pi^*$ transitions involve lone-pair electrons in heteroatoms (O, N, S) and are especially relevant in nanoparticle surface functionalization.

For metallic nanoparticles like Au, Ag, or Cu, absorption is mainly due to Localized Surface Plasmon Resonance (LSPR), which is the collective oscillation of conduction electrons excited by light. (Chen *et al.*, 2013) The fundamental quantitative relationship governing absorption is described by the Beer-Lambert law:

$$A = \varepsilon c l$$

Where A is absorbance (unitless), ε is the molar extinction coefficient ($Lmol^{-1}cm^{-1}$), c is the molar concentration ($molL^{-1}$), and l is the optical path length (cm). Absorbance is also directly related to transmittance T by:

$$A = -\log 10(T) = \log 10(I0/I)$$

Where I0 and I represent the intensities of incident and transmitted light, respectively. This connection between absorbance and transmittance is key for monitoring concentration and stability.(Chen *et al.*, 2013)

**Working and Instrumentation**

A UV–Vis spectrophotometer consists of UV (deuterium) and visible (tungsten–halogen) light sources, a monochromator for wavelength selection, a quartz or glass cuvette for the sample, and a detector such as a photodiode array or photomultiplier tube. In a double-beam configuration, reference and sample beams reduce baseline drift, while software converts transmitted intensity into absorbance spectra. For metallic nanofluids, surface plasmon resonance (SPR) peaks explained by Mie theory depend on nanoparticle size, shape, and surrounding medium; red shifts indicate particle growth or aggregation, and peak broadening reflects polydispersity (Chen *et al.*, 2013; Singh *et al.*, 2022).

**Applications to Nanofluids**

UV–Vis spectroscopy is a versatile tool for nanofluid characterization, widely used to monitor colloidal stability through time-dependent absorbance changes, where decreased absorbance indicates aggregation or sedimentation. It is also employed to evaluate surfactant effectiveness, determine nanoparticle concentration, and estimate particle size and shape via surface plasmon resonance (SPR) behavior; red shifts indicate aggregation, while sharp peaks suggest good dispersion (Mukesh Kumar *et al.*, 2020; Chen *et al.*, 2013, Singh *et al.*, 2022). For oxide and semiconductor nanofluids, UV–Vis-derived Tauc plots enable optical band-gap estimation. Overall, UV–Vis is a fast, non-destructive, cost-effective technique requiring minimal sample preparation, making it well-suited for routine nanofluid analysis.

**Limitations and Future Scope**

UV–Vis spectroscopy of nanofluids faces limitations such as light scattering from aggregates, overlapping absorption bands, and deviations from the Beer–Lambert law at high concentrations. It cannot replace imaging techniques for direct morphological analysis or provide mechanistic insight into interparticle forces or aggregation reversibility (Sadeghi *et al.*, 2014; Chen *et al.*, 2013; Rabbi *et al.*, 2020). Standardization remains a major challenge due to variations in sample preparation and measurement conditions. Future studies are expected to integrate UV–Vis with FTIR, DLS, and TEM, along with advanced SPR and bandgap modeling, to improve accuracy and reliability in nanofluid characterization.

**Table 5** Critical evaluation of UV-Vis for nanofluid characterization

| *Sr. No.* | *Characterization Parameter* | *Observation* |
|---|---|---|
| 1. | Obtained information | Dispersion uniformity, sedimentation behavior |
| 2. | Sensitivity | Moderate |
| 3. | Limitations | Limited particle size resolution |
| 4. | Application suitability | Stability monitoring and sedimentation analysis |
| 5. | Correlation with thermal conductivity | Indirect |
| 6. | Correlation with viscosity | Weak |
| 7. | Correlation with long-term stability | Strong |
| 8. | Overall data correlation strength | Moderate–Strong |

## 3.6 X-ray Diffraction (XRD) for Nanofluid Characterization

X-ray Diffraction (XRD) is used to determine the crystalline phase, crystal structure, and average crystallite size of nanoparticles in nanofluids, providing insight into crystallinity, lattice parameters, and phase purity. These structural features strongly influence nanofluid stability and thermophysical performance. XRD is essential for identifying nanoparticle phases, confirming crystallinity and purity, and monitoring structural changes during operation, making it a key technique for comprehensive nanofluid characterization (Kiani, 2023; Sonage and Mohanan, 2014).

**Working Principle of XRD**

XRD works on the principle of elastic scattering of X-rays by the regular arrangement of atoms in a crystalline material. When a beam of monochromatic X-rays, like Cu K$\alpha$ radiation, hits a crystalline sample, the incident waves scatter off the electron clouds around the atoms. Constructive interference happens when the path difference between waves scattered from successive lattice planes follows Bragg's law:

$$n\lambda = 2d \sin \theta$$

In this equation, n is the diffraction order, $\lambda$ is the wavelength of the incoming X-rays, d is the distance between crystallographic planes, and $\theta$ is the angle of incidence. This law is central to XRD, connecting the observed diffraction peaks to specific distances between planes in the crystal lattice. The diffraction condition can also be described in the context of the reciprocal lattice, where diffraction peaks correspond to points in the reciprocal lattice that intersect the Ewald sphere. The intensity of observed peaks is determined by the structure factor, which accounts for the types and positions of atoms within the unit cell. This creates XRD patterns that act as unique "fingerprints" for crystalline materials. Each reflection set, indexed with Miller indices (hkl), corresponds to specific lattice planes. Their relative positions and intensities carry both structural and compositional information (Kiani, 2023; Sonage and Mohanan, 2014, Mohan and Renjanadevi, 2016). The Critical evaluation of the XRD study for nanofluid characterization are summarised in Table 6.

For example, in silver nanoparticles, diffraction peaks at $2\theta = 38.1°$, $44.3°$, $64.5°$, and $77.3°$ correspond to the (111), (200), (220), and (311) planes, respectively, matching the standard JCPDS Card No. 07-0783. This indexing confirms the face-centered cubic structure of silver, while the absence of extra peaks verifies sample purity. (Mehta *et al.*, 2017)

**Working and Instrumentation**

A modern XRD system consists of an X-ray source (commonly Cu K$\alpha$ or Mo K$\alpha$), a goniometer, and a detector. X-rays are directed onto a dried nanofluid sample, and the goniometer varies the diffraction angle ($2\theta$) while the detector records diffracted intensity. Powder XRD is widely used for nanofluids, providing diffraction peaks that identify crystal structure and phases by comparison with standard databases (e.g., JCPDS). Crystallite size is estimated from peak broadening using the Scherrer equation $D = K\lambda/\beta \cos \theta$, while peak intensity analysis reveals crystallinity, preferred orientation, and phase composition (Kiani, 2023). Here, K is the shape factor (around 0.9), $\lambda$ is the X-ray wavelength, $\beta$ is the full width at half maximum (FWHM) of the diffraction peak in radians, and $\theta$ is the Bragg angle.

**Application to Nanofluids**

XRD is widely used to characterize nanofluids by identifying crystalline phases (e.g., anatase/rutile $TiO_2$, wurtzite ZnO, $Al_2O_3$), confirming nanopowder purity, and analyzing hybrid nanofluids such as Ag–graphene oxide or Cu–$TiO_2$ composites (Mohan and Renjanadevi, 2016; Kshirsagar and Venkatesh, 2021). It complements SEM/TEM by providing average crystallite size, lattice strain, and structural arrangement over large volumes. XRD also monitors structural changes during thermal cycling, aggregation, or phase transformations. Advantages include precision, non-destructive analysis, simultaneous size and phase determination, and reproducibility with reference databases (JCPDS). When combined with UV–Vis, FTIR, DLS, and TEM, XRD offers comprehensive insight into nanofluid structure, stability, and performance (Kiani, 2023).

**Limitations and Future Scope**

XRD analysis of nanofluids faces challenges due to sample drying, which can cause aggregation, and limited effectiveness for amorphous, very small (<2 nm), or highly diluted nanoparticles, resulting in weak or missing diffraction peaks (Jama *et al.*, 2016). Future approaches include using synchrotron sources for real-time analysis and combining XRD with TEM, SAXS, DLS, and FTIR, while Rietveld refinement and PDF analysis enhance interpretation for complex or partially amorphous systems.

**Table 6** Critical evaluation of XRD for nanofluid characterization.

| *S.No.* | *Characterization Parameter* | *Observation* |
|---|---|---|
| 1. | Obtained Information | Crystallinity, crystal structure, phase composition, |
| 2. | Sensitivity | High |
| 3. | Limitations | Dry-state analysis; no dispersion information |
| 4. | Application suitability | Phase identification and structural confirmation |
| 5. | Correlation with thermal conductivity | Moderate (phonon transport relevance) |
| 6. | Correlation with viscosity | Weak |
| 7. | Correlation with long-term stability | Weak |
| 8. | Overall data correlation strength | Weak–Moderate |

## *3.7 Particle Size and Zeta Potential Analysis*

Nanofluid properties strongly depend on nanoparticle stability, size, and surface charge. Poor dispersion can lead to aggregation and sedimentation, affecting thermal, optical, and rheological performance. Particle size analysis, commonly done via Dynamic Light Scattering (DLS), measures hydrodynamic diameter and indicates stability—monodisperse distributions suggest good stability, while polydispersity signals aggregation (Jing and Song, 2017). Particle size also influences thermal conductivity, viscosity, and dielectric properties, as shown in studies on $Al_2O_3$, $SiO_2$, $Fe_3O_4$, and hybrid nanofluids, highlighting its critical role in performance (Borode *et al.*, 2025).

**Zeta Potential Analysis**

Particle size affects nanofluid thermophysical properties, while zeta potential governs dispersion stability by reflecting the electrical potential at the slipping plane of nanoparticles. High zeta potential indicates strong electrostatic repulsion and better stability, with 25 mV often considered the threshold for moderate stability, 40–60 mV for stable colloids, and >60 mV for highly stable suspensions (Cacua *et al.*, 2019). Studies have shown its significance across nanofluids: $Al_2O_3$/water (Choudhary *et al.*, 2017), $SiO_2$/water (Augustyniak *et al.*, 2022), hybrid nanofluids (Razzaq *et al.*, 2025), and Cu/water systems (Li *et al.*). Stability is influenced by nanoparticle concentration, pH, and surfactants (e.g., CTAB 28.1 mV, SDBS – 43.8 mV, TX – 8.3 mV), where proper pH adjustment enhances electrostatic repulsion and prevents agglomeration. DLVO theory explains the balance of attractive and repulsive forces in stabilizing nanofluids (Dongsheng Zhu *et al.*; Haitao Zhu *et al.*). A critical evaluation of zeta potential for nanofluid characterization is summarized in Table 7.

AlMuhaysh *et al.* (AlMuhaysh *et al.*, 2025) investigated the colloidal stability of $Al_2O_3$–$H_2O$ nanofluids using zeta potential measurements, as illustrated in Figure 8. Zeta potential analysis was employed to quantify the electrostatic repulsive forces between suspended nanoparticles and to assess the influence of pH and nanoparticle concentration on dispersion stability. As shown in Figure 8, the magnitude of the zeta potential varies significantly with pH, reaching higher absolute values away from the isoelectric point, which indicates enhanced electrostatic repulsion and improved nanofluid stability. The results demonstrate that an optimal pH, around 4, leads to zeta potential values exceeding ±30 mV, signifying a highly stable, surfactant-free suspension. These findings highlight the critical role of surface charge regulation in controlling nanoparticle agglomeration and confirm zeta potential as a reliable and quantitative tool for evaluating both short- and long-term stability of nanofluids in heat transfer applications.

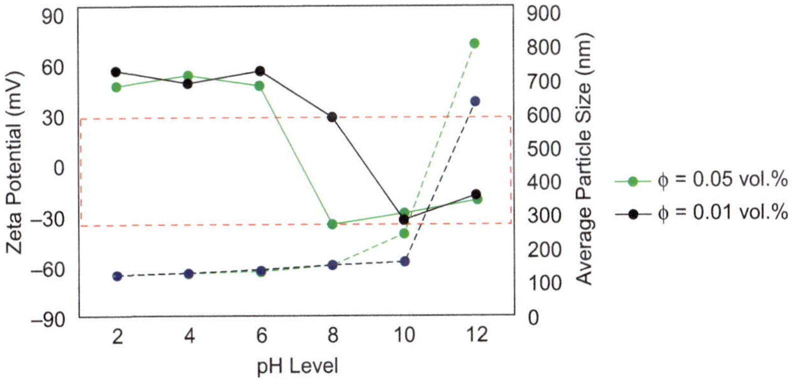


**Fig. 8** Relationship between zeta potential (solid line) and average particle size (dashed line) in $Al_2O_3$ NFs at different pH for specific concentrations. Adapted from (AlMuhaysh *et al.*, 2025) Springer Nature, licensed under CC BY 4.0.

## Applications

Zeta potential analysis is crucial for nanofluid applications, as it evaluates stability and optimizes thermal conductivity for heat exchangers and electronic cooling.

It also influences nanoparticle–cell interactions in drug delivery and enhances lubrication by reducing wear in mechanical systems (Liu *et al.*, 2018).

**Limitations**

Measurements can be affected by salts or ions that shield particle charges, pH sensitivity, and factors like sample concentration, temperature, and impurities, which may reduce accuracy.

**Table 7** Critical evaluation of for nanofluid characterization parameter and observation.

| *S.No.* | *Characterization Parameter* | *Observation* |
|---|---|---|
| 1. | Obtained information | Particle size distribution, surface charge |
| 2. | Sensitivity | High |
| 3. | Limitations | Strong dependence on pH and ionic strength |
| 4. | Application suitability | Stability prediction and dispersion optimization |
| 5. | Correlation with thermal conductivity | Moderate |
| 6. | Correlation with viscosity | Moderate |
| 7. | Correlation with long-term stability | Strong |
| 8. | Overall data correlation strength | Strong |

## 3.8 Raman Spectroscopy

Raman spectroscopy provides a non-destructive probe of the molecular vibrations and therefore an insight into the unique structure, stability, and thermophysical properties of nanofluids. The experimental verification was provided in 1928 by C.V. Raman and K.S. Krishnan, using sunlight or mercury lamps, which required lengthy exposures due to the weak signal, typically 1 in $10^7$ scattered photons. The replacement of these sources with lasers in the 1960s provided monochromatic, high-intensity excitation, turning it into a common analytical method that gained widespread use in vibrational spectroscopy. With improvements in detectors and optics, Raman spectroscopy became a popular technique for the identification of molecular structures owing to its 'chemical fingerprint (Rodriguez-Laguna *et al.*, 2019, Jorio and Saito, 2021).

**Principle**

Laser photons distort the electron cloud and induce a virtual transient state without real absorption. Rayleigh scattering returns the molecule to $v = 0$; Raman shifts involve $v = 0 \rightarrow v = 1$ (Stokes) or the reverse process for anti-Stokes. The anti-Stokes to Stokes intensity ratio $I_{aS}/I_S \propto \exp\left(-\frac{\Delta E}{k}T\right)$, where $I_{aS}$ is the intensity of the anti-Stokes Raman signal, $I_S$ is the intensity of the Stokes Raman signal,

$\Delta E$ is energy difference between the vibrational ground and first excited state, $k$ is Boltzmann's constant, and $T$ is the absolute temperature in Kelvin. This enables Raman thermometry, providing a route to map local temperatures in nanofluids and microchannels (Jorio and Saito, 2021).

Key factors includes laser power which elevates the signal linearly with $I_R$; shorter wavelength excitation results in stronger signals but increase the chances of blue excitation (fluorescence risk); concentration is reflected by the number of particles N. Wavelengths between 488 – 1064 nm balance signal intensity and sample safety. Key factors influencing the measurement include laser power, which increases the signal in an approximately linear fashion but may cause localized heating; excitation wavelength, where shorter wavelengths provide stronger signals but at greater risk of fluorescence; and particle concentration, represented by the number of scatterers (N). A critical evaluation of Raman Spectroscopy for nanofluid characterization is summarized in Table 8.

**Instrumentation**

A typical Raman setup includes a laser (e.g., $Ar^+$ 514.5 nm, Nd:YAG 1064 nm for FT-Raman), illumination optics, collection through a microscope objective, a notch or edge filter that suppresses Rayleigh scattering (~$10^3$-$10^6$ times suppression), a grating monochromator, and a CCD detector. Calibration is carried out using crystalline silicon (520 $cm^{-1}$) or a standard organic compound (e.g., indene).

**Table 8** Component, function, and examples of Raman spectra.

| *Component* | *Function* | *Examples* |
|---|---|---|
| Laser | Excitation | $Ar^+$ (488 nm), Nd:YAG (1064 nm) |
| Filter | Rayleigh rejection | Notch, edge |
| Disperser | Spectral resolution | Grating (1200-2400 grooves/mm) |
| Detector | Signal capture | CCD (thermally cooled), InGaAs (NIR) |

For nanofluids, additional factors to be taken into account include the employment of flow cells or microfluidic chips to minimize sedimentation as well as the installation of temperature-controlled stages that allow for differentiation between actual thermally induced spectral changes and artefacts caused by laser heating (Wu *et al.*, 2013, Jorio and Saito, 2021).

(AlMuhaysh *et al.*, 2025) investigated the molecular interactions in graphene-based nanofluids using Raman spectroscopy, revealing that graphene significantly modifies the vibrational modes of N, N-dimethylacetamide (DMAc) and N, N-dimethylformamide (DMF), indicating strong local interactions and π–π stacking between solvent molecules and graphene, which correlate with enhanced thermal conductivity. In contrast, N-methyl-2-pyrrolidinone (NMP)-based nanofluids showed minimal shifts in Raman bands (Figure 9, suggesting that the rigid cyclic structure and intrinsic molecular ordering of NMP limit the influence of graphene on solvent vibrations, explaining the negligible thermal conductivity enhancement.

Thus, Raman spectroscopy provides crucial insight into solvent–nanoparticle interactions and the mechanisms underlying nanofluid thermal behavior.

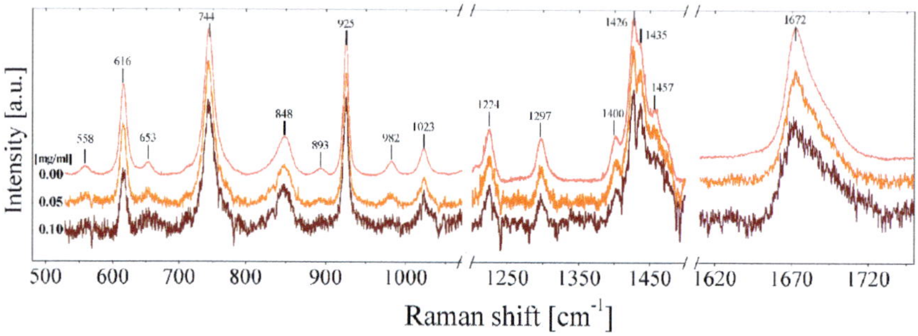


**Fig. 9** Raman spectra of NMP-based nanofluids for three different graphene concentrations. Adapted from Ref (AlMuhaysh *et al.*, 2025), MDPI, licensed under CC BY.

**Applications to Nanofluids**

In Raman analysis, dispersion stability can be gauged by repeated measurements of peak intensities and line shapes; aggregation can be determined by peak broadening and intensity loss, and interfacial interactions from systematic peak shifts or intensity changes.

Blue-shifts often signal 'hardening' effects associated with enhanced conductivity; for example, a ~4 $cm^{-1}$ shift has been reported in the N, N-Dimethylformamide (DMF) vibrational mode due to $\pi$-$\pi$ stacking in graphene-DMF nanofluids, and this correlated with increased thermal conductivity relative to the base fluid. Raman thermometry maps local temperature via $T = \dfrac{\Delta E}{k \ln\left(\dfrac{I_S}{I_{aS}} \cdot g\right)}$

where $g$ is the degeneracy, enabling visualization of temperature gradients in nanofluid-filled microchannels and devices (Wu *et al.*, 2013; Jorio and Saito, 2021). A critical evaluation and key features for nanofluid characterization are summarized in Table 9.

**Table 9** Types of nanofluids, key Raman feature, and insights.

| *Nanofluid Type* | *Key Raman Features* | *Insights* |
|---|---|---|
| $TiO_2$ (rutile/anatase) | Distinct polymorph bands | Phase identification |
| SWCNTs | RBM (100-300 $cm^{-1}$), $\omega_{RBM} \propto 1/d$ | Diameter, chirality |
| Graphene-DMF | DMF mode blue-shift (~4 $cm^{-1}$) | Orientation, stacking |

**Case Studies and Multi-Technique Correlation**

$TiO_2$ nanofluids can be classified as rutile (~610 $cm^{-1}$) or anatase (~144 $cm^{-1}$), and this classification is related to their different thermal conductivity responses when dispersed in water or glycols. The Radial Breathing Mode (RBM) in Raman spectroscopy can be used to estimate the size of the nanotube, through d(nm)=227/$\omega_{RBM}$. Graphene in N-methyl-2-pyrrolidone (NMP) does not show any peak shift or enhancement; graphene flakes mixed with DMF enhance thermal conductivity.

A representative multi-technique case study involves graphene-based nanofluids in polar organic solvents: Raman peak shifts and intensity changes of solvent bands reveal specific π–π or hydrogen-bond interactions, while THW measurements show that systems with strong Raman-detected interactions exhibit significantly higher thermal conductivity enhancements than systems with negligible spectral changes. By combining Raman (structure/interaction), DLS or zeta potential (stability), and THW (thermal conductivity), one can build a direct link from nanoscale dispersion and interfacial structure to macroscopic heat-transfer performance in simple test loops or heat exchangers (Zagabathuni *et al.*, 2016; Rodriguez-Laguna *et al.*, 2019).

## 3.9 Thermal Conductivity Characterization (THW Method)

The transient hot wire (THW) method is widely regarded as a reference technique for measuring the thermal conductivity of fluids and nanofluids. In THW, a thin metal wire immersed in the fluid acts simultaneously as a heater and a thermometer; after a step in electrical power, the temperature rise of the wire is recorded as a function of time, and the slope of temperature versus the logarithm of time is related to the fluid's thermal conductivity. THW offers absolute measurements with relatively low sensitivity to convection when an appropriate short time window is used, making it a standard in benchmark studies and inter-laboratory comparisons of nanofluid thermal conductivity (Zagabathuni *et al.*, 2016). A critical evaluation of the THW method for nanofluid characterization is summarized in Table 10.

**Advantages and Limitations**

Advantages: non-destructive, in-situ analysis, compatible with aqueous environments, and polarizability-sensitive measurements. Limitations: weak signal necessitating lengthy integration times, fluorescence background alleviated by near-infrared excitation, and heating due to illumination that can be compensated for using low-power heating. Enhancements: surface-enhanced Raman spectroscopy (SERS ) realizes a signal enhancement factor of roughly $10^6$–$10^8$ times by the use of Ag or Au substrates.

**Table 10** Summary of overall advantages and limitations with respective mitigation.

| *Aspect* | *Advantage* | *Limitation* | *Mitigation* |
|---|---|---|---|
| Signal | Chemical specificity | Weak ($10^{-7}$) | Lasers, SERS |
| Sample | Non-destructive, aqueous | Fluorescence | NIR, FT-Raman |
| Spatial | Micron resolution | Heating | Low power, objectives |

Portable Raman instruments enable field nanofluid monitoring. Microfluidic integration promises a full-volume analysis with reduced risk of clogging. The

**Table 11** Comparative Study of Characterization Techniques.

| *S. No.* | *Techniques* | *Role in Nanofluid Studies* | *Critical Characteristics Analyzed* | *Instruments Used* | *Representative Nanofluids Studied* | *Resolution / Sensitivity* | *Sample Requirement* | *Limitations* |
|---|---|---|---|---|---|---|---|---|
| **1.** **a.** | **Microscopic Techniques** Transmission Electron Microscopy (TEM) | Used for the direct imaging of nanoparticles | Particle size, shape, morphology | JEOL JEM-2100, FEI Tecnai G2 | $TiO_2$–water (size confirmation), CNT–water (dispersion) | Very high (<0.1 nm) | Dried nanofluid drop, thin film | Sample prep can alter morphology, expensive |
| **b.** | **SEM** | It is used for Surface morphology and particle shape | Surface texture, aggregation | ZEISS Sigma 300, FEI Quanta 250 | $Fe_3O_4$–water (aggregation), GO–EG (sheet morphology) | High (1–10 nm) | Dried solid sample | Conductive coating needed, only surface info |
| **c.** | **FESEM** | Give information on High-resolution morphology | Nanostructure details | Hitachi SU8230, JEOL JSM-7610F | $SiO_2$–water (nanostructure), ZnO–water (crystallite size) | Very high (<1 nm) | Solid dried sample | Costly, limited to surface |
| **2.** | **XRD** | Detailed information on the Crystal phase analysis | Crystallinity, crystallite size | Rigaku SmartLab, Bruker D8 Advance | $Al_2O_3$–water (crystal phases), CuO–EG (crystallinity) | High for crystalline phases | Powder or dried nanofluid | Cannot detect amorphous structures |

*Contd.*

**Table 11** *Contd.*

| *S. No.* | *Techniques* | *Role in Nanofluid Studies* | *Critical Characteristics Analyzed* | *Instruments Used* | *Representative Nanofluids Studied* | *Resolution / Sensitivity* | *Sample Requirement* | *Limitations* |
|---|---|---|---|---|---|---|---|---|
| **3. a.** | **Spectroscopic Techniques** Fourier Transform Infrared Spectroscopy (FTIR) FTIR | Functional group analysis and Surface chemistry & bonding | Functional groups | Bruker Tensor II, PerkinElmer Spectrum 100 | GO–water (oxygen groups), CNT–water (carboxylation) | High (bond-specific) | Dried nanofluid film/ pellet | Overlapping peaks can limit interpretation |
| **b.** | **UV-Vis** | Optical spectroscopic details like absorption and band gap | Absorption peaks, band gap | Shimadzu UV-2600, Agilent Cary 5000 | Ag–water (plasmon peak), $TiO_2$–EG (band gap) | High (200–800 nm) | Small liquid sample | Only optical-active species were detected |
| **4.** | **Zeta Potential** | Surface charge and Stability evaluation | Surface charge, repulsion | Malvern Zetasizer Nano ZS, Horiba SZ-100 | $Al_2O_3$–water (±30 mV stability), $Fe_3O_4$–EG (colloidal stability) | High (±1 mV) | Liquid nanofluid suspension | Sensitive to ionic strength & pH |

hybrid (Raman-AFM) technique outperforms the diffraction limits for single-particle statistical analyses. Optical tweezers can trap/isolate nanoparticles in fluid suspensions. High-throughput tracking of the reaction processes will advance the design of thermal-fluid systems.

## 4. Best Practices for Reproducibility and Accuracy

Reproducible nanofluid characterization requires careful control at all stages, including sample synthesis, preparation, measurement, and instrument calibration. Key practices include reporting nanoparticle size, morphology, purity, and synthesis parameters; maintaining constant environmental conditions; cleaning all equipment; following standard protocols; calibrating instruments; and performing at least three independent samples with multiple measurements. Transparent documentation and adherence to these practices ensure reliable, accurate, and repeatable results, reducing inconsistencies in nanofluid research.

## 5. Conclusion

Nanofluid properties are analyzed using microscopic (SEM, FESEM, TEM), spectroscopic (FTIR, Raman, UV–Vis), XRD, and zeta potential techniques to determine particle size, morphology, crystallinity, surface functionalization, aggregation, optical properties, and stability. No single method provides complete insight; combining complementary techniques links nanoparticle features to bulk nanofluid performance. Standardized, reproducible characterization is essential for reliable data and optimization of nanofluids in heat transfer, energy, biomedical, and environmental applications. Future research must focus on integrated protocols, minimizing sample errors, and correlating lab-scale results with real-world performance to advance practical applications.

## References


[1] Almuhaysh, K.A., Sergis, A. and Hardalupas, Y. 2025. Effects of Ph and Nanoparticle Concentration On Al2o3–H2o Nanofluid Stability. *International Journal Of Thermophysics,* 46.

[2] Augustyniak, J., Zglobicka, I., Kurzydlowski, K., Misiak, P., Wilczewska, A.Z., Gluch, J., Liao, Z. and Perkowski, D. M. 2022. Characterization Of Nanofluids Using Multifractal Analysis of a Liquid Droplet Trace. *Sci Rep,* 12, 11111.

[3] Chen, Z., Deutsch, T.G., Dinh, H.N., Domen, K., Emery, K., Forman, A.J., Gaillard, N., Garland, R., Heske, C., Jaramillo, T.F., Kleiman-Shwarsctein, A., Miller, E., Takanabe, K. and Turner, J. 2013. Uv-Vis Spectroscopy. *In:* Chen, Z., Dinh, H. N. & Miller, E. (Eds.) *Photoelectrochemical Water Splitting: Standards, Experimental Methods, and Protocols.* New York, NY : Springer New York.

[4] Das, L., Habib, K., Saidur, R., Aslfattahi, N., Yahya, S.M. and Rubbi, F. 2020. Improved Thermophysical Properties and Energy Efficiency of Aqueous Ionic Liquid/Mxene Nanofluid in a Hybrid Pv/T Solar System. *Nanomaterials (Basel),* 10.

[5] Gupta, M., Singh, V., Kumar, R. and Said, Z. 2017. A Review on Thermophysical Properties of Nanofluids and Heat Transfer Applications. *Renewable and Sustainable Energy Reviews,* 74, 638–670.

[6] Jorio, A. and Saito, R. 2021. Raman Spectroscopy for Carbon Nanotube Applications. *Journal of Applied Physics,* 129.

[7] Khullar, V., Bhalla, V. and Tyagi, H. 2018. Potential Heat Transfer Fluids (Nanofluids) for Direct Volumetric Absorption-Based Solar Thermal Systems. *Journal of Thermal Science and Engineering Applications,* 10.

[8] Kiani, D. 2023. X-Ray Diffraction (Xrd). *In:* Wachs, I.E. and Bañares, M.A. (Eds.) *Springer Handbook of Advanced Catalyst Characterization.* Cham: Springer International Publishing.

[9] Mehta, B.K., Chhajlani, M. and Shrivastava, B.D. 2017. Green Synthesis of Silver Nanoparticles and Their Characterization By Xrd. *Journal of Physics: Conference Series,* 836.

[10] Mohan, A.C. and Renjanadevi, B. 2016. Preparation of Zinc Oxide Nanoparticles and its Characterization using Scanning Electron Microscopy (Sem) and X-Ray Diffraction(Xrd). *Procedia Technology,* 24, 761–766.

[11] Mohd Saidi, N., Norizan, M.N., Abdullah, N., Janudin, N., Kasim, N.A.M., Osman, M.J. and Mohamad, I.S. 2022. Characterizations of Mwcnts Nanofluids on the Effect of Surface Oxidative Treatments. *Nanomaterials (Basel),* 12.

[12] Rodriguez-Laguna, M.D.R., Gomez-Romero, P., Sotomayor Torres, C.M. and Chavez-Angel, E. 2019. Modification of the Raman Spectra in Graphene-Based Nanofluids and its Correlation with Thermal Properties. *Nanomaterials (Basel),* 9.

[13] Rubaiee, S., Yahya, S.M., Fazal, M.A. and Danish, M. 2023. Characterization of $Al_2O_3$, $TiO_2$, Hybrid $Al_2O_3$-$TiO_2$ and Graphene Oxide Nanofluids and their Performance Evaluations in Photovoltaic Thermal System. *Journal of Thermal Analysis and Calorimetry,* 148, 11467–11477.

[14] Safiei, W., Rahman, M.M., Kulkarni, R., Ariffin, M.N. and Abd Malek, Z.A. 2020. Thermal Conductivity and Dynamic Viscosity of Nanofluids: A Review. *Journal of Advanced Research in Fluid Mechanics and Thermal Sciences,* 74, 66–84.

[15] Sahamifar, S., Naylor, D. and Friedman, J. 2024. An Interferometric Method to Visualize and Quantify Nanofluid Stability. *International Journal of Heat and Mass Transfer,* 235.

[16] Shanbedi, M., Zeinali Heris, S. and Maskooki, A. 2015. Experimental Investigation of Stability and Thermophysical Properties of Carbon Nanotubes Suspension in the Presence of Different Surfactants. *Journal of Thermal Analysis and Calorimetry,* 120, 1193–1201.

[17] Sonage, B.K. and Mohanan, P. 2014. Characterization of Zinc Oxide Nanoparticles used for Preparation of Nanofluids. *Procedia Materials Science,* 5, 1160–1164.

[18] Thuan, N.D., Cuong, H.M., Nam, N.H., Lan Huong, N.T. and Hong, H.S. 2024. Morphological Analysis of Pd/C Nanoparticles Using Sem Imaging and Advanced Deep Learning. *Rsc Adv,* 14, 35172–35183.

[19] Wu, J., Kwok, T.Y., Li, X., Cao, W., Wang, Y., Huang, J., Hong, Y., Zhang, D. and Wen, W. 2013. Mapping three-Dimensional Temperature In Microfluidic Chip. *Sci Rep,* 3, 3321.

[20] Yu, W., Xie, H. and Liu, L.-H. 2011. A Review on Nanofluids: Preparation, Stability Mechanisms, and Applications. *Journal of Nanomaterials,* 2012.

[21] Zagabathuni, A., Ghosh, S. and Pabi, S.K. 2016. The Difference in the Thermal Conductivity of Nanofluids Measured by Different Methods and its Rationalization. *Beilstein J Nanotechnol,* 7, 2037–2044.

# 6 Nanofluid Thermophysical Properties Prediction Using Machine Learning and Artificial Intelligence

Ali Alsayegh[1*], Hawraa Alsayegh[2] and Annisa Fatharani[3]

Nanofluids exhibit complex thermophysical behaviours that traditional empirical correlations and effective-medium theories fail to predict accurately due to oversimplified assumptions about particle distribution and interfacial effects. This chapter critically examines the application of artificial intelligence and machine learning techniques for predicting nanofluid thermal conductivity, viscosity, specific heat capacity, and density, demonstrating how data-driven approaches overcome the limitations of conventional models. The analysis traces the evolution from early algorithms such as artificial neural networks and support vector machines to advanced ensemble methods, including random forests, XGBoost, and LightGBM, highlighting their superior predictive capabilities across diverse experimental conditions. Particular attention is given to physics-informed neural networks, which embed conservation laws directly into model architectures to ensure physically plausible predictions whilst maintaining data-driven flexibility. The chapter further explores emerging generative AI techniques that address persistent challenges of limited experimental data and inconsistent reporting standards through synthetic data generation and natural language processing for metadata extraction. However, significant obstacles remain, including data quality issues, computational requirements, and model interpretability concerns, which are systematically analysed alongside validation protocols for synthetic data. The chapter concludes by identifying high-impact research opportunities in explainable AI frameworks, standardised data-sharing platforms, and autonomous experimental systems that integrate prediction with real-time validation, thereby accelerating nanofluid innovation whilst ensuring reproducible, scientifically rigorous property predictions.

## 1. Introduction

The introduction of nanoparticles into conventional base fluids, first proposed by Choi and Estman (1995), opened a new frontier in thermal sciences. Researchers

[1] Department of Design, Manufacturing, and Engineering Management, University of Strathclyde, Glasgow, United Kingdom.

[2,3] School of Cancer and Pharmaceutical Sciences, King's College London, London, United Kingdom.

* Corresponding author: a.alsayegh@strath.ac.uk

DOI: 10.1201/9781003686002-6

have discovered that even tiny amounts of solid particles added to fluids can significantly enhance their ability to transfer heat, particularly by increasing their thermal conductivity (Choi and Eastman, 1995; Yu *et al.*, 2008). These findings quickly caught the attention of industries like energy, manufacturing, and electronics cooling, where managing heat efficiently is a critical challenge.

As research progressed, it became evident that the nanofluids' thermophysical properties are affected by a complex interplay of factors. These include particle size, shape, and concentration, as well as the chemistry of the base fluid, operating temperature, and shear rate (Angayarkanni and Philip, 2015). Significantly, preparation variables such as surfactant type, sonication energy, pH level, and settling or shelving time have a substantial impact on nanofluid dispersion stability, which, in turn, influences their bulk properties. Traditional empirical correlations and effective medium theories have struggled to fully capture these non-linear relationships, often yielding inconsistent results outside the conditions for which they were initially developed. This limitation underscores the need for more adaptable, data-driven approaches.

Artificial intelligence and machine learning (AI-ML) have emerged as promising solutions capable of revealing patterns hidden in experimental datasets without requiring explicit function assumptions. As shown in Figure 1, the number of publications focusing on nanofluid thermophysical property prediction using AI/ML techniques has grown steadily over the past decade, with a particularly sharp increase in recent years. This trend reflects the growing recognition of AI/ML as essential tools for tackling the complex, nonlinear relationships governing nanofluid behaviour. Early applications of ML to nanofluids focused on algorithms. Artificial neural networks (ANNs) and support vector machines (SVMs) are the

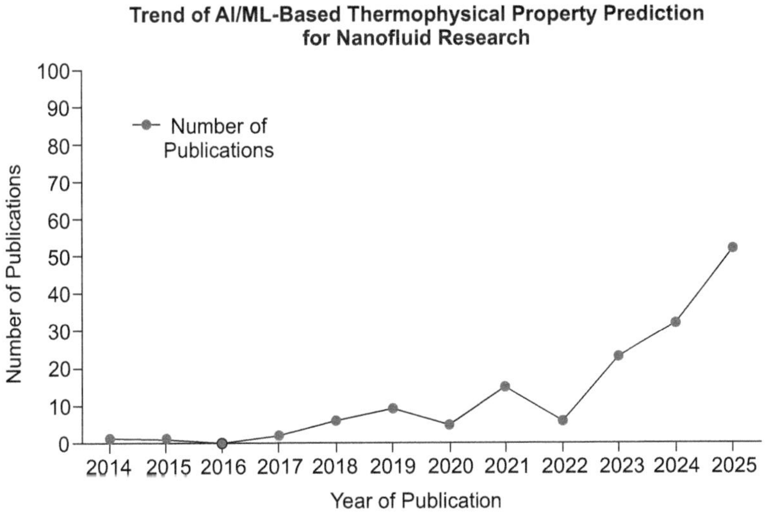


**Fig. 1** Annual publication trend of material science research articles focused on applying AI/ML techniques to predict nanofluid thermophysical properties.

ones that are commonly applied in this field. In recent years, ensemble techniques in random forests (RF) and gradient-boosting frameworks, such as XGBoost and LightGBM, have gained traction for their accuracy and ability to identify the relative importance of different input features.

Beyond purely data-driven models, a growing body of work integrates physical principles into the learning process. Physics-informed and physics-guided models incorporate constraints derived from conservation laws, constitutive trends, or low-fidelity theoretical models into the algorithm design, improving interpretability and ensuring that predictions remain physically plausible. Meanwhile, recent breakthroughs in generative AI have opened new possibilities for tackling two persistent challenges in nanofluid research: limited data availability and inconsistent reporting. Natural language processing (NLP) in large language models (LLMs) can extract experimental metadata from the literature, including preparation methods, sonication protocols, and measurement techniques, thereby enabling the creation of richer, machine-readable datasets. Generative approaches, such as conditional tabular generative adversarial networks (CTGANs), while not yet applied in nanofluid research, can be used to produce synthetic but physically consistent data that help balance datasets and explore under-sampled regions of the design space.

Building on the previously covered literature, this chapter serves as a literature review that critically analyses the application of AI/ML to predict the thermophysical properties of nanofluids. The chapter also integrates original interpretive analysis, including a comparative summary of key studies (Table 4) and representative case studies that illustrate the practical use and performance of popular AI/ML models. The intention is to bridge the gap between literature synthesis and applied understanding, providing both researchers and practitioners with an organised perspective on methodological trends, limitations, and future research opportunities. By drawing on insights from computational modelling, materials science, and fluid mechanics, it examines both the considerable progress made so far and the hurdles that must be addressed to create dependable, scalable methods for future nanofluid applications research.

## 2. Role of Artificial Intelligence and Machine Learning

The thermophysical properties of nanofluids arise from intricate, multiscale interactions among particles, base fluids, and preparation conditions. These interactions are highly nonlinear and context-dependent, making it challenging to describe them using traditional empirical correlations or effective-medium theories. Traditional models often rely on simplifying assumptions, such as uniform particle distribution or negligible interfacial effects, which rarely hold in real-world systems. As a result, their predictive accuracy declines when applied to new systems or operating conditions, yielding inconsistent or even contradictory results across laboratories.

AI/ML provides a fundamentally different approach by learning patterns directly from data rather than imposing a fixed mathematical form. This makes them particularly suitable for nanofluids, where property variations depend on multiple interacting variables in particle concentration, particle size and morphology, temperature, base fluid chemistry, and stability factors. After training, AI/ML models can predict key thermophysical properties, such as viscosity ($\mu$), specific heat capacity ($Cp$), density ($\rho$), and thermal conductivity ($k$) across a range of operating conditions. Thereby, supporting both scientific understanding and engineering design. The integration of AI/ML into nanofluid property prediction can be seen in Figure 2.

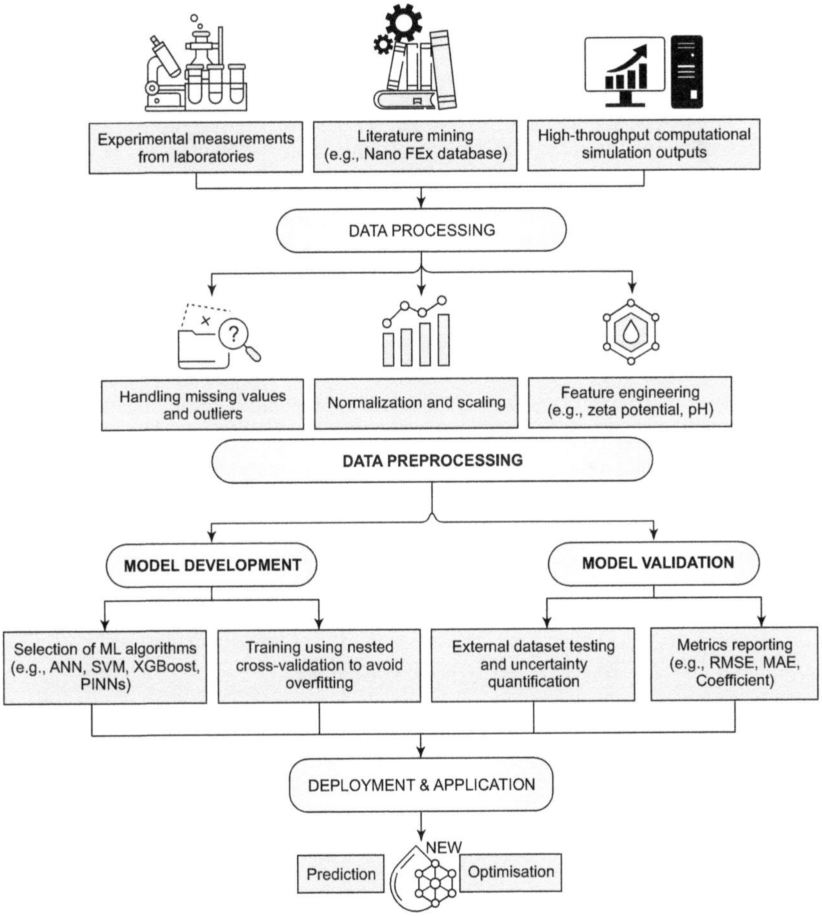


**Fig. 2** Workflow for ML-Based Nanofluid Property Prediction.

Early applications of ML in this field focused on ANNs and SVMs. ANNs are particularly effective at capturing complex, non-linear relationships. In contrast,

SVMs are more suitable for smaller datasets and noisy measurements, thanks to their strong generalisability performance. Numerous studies have demonstrated that these algorithms can significantly outperform traditional empirical models, providing more reliable predictions for viscosity and thermal conductivity properties.

As datasets grew and computing power increased, ensemble learning techniques like random forests (RF) and gradient boosting were adopted. These models combine multiple learners to improve accuracy and robustness while providing interpretable measures of variable importance, helping researchers identify which experimental factors most strongly influence nanofluid behaviour (Breiman, 2001; Chen and Guestrin, 2016; Friedman, 2001; Gholizadeh *et al.*, 2020; Ke *et al.*, 2017; Onyiriuka, 2023). For example, studies using ensembles have highlighted the significant impact of preparation parameters, such as sonication energy and dispersant type, on viscosity and thermal conductivity (Said *et al.*, 2023).

A recent and transformative development has been the integration of physical principles into ML models. Physics-informed neural networks (PINNs) embed conservation laws and constitutive relationships directly into the training process, preventing models from producing non-physical predictions and improving their performance in data-scarce regimes (Bhaumik *et al.*, 2022; Peng *et al.*, 2023; Raissi *et al.*, 2019; Wang *et al.*, 2022). Hybrid approaches can also combine low-fidelity theoretical models with high-fidelity experimental data in multi-fidelity learning architectures, further enhancing extrapolation capability.

AI/ML has evolved from simple curve-fitting to a comprehensive, interdisciplinary framework in nanofluid research. Instead of replacing experimental and theoretical methods, these tools complement them, with experiments providing crucial data, physics providing guiding principles, and ML connecting the two by recognising complex patterns and enabling reliable, scalable prediction. This paves the way for more reproducible and generalisable models of nanofluid thermophysical properties, ultimately accelerating both scientific discovery and practical applications.

## 3. Data Preparation and Input Variables

The predictive ability of any AI/ML model is primarily limited by the quality, completeness, and consistency of the data on which it is trained. In nanofluid research, this presents a particular challenge, as experimental studies often vary significantly in their preparation methods, measurement techniques, and reporting standards, leading to disjointed and sometimes incompatible datasets. Without careful attention to data preparation, even the most advanced models risk producing misleading or non-generalisable results and predictions (Esfahani, 2023; Jamei and Said, 2022; Sharma *et al.*, 2022).

Most published nanofluid studies report only a minimal set of experimental variables, typically temperature, nanoparticle volume fraction, and occasionally particle size or base fluid type. While these are essential, they are insufficient to capture the full range of factors governing thermophysical behaviour. Stability-related descriptors, such as zeta potential, pH, dispersant type, and sonication energy, strongly influence nanoparticle dispersion and aggregation, which, in turn, affect bulk properties (Ali *et al.*, 2019). Their omission leads to hidden variability, making it hard to compare results across laboratories. For instance, two experiments with the same temperature and nanoparticles pKa values may still produce very different viscosities if their sonication methods differ (Aljuwayhel *et al.*, 2023; Misbah *et al.*, 2022).

Table 1 summarises the input variables primarily used in nanofluid AI/ML modelling, grouped by category, highlighting their reported frequency, physical relevance, and potential issues when omitted. This information shows why stability descriptors and preparation metadata must be prioritised alongside basic thermophysical parameters for reliable AI/ML models. Preparing a high-quality dataset, then, involves three interconnected steps:

1. Data Cleaning and Harmonisation: Raw experimental data must be examined for outliers, duplicate entries, and inconsistent units. It is imperative to standardise variable names and units across studies when constructing integrated datasets, particularly when aggregating data from multiple sources or publications (Amoo *et al.*, 2024).
2. Feature Selection and Augmentation: Selecting relevant features goes beyond statistical correlation, where it requires physical insight into which factors are mechanistically linked to nanofluid stability and thermophysical properties. When key features are missing from the literature, they can sometimes be reconstructed using NLP pipelines. LLMs are increasingly utilised to extract experimental metadata from text, tables, and supplementary files, thereby reducing human effort and enabling the creation of more comprehensive datasets (Göpfert *et al.*, 2024; Jiang *et al.*, 2025).
3. Normalisation and Scaling: These are crucial processes, as many AI/ML algorithms exhibit sensitivity to discrepancies in the scale of input variables. Standardising features, through techniques such as min-max scaling or z-score normalisation, guarantees stable training and prevents variables with larger scales from dominating the optimisation process (Adun *et al.*, 2022; Onyiriuka, 2023).

Recent studies in curated databases are addressing the persistent problem of fragmented data. For instance, the NanoFEx database provides a structured repository of thermophysical measurements, including metadata on preparation and measurement conditions, to facilitate reproducible ML workflows (Amoo *et al.*, 2024). These resources, combined with new reporting guidelines, are crucial for reducing bias and improving the generalisability of AI/ML models. Standardisation efforts also help address dataset imbalance, where certain particle-based fluid combinations are overrepresented, while others remain sparse, a common source of overfitting in nanofluid research (Zhu *et al.*, 2024).

**Table 1** Main input variables for AI/ML approach in nanofluid thermophysical property prediction.

| *Category* | *Variable* | *Typical Reporting Frequency* | *Physical Relevance* | *Common Issues When Omitted* |
|---|---|---|---|---|
| Basic Parameters | Temperature (T) | Very High | Directly affects all thermophysical properties (k, μ, Cp). | Models cannot accurately capture temperature dependence. |
| Basic Parameters | Nanoparticle volume fraction (φ) | Very High | Determines particle interactions and clustering. | No distinction between dilute and concentrated systems. |
| Basic Parameters | Particle size (nm) | Medium | Influences Brownian motion and surface area effects. | Ignores size-driven thermal conductivity changes. |
| Base Fluid Info | Base fluid type (e.g., water, ethylene glycol) | Medium | Governs viscosity and dispersion behaviour. | Cannot generalise across different fluid systems. |
| Stability Descriptors | Zeta potential (mV) | Low | Key indicator of dispersion stability. | Hidden variability in viscosity and thermal conductivity results. |
| Stability Descriptors | pH level | Low | Affects surface charge and nanoparticle interactions. | Poor reproducibility between labs. |
| Stability Descriptors | Sonication energy (J) | Low | Determine the s degree of de-agglomeration of the same. | The same φ and T may yield different results due to inconsistent preparation. |
| Stability Descriptors | Surfactant type/concentration | Very Low | Controls stability and aggregation prevention. | Misleading comparisons between datasets. |
| Measurement Context | Measurement technique (e.g., transient hot-wire, rotational viscometer) | Low | Method differences directly influence reported values. | Dataset noise and inconsistent accuracy. |

Choosing and using the right input variables significantly affects how well AI/ML models predict nanofluid thermophysical properties. While traditional studies focus on core variables such as temperature and nanoparticle concentration, stability-related features are often underreported, limiting model generalisation. Figure 3 illustrates the broader feature space, grouping variables into the three most important categories in data preparation.

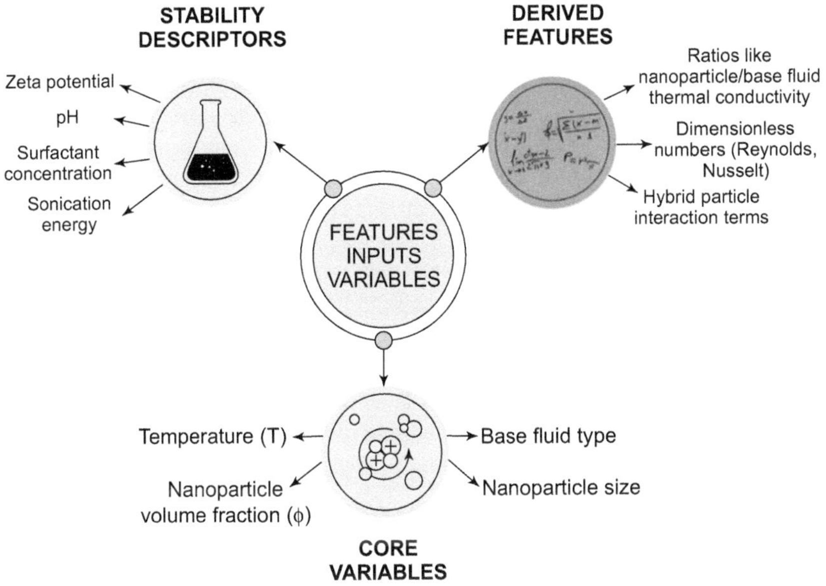


**Fig. 3** Feature Space for Nanofluid Thermophysical Property Prediction.

The success of AI/ML in predicting nanofluid properties depends not only on the selection of the algorithm but also on the discipline applied to data preparation and analysis. Using a meticulously curated dataset, supplemented with stability descriptors and experimental context, lays a robust groundwork for models that are both accurate and interpretable. On the other hand, models trained on incomplete or inconsistent data may show impressive metrics on paper but struggle when applied to new systems or in real-world scenarios.

## 4. Popular Algorithms and Techniques

Choosing the correct AI/ML algorithm is crucial to determining the accuracy, generalisability, and interpretability of predictions for nanofluid properties. In the past decade, several algorithms have been used, each with its own strengths and disadvantages. There are four categories of popular algorithms and techniques in the AI/ML approach to predicting nanofluid properties: ANNs, SVMs, ensemble and hybrid models, and emerging generative AI techniques. Early research primarily utilised traditional algorithms such as SVMs and ANNs, which were effective for small experimental datasets but lacked scalability. Ensemble methods, including RF and gradient boosting algorithms, have gained prominence, leading to improved robustness and accuracy. Recently, physics-guided deep learning models have addressed the limitations of data-driven approaches by directly incorporating physical laws into neural networks.

## *4.1 Artificial Neural Networks (ANN)*

ANNs are one of the most commonly used ML techniques for modelling the thermophysical properties of nanofluids. Modelled on the structure of biological neurons, ANNs are well-suited for analysing the complex, non-linear relationships between experimental input variables and target outputs. As shown in Figure 4, this makes them highly effective for predicting complex properties such as thermal conductivity ($k$), dynamic viscosity ($\mu$), and specific heat capacity ($Cp$), which are influenced by multiple interdependent factors, such as temperature, nanoparticle concentration ($\varphi$), particle size, and base fluid type (Alade *et al.*, 2020; Heidari *et al.*, 2016; Hemmat Esfe *et al.*, 2019). These inputs pass through multiple hidden layers, where weighted connections and activation functions allow the network to learn complex patterns. The final output layer produces predicted properties, such as viscosity or thermal conductivity, enabling researchers to forecast performance under different experimental conditions.

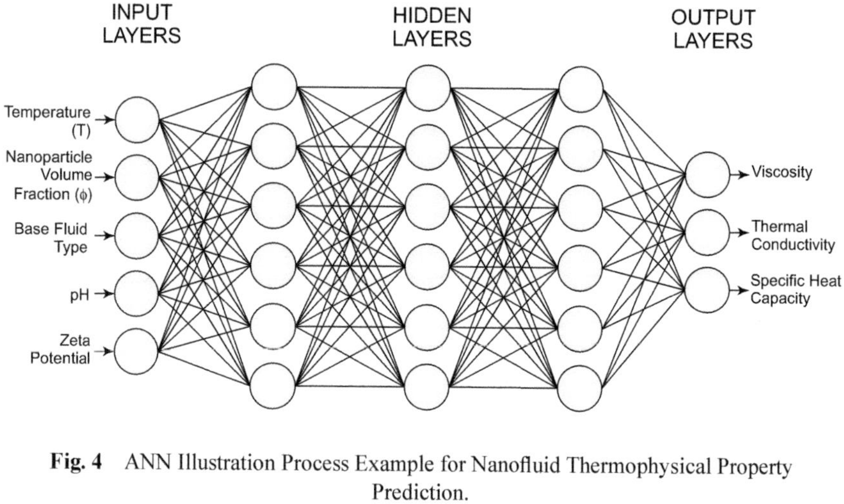


**Fig. 4** ANN Illustration Process Example for Nanofluid Thermophysical Property Prediction.

Since an ANN consists of an input layer, one or more hidden layers, and an output layer, each neuron processes incoming signals by applying a weighted sum of the inputs, followed by a non-linear function. The mathematical representation of a single neuron is given in Eq. 1 (Haykin, 2009):

$$y = f\left(\sum_{i=1}^{n} w_i x_i + b\right) \qquad ...(1)$$

where $x_i$ represents the input variables (temperature, particle volume fraction, or stability descriptors); $w_i$ are the weights that determine the contribution of each input; $b$ is the bias term; $f(\cdot)$ is the activation function (the sigmoid), ReLU, or tanh function; and $y$ is the predicted thermophysical property.

Throughout the training process, the network optimises its weights and biases through iterative updates using algorithms such as backpropagation to reduce the error between predicted and actual values. This makes ANNs highly flexible in modelling complex non-linear behaviour where traditional regression techniques fail. Researchers in nanofluidics have successfully applied ANNs to simultaneously predict multiple properties. Heidari *et al.* (2016) showed that a multi-layer perceptron ANN could accurately predict viscosity across a range of temperatures and concentrations, outperforming empirical method correlations. Similarly, Esfe H. *et al.* (2019) used ANNs to model both the thermal conductivity and viscosity of hybrid nanofluids, demonstrating their ability to handle multi-output problems.

Despite their effectiveness, ANNs face some challenges. First, they require large, high-quality datasets to prevent overfitting, a recurring issue in nanofluid studies where experimental data are limited and often inconsistent across laboratories (Sharma *et al.*, 2022). Secondly, ANNs are frequently criticised for being a "black box", which limits interpretability. Researchers have addressed this by integrating Shapley Additive Explanations (SHAP), an explainable AI (XAI) technique, to quantify the influence of individual features, or by embedding physical constraints directly into the network architecture to ensure predictions remain physically meaningful (Bhaumik *et al.*, 2022; Esfahani, 2023).

Overall, ANNs remain a foundation of AI/ML applications in nanofluid research, offering high predictive power for complex thermophysical properties. When coupled with robust data preparation (Section 3) and explainability tools, they form the foundation for building reliable and generalisable predictive models.

## 4.2 Support Vector Machines (SVM)

SVM is another widely used AI/ML approach in nanofluid research, particularly suited for situations where datasets are small or noisy, which is often the case in experimental nanofluid studies due to the high cost and complexity of data generation. Unlike ANNs, which are highly adaptable but require a large amount of data, SVMs perform well with limited data, making them suitable for predicting thermophysical properties (Adun *et al.*, 2022; Ahmadi *et al.*, 2019). As shown in Figure 5, SVMs identify the best possible hyperplane that separates data points (such as low- and high-viscosity samples) with the most significant margin. The support vectors located closest to this hyperplane play an important role in defining the decision lane.

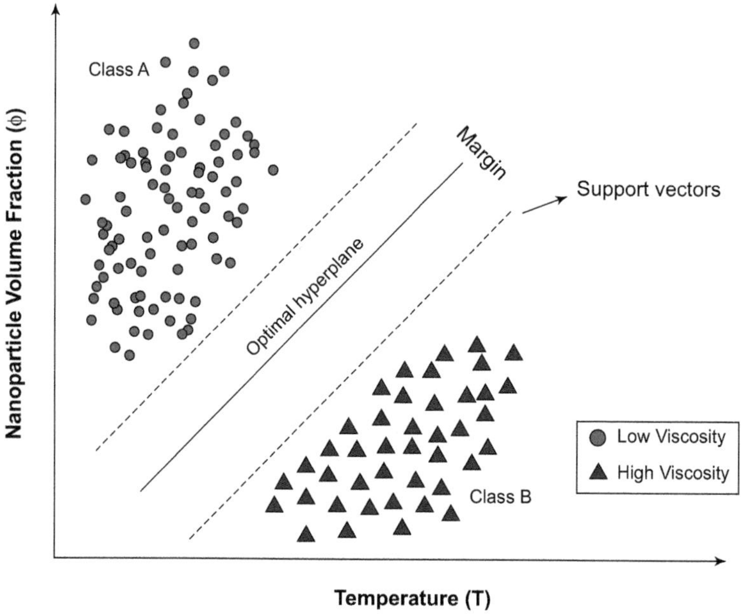


**Fig. 5** SVM Illustration Example for Nanofluid Property Classification and Prediction.

For regression tasks, SVMs are implemented as Support Vector Regression (SVR) models. At its core, SVR aims to identify a function $f(x)$ that closely represents the relationship between input variables and target outputs. The optimisation objective for SVR is outlined in Eq. 2 (Smola and Schölkopf, 2004; Vapnik, 2000):

$$= \frac{1}{2} | w |^2 + C \sum_{i=1}^{n} (\xi_i + \xi_i^*) \qquad ...(2)$$

where $x_i$ are the input features (temperature, particle concentration, particle size, and stability descriptors); $y_i$ represents the measured thermophysical property; $w$ and $b$ are the weight vector and bias term, respectively; $\epsilon$ defines the margin of tolerance within which prediction errors are not penalised; $\xi_i$ and $\xi_i^*$ are slack variables allowing for soft margin violations; and $C$ is the regularisation parameter trade-off between model complexity and error tolerance.

SVR delivers impressive results by utilising kernel functions that transform the original input data into a higher-dimensional feature space, enhancing its ability to model complex patterns in space. This enables the modelling of complex, non-linear relationships without explicitly increasing the problem's dimensionality. Consequently, SVMs have proven to be effective in predicting various properties of nanofluids. Ahmadi *et al.* (2019) optimised SVR hyperparameters using genetic algorithms, resulting in reliable predictions of nanofluid thermal conductivity across a broad range of nanoparticle-based fluid systems. Similarly, Adun *et al.* (2022) demonstrated that SVR outperformed several other ML methods in estimating the viscosity of hybrid nanofluids, particularly when the training data were limited.

However, SVMs also present certain limitations. Their performance depends significantly on selecting the correct kernel and fine-tuning parameters, which can often require time-consuming cross-validation. While they provide accurate predictions, SVMs are somewhat less easy to interpret than tree-based ensemble methods, as they do not naturally indicate which variables are most important (Onyiriuka, 2023). Even with these challenges, SVMs are a valuable tool for nanofluid research, especially in the early stages when experimental data may be limited. When paired with thoughtful data preparation and feature engineering (Section 3), they can be a reliable baseline, paving the way for more advanced methods in hybrid or ensemble approaches.

### 4.3 Hybrid and Ensemble Models

As nanofluid research has advanced and datasets have become more complex, there has been a significant shift from using single AI/ML algorithms to ensemble and hybrid approaches. These techniques enhance both prediction accuracy and generalisation by combining multiple models or directly incorporating physical laws into the learning process. They particularly help address two common challenges in nanofluid modelling: limited and imbalanced data, and the need for predictions that make physical sense. Ensemble methods combine the results of multiple base learners, thereby strengthening the overall model and reducing its susceptibility to the weaknesses of individual models. Two popular ensemble strategies are bagging and boosting.

Bagging (bootstrap aggregation) trains multiple models on random subsets of the dataset and calculates their average predictions. RF, introduced by Breiman (2001) as a technique, is the most common bagging-based method and has been successfully applied to predict nanofluids' properties (Gholizadeh *et al.*, 2020; Jamei and Said, 2022). Boosting, on the other hand, builds models one after the other to complement the predecessors. Gradient boosting frameworks, such as XGBoost and LightGBM, have demonstrated exceptional performance in predicting nanofluid properties using this method (Chen and Guestrin, 2016; Ke *et al.*, 2017). The iterative update process of boosting is mathematically described in Eq.3 (Chen and Guestrin, 2016; Friedman, 2001; Hastie *et al.*, 2001):

$$F_m(x) = F(m-1)(x) + \gamma h_m(x) \qquad ...(3)$$

where $F_m(x)$ is the updated ensemble model at iteration $m$; $F_{m-1}(x)$ is the iteration's model; $h_m(x)$ is the newly added weak learner, such as a shallow decision tree; $\gamma$ is the learning rate, which controls the contribution of each new model.

Recent studies have shown the practical advantages of boosting methods for nanofluids. Zhu *et al.* (2024) utilised XGBoost to predict the thermal conductivity of $Al_2O_3$-ethylene glycol nanofluids, revealing that preparation variables, such as sonication energy and zeta potential, can be as influential as temperature and volume fraction. This provides insights that simpler models would have overlooked. These interpretability features make ensemble methods especially helpful, not just for delivering accurate predictions but also for guiding the design of experimental efforts.

Although ensemble methods primarily focus on predictive accuracy, hybrid models take it a step further by explicitly incorporating physical laws and domain knowledge into the learning process. A notable example is the PINN, which integrates governing equations, such as the heat conduction equation, directly into the network's loss function. This approach ensures that the model's predictions remain physically plausible, even when trained on sparse or noisy data datasets. The general form of a PINN loss function is shown in Eq. 4 (Raissi *et al.*, 2019):

$$L = L_{\text{data}} + \lambda L_{\text{physics}} \qquad \text{...(4)}$$

where $L_{\text{data}}$ represents the conventional data loss (mean squared error between the values of predicted and measured); $L_{\text{physics}}$ penalises violations of known physical constraints or governing equations; and $\lambda$ is a weighting factor trade-off between data fidelity and physical consistency.

Researchers in nanofluid studies, such as Bhaumik *et al.* (2022), have shown the potential of physics-guided deep neural networks by accurately predicting viscosity while maintaining thermodynamic consistency. Likewise, Peng *et al.* (2023) used PINNs to model heat conduction in nanofluids, highlighting that incorporating physical principles directly into the learning process greatly improves generalisation and prevents the model from producing physically unrealistic results. These combined approaches have significant benefits. Ensemble methods reduce overfitting, enhance robustness, and provide clear insights into which experimental factors have the most significant impact on nanofluid properties. Hybrid models, by contrast, directly address the interpretability issue faced by purely data-driven algorithms. Overall, these techniques are particularly effective for sparse or imbalanced datasets, where traditional ANNs or SVMs may struggle to generalise effectively.

Ensemble methods, especially boosting algorithms, can be quite complex and often require substantial computational resources to fine-tune hyperparameters. Creating hybrid models also requires expertise in both physical sciences and ML, which can be a challenge for some research teams. Additionally, if the embedded physical laws are not comprehensive or overly simplified, they could overly restrict the model, resulting in biased or less accurate outcome predictions.

### *4.4 Generative Artificial Intelligence Model*

One of the most significant challenges in applying ML methods to nanofluid research is the scarcity of high-quality data. Experimentally measuring nanofluids' thermophysical properties is costly, time-consuming, and prone to inconsistencies, as variations in preparation and measurement techniques between laboratories often produce conflicting results (Esfahani, 2023). Consequently, many combinations of nanoparticle type, base fluid, and operating conditions remain underrepresented in current datasets, creating severe class imbalances. This limits the ability of conventional ML and deep learning models to generalise effectively, often leading to overfitting and reduced predictive performance.

After a comprehensive review of the current literature, no published studies have reported the direct application of GenAI to predict nanofluid thermophysical properties. Therefore, this section is presented as a prospective overview rather than a report of existing work, as it holds significant potential for addressing challenges in data scarcity and experimental variability. By learning the statistical patterns of real experimental data, such as GANs, one could, in principle, generate plausible synthetic datasets of thermophysical properties, thereby expanding the design space and alleviating related issues of imbalance.

Recent advances in GANs and broader GenAI approaches have demonstrated success in other scientific and engineering domains for creating synthetic yet realistic datasets (Dan *et al.*, 2020; Xu *et al.*, 2019).GAN comprises two competing neural networks. The generator (G) generates synthetic samples, while the discriminator (D) endeavours to differentiate the authentic and generated data. The generator improves its performance by attempting to deceive the discriminator, while the discriminator concurrently enhances its ability to identify counterfeit data. Through an iterative feedback loop, the generator becomes progressively more proficient in producing realistic data, whilst the discriminator improves its ability to evaluate authenticity. This dynamic process has been shown to produce high-quality synthetic datasets across fields such as materials science and healthcare (Trabassi *et al.*, 2024; Xu *et al.*, 2019).

Beyond dataset augmentation, GenAI could also serve a transformative role in inverse design (Channi *et al.*, 2025). In nanofluid research, by sampling from a trained generator, researchers may theoretically investigate novel combinations of nanoparticle types, base fluids, and preparation parameters to achieve targeted outcomes such as maximised thermal conductivity or minimised viscosity. While direct applications in nanofluids remain unexplored, recent work has introduced NANOGPT, a large language model framework specifically designed to accelerate nanotechnology research (Chandrasekhar *et al.*, 2025). Although NANOGPT primarily focuses on literature mining and knowledge synthesis, its development signals growing interest in leveraging generative AI to address data challenges across nanofluid domains.

The application of the GenAI approach to nanofluids, hence, remains an area for future research, with experimental validation yet to be demonstrated. When utilising synthetic data in scientific investigations, rigorous validation protocols are imperative to ensure the credibility and utility of the generated datasets (Dan *et al.*, 2020; Miletic and Sariyar, 2024). Recommended protocols include:

1. Train-on-Synthetic, Test-on-Real (TSTR): Involves training AI/ML models with synthetic data and then evaluating their performance on real-world data, providing a good sense of their practical applicability.
2. Train-on-Real, Test-on-Synthetic (TRTS): Helps evaluate whether the synthetic data used genuinely mirrors the distributions of real-world data, ensuring that the synthetic datasets are realistic and provide accurate measurements.

These validation strategies are essential to ensure that synthetic datasets are both realistic and informative. While GenAI applications remain conceptual at this stage, their potential impact is considerable. Future research may combine GenAI-based data augmentation with physics-informed constraints to ensure that generated samples remain physically consistent, bridging data-driven innovation with thermodynamic realism. Nevertheless, the generation process may introduce unforeseen biases, especially if the training data are unbalanced or contain systematic experimental errors (Shahul Hameed *et al.*, 2024).

Selecting the appropriate AI/ML techniques has a substantial impact on model performance, interpretability, and data requirements for predicting nanofluid thermophysical properties. Table 2 provides a comparison of their characteristics, strengths, weaknesses, and suggested applications.

## 5. Model Validation and Accuracy Assessment

For AI/ML models to perform well in nanofluid research, it is not just about choosing the correct algorithm. Robust validation and assessment procedures are also crucial. If validation is not done thoroughly, even highly accurate models may not be able to generalise beyond the specific experimental datasets they were trained on, resulting in misleading or physically impossible results predictions. Model validation serves the purposes of evaluating accuracy and generalisability, ensuring that predictions remain reliable for unseen data, and identifying overfitting or underfitting, which helps researchers refine model complexity and prevent spurious relationships from being learned. The following are the current methods used for validation and accuracy assessment of AI/ML models for nanofluid thermophysical property predictions.

### a. Cross-Validation Strategy

For nanofluid datasets, which are often small and heterogeneous, cross-validation (*CV*) is essential. One of the most commonly used strategies is *k*-fold cross-validation. The model is trained on *k*–1 folds and tested on the remaining fold, with this process repeated *k* times to ensure every data point is used for both training and testing. The average performance across all folds provides a reliable estimate of the model's generalisability. Mathematically, the *k*-fold CV error is calculated as shown in Eq. 5 (Hastie *et al.*, 2001):

$$E_{CV} = \frac{1}{k}\sum_{i=1}^{k} E_i \quad ...(5)$$

**Table 2** Comparison of commonly used AI/ML algorithms for nanofluid thermophysical property prediction.

| *Algorithm(s)* | *Features* | *Strengths* | *Weaknesses* | *Recommended Use Cases* |
|---|---|---|---|---|
| Artificial Neural Networks (ANN) | Multi-layer network capable of capturing non-linear relationships. | Excellent for complex and non-linear mappings; flexible architecture. | Prone to overfitting; limited interpretability. | Viscosity prediction with diverse experimental parameters. |
| Support Vector Machines (SVM) | Kernel-based regression using support vectors. | Strong generalisation for smaller datasets; robust to overfitting. | Requires careful parameter tuning; limited scalability. | Early-stage experiments with limited data. |
| Random Forest (RF) | Ensemble of decision trees using bagging. | Interpretable through feature importance; handles mixed data types well. | Limited extrapolation beyond training range. | Identifying critical experimental variables; Cp prediction. |
| XGBoost | Gradient boosting with regularisation. | High accuracy; effective for tabular datasets; relatively interpretable. | Sensitive to hyperparameter tuning; higher computational cost. | Predicting thermal conductivity across large datasets. |
| Physics-Informed Neural Networks (PINNs) | Neural networks incorporating physical laws in the loss function. | Ensures physically consistent predictions; reduces data requirements. | Complex implementation; slower training. | Sparse datasets where physical laws are known. |
| Generative Adversarial Networks (GANs) | Generates synthetic data through adversarial learning. | Addresses data scarcity; enables inverse design. | Risk of bias amplification; validation required. | Creating synthetic datasets; exploring design spaces. |

where $E_{CV}$ is the average cross-validation error; $E_i$ is the error for fold $i$; and $k$ is the total number of folds, typically set between 5 and 10 for nanofluid datasets.

For small datasets, leave-one-out cross-validation (LOOCV) can be used, where each data point serves as a test case exactly once (Lumumba *et al.*, 2025).

**b. Performance Metrics**

Various statistical metrics are used to assess the performance of AI/ML models in nanofluids research. The most common include:

- Mean Absolute Error (MAE) (Chai and Draxler, 2014):

$$MAE = \frac{1}{n}\sum_{i=1}^{n} |y_i - \hat{y}_i| \qquad ...(6)$$

where $n$ is the number of data points; $y_i$ is the experimental value of the thermophysical property; and $\hat{y}_i$ is the model-predicted value. This metric calculates the average size of errors, disregarding their direction. It helps compare models when significant outliers are not dominant.

- Root Mean Squared Error (RMSE) (Chai and Draxler, 2014):

$$RMSE = \sqrt{\frac{1}{n}\sum_{i=1}^{n} (y_i - \hat{y}_i)^2} \qquad ...(7)$$

where larger deviations are penalised more heavily, making RMSE particularly sensitive to outliers. This metric places greater emphasis on larger errors, making it more sensitive to outliers and particularly suitable when significant deviations in thermophysical property predictions are particularly problematic, such as with viscosity modelling.

- Coefficient of Determination ($R^2$) (Hastie *et al.*, 2001; Karimipour *et al.*, 2019):

$$R^2 = 1 - \frac{\sum_{i=1}^{n} (y_i - \hat{y}_i)^2}{\sum_{i=1}^{n} (y_i - \underline{y})^2} \qquad ...(8)$$

where $\hat{y}$ is the mean of experimental values. A value of $R^2$ close to 1.0 indicates that the model explains most of the variance in the experimental data, representing strong predictive performance. This metric indicates the proportion of variance in the experimental data that can be attributed to the model.

While individual metrics offer helpful insights into model performance, choosing the right metric depends on the study's aims and the data's unique features. Finding the right metric ensures that the evaluation truly reflects the model's strengths and areas for improvement in the dataset. For instance, RMSE is often preferred for viscosity modelling (Ahmadi *et al.*, 2020; Zhu *et al.*, 2024), where large prediction errors can have a pronounced impact on system performance. In contrast, MAE offers a more interpretable measure when datasets contain few extreme outliers. Similarly, $R^2$ is widely used for comparing the explanatory power of different models, but should always be considered alongside error-based metrics to avoid misleading conclusions (Ahmadi *et al.*, 2020). Table 3 below outlines the primary purposes, benefits, limitations, and common uses of these performance metrics within the context of the AI/ML approach to predicting nanofluid thermophysical properties.

**Table 3** Comparison of performance evaluation metrics used in AI/ML models for nanofluid thermophysical property predictions.

| *Metric(s)* | *Purpose* | *Advantages* | *Limitations* | *Typical Use Cases* |
|---|---|---|---|---|
| **Mean Absolute Error (MAE)** | Measures the average magnitude of errors without considering direction. | Simple to interpret; not overly influenced by outliers. | Does not reflect the severity of significant errors. | General benchmarking of viscosity or thermal conductivity predictions where outliers are minimal. |
| **Root Mean Squared Error (RMSE)** | Penalises larger errors more heavily, making it sensitive to outliers. | Provides higher sensitivity to large deviations, ideal for critical safety-related predictions. | Can be skewed by outliers; harder to interpret directly. | Viscosity modelling or stability predictions where large deviations are particularly problematic. |
| **Coefficient of Determination ($R^2$)** | Indicates the proportion of variance in the experimental data explained by the model. | Gives a standardised measure of overall model fit; widely recognised in literature. | High $R^2$ does not guarantee low error; may be misleading when data range is narrow. | Comparing multiple models for thermal conductivity or specific heat capacity datasets. |

## c. Train-Test Splits and External Validation

When developing AI/ML techniques, it is a common practice to divide the dataset into 3 categories: training, validation, and test set. The training set is used to fit the model, while the validation set helps fine-tune the hyperparameters and prevent overfitting. The test set was kept entirely separate to manage an unbiased estimate of performance on new, unseen data. For nanofluid research, external validation using independent experimental datasets is particularly valuable. This type of validation demonstrates that the model can generalise beyond a single laboratory or set of experimental conditions, addressing concerns about reproducibility (Esfahani, 2023).

Beyond the three validation and accuracy evaluations above, assessing uncertainty is also crucial and needs to be implemented, especially when ML models inform experimental design or industrial decisions. As AI/ML methods gain momentum in nanofluid research, there's a growing focus on transparent and reproducible validation procedures. Reporting only high-accuracy metrics without detailing the validation process can lead to reproducibility crises, where models fail when applied to new systems or experimental protocols.

## 6. Case Studies

Over the last decade, AI/ML techniques have significantly advanced nanofluid research, driven by the need to understand complex, nonlinear interactions among variables in nanofluids' thermophysical properties. These approaches are essential for predicting thermophysical behaviours that traditional models struggle to capture. Table 4 presents research studies that employ popular AI/ML models to forecast nanofluid thermophysical properties, highlighting their performance metrics and comparing experimental results with predictions. This demonstrates the growing role of AI/ML in enhancing accuracy, efficiency, and scientific insight into nanofluid systems.

As outlined in Section 6.1, traditional nanofluid research, which depends on empirical correlations and purely physics-based models, faces notable hurdles. Experimental data are often limited, expensive to produce, and vary significantly due to differences in nanoparticle preparation and measurement techniques across laboratories. These issues hinder a comprehensive understanding of the complex, nonlinear behaviour of nanofluid systems. To overcome these challenges, AI/ML methods have gained considerable momentum recently. Their capacity to model complex, nonlinear relationships and to incorporate diverse datasets provides a robust alternative to traditional methods. The following case studies exemplify how various AI/ML techniques have been successfully used to predict nanofluid thermophysical properties and enhance understanding of nanofluid behaviour, highlighting their importance as essential tools for research and for the design of practical thermal system applications.

### a. Case Study 1: ANN-Based Viscosity Prediction Across Diverse Nanofluids

Heidari *et al.* (2016) were among the first to utilise machine learning in nanofluid research, creating an MLP-ANN model to predict the relative viscosity of various nanofluids, such as $Al_2O_3$, $TiO_2$, $SiO_2$, and CuO dispersed in different base fluids. Unlike typical studies that focus on a single nanofluid, this pioneering work included five key variables: temperature, nanoparticle volume fraction, particle size, nanoparticle density, and base fluid viscosity. The dataset comprised 1,490

**Table 4** Example of some research studies applying popular AI/ML techniques to nanofluid thermophysical property prediction with the performance metrics and experimental vs prediction results.

| *Study* | *Nanofluid Type* | *Input Variables* | *Property Prediction* | *AI/ML Model Used* | *Performance/Validation Metrics* | | | *Experimental vs. Prediction Results* |
|---|---|---|---|---|---|---|---|---|
| | | | | | *RMSE* | *MAE* | *$R^2$* | |
| (Vaferi *et al.*, 2014) | $Al_2O_3 + H_2O$, EG, PG, EG/$H_2O$ mixtures | T (K), φ (%), Dp (nm), Tc (K), Pc (MPa), ω | μ (mPa·s) | MLPNN, CFNN, RBFNN, LS-SVM | 0.3797 | – | 0.99947 | MLPNN outperformed other models and empirical correlations, showing very high accuracy in predicting μ of various alumina-based nanofluids. |
| (Heidari *et al.*, 2016) | Multiple nanofluids ($Al_2O_3$, $TiO_2$, $SiO_2$, CuO in various base fluids) | T, Nanoparticle Size, φ, ρ, μ | Relative Viscosity ($\mu_{nf}/\mu_{bf}$) | MLP-ANN (Levenberg–Marquardt algorithm) | MSE = $9.7274 \times 10^{-5}$ (Training/ Test, 0.0185 (Validation) | AARD = 0.41% (Training/ Test), 5.47% (Validation) | 0.99998 (Training/ Test) 0.9896 (Validation) | ANN model accurately predicted relative viscosity with <1% deviation for most cases, significantly outperforming classical theoretical models and correlations. |
| (Gholami *et al.*, 2018) | $Al_2O_3$ + various base fluids ($H_2O$, EG, PG, EG/ $H_2O$) | T (K), φ (%), Dp (nm), Tc (K), Pc (MPa), ω | μ (mPa·s) | MLPNN, CFNN, RBFNN, LS-SVM | 0.3797 | – | 0.99947 | MLPNN was identified as the best-performing model for predicting μ, surpassing other AI methods and empirical correlations. |

*Contd.*

**Table 4** *Contd.*

| *Study* | *Nanofluid Type* | *Input Variables* | *Property Prediction* | *AI/ML Model Used* | *Performance/Validation Metrics* | | | *Experimental vs. Prediction Results* |
|---|---|---|---|---|---|---|---|---|
| | | | | | *RMSE* | *MAE* | $R^2$ | |
| (Hemmat Esfe *et al.*, 2018a) | DWCNT–$SiO_2$/ Ethylene Glycol Hybrid Nanofluid | T, φ | TCR | ANN | 0.00000083 | 0.0294 | 0.9991 | ANN provided excellent prediction of TCR with max deviation <0.5%; highest enhancement was 38% over base fluid at φ=1.71%, T=50 °C. |
| (Hemmat Esfe *et al.*, 2018c) | F-SWCNT/EG | T (°C), φ (%) | k (W/m·K) | ANN | $3.67\times10^{-5}$ | <0.009 | 0.9989 | ANN showed very high accuracy in predicting thermal conductivity compared to traditional correlations, with minimal deviation from experimental data. |
| (Hemmat Esfe *et al.*, 2018b) | $ZrO_2$-MWCNT/10W40 hybrid nano-lubricant | T (°C), φ (%), γ ($s^{-1}$) | μ (Pa·s) | ANN (MLP) | $7.0631\times10^{-5}$ | AARD = 0.0051 | 0.9905 | ANN provided superior prediction of viscosity behavior compared to theoretical models, effectively capturing non-Newtonian pseudoplastic trends. |

*Contd.*

**Table 4** *Contd.*

| *Study* | *Nanofluid Type* | *Input Variables* | *Property Prediction* | *AI/ML Model Used* | *Performance/Validation Metrics* | | | *Experimental vs. Prediction Results* |
|---|---|---|---|---|---|---|---|---|
| | | | | | *RMSE* | *MAE* | $R^2$ | |
| (Karimipour *et al.*, 2019) | MWCNT-CuO/ Water Hybrid Nanofluid | T, φ | k | ANN SVR | – | – | ANN ≈ 0.99997 SVR ≈ 0.9979 | Both models predicted k accurately, but SVR demonstrated better generalisation and resistance to overfitting in small datasets |
| (Ahmadi *et al.*, 2019) | $Al_2O_3$/EG nanofluid | T, φ, Particle Size | $k/k_0$ | LSSVM optimized by Genetic Algorithm (GA) | $8.69\times10^{-5}$ | – | 0.9902 | The hybrid GA-LSSVM model accurately predicted the k ratio of $Al_2O_3$/ EG nanofluid, showing strong correlation with experimental data and low error, proving its reliability for modeling k with size, concentration, and T as inputs. |
| (Akhgar *et al.*, 2019) | MWCNT-$TiO_2$/ Water-Ethylene Glycol (50:50) hybrid nanofluid | T, φ | k | ANN | – | – | – | ANN outputs showed maximum deviation of 2.1% compared to experimental data, outperforming correlation model (2.72% max error), demonstrating reliable predictive capability. |

*Contd.*

**Table 4** *Contd.*

| *Study* | *Nanofluid Type* | *Input Variables* | *Property Prediction* | *AI/ML Model Used* | *Performance/Validation Metrics* | | | *Experimental vs. Prediction Results* |
|---|---|---|---|---|---|---|---|---|
| | | | | | *RMSE* | *MAE* | *$R^2$* | |
| (Ghaffarkhah *et al.*, 2019) | COOH-Functionalized MWCNTs (20 vol%) + Oxide nanoparticles (80 vol%) in SAE 40 engine oil | T, φ | Dynamic μ | Decision Tree<br>Random Forest<br>SVM<br>RBF-ANN | – | – | – | Decision Tree and SVM achieved highest prediction accuracy; Random Forest moderate; RBF-ANN lowest accuracy. Available theoretical models (Batchelor, Wang, Esfe) failed to predict μ effectively. |
| (Alade *et al.*, 2020) | CuO/Water Nanofluid | T, φ, Cp of CuO nanoparticles | Cp | SVR<br>ANN | SVR: 0.0023<br>ANN: 0.0025 | – | 0.999 (both models) | Both SVR and ANN accurately predicted Cp; SVR slightly outperformed ANN and showed clear superiority over traditional empirical models. |
| (Gholizadeh *et al.*, 2020) | Various nanofluids ($Al_2O_3$, $SiO_2$, $TiO_2$, CuO, ZnO, MgO, $Fe_3O_4$, etc.) | T, φ, μ, Nanoparticle Size, ρ | μ | RF<br>MLP<br>SVR | 0.139 | – | 0.989 | The RF model provided the most accurate prediction of nanofluid μ, outperforming MLP and SVR. Nanoparticle φ and ρ were identified as the most significant factors affecting μ. |

*Contd.*

**Table 4** *Contd.*

| *Study* | *Nanofluid Type* | *Input Variables* | *Property Prediction* | *AI/ML Model Used* | *Performance/Validation Metrics* | | | *Experimental vs. Prediction Results* |
|---|---|---|---|---|---|---|---|---|
| | | | | | *RMSE* | *MAE* | $R^2$ | |
| (Shahsavar *et al.*, 2020) | Magnetite-water ferrofluid | γ, φ, Magnetic Field (MF) Strength | μ | GMDH-type Neural Network | – | – | 0.9692-0.9772 | The developed ANN model accurately predicted μ under varying magnetic fields and concentrations, demonstrating its capability to capture non-Newtonian behavior of ferrofluids. |
| (Yang *et al.*, 2021) | Water/EG (80:20) + MWCNTs + $TiO_2$ + ZnO | φ, T, NF type | $k_{nf}$ | ANN | 2.04E-05 | – | – | ANN accurately predicted k_nf with max absolute error <0.018, demonstrating excellent agreement with experimental data. |
| (Hemmat Esfe *et al.*, 2022) | 10W40 + MWCNT + MgO (1:1) | φ, T, SR | μ, μr | GA-RBF, LS-SVM, GEP | – | – | 0.9963 (for μ) | ML models reliably predicted viscosity; GA-RBF had best performance, with deviations within ±12%. |

*Contd.*

**Table 4** *Contd.*

| *Study* | *Nanofluid Type* | *Input Variables* | *Property Prediction* | *AI/ML Model Used* | *Performance/Validation Metrics* | | | *Experimental vs. Prediction Results* |
|---|---|---|---|---|---|---|---|---|
| | | | | | *RMSE* | *MAE* | $R^2$ | |
| (Sepehrnia *et al.*, 2022) | 5W30 + ZnO + MWCNT (30:70) | φ, T, $\dot{\gamma}$ | μ | ANN | – | – | – | ANN demonstrated superior prediction capability for μ compared to the novel correlation, capturing non-Newtonian behavior precisely. |
| (Adun *et al.*, 2022) | Various hybrid nanofluids (e.g., $Al_2O_3$/MWCNT, ZnO/MWCNT) | T, φ, ρ, Nanoparticle Size, Mixture Ratio, Acentric Factor | μ | O-GPR<br>BTR<br>ANN<br>SVR | 0.015975 (O-GPR test) | 0.0084404 (O-GPR test) | 0.999998 (O-GPR test) | The O-GPR model provided the most accurate predictions, significantly outperforming ANN, SVR, BTR, and traditional theoretical models. This high accuracy enables precise viscosity estimation for hybrid nanofluids, which is crucial for optimising the thermal efficiency and reducing the pumping power requirements in solar thermal systems. |

*Contd.*

**Table 4** *Contd.*

| *Study* | *Nanofluid Type* | *Input Variables* | *Property Prediction* | *AI/ML Model Used* | *Performance/Validation Metrics* | | | *Experimental vs. Prediction Results* |
|---|---|---|---|---|---|---|---|---|
| | | | | | *RMSE* | *MAE* | *$R^2$* | |
| (Bhaumik *et al.*, 2022) | Water-based nanofluids with $Al_2O_3$, CuO, $TiO_2$ | T, φ | k | Physics-aware Neural Network (Physics-guided MLP) compared with Ridge Regression. Lasso Regression Random Forest XGBoost MLP | 0.0042 W/mK | 0.7075% (MAPE) | 0.9525 | The physics-guided model provided superior accuracy compared to traditional and black-box models, effectively predicting k of nanofluids. |
| (AbuShanab *et al.*, 2023) | Polyalpha-Olefin - hexagonal Boron Nitride (PAO-hBN) | T, γ, φ | Dynamic μ | SVR | 0.0015979 | – | 0.98 | SVR model achieved $R^2 = 0.98$ but had reduced accuracy at lower μ ranges. |
| (AbuShanab *et al.*, 2023) | PAO-hBN | T, γ, φ | Dynamic μ | ANN | 4.50306e−06 (Training) 9.14446e−06 (Validation) 8.74922e−06 (Testing) | – | 0.99994 | ANN model outperformed SVR and correlation model, showing superior predictive accuracy across most ranges. |

*Contd.*

**Table 4** *Contd.*

| *Study* | *Nanofluid Type* | *Input Variables* | *Property Prediction* | *AI/ML Model Used* | *Performance/Validation Metrics* | | | *Experimental vs. Prediction Results* |
|---|---|---|---|---|---|---|---|---|
| | | | | | *RMSE* | *MAE* | *$R^2$* | |
| (AbuShanab *et al.*, 2023) | PAO-hBN | T, γ, φ | Dynamic μ | ANFIS | 0.000782 (Training)<br>0.00123230 (Validation) | – | 1 | ANFIS performed equivalently to ANN but required more computational resources. |
| (AbuShanab *et al.*, 2023) | PAO-hBN | T, φ (Without SR) | Dynamic μ | ANN Optimized | – | – | 0.99994 | Removing SR improved ANN accuracy, reducing absolute relative error to less than 1.89% across full T range. |
| (Borode *et al.*, 2023) | GNP-$Al_2O_3$ / Water Hybrid Nanofluid | T, φ | μ, λ, Electrical Conductivity (σ) | ANFIS (Adaptive Neuro-Fuzzy Inference System) | μ: 0.00305<br>λ: 0.00234<br>σ: - | – | μ: 0.99965<br>λ: 0.99424<br>σ: 0.99995 | ANFIS accurately predicts thermophysical properties; optimal nanoparticle concentration for heat transfer was 0.1-0.2 vol% due to balanced μ increase and thermal performance. |

*Contd.*

**Table 4** *Contd.*

| *Study* | *Nanofluid Type* | *Input Variables* | *Property Prediction* | *AI/ML Model Used* | *Performance/Validation Metrics* | | | *Experimental vs. Prediction Results* |
|---|---|---|---|---|---|---|---|---|
| | | | | | *RMSE* | *MAE* | $R^2$ | |
| (Esfahani, 2023) | Various nanofluids (19 classes) | Nanoparticle diameter, ρ, φ, T | Relative viscosity (μr) | Multi-Fidelity Neural Network (MFNN) | 0.0051 (Adam optimizer) | – | 0.997 (training and test), 0.991 (extrapolation) | MFNN significantly outperforms traditional ANN by incorporating theoretical models and experimental data, providing accurate μ predictions and strong extrapolation capability. |
| (Ganga *et al.*, 2023) | Water-based nanofluids with various nanoparticles | T, φ, k (for k prediction) | μ, k | Decision Tree<br>Random Forest<br>Extra Tree<br>KNN<br>Polynomial Regression | – | – | μ (3-input model, Extra Trees): 0.9771<br><br>μ (2-input model, Decision Tree): 0.9678<br><br>k (Extra Trees): 0.9403 | Extra Trees algorithm provided the most accurate predictions for both μ and k; Decision Tree best for simplified two-input μ model. |
| (Li *et al.*, 2023) | $TiO_2$/Water | T, φ | $k_{nf}$, $\mu_{nf}$ | LS-SVM | 0.00104 ($k_{nf}$), 0.0009 ($\mu_{nf}$) | 0.00092 ($k_{nf}$), 0.00074 ($\mu_{nf}$) | 0.99864 ($k_{nf}$), 0.99995 ($\mu_{nf}$) | LS-SVM provides highest accuracy for predicting both thermal conductivity and viscosity. |

*Contd.*

**Table 4** *Contd.*

| *Study* | *Nanofluid Type* | *Input Variables* | *Property Prediction* | *AI/ML Model Used* | *Performance/Validation Metrics* | | | *Experimental vs. Prediction Results* |
|---|---|---|---|---|---|---|---|---|
| | | | | | *RMSE* | *MAE* | $R^2$ | |
| (Hemmat Esfe *et al.*, 2023) | MWCNT-$Al_2O_3$ (40:60)/Oil 5W50 | T, φ, SR | $\mu_{nf}$ | ANN-MLP | – | – | 0.98 (Train), 0.978 (Test), 0.925 (Validation) | ANN effectively models dynamic viscosity with errors <2.6%. |
| (Said *et al.*, 2023) | f-MWCNTs (functionalized Multi-Walled Carbon Nanotubes) / Water | φ, T, J time, Number of days | k | XGBoost, GPR | XGBoost: 0.0017 (train), 0.001 (test)<br><br>GPR: 0.00255 (train), 0.0022 (test) | XGBoost: 0.0014 (train), 0.0008 (test)<br><br>GPR: 0.002 (train), 0.0018 (test) | XGBoost: 0.9901 (train), 0.9942 (test)<br><br>GPR: 0.9847 (train), 0.9941 (test) | XGBoost outperformed GPR with lower errors and higher $R^2$. Optimal J: 0.10 vol.% concentration for 80 min for stable nanofluid with enhanced k. |
| (Kanti *et al.*, 2024) | Water-based nanofluids containing GO, $SiO_2$, $TiO_2$, GO-$SiO_2$, GO-$TiO_2$ | T, φ | k | DNN, XGBoost | 0.0010 (XGBoost) | 0.0171 (XGBoost) | 0.8616 (XGBoost) | XGBoost provided higher accuracy and lower error compared to DNN; concentration had the strongest effect on TC. |
| (Kanti *et al.*, 2024) | Water-based nanofluids containing GO, $SiO_2$, $TiO_2$, GO-$SiO_2$, GO-$TiO_2$ | T, φ | μ (VST) | DNN, XGBoost | 0.0110 (XGBoost) | 0.0760 (XGBoost) | 0.9122 (XGBoost) | XGBoost outperformed DNN in μ prediction; higher concentration increased VST while higher T decreased it. |

*Contd.*

**Table 4** *Contd.*

| *Study* | *Nanofluid Type* | *Input Variables* | *Property Prediction* | *AI/ML Model Used* | *Performance/Validation Metrics* | | | *Experimental vs. Prediction Results* |
|---|---|---|---|---|---|---|---|---|
| | | | | | *RMSE* | *MAE* | $R^2$ | |
| (Zhu *et al.*, 2024) | EG-$Al_2O_3$ nanofluids | Particle size, T, φ | k | ANN, DT, GBDT, KNN, MLP, XGBoost | 0.0026 (training), 0.0033 (testing) | 0.0009 (training), 0.0028 (testing) | 0.9974 (training), 0.9958 (testing) | GBDT achieved the best performance; T was identified as the most important feature using SHAP analysis |
| (Amoo *et al.*, 2024) | Multiple mono and hybrid nanofluids (e.g., Water + $Al_2O_3$, Water + $Fe_3O_4$ + $SiO_2$) | T, φ, Pressure | ρ, k, Cp | GPR Feed-Forward Neural Network (Bayesian Regularization) | – | – | 0.93 | The developed NanoFEx tool demonstrated over 93% accuracy in predicting thermophysical properties of nanofluids, providing a comprehensive database for both mono and hybrid nanofluids and enabling accurate property estimation where experimental data is lacking. |
| (Kumar K *et al.*, 2024) | GO/MXene (90:10) in vanadium electrolyte | T, φ | k, σ, μ (VST) | LSBoost | 0.00089 (TC) 5.55 (EC) 0.094 (VST) | 0.00069 (TC) 4.09 (EC) 0.061 (VST) | 0.9981 (TC) 0.990 (EC) 0.9954 (VST) | LSBoost with Bayesian optimization provided a highly accurate prediction for thermophysical properties of hybrid nanofluid, demonstrating potential for VRFB applications. |

*Contd.*

**Table 4** *Contd.*

| *Study* | *Nanofluid Type* | *Input Variables* | *Property Prediction* | *AI/ML Model Used* | *Performance/Validation Metrics* | | | *Experimental vs. Prediction Results* |
|---|---|---|---|---|---|---|---|---|
| | | | | | *RMSE* | *MAE* | *$R^2$* | |
| (Qu *et al.*, 2024) | $SiO_2$-$Al_2O_3$-MWCNT / $H_2O$ | φ, T | μ | Curve Fitting | 0.08384 | – | 0.9945 | Developed a nonlinear polynomial model with high accuracy to predict μ of hybrid nanofluid; μ↑ with φ and μ↓ with T |
| (Gao *et al.*, 2024) | MgO / SAE 5W30 Oil | φ, T, γ | μ | RBF-ANN | – | – | 1 | RBF-ANN successfully modeled μ behavior with high prediction accuracy; μ↓ with T and γ, μ↑ with φ |
| (Bagholizadeh *et al.*, 2024) | MWCNT-MgO / SAE 50 Oil | φ, γ, T | μ, τ | MLR, D-Tree, Ridge, PLR, SVM, Lasso, ECR, GPR, MPR, NSGA-II | 7.6639 | – | 0.9983 | MPR with NSGA-II achieved highly precise prediction and optimization of μ and τ for hybrid nano-lubricant |
| (Alyousef *et al.*, 2025) | Various nanofluids (over 110 types) | T, nanoparticle size, φ, base fluid k, nanoparticle k, nanoparticle ρ, nanoparticle shape, nanoparticle specific surface area | k | MLP, XGBOOST, LGBM | Not specified | Not specified | Not specified | MLP achieved the best accuracy. A combination of seven key variables (excluding specific surface area) yielded the most accurate prediction. k of the base fluid was the most influential feature. |

*Contd.*

**Table 4** *Contd.*

| *Study* | *Nanofluid Type* | *Input Variables* | *Property Prediction* | *AI/ML Model Used* | *Performance/Validation Metrics* | | | *Experimental vs. Prediction Results* |
|---|---|---|---|---|---|---|---|---|
| | | | | | *RMSE* | *MAE* | $R^2$ | |
| (Jamei *et al.*, 2025) | $Mg(OH)_2$ + MWCNT / Engine Oil Hybrid Nanofluid | T, φ | k (THC), μ, Heat Transfer Performance under Laminar (HTPL) and Turbulent Flow (HTPT) | ANFIS-TLBO<br>ANN<br>ANFIS<br>MARS | 0.00086 (THC)<br>3.15031 (μ)<br>0.03026 (HTPL)<br>0.00419 (HTPT) | – | 0.99888 (THC)<br>0.99959 (μ)<br>0.99580 (HTPL)<br>0.99306 (HTPT) | ANFIS-TLBO provided the most accurate predictions for thermophysical properties and HTP, outperforming ANN, ANFIS, and MARS. |
| (Aktary *et al.*, 2025) | Cu-MgO/$H_2O$ | Rayleigh Number (Ra), Hartmann Number (Ha), φ | $Nu_{av}$ | ANN, RSM | – | – | 0.9843 | ANN and RSM effectively modeled natural convection in a trapezoidal cavity, with ANN showing strong predictive ability for $Nu_{av}$. |
| (Kholoud *et al.*, 2025) | MWCNT-$Al_2O_3$/ Water-EG (80:20) | φ, T, γ | μ | CFNN | – | 0.4992 | 0.999 (training), 0.997 (testing) | CFNN successfully predicted dynamic viscosity across varying conditions, demonstrating high accuracy and minimal error. |

experimental data points from different literature sources, offering a thorough view of nanofluid behaviour. The model achieved high predictive accuracy, with an $R^2$ of 0.99998 for both training and testing datasets and a mean absolute relative deviation (MARD) of 0.41%, greatly exceeding traditional empirical correlations. Even on unseen data, the model performed well ($R^2 = 0.9896$). These findings demonstrate how ANN models effectively capture the complex nonlinear, multivariate interactions between nanoparticles and base fluids, especially at higher concentrations where conventional regression methods often fall short fail.

Further research broadened the scope and enhanced the capabilities of ANN-based methods for predicting viscosity. Hemmat Esfe *et al.* (2022, 2023a, 2023b) utilised ANN and other AI/ML techniques, such as GA-RBF, LS-SVM, and GEP, to estimate the dynamic and relative viscosities of engine oil-based hybrid nanofluids containing MWCNT, MgO, and $Al_2O_3$. These models achieved $R^2$ values over 0.996, showing excellent alignment with experimental results across various shear rates, temperatures, and nanoparticle concentrations. Likewise, Sepehrnia *et al.* (2022) demonstrated that ANN models can effectively model the non-Newtonian behaviour of multi-component nanofluids, thereby enhancing predictive accuracy over newly developed empirical method correlations.

Recent studies highlight the scalability and versatility of ANN models. For instance, Li *et al.* (2023) utilised an LS-SVM approach to simultaneously predict the thermal conductivity and dynamic viscosity of $TiO_2$/water nanofluids, achieving $R^2$ values of 0.99995 for viscosity and 0.99864 for thermal conductivity. These multi-output models exemplify how ANN frameworks extend beyond single-property predictions to facilitate integrated thermal management design. In advanced applications, Kholoud *et al.* (2025) employed a CFNN to model the rheological behaviour of MWCNT-$Al_2O_3$/water-EG hybrid nanofluids, attaining near-perfect predictions with $R^2 = 0.999$ for training and $R^2 = 0.997$ for testing. This demonstrates the capability of modern ANN architectures to deliver both accuracy and stability, even with complex hybrid systems of nanofluids.

**b. Case Study 2: LSSVM for Accurate Thermal Conductivity Prediction with Moderate Datasets**

Obtaining large, high-quality experimental datasets for nanofluid property prediction is often difficult due to high costs, lengthy procedures, and the complexity of measuring variables such as thermal conductivity and viscosity. In such situations, machine learning models that can effectively generalise from limited data are essential. Recent research has highlighted the robustness of LSSVM for this application. For instance, Li *et al.* (2023) employed LSSVM to predict thermal conductivity and dynamic viscosity of $TiO_2$/water nanofluids, using a moderately sized experimental dataset. The model utilised temperature and nanoparticle volume fraction as inputs, successfully capturing the nonlinear relationships between these factors and the properties. Their results showed that LSSVM achieved $R^2$ values of 0.99864 for thermal conductivity and 0.99995 for viscosity, with low RMSE values (0.00104 and 0.0009). This performance greatly exceeded traditional correlations and was comparable to more complex ANN models, illustrating that LSSVM is highly effective for moderate datasets where deep learning risks overfitting underperform.

Other studies, like those by Esfe *et al.* (2018, 2022, 2023), have integrated LS-SVM with hybrid modelling approaches, comparing its performance to other AI/ML techniques such as ANN, GA-RBF, and GEP. While ensemble or deep learning models sometimes showed slightly better results, LS-SVM consistently proved to be robust in generalisation and computational efficiency, making it particularly suitable for situations with limited data. As outlined in Table 4, LSSVM has become an important tool in nanofluid research, filling the gap between traditional empirical models and data-driven deep learning methods. Using kernel functions, LSSVM maps input variables like temperature, nanoparticle size, and volume fraction into a higher-dimensional space, capturing the complex nonlinear behaviours typical of nanofluids. This allows researchers to make reliable property predictions early in the research process, aiding in experimental design and resource optimisation use.

**c. Case Study 3: Ensemble Methods for High Accuracy and Input Factor Analysis**

While ANN and SVR models are well-known for their strong predictive performance, their black-box nature often limits their interpretability. Recent studies have turned to ensemble learning algorithms, which not only deliver high predictive accuracy but also provide insights into feature importance, offering researchers guidance on which factors most influence nanofluid thermophysical properties. Zhu *et al.* (2024) exemplified this approach by evaluating six machine learning models (ANN, DT, GBDT, KNN, MLP, and XGBoost) to predict the thermal conductivity of $Al_2O_3$/ethylene glycol nanofluids. Their dataset consisted of 94 experimental data points with temperature, nanoparticle volume fraction, and particle size as input features.

Among these models, GBDT achieved the highest predictive performance, with an $R^2$ of 0.9958, an RMSE of 0.0033, and an MAE of 0.0028, outperforming both the ANN and traditional regression models. Beyond prediction, the study employed SHAP to interpret model outputs. The analysis revealed that temperature was the primary influence on thermal conductivity, along with volume fraction and particle size. This interpretability is essential for experimental design, as it highlights which variables should be prioritised in future research to optimise nanofluid properties. Other ensemble-based studies listed in Table 4 show similar advancements. Kumar *et al.* (2024) utilised LSBoost, an ensemble boosting technique, to predict multiple properties such as thermal conductivity, electrical conductivity, and viscosity for a GO/MXene hybrid nanofluid. Their model demonstrated very high accuracy ($R^2 > 0.99$ for all outputs) and supported multi-output prediction, showcasing the versatility of ensemble methods for complex, multi-parameter systems. Furthermore, Baghoolizadeh *et al.* (2024) combined ensemble learning with multi-objective optimisation (NSGA-II) to predict and optimise dynamic viscosity and torque in MWCNT-MgO hybrid nano-lubricants.

This approach not only achieved excellent accuracy ($R^2 = 0.9983$) but also helped identify the best experimental conditions for desired performance metrics.

**d. Case Study 4: Physics-Aware Neural Networks for Hybrid Modelling**

Purely data-driven machine learning models often struggle to produce physically consistent predictions, especially when experimental data is limited, noisy, or sparsely distributed across the input space. This issue is particularly critical in nanofluid research, where thermophysical properties such as viscosity and thermal conductivity depend on complex, nonlinear interactions that must still obey fundamental thermodynamic and physical laws. To address these challenges, Bhaumik *et al.* (2022) proposed a PANN framework for predicting nanofluid viscosity. This hybrid method blends the flexibility of deep learning with domain knowledge by explicitly incorporating thermodynamic relationships into the loss function. The PANN model simultaneously minimises two objectives: data loss, which reduces the difference between predicted and experimental viscosity for statistical accuracy; and physics loss, which discourages violations of thermodynamic principles, ensuring physically consistent predictions even outside the tested conditions regions.

The study showed that PANN outperformed a traditional ANN, with an $R^2$ of 0.996 versus 0.989 for the standard ANN. This advantage was most noticeable at low nanoparticle concentrations and extreme temperatures, where experimental data were scarce. While the standard ANN drifted from expected asymptotic trends, the PANN exhibited physically consistent behaviour, highlighting its better generalisation. These findings are supported by other recent studies listed in Table 4. Kumar *et al.* (2024) used ensemble learning (LSBoost) with domain-specific constraints to accurately predict various thermophysical properties (thermal conductivity, electrical conductivity, and viscosity) of GO/MXene-based hybrid nanofluids, achieving $R^2$ values above 0.99. Similarly, Shirmahd *et al.* (2025) combined ANN-MLP with Monte Carlo simulation data to model the magnetic properties of multi-layered nanostructures, showing how physics-guided ML can go beyond fluid properties to nanomaterial design. Integrating physics-based knowledge with data-driven models, PANNs, and other hybrid approaches paves the way for next-generation predictive tools in nanofluids research.

The case studies discussed above collectively emphasise the varied strengths and complementary roles of modern AI/ML techniques in nanofluid research. No single algorithm is universally the best; the optimal choice depends on factors such as dataset size, complexity, prediction objectives, and the need for physical interpretability. Generally, ANNs excel with large, diverse datasets, while models such as LSSVM are more suitable for moderate-sized datasets where deep learning might overfit. Ensemble methods strike a balance between accuracy and interpretability, enabling researchers to identify which experimental factors most significantly influence thermophysical properties. Meanwhile, physics-aware or hybrid networks combine data-driven flexibility with physical consistency, ensuring realistic predictions even with limited or extrapolated data. Across all approaches, the studies summarised in Table 4 highlight the importance of thorough validation and error analysis (Section 5) to ensure reliability and scientific credibility, especially when working with noisy or limited experimental data. Additionally, although emerging techniques like GANs (Section 4.4) are not yet explored for nanofluids, they show potential for future applications in synthetic data generation, dataset balancing, and addressing data scarcity issues. Integrating these methods with experimental nanoscience can foster the development of smarter, more efficient thermal systems and expand the horizons of nanofluid research.

## 7. Challenges and Opportunities

Although AI/ML approaches have made significant progress in predicting nanofluid thermophysical properties, particular challenges continue to restrict the full potential of AI/ML in this area. At the same time, emerging trends and technological advancements provide opportunities to overcome these limitations and transform the way nanofluids are researched, designed, and optimised.

### *7.1 Challenges*

a. Data Scarcity: One significant challenge is the shortage of high-quality, standardised datasets. The case studies in Section 6.6 demonstrate that many nanofluid experiments are costly and time-consuming, yielding small datasets that are often highly uneven across nanoparticle types, base fluids, and operating conditions. This scarcity makes it challenging for data-hungry algorithms, such as deep ANNs or gradient boosting models, to generalise effectively.
b. Overfitting and Lack of Generalisability: Overfitting remains a significant problem in nanofluid AI/ML modelling. Numerous studies, particularly those using high-capacity models such as deep artificial neural networks, produce outstanding results on training data but struggle to generalise to new experimental conditions. This was evident in several early studies where models performed well under controlled conditions but failed when applied to different nanoparticle formulations or temperature ranges. However, such practices are not always consistently employed in research, leading to exaggerated performance claims.
c. Lack of Interpretability: Many powerful algorithms, including ANNs and deep learning architectures, are often labelled as "black boxes." Although they excel at capturing complex, non-linear patterns, they provide little insight into why a prediction is made. This lack of transparency hinders scientific understanding and erodes trust in AI-driven results. Without interpretable models, there's a risk that researchers may prioritise solely predictive accuracy over the underlying physical relationship mechanisms.

d. Computational and Resource Constraints: Advanced AI/ML methods, rigorous learning, and hybrid models such as PINNs, as well as generative approaches like GANs, are computationally intensive. Training these models requires high-performance computing resources and specialist expertise, which are not available to many research laboratories. It requires a complex implementation and lengthy training times. Therefore, this kind of computational barrier may limit widespread adoption and contribute to disparities.
e. Ethical and Validation Concerns for Synthetic Data: AI/ML approaches raise ethical and scientific concerns. For example, if using synthetic data that is not thoroughly validated, this can introduce hidden biases or unrealistic patterns into downstream models. Furthermore, synthetic datasets may unintentionally conceal gaps in experimental knowledge, leading to overconfidence in models trained on artificially balanced data. These concerns underscore the importance of transparency and reproducibility when utilising synthetic data in nanofluid research.

## 7.2 *Opportunities*

a. Integration of Physics-Based Knowledge: One of the most transformative trends is the development of hybrid models that integrate data-driven learning with physical laws. ANNs and gradient boosting frameworks excel at capturing complex nonlinear relationships. However, they are susceptible to generating physically implausible predictions, particularly when data are sparse or noisy. This limitation can be mitigated by incorporating domain knowledge directly into the AI/ML architecture, thereby ensuring that outputs are both accurate and consistent with established scientific principles. Future developments could build on these methods to model multi-physics systems, combining heat transfer, mass diffusion, and fluid dynamics.
b. Standardisation and Open Data Initiatives: Developing standardised, high-quality datasets is crucial for advancing AI/ML applications in nanofluid research. Currently, inconsistencies hinder progress due to variations in experimental protocols, measurement techniques, and incomplete reporting of key variables. These issues introduce noise and fragmentation into datasets, limiting the ability of AI/ML models to generalise effectively. Addressing these challenges could be achieved by the research community adopting standardised reporting guidelines for experimental variables. Standardisation would allow for the creation of benchmark datasets featuring a range of nanofluid systems under controlled conditions, enabling fair comparisons between algorithms and promoting reproducibility. Furthermore, quality control measures such as replication studies and metadata tracking would enhance reliability and transparency.
c. XAI for Scientific Insight: Whilst predictive accuracy is crucial, the real value of AI/ML lies in its capacity to generate scientific understanding.

Many of the most powerful models, such as ANNs, are often criticised for their lack of transparency, which reduces their usefulness for hypothesis development and experimental planning. The growing field of XAI provides tools to address this issue by making model outputs more transparent and understandable (Kanti *et al.*, 2024). Techniques like SHAP offer insights into how specific input variables affect predictions. This enables researchers to identify key experimental variables for optimisation, spot systematic errors or biases within datasets, and develop mechanistic hypotheses regarding nanoparticle interactions and stability.

d. GenAI for Accelerated Discovery: GenAI, although not yet adopted in nanofluid thermophysical property prediction studies, presents a promising solution to one of the most persistent challenges in nanofluid research. New GenAI models, such as GANs, have the potential to generate realistic synthetic datasets, thereby greatly expanding the range and diversity of data available for AI/ML applications. By supplementing limited experimental measurements with carefully validated synthetic data, these techniques could help reduce overfitting and improve the robustness of downstream predictive algorithms, such as gradient boosting models or hybrid neural networks. GenAI also opens up opportunities for inverse design. This approach lets researchers virtually explore untested combinations of nanoparticle types, base fluids, and preparation conditions to achieve specific performance targets.
e. Industrial Applications and Scale-Up: For the success of AI-driven nanofluid research, it must have a significant impact on large-scale industrial applications. Sectors such as electronic thermal management, renewable energy systems, and advanced manufacturing stand to gain substantially from the predictive and optimisation capabilities of AI/ML models. These tools facilitate bridging laboratory experimentation with real-world applications by forecasting performance across varied operational conditions, optimising formulations for cost efficiency and enduring stability, and reducing both time and costs associated with iterative industrial process testing. Upon the generation of industrial-scale datasets, transfer learning techniques can be employed to adapt academic models for commercial deployment, thereby ensuring that the advantages of AI/ML are realised throughout the innovation of nanofluid research.

As discussed in Section 1, the field of nanofluid research is currently at a pivotal stage. The initial successes in case studies (Section 6) demonstrate that AI/ML approaches can achieve cutting-edge predictive performance. However, to fully realise this potential, researchers must adopt a more systematic, collaborative approach that incorporates standardisation, explainability, and hybrid modelling. Table 5 summarises the main challenges in applying AI/ML models to nanofluid property prediction, along with actionable opportunities and relevant techniques that could be adopted.

**Table 5** Summary of the main challenges encountered when applying AI/ML methods to nanofluid thermophysical property prediction, along with corresponding opportunities and recommended future techniques

| *Main Challenges* | *Impact* | *Opportunities* | *Relevant Implementable AI/ ML Techniques* |
|---|---|---|---|
| Data Scarcity | Limited ability to train generalisable ML models; overfitting risk. | Use synthetic datasets via generative models and expand curated databases. | GANs, CTAB-GAN, NanoFEx database initiatives |
| Overfitting and Poor Generalisation | Models perform well on training data but fail on unseen datasets. | Nested cross-validation, external validation, and physics constraints in models. | PINNs, group-split CV, physics-guided DNNs |
| Lack of Interpretability | Stakeholders cannot trust or deploy models confidently. | Use XAI tools to reveal feature importance and decision logic. | SHAP values, XGBoost feature importance |
| Inconsistent Experimental Reporting | Difficult to reproduce datasets and replicate models. | Adopt FAIR (Findable, Accessible, Interoperable, Reusable) principles and standardised metadata protocols. | FAIR-compliant reporting templates, centralised repositories |
| Bias Amplification in Synthetic Data | GAN-generated data may replicate or amplify existing biases. | Regular audits and validation of generated data; diversity checks. | CTAB-GAN, statistical bias detection algorithms |
| Integration of Classical Physics | Purely data-driven models lack physical realism. | Combine physics-based constraints with AI/ML algorithms. | PINNs |

## 8. Conclusion

AI/ML methodologies have rapidly transitioned from emerging computational techniques to influential catalysts of innovation within the field of nanofluid research. These methodologies have demonstrated their capacity to model highly intricate, non-linear interactions among nanoparticle characteristics, base fluid properties, and preparation variables, providing predictive capabilities that markedly exceed those of traditional empirical correlations and analytical models. Through data-driven modelling, AI/ML not only offers precise property predictions but also reveals latent patterns and relationships within experimental data. By employing various algorithms, including ANN, SVM, ensemble approaches such as RF and XGBoost, as well as hybrid models informed by physical principles, the understanding of nanofluid behaviour has been substantially enhanced. The advent of PINNs serves to bridge the divide between opaque AI/ML models and scientific rigour, thereby ensuring that predictions adhere to fundamental thermodynamic and fluid dynamic principles. Concurrently, GenAI models extend opportunities to mitigate data scarcity by generating synthetic yet realistic datasets, facilitating the exploration of untested formulations and supporting inverse design methodologies that expedite innovation.

Although notable progress has been made, challenges persist. The deficiency of comprehensive, high-quality datasets remains a significant obstacle, compounded by inconsistent reporting standards and variations in experimental methodologies. Additionally, the computational requirements pose substantial difficulties, especially for advanced model architectures and hybrid frameworks, which demand considerable resources and expertise. Moreover, the lack of transparency in many AI/ML models raises concerns regarding their interpretability, whilst the ethical considerations associated with the generation of synthetic data warrant careful deliberation and rigorous validation protocols. Overcoming these challenges will require international collaboration on standardisation, open data sharing, and the promotion of reproducible research practices.

Nonetheless, opportunities exist beyond conventional boundaries. As collaborative frameworks expand, AI-driven nanofluid research is poised to transition from isolated proof-of-concept studies to fully integrated design pipelines. The prospective future may encompass autonomous research ecosystems in which AI/ML models suggest novel nanofluid formulations, and predictive algorithms assess their properties. Automated laboratories can conduct validation experiments in real-time. Such systems would substantially accelerate discoveries in nanofluid research while simultaneously reducing costs and environmental impact. Furthermore, the application of XAI techniques can enhance transparency and diminish bias, thereby transforming AI/ML from a mere predictive instrument into a catalyst for fundamental scientific advancement and insight.

AI/ML algorithms represent a considerable departure from conventional approaches, rather than incremental advancements, and are poised to fundamentally transform research on nanofluids. By integrating data-driven intelligence with physics-based insights and adhering to rigorous scientific ethics, researchers can achieve substantial breakthroughs in thermal management, renewable energy, and advanced manufacturing. This development has the potential to yield optimised, sustainable solutions that address the requirements of contemporary technology while also deepening the comprehension of the scope of nanofluid applications behaviours.

## References


[1] AbuShanab, Y., Al-Ammari, W.A., Gowid, S. and Sleiti, A.K., 2023. Accurate prediction of dynamic viscosity of polyalpha-olefin boron nitride nanofluids using machine learning. *Heliyon* 9, e16716.

[2] Adun, H., Wole-Osho, I., Okonkwo, E.C., Ruwa, T., Agwa, T., Onochie, K., Ukwu, H., Bamisile, O. and Dagbasi, M. 2022. Estimation of thermophysical property of hybrid nanofluids for solar Thermal applications: Implementation of novel Optimizable Gaussian Process regression (O-GPR) approach for Viscosity prediction. *Neural Comput Appl* 34, 11233–11254.

[3] Ahmadi, M.H., Ahmadi, M.A., Nazari, M.A., Mahian, O. and Ghasempour, R. 2019. A proposed model to predict thermal conductivity ratio of Al2O3/EG nanofluid by applying least squares support vector machine (LSSVM) and genetic algorithm as a connectionist approach. *J Therm Anal Calorim* 135, 271–281.

[4] Ahmadi, M.H., Mohseni-Gharyehsafa, B., Ghazvini, M., Goodarzi, M., Jilte, R.D. and Kumar, R. 2020. Comparing various machine learning approaches in modeling the dynamic viscosity of CuO/water nanofluid. *J Therm Anal Calorim* 139, 2585–2599.

[5] Akhgar, A., Toghraie, D., Sina, N. and Afrand, M. 2019. Developing dissimilar artificial neural networks (ANNs) to prediction the thermal conductivity of MWCNT-TiO2/Water-ethylene glycol hybrid nanofluid. *Powder Technol* 355, 602–610.

[6] Aktary, R., Islam, S., Ray, S.C., Khatun, S., Kundu, R. and Biswas, A. 2025. Thermal performance analysis of Cu-MgO-HO hybrid nanofluid in a partially heated trapezoidal cavity: FEM, RSM and ANN approaches. *Thermal Advances* 4, 100062.

[7] Alade, I.O., Abd Rahman, M.A., Abbas, Z., Yaakob, Y. and Saleh, T.A. 2020. Application of support vector regression and artificial neural network for prediction of specific heat capacity of aqueous nanofluids of copper oxide. *Solar Energy* 197, 485–490.

[8] Ali, N., Teixeira, J.A. and Addali, A. 2019. Effect of Water Temperature, pH Value, and Film Thickness on the Wettability Behaviour of Copper Surfaces Coated with Copper Using EB-PVD Technique. *Journal of Nano Research* 60, 124–141.

[9] Aljuwayhel, N.F., Ali, N. and Bahman, A.M. 2023. Experimental evaluation of split air conditioning performance using nanodiamonds particles in compressor polyester lubricant oil. *Appl Therm Eng* 231, 120961.

[10] Alyousef, B.A., Abdulhay, E.W. and Khnouf, R.E. 2025. Modeling and Predicting of Thermal Conductivity of Nanofluids Using an Ensemble of Different Tree-based Gradient Boosting Algorithms, *Jordanian Journal of Engineering and Chemical Industries.*

[11] Amoo, O.M., Ajiboye, A. and Oyewola, M.O. 2024. Analysis of thermophysical and transport properties of nanofluids using machine learning algorithms. *International Journal of Thermofluids* 21, 100566.

[12] Angayarkanni, S.A. and Philip, J. 2015. Review on thermal properties of nanofluids: Recent developments. *Adv Colloid Interface Sci* 225, 146–176.

[13] Baghoolizadeh, M., Pirmoradian, M., Sajadi, S.M., Salahshour, S. and Baghaei, S. 2024. Prediction and extensive analysis of MWCNT-MgO/oil SAE 50 hybrid nano-lubricant rheology utilizing machine learning and genetic algorithms to find ideal attributes. *Tribol Int* 195.

[14] Bhaumik, B., Changdar, S. and De, S. 2022. An Expert Model Based on Physics-Aware Neural Network for the Prediction of Thermal Conductivity of Nanofluids. *J Heat Transfer* 144.

[15] Borode, A., Tshephe, T., Olubambi, P., Sharifpur, M. and Meyer, J. 2023. Experimental study and ANFIS modelling of the thermophysical properties and efficacy of GNP-$Al_2O_3$ hybrid nanofluids of different concentrations and temperatures. *SN Appl Sci* 5, 337.

[16] Breiman, L. 2001. Random Forests. *Mach Learn* 45, 5–32.

[17] Chai, T. and Draxler, R.R. 2014. Root mean square error (RMSE) or mean absolute error (MAE)? – Arguments against avoiding RMSE in the literature. *Geosci Model Dev* 7, 1247–1250.

[18] Chandrasekhar, A., Farimani, O.B., Ajenifujah, O.T., Ock, J. and Farimani, A.B. 2025. NANOGPT: A Query-Driven Large Language Model Retrieval-Augmented Generation System for Nanotechnology Research.

[19] Channi, H.K., Kaur, A. and Kaur, S. 2025. AI-Driven Generative Design Redefines the Engineering Process, in: Generative Artificial Intelligence in Finance. *Wiley*, pp. 327–359.

[20] Chen, T. and Guestrin, C. 2016. XGBoost, in: Proceedings of the 22nd ACM SIGKDD International Conference on Knowledge Discovery and Data Mining. *ACM, New York*, NY, USA, p. 785

[21] Choi S.U.S and Eastman, J. 1995. Enhancing thermal conductivity of fluids with nanoparticles. *ASME International Mechanical Engineering Congress & Exposition,*

[22] Dan, Y., Zhao, Y., Li, X., Li, S., Hu, M. and Hu, J. 2020. Generative adversarial networks (GAN) based efficient sampling of chemical composition space for inverse design of inorganic materials. *NPJ Comput Mater* 6, 84.

[23] Esfahani, C.I. 2023. A data-driven physics-informed neural network for predicting the viscosity of nanofluids. *AIP Adv* 13.

[24] Friedman, J.H. 2001. Greedy Function Approximation: A Gradient Boosting Machine, Source: The Annals of Statistics.

[25] Ganga, S., Uddin, Z., Asthana, R., Hassan, H. and Bhardwaj, A. 2023. Modelling of Viscosity and Thermal Conductivity of Water-Based Nanofluids using Machine-Learning Techniques. *International Journal of Mathematical, Engineering and Management Sciences* 8, 817–840.

[26] Gao, J., Jasim, D.J., Sajadi, S.M., Eftekhari, S.A., Hekmatifar, M., Salahshour, S., Shahdost, F.T. and Toghraie, D., 2024. An RBF-based artificial neural network for prediction of dynamic viscosity of MgO/SAE 5W30 oil hybrid nano-lubricant to obtain the best performance of energy systems. *Mater Today Commun* 38.

[27] Ghaffarkhah, A., Bazzi, A., Azimi Dijvejin, Z., Talebkeikhah, M., Keshavarz Moraveji, M. and Agin, F. 2019. Experimental and numerical analysis of rheological characterization of hybrid nano-lubricants containing COOH-Functionalized MWCNTs and oxide nanoparticles. *International Communications in Heat and Mass Transfer* 101, 103–115.

[28] Gholami, E., Vaferi, B. and Ariana, M.A. 2018. Prediction of viscosity of several alumina-based nanofluids using various artificial intelligence paradigms – Comparison with experimental data and empirical correlations. *Powder Technol* 323, 495–506.

[29] Gholizadeh, M., Jamei, M., Ahmadianfar, I. and Pourrajab, R. 2020. Prediction of nanofluids viscosity using random forest (RF) approach. *Chemometrics and Intelligent Laboratory Systems* 201, 104010.

[30] Göpfert, J., Weinand, J.M., Kuckertz, P. and Stolten, D. 2024. Opportunities for large language models and discourse in engineering design. Energy and AI 17, 100383.

[31] Hastie, T., Friedman, J. and Tibshirani, R. 2001. *The Elements of Statistical Learning*. Springer New York, New York, NY.

[32] Haykin, S.S., 2009. *Neural networks and learning machines*. Pearson.

[33] Heidari, E., Sobati, M.A. and Movahedirad, S. 2016. Accurate prediction of nanofluid viscosity using a multilayer perceptron artificial neural network (MLP-ANN). *Chemometrics and Intelligent Laboratory Systems* 155, 73–85.

[34] Hemmat Esfe, M., Abbasian Arani, A.A., Shafiei Badi, R. and Rejvani, M. 2018a. ANN modeling, cost performance and sensitivity analyzing of thermal conductivity of DWCNT–SiO2/EG hybrid nanofluid for higher heat transfer. *J Therm Anal Calorim* 131, 2381–2393.

[35] Hemmat Esfe, M., Esmaily, R., Khabaz, M.K., Alizadeh, A., Pirmoradian, M., Rahmanian, A. and Toghraie, D., 2023. A novel integrated model to improve the dynamic viscosity of

MWCNT-Al2O3 (40:60)/Oil 5W50 hybrid nano-lubricant using artificial neural networks (ANNs). *Tribol Int* 178.

[36] Hemmat Esfe, M., Motallebi, S.M. and Toghraie, D. 2022. Investigation of thermophysical properties of MWCNT-MgO (1:1)/10W40 hybrid nanofluid by focusing on the rheological behavior: Sensitivity analysis and price-performance investigation. *Powder Technol* 405.

[37] Hemmat Esfe, M., Rostamian, H., Rejvani, M. and Emami, M.R.S. 2018b. Rheological behavior characteristics of ZrO2-MWCNT/10w40 hybrid nano-lubricant affected by temperature, concentration, and shear rate: An experimental study and a neural network simulating. *Physica E Low Dimens Syst Nanostruct* 102, 160–170.

[38] Hemmat Esfe, M., Sarmasti Emami, M.R. and Kiannejad Amiri, M. 2019. Experimental investigation of effective parameters on MWCNT–TiO2/SAE50 hybrid nanofluid viscosity. *J Therm Anal Calorim* 137, 743–757.

[39] Hemmat Esfe, M., Tatar, A., Ahangar, M.R.H. and Rostamian, H. 2018c. A comparison of performance of several artificial intelligence methods for predicting the dynamic viscosity of TiO2/SAE 50 nano-lubricant. *Physica E Low Dimens Syst Nanostruct* 96, 85–93.

[40] Jamei, M., Alkouh, A.B., Karbasi, M. and Yaseen, Z.M. 2025. Thermo-physical properties estimation of an oil-based hybrid nanofluid: application of a new hybrid neurocomputing approach. *J Therm Anal Calorim* 150, 2295–2318.

[41] Jamei, M. and Said, Z. 2022. Recent advances in the prediction of thermophysical properties of nanofluids using artificial intelligence, in: *Hybrid Nanofluids*. Elsevier, pp. 203–232.

[42] Jiang, X., Wang, W., Tian, S., Wang, H., Lookman, T., Su, Y., 2025. Applications of natural language processing and large language models in materials discovery. NPJ Comput Mater 11, 79.

[43] Kanti, P.K., Prabhakar Sharma, Wanatasanappan, V.V. and Said, N.M. 2024. Explainable machine learning techniques for hybrid nanofluids transport characteristics: an evaluation of shapley additive and local interpretable model-agnostic explanations. *J Therm Anal Calorim* 149, 11599–11618.

[44] Karimipour, A., Bagherzadeh, S.A., Taghipour, A., Abdollahi, A. and Safaei, M.R. 2019. A novel nonlinear regression model of SVR as a substitute for ANN to predict conductivity of MWCNT-CuO/water hybrid nanofluid based on empirical data. *Physica A: Statistical Mechanics and its Applications* 521, 89–97.

[45] Ke, G., Meng, Q., Finley, T., Wang, T., Chen, W., Ma, W., Ye, Q. and Liu, T.-Y. 2017. LightGBM: A Highly Efficient Gradient Boosting Decision Tree. *Long Beach*, CA, USA.

[46] Kholoud, M.J., Mahmoudi Soumar, N., Torabi, A., Ghahderijani, M.J. and Dehkordi, M.H.R. 2025. Rheological modeling of MWCNT-$Al_2O_3$ hybrid nanofluid using cascade forward neural network method. *Diam Relat Mater* 151.

[47] Kumar, K.P., Deepthi Jayan, K., Sharma, P. and Alruqi, M. 2024. Thermo-electro-rheological properties of graphene oxide and MXene hybrid nanofluid for vanadium redox flow battery: Application of explainable ensemble machine learning with hyperparameter optimization. FlatChem 43.

[48] Li, J., Deng, W., Qing, S., Liu, Y., Zhang, H. and Zheng, M. 2023. Prediction and Optimization of the Thermal Properties of TiO2/Water Nanofluids in the Framework of a Machine Learning Approach. *Fluid Dynamics and Materials Processing* 19, 2181–2200.

[49] Lumumba, V.W., Kiprotich, D., Lemasulani Mpaine, M., Grace Makena, N. and Daniel Kavita, M. 2025. Comparative Analysis of Cross-Validation Techniques: LOOCV, K-folds Cross-Validation, and Repeated K-folds Cross-Validation in Machine Learning Models. *SSRN Electronic Journal.*

[50] Miletic, M. and Sariyar, M. 2024. Challenges of Using Synthetic Data Generation Methods for Tabular Microdata. *Applied Sciences* 14, 5975.

[51] Misbah, B., Sedaghat, A., Rashidi, M., Sabati, M., Vaidyan, K., Ali, N., Omar, M.A.A. and Hosseini Dehshiri, S.S. 2022. Friction reduction of $Al_2O_3$, $SiO_2$ and $TiO_2$ nanoparticles added to non-Newtonian water based mud in a rotating medium. *J Pet Sci Eng* 217, 110927.

[52] Onyiriuka, E. 2023. Modeling nanofluid viscosity: comparing models and optimizing feature selection—a novel approach. *Bull Natl Res Cent* 47, 139.

[53] Peng, J.-Z., Hua, Y., Li, Y.-B., Chen, Z.-H., Wu, W.-T. and Aubry, N. 2023. Physics-informed graph convolutional neural network for modeling fluid flow and heat convection. *Physics of Fluids* 35.

[54] Qu, M., Jasim, D.J., Alizadeh, A., Ali Eftekhari, S., Nasajpour-Esfahani, N., Zekri, H., Salahshour, S. and Toghraie, D. 2024. A new model for viscosity prediction for silica-alumina-MWCNT/Water hybrid nanofluid using nonlinear curve fitting. *Engineering Science and Technology, an International Journal* 50.

[55] Raissi, M., Perdikaris, P. and Karniadakis, G.E. 2019. Physics-informed neural networks: A deep learning framework for solving forward and inverse problems involving nonlinear partial differential equations. *J Comput Phys* 378, 686–707.

[56] Said, Z., Sharma, P., Bora, B.J. and Pandey, A.K., 2023. Sonication impact on thermal conductivity of f-MWCNT nanofluids using XGBoost and Gaussian process regression. *J Taiwan Inst Chem Eng* 145, 104818.

[57] Sepehrnia, M., Mohammadzadeh, K., Veyseh, M.M., Agah, E. and Amani, M. 2022. Rheological behavior of engine oil based hybrid nanofluid containing MWCNTs and ZnO nanopowders: Experimental analysis, developing a novel correlation, and neural network modeling. *Powder Technol* 404.

[58] Shahsavar, A., Khanmohammadi, S., Afrand, M., Shahsavar Goldanlou, A. and Rosatami, S., 2020. On evaluation of magnetic field effect on the formation of nanoparticles clusters inside aqueous magnetite nanofluid: An experimental study and comprehensive modeling. *J Mol Liq* 312, 113378.

[59] Shahul Hameed, M.A., Qureshi, A.M. and Kaushik, A. 2024. Bias Mitigation via Synthetic Data Generation: A Review. Electronics (Basel) 13, 3909.

[60] Sharma, P., Said, Z., Kumar, A., Nižetić, S., Pandey, A., Hoang, A.T., Huang, Z., Afzal, A., Li, C., Le, A.T., Nguyen, X.P. and Tran, V.D. 2022. Recent Advances in Machine Learning Research for Nanofluid-Based Heat Transfer in Renewable Energy System. *Energy & Fuels* 36, 6626–6658.

[61] Smola, A.J. and Schölkopf, B. 2004. A tutorial on support vector regression. *Stat Comput* 14, 199–222.

[62] Trabassi, D., Castiglia, S.F., Bini, F., Marinozzi, F., Ajoudani, A., Lorenzini, M., Chini, G., Varrecchia, T., Ranavolo, A., De Icco, R., Casali, C. and Serrao, M. 2024. Optimizing Rare Disease Gait Classification through Data Balancing and Generative AI: Insights from Hereditary Cerebellar Ataxia. *Sensors* 24, 3613.

[63] Vaferi, B., Samimi, F., Pakgohar, E. and Mowla, D. 2014. Artificial neural network approach for prediction of thermal behavior of nanofluids flowing through circular tubes. *Powder Technol* 267, 1–10.

[64] Vapnik, V.N. 2000. *The Nature of Statistical Learning Theory*. Springer New York, New York, NY.

[65] Wang, Q., Zhou, W., Yang, L. and Huang, K. 2022. Comparison between conventional and deep learning-based surrogate models in predicting convective heat transfer performance of U-bend channels. *Energy and AI* 8, 100140.

[66] Xu, L., Skoularidou, M., Cuesta-Infante, A. and Veeramachaneni, K. 2019. Modeling Tabular Data using Conditional GAN. Vancouver, Canada.

[67] Yang, X., Boroomandpour, A., Wen, S., Toghraie, D. and Soltani, F. 2021. Applying Artificial Neural Networks (ANNs) for prediction of the thermal characteristics of water/ethylene glycol-

based mono, binary and ternary nanofluids containing MWCNTs, titania, and zinc oxide. *Powder Technol* 388, 418–424.

[68] Yu, W., France, D.M., Routbort, J.L. and Choi, S.U.S. 2008. Review and Comparison of Nanofluid Thermal Conductivity and Heat Transfer Enhancements. *Heat Transfer Engineering* 29, 432–460.

[69] Zhu, T., Mei, X., Zhang, J. and Li, C. 2024. Prediction of Thermal Conductivity of EG–Al2O3 Nanofluids Using Six Supervised Machine Learning Models. *Applied Sciences* 14, 6264.

# 7 Environmental and Economic Considerations of Nanofluid Deployment

Sanatan Ramnath Iyer[1], Kaajal Anil Rai[2], Sonali Biswas[3] and Ganesh Moorthy I[4*]

The environmental outlook on the efficiency of thermal systems is a significant topic for discussion from a global perspective on resource utilisation, with an underlying concern and motive towards sustainability. This chapter focuses on the environmental and economic analysis related to the integration of nanofluids in thermal systems. It provides a promising approach in analysing recent advancements contributing to the global climate goals by focusing on lifecycle analysis, renewable energy integration, and economic viability. The industrial applications of nanofluid-based thermal systems are known to be important in contributing towards measures like heat transfer efficiency and reduced power consumption. Due to several regulatory frameworks being implemented, environmental considerations are of very significant value when understanding nanofluid integration in thermal systems. This chapter explores how the process of optimizing thermal systems with efficient nanofluid selection plays a crucial and environmentally beneficial role. Nanoparticle production methods consider environmental and greener options, while practical insights evaluate ROI, market trends, and material selection, making nanofluids economically beneficial across various sectors. Limitations such as nanoparticle toxicity and preparation costs are addressed along with strategies to combat them. Research, backing claims on successful integration with nanofluids and thermal systems, helps us to transition from experimental use to real-world applications. The integration of artificial intelligence and machine learning are emerging as an important tool to optimize the thermophysical properties of nanofluids and help in performance prediction, which is analysed and presented as a fresh perspective. While we gain insights into the global movement towards sustainable approaches, vast possibilities align with the advancing research on environmental and economic feasibility throughout the production line for nanofluids as part of the solution.

[1,2,4] School of Chemical Engineering, Vellore Institute of Technology, Tamilnadu, Vellore 632014, India.

[3] School of Bio-Sciences and Technology, Vellore Institute of Technology, Tamilnadu, Vellore 632014, India.

* Corresponding author: ganeshmoorthy.i@vit.ac.in

DOI: 10.1201/9781003686002-7

## Glossary

| *Term* | *Definition* |
|---|---|
| Energy Efficiency | A measure of how well energy input is converted into useful output, minimizing energy waste. Defined by the ratio of useful energy output to total energy input. |
| Thermal Conductivity | The property of a material to conduct heat, defined as the amount of heat transferred per unit time per unit area per unit temperature gradient, typically in W/(m·K)W/(m·K). |
| Toxicity | The degree to which a substance can damage an organism, measured by effects on cells or whole organisms, is often quantified by lethal dose (LD50) or reference dose (RfD). |
| Zeta Potential | The electrokinetic potential in colloidal dispersions indicates particle stability in suspension; a higher magnitude zeta potential implies greater stability. |
| Life Cycle Assessment (LCA) | A technique to assess environmental impacts associated with all the stages of a product's life, from raw material extraction through disposal. |
| Nanofluid | A fluid containing nanometer-sized particles (nanoparticles), designed to enhance thermal properties such as thermal conductivity and heat transfer. |

## Quantity and Symbols

| *Quantity* | *Current Units* | *Conversion Notes for alternate units* |
|---|---|---|
| Temperature | °C (degrees Celsius) | To read in Kelvin (K): $K = °C + 273.15$ |
| Pressure | bar | To read in (Pa): $1\ \text{bar} = 10^5\ \text{Pa}$ |
| Energy | kWh | To read in Joules (J): $1\ \text{kWh} = 3.6 \times 10^6\ \text{J}$ |

## Symbols and Descriptions

| *Symbol* | *Full Form / Meaning* | *Equation or Description* |
|---|---|---|
| $\nabla \cdot (\rho_{nf} u) = 0$ | Continuity Equation (mass conservation) | Divergence of nanofluid density and velocity vector equals zero |
| $\rho_{nf}(u \cdot \nabla)u = -\nabla p + \mu_{nf}\nabla^2 u$ | Navier-Stokes Equation (momentum conservation) | Fluid motion under pressure and viscous forces |
| $(\rho Cp)n_{nf}(u \cdot \nabla T) = k_{nf}\nabla^2 T$ | Energy Equation (heat transfer) | Convective and conductive heat transfer within a nanofluid |
| $u = u_{in}$ | Inlet velocity | Velocity of fluid entering the system |

| | | |
|---|---|---|
| $T = T_{in}$ | Inlet temperature | Temperature of fluid entering the system |
| $u = 0$ | No-slip condition | Fluid velocity at the solid boundary is zero |
| $q'' =$ constant | Constant heat flux | Heat flux applied to the boundary surface |
| $\partial u/\partial x = 0$ | Fully developed flow velocity gradient | The velocity gradient along the flow direction is zero |
| $\partial T/\partial x = 0$ | Fully developed temperature gradient | Temperature gradient along the flow direction is zero |
| Nu | Nusselt number | Dimensionless number describing convective heat transfer |
| $\mu_{nf}$ | Effective dynamic viscosity | Viscosity of nanofluid |
| $k_{nf}$ | Effective thermal conductivity | Thermal conductivity of nano-fluid |
| Cp | Specific heat capacity | Heat capacity at constant pressure |
| $\rho$ | Density | Mass per unit volume |
| LD50 | Lethal Dose 50 | Dose killing 50% of test animals |
| RfD | Reference Dose | Safe daily intake without adverse effects |
| mV | Millivolt | Unit for zeta potential |

## 1. Introduction

The need for a useful system for the management of sustainable energy has significantly increased over the past few decades. This occurred due to the increase in energy demands and consumption, the need for fossil fuels for transportation and luxury, and the emission of harmful greenhouse gases. The environmental footprint and the lower efficiency of management in energy production have led to the search for advanced systems that have the capability to perform with increased efficiency and lower ecological impacts. Thermal systems play a significant role in industrial operations to increase efficiency in processes, such as power generation, heating, and cooling operations. Thermal systems can help in efficient heat transfer. (Eze *et al.*, 2024). Energy conservation stands to be of a greater priority as the world moves towards sustainability, especially when the focus is on climate, continuous increase in population, and ecological stability. Therefore, a run towards finding clever and efficient optimization techniques for thermal systems with increased emphasis on research and development is a keynote these days. Similarly, it is not just the design of these thermal systems that contributes to the solution to the ongoing scientific debates over efficiency, but the sustainable outlook to the problems and challenges posed by thermal systems and for achieving net-zero thermal systems (Kim, 2022), can help understand the need of the hour of reducing environmental impacts and encouraging energy conservation. This is where nanofluids come in. The addition of normal fluids into nanoparticles allows nanofluids to offer broader advantages without altering or compromising the original integrity and properties of a system. It is important to note that nanoparticles in the nanofluids must be stable in nature. The instability of nanoparticles can cause fouling of equipment and a greater requirement for power, which will fundamentally increase the negative environmental impact. The stability will help in managing the thermochemical properties, preventing aggregation and alteration of the original chemistry of systems. This is achieved mainly by using various carbon compositions, metals like copper and gold, metal oxides like $Al_2O_3$, $SiO_2$, and more (Suneetha, Subbarayudu, and Reddy, 2022). Traditional fluids do not show as much thermal energy efficiency potential as nanofluids do, and more so with hybrid nanofluids. It is a very effective and attainable stability means. Table 1 below provides a glance at the comparison summary between nanofluids and traditional fluids.

**Table 1** Comparison between Nanofluids and Traditional Fluids.

| *Aspect* | *Nanofluids* | *Traditional Fluids* | *Reference* |
|---|---|---|---|
| Thermal Conductivity | High | Low to moderate | (Manimaran *et al.*, 2025) |
| Heat Transfer Rate | Enhanced | Standard | (Borode *et al.*, 2021) |
| Viscosity | Higher | Lower | (Rahman *et al.*, 2024) |
| Environmental Risk | Potential nanoparticle toxicity | Generally lower toxicity | (Elsaid *et al.*, 2020) |
| Cost | Higher | Lower | (Razzaq *et al.*, 2025) |
| Stability | Possible sedimentation | Generally stable | (Mukherjee *et al.*, 2022) |
| Energy Efficiency | Potential energy savings | Standard | (Samylingam *et al.*, 2024) |

However, this includes the setbacks that fall in relation to viscosity and pressure drop conditions from hybrid nanofluid utilization over mono nanofluids (Tao *et al.*, 2023). There is an increasing demand for energy due to the accelerated growth of countries. Solar thermal energy is emerging as a prominent solution for the expanding energy usage. However, more analysis and improvements are needed due to limitations like efficiency, cost, and poor design (Sharma *et al.*, 2025). Several types of nanofluids, such as molten salt nanofluids, address parameters such as increasing the specific heat capacity (Akanda and Shin, 2023). The increased heat transfer and capacity of smaller systems further facilitate the implementation of nanofluids in applications such as high-power fuel cell vehicles, where they act as an alternative coolant due to the large amount of heat dissipation found in conventional ones (Tao *et al.*, 2023). Certain metallic oxide nanomaterials, such as MgO, $TiO_2$, and ZnO, also improve factors such as the heat transfer coefficient, which are significant for various thermal applications, such as a double pipe heat exchanger (Louis *et al.*, 2022b). Further on, Figure 7.1 shows a few other types of nanofluids and their applications in brief.

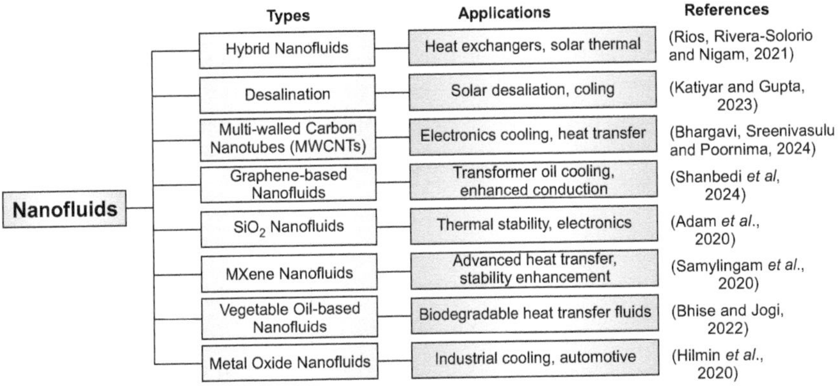


**Fig. 1** Types of Nanofluids and Their Applications.

Industries are, however, responsible for more than just efficiency targets in their processes. They are obliged to meet environmental well-being regulations and are therefore encouraged to use and develop technologies that not only enhance performance but also minimize greenhouse gas emissions as well as waste generation. This can be achieved by aligning their production with the 3Rs of reduce, reuse, and recycle and inculcating sustainable manufacturing goals (Hermawan *et al.*, 2023). There is an increasing number of regulatory frameworks being applied for various industries, right from energy and transport to agriculture. The 2030 Agenda for Sustainable Development is one of them, which tackles global challenges. As a result, nanofluids are drawing significant interest as industries are adopting such standards, transforming their operations and boosting efficiency by integrating sustainable resources and innovations (Lella, Oses-Eraso, and Stamos, 2024; Klimczuk *et al.*, 2024). Energy auditing is an ideal model for practical studies of the process efficiencies by industries, which can culminate data from decades' worth of resources and help set up measures for environmental benefits with a comparative measure to these audits (Kapp, Choi, and Kissock, 2021). This is a way of deciding emission standards at local levels, which are formed based on these audit data. There are several economic barriers, such as the production costs and stability of nanofluids, which limit the adoption of nanofluids into industrial applications (Bacha *et al.*, 2024). There are additional developments and innovative strategies that offset these costs, as discussed in this chapter. Sustainable economic development can be achieved through innovation in design and technology, and this can be led through more partnerships between industries and universities with goals to enhance production process efficiency through market research, labor productivity, financial system developments, and more (Zhang *et al.*, 2021). Total Cost Analysis or TCA is a method that considers all relevant costs with respect to purchasing and supply chain management, including making decisions between alternatives at the point of time (Oskarsson, 2024). However, its application is not widely established through research and has limitations that prevent a clear understanding. Now, to understand the thermodynamics of nanofluids better, we can identify the fundamental conservation laws for mass, momentum, and energy. The study for the $Al_2O_3$-water nanofluid gives us the Continuity Equation for complete mass conservation, stating that the divergence of the product of the nanofluid density ($\rho_{nf}$) and velocity vector (u) is equal to zero.

**Equation of Continuity**

$$\nabla \cdot (\rho nf\, u) = 0 \qquad ...(1)$$

Also, the Momentum Equation, famously known as the Navier-Stokes Equation, is used to describe the motion of the fluid under the pressure and viscous forces influence. This can be observed by the following equation.

Navier Stokes Equation:

$$\rho_{nf}\,(u \cdot \nabla)u = -\,\nabla p + \mu_{nf}\,\nabla^2 u \qquad ...(2)$$

Here, the equation has inertial forces represented on the left side and a negative pressure gradient ($-\nabla p$) on the right side, as well as viscous diffusion term involving the effective dynamic viscosity ($\mu_{nf}$).

Lastly, the Energy Equation governing the heat transfer within the fluid can be represented for nanofluids as:

$$(\rho Cp)_{nf}(u \cdot \nabla T) = k_{nf}\,\nabla^2 T \qquad ...(3)$$

The convective heat transport of thermal energy is represented by this equation with the left-hand side involving the effective heat capacity (ρCp)nf) to the heat diffusion governed by the effective thermal conductivity (knf) and the Laplacian of the temperature (T).

Now, focusing on the boundary conditions, it is important to first understand the purpose of setting boundary conditions. Any solution domain is typically constrained by certain, problem-specific boundary conditions in order to validate the mathematical model. Here, at the inlet point, the fluid enters with a uniform velocity (u = uin) and at a uniform temperature (T = Tin). The microchannel solid walls have the no-slip condition, requiring the fluid velocity to be zero (u = 0), whereas the constant heat flux (q" = constant) is applicable to simulate the heating element. When looking at the outlet of our system, we assume a fully developed flow, which in turn implies that the velocity and temperature gradients in the flow direction are negligible :

$$\partial u / \partial x = 0 \qquad ...(4)$$

$$\partial T / \partial x = 0 \qquad ...(5)$$

The Uncertainty Analysis is used to validate the numerical predictions. Performing grid independence tests (a fundamental procedure in Computational Fluid Dynamics) ensures that the results are not manipulated or influenced by the mesh size (with deviations usually less than 2%). Along with this, the uncertainty in Nusselt Number (Nu) prediction is mainly quantified from the empirical correlated properties that might resolve in 3-5% error, based on the nanoparticle loadings. By setting strict convergence criteria, we can confirm the numerical scheme's stability with residuals below $10^{-6}$ (Ma *et al.*, 2023; Bahrami, Allahdadi, and Saffari, 2025).

## 2. Life-Cycle Assessment of Nanofluids

LCA plays an important role in understanding the environmental and economic effects of using nanofluids by assessing different impact categories across all phases of its lifecycle. A large amount of energy is used for nanomaterial production through methods such as vapour deposition and liquid-phase synthesis, which release pollutants into the air, water, and soil. Life Cycle Assessment (LCA) is described through ISO 14040, which offers a holistic tool to assess these impacts right from raw material extraction to disposal (Ramadan et al., 2024). Lifecycle Assessment being applied to nanomaterials has increased significantly, but most studies have been limited to cradle-to-gate stages and don't focus much on the disposal and long-term impacts. Although LCA has numerous benefits, right from identification of environmental effects to energy usage and emissions, there are some struggles involved, such as poor data transparency and limited methods for characterising emissions, resulting in uncertainty. Nevertheless, many studies prove that nanomaterials can reduce emissions and improve energy efficiency over conventional options, highlighting the fact that LCA plays a vital role in thermal nanofluid applications (Nizam, Hanafiah, and Woon, 2021). Many studies on nanomaterials highlight that their production is considerably more energy-intensive than conventional materials. This is mainly due to the energy-intensive processes, such as high-temperature processing or the use of specialized precursors. As a result, there is a direct relationship between manufacturing choices and their associated environmental impacts and energy consumption. LCA was done on $TiO_2$ nanoparticles from cradle to gate, where 5 different precursors were tested right from titanium oxysulphate to titanium bis(ammonium lactato) dihydroxide, where nanoparticle characteristics such as crystallinity and size distribution, as well as environmental impacts such as global warming potential and energy demand were measured. It was found out that precursor choice plays a significant role in driving most of the environmental impacts, where simple inorganics like TiOS had lower impacts, contributing to less carbon emissions and energy usage. Process conditions, such as using water as a solvent (TiOS), reduced impacts. This shows how LCA provides guidance for making better choices in material design and production (Caramazana-González *et al.*, 2017).

The review notes that nanofluid production methods strongly shape environmental outcomes: scaling up nanoparticle synthesis can reduce emissions by up to 90%, while continuous flow techniques lower hazardous waste compared to batch methods. At the same time, the end-of-life stage remains a neglected hotspot, as nanofluids often enter waste streams without clear disposal strategies. LCA plays a dual role here by helping in identifying cleaner, less resource-intensive production routes and highlighting the importance of safe disposal and recycling of nanoparticles to achieve circularity and reduce long-term environmental risks (Rehman and Park, 2024). Studies on nanofluid-based CPV/T-TEG systems indicate that a substantial amount of energy is utilized due to thermoelectric materials and nanofluid preparation, but over the span of 25 years, there is a much higher exergy output and reduced environmental impact in the long run. This shows that despite the initial costs and energy usage, nanofluid deployment enhances long-term sustainability in thermal applications (Lekbir, Hassani, and Mekhilef, 2024). A study shows that $TiO_2$-SiC/water hybrid nanofluid in a shell and tube heat exchanger provides economic and environmental benefits.

At 0.5 vol% concentration, there was a significant increase of around 65% in Nusselt number and a slight increase in pressure drop of around 8%. The LCA identifies that most of the energy (around 89.5%) is consumed during its operation, and around 75.7% of $CO_2$ emissions were accounted for, but despite the initial costs of nanofluid integration, there is a long-term efficiency validated by ANN cost modelling (Malika and Sonawane, 2022). Another LCA study revealed a 12% decrease in carbon emissions, 15% shorter payback period, as well as 10% annual operational savings compared to water-based flat plate solar collector (FPSC) systems. $Si_3\,N_4$ nanofluids gave 8% higher thermal efficiency and 33% higher thermal conductivity, reducing the size of the collector plate by 6.1%. Therefore, despite the huge amount of energy usage and emissions during the operational phase, LCA validates the economic and environmental benefits of $Si_3\,N_4$ nanofluids in FPSC systems (Rahman *et al.*, 2025). Due to insufficient data, specific features aren't taken into account by LCA regarding the toxicity of nanoparticles, which are detrimental to human life and the environment. Through LCA studies, viscosity and method of production are dominant criteria in deciding whether the addition of nanoparticles is valuable (Grosu *et al.*, 2020). There are some limitations found when LCA studies are applied. The first challenge that needs to be addressed is the lack of lifecycle inventory(LCI) data, which contains inadequate inputs such as energy and raw materials, along with outputs such as emissions and wastes. The other concern is the insufficient amount of manufactured nanomaterials' specific characterization factors, which helps in the computation of environmental effects such as global warming potential and toxicity factors (Salieri *et al.*, 2017; Miseljic and Olsen, 2014).

## 3. Environmental Impact and Toxicity Concerns

Advancements in nanofluids have greatly enhanced their thermophysical properties, influencing their use in diverse applications such as refrigeration, solar thermal systems, electronics cooling, biomedical processes, and automotive heat exchangers. However, the environmental and human health impacts of nanofluids are significant due to the type, shape, concentration, and dispersion quality of nanoparticles. Nanofluids often deal with difficulties such as sedimentation and agglomeration, which lead to a reduction in stability and their performance in various thermal integrated applications. Destabilization occurs due to various attractive and repulsive forces acting among suspended nanoparticles that result from the interplay between attractive (van der Waals, gravitational) and repulsive (electrostatic, buoyant) forces. Due to the tendency of nanoparticles to form aggregates and larger clusters, it is essential to understand theories such as DLVO theory and the concept of zeta potential, which constitute the basis for assessing the behaviour of nanofluid stability. The DLVO theory, named after Derjaguin, Landau, Verwey, and Overbeek, shows how particles in a colloid attract or repel each other from two effects, which are van der Waals attraction (VA) and electrostatic repulsion (VR). The total energy between two particles, $V_T + V_R$, determines the system's stability, which also involves factors such as ion type, concentration, surface potential value, and particle size. Van der Waals interactions cause attraction, whereas repulsion results from the overlap of electrical double layers surrounding each particle. Therefore, aggregation is prevented when repulsive forces dominate, whereas destabilization is observed when attractive forces dominate (Manimaran et al., 2025). The DLVO theory has a few limitations in explaining nanoparticle suspensions. Its first drawback is assuming interactions similar to those in colloidal systems, despite nanoparticles having larger interaction ranges and a higher diffusion coefficient. Its second limitation is assuming only two types of forces and neglecting other forces such as Born, hydration, hydrophobic, and steric interactions that play vital roles in nanoparticle stability (Spasojević, Ivanišević, and Sikirić, 2025). Previous research has explored several techniques that enhance nanofluid stability, such as ultrasonic agitation and surfactant addition. Ultrasonication is widely used across industries and applies sound waves above 20 kHz to disperse particles. In nanofluids, the zeta potential is another key measurable indicator of dispersion stability. It is the potential difference between the particle surface and the layer of fluid moving alongside it. By checking this value, we can understand whether a nanofluid will stay well-dispersed or start forming aggregates. Low values (below ±10 mV) result in rapid coagulation, while higher values (above ±40 mV) indicate good to excellent stability, with ±61 mV or more representing highly stable dispersions (Zainon and Azmi, 2021). The pH of a nanofluid significantly affects its zeta potential. A positive zeta potential was attained due to the absorption of H+ ions, whereas a negative value was observed due to the absorption of OH- ions. The isoelectric point is where the zeta potential is zero at a certain pH. The nanofluid is considered to be stable if the zeta potential shifts away from the IEP.

In one study, there was a significant increase in the thermal conductivity of CuO/water and $Al_2O_3$/water nanofluids when ultrasonication was done for 1-4 hrs and SDBS surfactant was added (Gangadevi, Vinayagam, and Senthilraja, 2018). Both $Al_2O_3$/water and CuO/deionized-water had zeta potentials of around –72.2

mV and −85.1 mV, confirming strong repulsion between the particles and ensuring good stability for more than a week with surfactant use (Khairul et al., 2016). We can thus conclude that both sonication and surfactant use play vital roles in the stability of nanofluids.

Conventional methods often involve energy-intensive procedures, hazardous byproducts, and toxic chemicals, causing environmental burdens and health hazards. Green biological and chemical synthesis methods, which use mild reaction conditions, renewable resources, and environmentally friendly techniques, have become a viable alternative. Green synthesis lowers the carbon footprint of producing nanoparticles while reducing the use of dangerous chemicals. Green-synthesized nanoparticles improve thermophysical characteristics without compromising environmental sustainability. Bioinspired or plant-based extracts can improve thermal conductivity and surface functionalization, increasing stability during dispersion and increasing heat transfer efficiency. Green synthesis promotes both technological innovation and environmental stewardship, allowing the integration of nanofluids into heating and cooling applications while striking a balance between performance and ecological responsibility. (Elsaid *et al.*, 2020). As exposure to nanomaterials is rising, ecosystem health is becoming a primary concern that must be tackled. This can be managed by gaining insight into the behaviour of manufactured nanomaterials. Nanomaterials can settle on land, water, and exposure via inhalation is possible when particles are suspended in the air. Accumulation of nanoparticles changes their properties and can influence their characteristics in internal and external environments. Nanomaterials such as $TiO_2$, $Al_2O_3$, and carbon black may have a higher toxicity than micron-sized particles and can penetrate through the skin as well as reach the brain via the nasal epithelium olfactory neurons (Atofarati, Sharifpur, and Huan, 2024). This highlights the diverse risks that nanoparticles may present (Ray, Yu, and Fu, 2009). Cytotoxicity is the ability of the substance to cause damage to the cells (Hossain *et al.*, 2022). A study was done between ZnO and $SiO_2$ particles, where results found both nano and micron ZnO and $SiO_2$ to be toxic. However, nanoparticles were generally more harmful. ZnO toxicity was similar for L-132 cells but stronger for nano in THP-1 cells, while $SiO_2$ toxicity was clearly higher in the nano form for both lines. This study shows that nanoparticle toxicity is influenced by size, surface properties, and cell reactivity (Sahu *et al.*, 2016). Due to the extensive use of nanoparticles in surfactant-stabilised formulations, their release into the environment and wastewater cannot be avoided (Huang *et al.*, 2016). This study investigated the nanoparticle (silica) -surfactant combination to form nanofluids. Results showed that nanoparticles can alter surfactant performance and depend on the concentration. Biodegradation is the main pathway for removing organic compounds in wastewater. Lower concentrations of nanoparticles and surfactants (around 25mg/l) showed higher biodegradability. These findings highlight the importance of assessing biodegradability when designing nanofluid formulations to reduce environmental implications (Lechuga *et al.*, 2022). It is important to analyze the synthesis processes of nanoparticles, which are potentially adverse to ecosystems and human health. After utilization of nanofluids, it is essential to recover the primary fluids and the nanoparticles associated with it as they can leak into the environment and contaminate water sources. There are various methods used to reduce these harmful effects, such as centrifugation followed by evaporation of the fluid, leaving behind the nanoparticles for recovery. However, more advanced materials and chemicals are a primary concern for environmental hazards. Such nanomaterials enter the human body through the respiratory tract, affecting the cell's backbone and pushing the body to produce inflammatory signals (Razzaq *et al.*, 2025).

Multiple factors, such as heat transfer coefficient and thermal conductivity, must be considered while evaluating the ecological impact of heat transfer fluids. Different studies use different techniques, so there is no single, standard method for testing and comparing toxicity. Another key factor of 2D nanomaterials, apart from size and composition, that can increase toxicity is their shape (Shaik *et al.*, 2022). One study found that the cytotoxicity of gold nanoparticles (AuNPs) strongly depends on their shape. Within the analysed particle shapes, AuNP stars (around 215 nm) showed the highest cytotoxicity, particularly against cancer cells, whereas spheres (around 6.3 nm) were the least toxic but also had the lowest anti-cancer potential (Steckiewicz *et al.*, 2019). As discussed, green synthesis has proven to be a sustainable choice over conventional techniques. In a study where nanofluids can be employed as a conventional heat transfer fluid in solar thermal systems, mitigating the issue of poor thermal conductivity can be further developed to improve and address environmental issues. Among the various approaches, green synthesis has become a promising alternative as a cleaner method to produce nanomaterials, helping to reduce environmental degradation and health risks.

Green synthesis relies on eco-friendly chemicals along with sustainable sources such as plant-derived compounds, bacteria, and fungi, to produce nanoparticles (Cardoso *et al.*, 2025). A recent study found that $CuFe_2O_4$ nanoparticles were synthesized using *Chlorella vulgaris* extract through green methods, leading to a 4.8% increase in thermal conductivity and a 0.5°C rise in outlet temperature in a serpentine heat exchanger. This points to the encouraging role of such methods for sustainable heat transfer functions (B. Cardoso *et al.*, 2025). There is a growing demand for advanced insulators with respect to cooling capacity, dielectric strength, environmental sustainability, and reliable operation in high voltage applications. In another study, green synthesis is done using apple leaf extracts, which are dispersed in vegetable oils. $TiO_2$, $SiO_2$, and $Al_2O_3$ nanoparticles were used. The results revealed that 0.02% $TiO_2$ nanofluid sample in 275 mL of oil indicated greater insulating properties along the AC/DC breakdown voltage. Therefore, vegetable oil-based nanofluids are a suitable replacement for traditional mineral oil, showcasing the importance of green synthesis routes for thermal applications (Siddique *et al.*, 2024). Nanofluid-based solar heaters and PV/T systems improve energy efficiency and can considerably bring down emissions compared to conventional technologies. Even if bulk materials are considered safe, reducing them to nanoparticles makes them more reactive and potentially more hazardous.

The toxicity of nanoparticles also depends on the parent material. For instance, titania causes ecotoxicity, while CuO is highly toxic to aquatic systems (Said *et al.*, 2023). Furthermore, a summary of environmental toxicity and economic value of nanofluids for different nanoparticle types has been graphically represented in Figure 2. Porous carbon nanomaterials are attracting increasing attention for several mechanical and energy applications. A study shows the utilization of these nanomaterials in flat plate solar collectors (FPSCs). The synthesis process was optimized and prepared in such a way as to minimize environmental effects. Environmental assessment also revealed that PCNF reduces about 260 kg of $CO_2$ emissions as well as other associated hazardous emissions (Khan *et al.*, 2025). Therefore, understanding and carefully evaluating toxicity and environmental impacts will lead to smoother adoption of nanomaterials for sustainable applications with high operational efficiency.

The use of quantitative toxicity data, such as $LD_{50}$ values or dose–response curves play an important part in assessing their environmental impact. These studies are essential to understand safe operational and disposal limits when exposed to real-world conditions in thermal systems.

The $LD_{50}$ value aids in measuring how much of a substance is required to kill 50% of test animals. Substances with $LD_{50} < 25$ mg/kg are highly toxic, whereas if $LD_{50} > 5000$ mg/kg, it is considered less toxic. One experiment used Karber's method for m-$TiO_2$ nanoparticles, where $LD_{50} > 5000$ mg/kg b.w., as all mice survived even at the highest dose. Hence, m-$TiO_2$ NPs exhibit very low acute toxicity (Basante-Romo, Gutiérrez-M, and Camargo-Amado, 2021). Using Karber's method, $LD_{50}$ exceeded 2000 mg/kg for all oxides, placing them in the slightly toxic band. Comparatively, CuO showed the lowest $LD_{50}$ (2012 mg/kg), while $SiO_2$ showed the highest $LD_{50}$ (3000 mg/kg). The estimated lethal doses for ZnO and TiO2 were found to be 2225 and 2775, respectively (Shaker et al., 2017). The reference dose (RfD) indicates the amount of nanoparticles that can be safely taken in each day without harmful effects. $SiO_2$ (2 mg/kg/day) and $C_{60}$ (1 mg/kg/day) are the safest. Ni (0.008) and $Al_2O_3$ (0.006) are the most toxic. CuO (0.026) and ZnO (0.03125) are of moderate risk. Among the three emission routes, freshwater emissions ($1.55 \times 10^3$) had the highest ecotoxic potential, followed by air ($6.05 \times 10^2$) and soil ($1.19 \times 10^0$), indicating that $TiO_2$ nanoparticles pose the greatest risk to aquatic environments. Hence, it is essential to carefully select nanofluids for practical deployment (Bahrami, Allahdadi, and Saffari, 2025).

## 4. Recycling and Waste Management

It is essential to discuss recycling and waste management, as they play a very crucial role and a deciding factor in many environmental issues. To conserve natural resources, minimise pollution levels , and mitigate climate change effects, we must dive deeper and study recycling and waste management (Fletcher, St Clair,

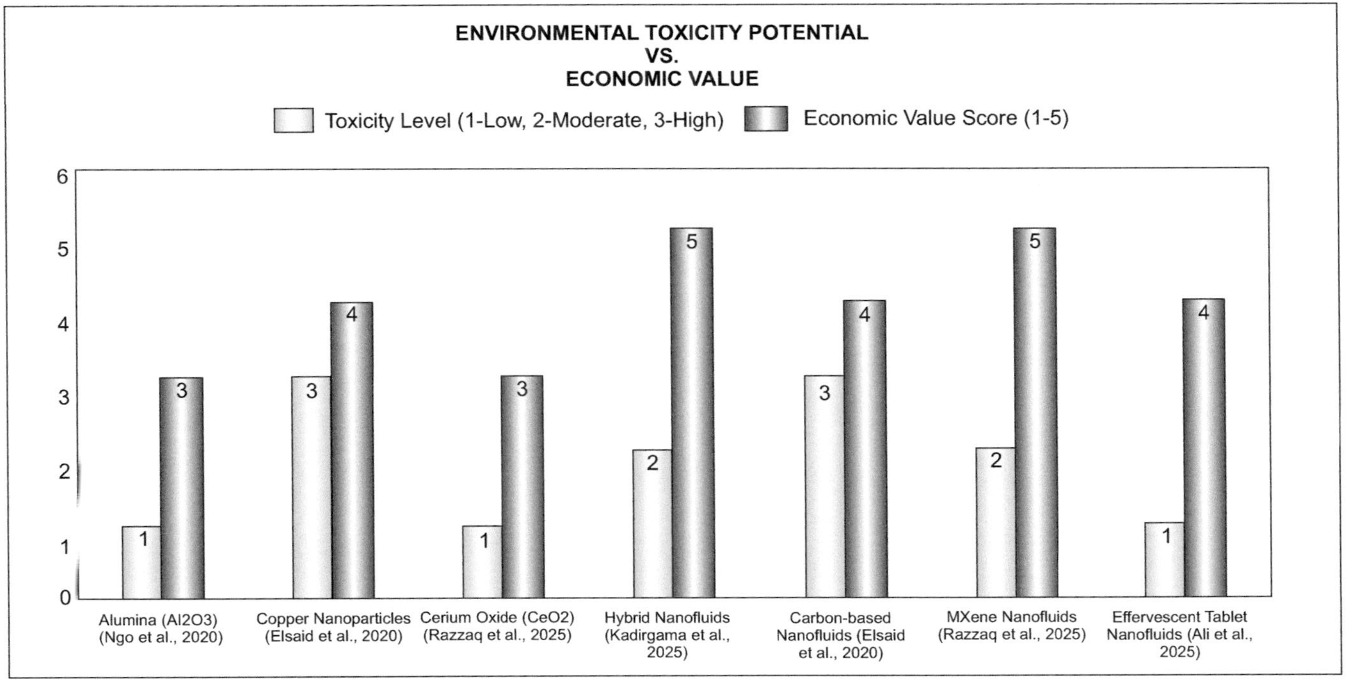


**Fig. 2** Comparative summary of Environmental Toxicity vs. Economic Value of Nanofluids for different nanoparticle types.

and Sharmina, 2021). By correctly managing garbage, we may limit the quantity of waste that is produced and put in landfills, which would lower greenhouse gas emissions as well as soil and groundwater contamination (Hajam, Kumar, and Kumar, 2023). Nanofluids are engineered with nanoparticles, which require their own specific waste disposal techniques because of their respective compositional challenges. This is why studying ways to manage waste generated from nanofluids is of vital importance and can be seen as an opportunity to help improve the environmental aspects and for risk mitigation from toxic materials that have the potential to negatively impact the environment at the end of their lifecycle .

To begin with, critical surveys highlight that nanofluids have a significant potential for wasted heat recovery in large quantities. This includes sources like vehicles, chimneys, solar energy systems, etc. It has been proven that nanofluids have an efficient conversion rate for enhancing energy recovery greater than 20% (Pereira, Moita, and Moreira, 2023). Industries focusing on steelwork systems also utilise nanofluids to improve energy capacities up to three times. Heat-to-power systems are another important application of nanofluid utilization. Heat-to-power systems capture unwanted heat from an industrial process, engine, or facility to generate electricity and/or useful thermal energy. Silica nanofluid, supported by some studies, shows that it can improve the power outputs of hybrid photovoltaic/thermoelectric generator systems by nearly 54%, while cobalt oxide aqueous nanofluids can increase electrical power generation by almost 11% when compared to water-based fluids (Corporation, 2022).

Moving towards more practical applications, there's a study that serves as the perfect example. The report, "Analysis of Nanofluids Behaviour in a PV-Thermal-Driven Organic Rankine Cycle with Cooling Capability", investigated various nanofluids like $Fe_3O_4$, $Al_2O_3$, CuO, and $SiO_2$ to check their performance-enhancing efficiencies in hybrid systems of a Photovoltaic (PV) thermal collector, an Organic Rankine Cycle (ORC), and a cooling coil. Copper oxide nanofluid proved to be of the highest efficiency, and the hybrid system's efficiency on average was 17% higher with CuO nanofluid than with water as the heat transfer fluid. Also, the cooling effect was better. This form of recycling demonstrates a good waste thermal energy recovery alternative to combat the otherwise lost energy from the PV system (Sami, 2020).

It is also imperative to understand how nanomaterials themselves need to be handled and disposed of after their utilisation. Although there isn't an official definition for "nanowaste", a proposed definition defines nanowaste as material that is "separately collected or collectable waste materials which are or contain engineered nanomaterials." (MANTRA, 2024) The sources of nanowastes are broader, like households, industrial, research institutes, and hospitals. The currently undertaken techniques for dealing with nanowastes aren't different from normal waste disposal. Remediation techniques for several nanomaterials are mentioned in Table 2 below. These techniques include recycling, landfilling, and incineration, which leads us to focus on the environmental concerns and risks with this way of handling nanowastes.

For example, mechanical processes like crushing and shredding potentially expose the environment to nanomaterials. Landfilling, on the other hand, depends on natural decomposition or disintegration of the waste, which can aid in the leaching of these particles into soil and water sources. Whereas incineration at high temperatures can cause volatility in nanowastes and make them an airborne threat. It is important to acknowledge the facts about significant knowledge gaps related to the lack of available data on questions like what the characteristics and amounts of nanomaterial-containing wastes are, the rate of release of nanowastes in landfills and their fate post-incineration, and, most importantly, the environmental and human exposure to the toxins (Puisney *et al.*, 2018) released by them during recycling operations. With respect to the manufacturing of nanomaterials, there are a few explored alternatives that could be processed to prepare nanomaterials. These alternatives include recycled wastes mainly from agricultural wastes (vegetables, fruits, leaves, etc.), animal wastes like waste from marine animals, human hair, cattle waste, and also wastes from plastic and industries such as silk and battery manufacturing industry byproducts , which are all very sustainable alternatives and they prevent waste accumulation (Ali and Makhlouf, 2021).

**Table 2** Nanomaterials and their remediation techniques.

| *Nanomaterial* | *Remediation Technique* | *References* |
|---|---|---|
| Aluminium Oxide ($Al_2O_3$) | Biosorption (algae composites) | (Al-Ahmari, Alghanmi, and Hamouda, 2025) |
| Copper Oxide (CuO) | PAC-CA coagulation (charge neutralisation, adsorption bridging) | (Sun *et al.*, 2021) |
| Carbon Nanotubes (CNTs) | Chemical oxidation (NaClO) | (Yang, Okazaki and Zhang, 2021) |
| Titanium Dioxide ($TiO_2$) | Coagulation & flocculation ($FeCl_3$, $FeSO_4$, $Al_2(SO_4)_3$) | (Honda *et al.*, 2014) |
| Zinc oxide (ZnO) | Wastewater treatment and antimicrobials | (Sonawane *et al.*, 2022) |
| Silver nanoparticles (AgNPs) | Bactericidal | (Krishna *et al.*, 2024) |
| Nickel oxide (NiO) | Adsorption of metal ions like copper, chromium, and zinc | (Dhanapal *et al.*, 2024) |

## 5. Economic Feasibility and Cost Analysis

Economic feasibility and cost analysis are very important to consider when incorporating into thermal systems. Despite their ability to maximise thermal conductivity, heat transfer efficiency, and system performance, there are challenges, such as nanoparticle cost, preparation techniques, and long-term stability. Associated costs such as pumping power, material prices, and maintenance can outweigh the thermal benefits, making cost-effectiveness an important parameter to analyse, preventing nanofluids from becoming less attractive.

A study was done to analyse the financial viability of MWCNT–water and CuO–water nanofluids. Results showed a reduction in pumping costs by 41% for MWCNT nanofluids, but were surpassed by their high purchase price ($382/kg) and short-term stability (180 days), which resulted in an annual loss of around $20.57/m$^2$. CuO nanofluids, due to higher concentration needs and higher cost ($328/kg), were even less feasible, with losses over $3900/m$^2$ per year. In some rare instances, stability extended to 5 years, resulting in positive returns (Karthikeyan, Raphael, and Tavares, 2021). In various experiments, nanofluids improved heat transfer by up to 50%. Yet, they have not been commercialized to a reasonable extent even after 2 decades of nanofluid research. Cost analysis shows nanoparticle costs, along with their manufacturing methods (e.g., silver, diamond-based nanofluids), lead to significantly higher costs even at low dosages (2 to 3 times the cost of base fluids). The method of preparation also plays a major role, where one-step methods are much more expensive than two-step methods.

Currently, there are limited cost-benefit tools, such as Price Performance Factor (PPF), Price Performance Index (PPI), and Economic Performance Index (EPI), which remain limited and do not yield a conclusive understanding (Mukherjee *et al.*, 2023). Due to the disadvantages found in the two-step and one-step approaches, such as higher electrical energy and careful stabilization, there are a few new techniques introduced, such as the effervescent tablet-based method (Ali *et al.*, 2025). The tablet approach lowers the capital investment due to a simple compression procedure. The increasing interest rates of 20-30% over the overall lifetime of around 20 years significantly affected the annual cost, but the effervescent method still outshone the conventional process with positive financial indicators such as payback period and IRR (Ali *et al.*, 2023). Another work assessed a three-fluid heat exchanger (HETF) with graphene-water-based nanofluid used in residential heating systems. Techno-economic results such as the increase in the benefit–cost ratio (90% and 92.3%) prove its suitability and high economic competitiveness (Bargah *et al.*, 2025).

Graphene has attracted lots of researchers due to its unique properties. A study looked into how graphene nanoplatelets (GNPs) can help in the profitability as well as heat transfer when utilized as nanofluids in heat pipes. The size of the heat transfer exchanger and its power requirement could be reduced by certain amounts depending on the concentration of nanoparticles relative to the base fluid. This helps in analyzing a suitable concentration that gives a highly efficient performance with reduced operating costs. Factors such as reduced dimensions and compact design play a vital role in the economic feasibility of such thermal applications. As per the study, the heat exchanger could be decreased by 16.10% at 0.025 wt%, 21.92% at 0.05 wt%, 25.37% at 0.075%, and 29.35% at 0.1% concentrations. The energy requirements of the base fluid were reduced from 422W to 354W, 326W, 315W, and 298W at the same concentrations, respectively. Even with a higher initial capital outlay, the operational savings make it economically feasible over the course of time (Mohammad *et al.*, 2022). Nowadays, nanofluids and phase change materials help in increasing the thermal efficiency of parabolic trough collectors (PTCs), which are commonly employed in solar power systems. By using Mxene/syltherm 800 hybrid nanofluid, the costs of producing energy appeared more economical, reaching 0.1324 $/kWh, standing out as a competitive option compared to conventional fluids economically (Nagar and Sreenivasa, 2025). Thermoelectric air-conditioners can be integrated with nanofluids, too, to improve energy optimization and cut emissions. The OPEX was brought down by 91.16 USD in comparison with conventional 1-ton split air conditioners. As CAPEX is a primary concern with most of the nanofluids, the payback period was found to be 39.74 years to make up for the high initial outlay (Bakthavatchalam *et al.*, 2025). In conclusion, the deployment of nanofluids depends on various factors. The economic feasibility depends on the various tools required to analyze various operational and financial conditions.

**Table 3** Reported Global Warming Potential (GWP) of Selected Nanoparticles: $CO_2$-Equivalent Emissions per Kilogram Produced.

| *Nanoparticle/System* | *Reported GWP (kg $CO_2$-eq per kg nanoparticle)* | *Reference(s)* |
|---|---|---|
| Carbon black (CB) | 2.368 kg $CO_2$-eq $kg^{-1}$ | (Lin, Chien and Chiu, 2016); (Spahr and Rothon, 2016) |
| Carbon nanotubes (CNTs) | $10^1$–$10^2$ kg $CO_2$-eq $kg^{-1}$ (industrial scale) to 0.48–28.55 kg $CO_2$ -eq $g^{-1}$ (high-quality SWCNT) | (Teah et al., 2020); (Roes et al., 2009) ; (Singh et al., 2008) |
| Carbon nanofibres (CNFs) | Around 500–700 kg $CO_2$-eq $kg^{-1}$ | (Khanna, Bakshi and Lee, 2008) |
| $TiO_2$ nanoparticles | Around 16.52 kg $CO_2$-eq $kg^{-1}$ | (Wu, Zhou and Hicks, 2019); (Grubb and Bakshi, 2010) |
| Ag nanoparticles | Hundreds of kg $CO_2$-eq $kg^{-1}$ | (Pourzahedi and Eckelman 2015); (Temizel-Sekeryan and Hicks 2020); (Zhang et al., 2016) |

Nanofluids, when added to base fluids or common fluids, have been responsible for significant improvements in the heat transfer ability of the base fluids. Particles like $TiO_2$, $Al_2O_3$, $Fe_3O_4$, and graphene have evidently achieved increased thermal performance with enhancement from 7% (low) to 47.5% (high), this is with Graphene Oxide in water at 40°C. All these compounds were tested in water or ethylene glycol-like fluids, and the experimental data results prove that maximum enhancement is observed at low concentrations. This can be backed by $Fe_3O_4$, achieving around 25–48% enhancement at a low 2 vol% concentration, while Graphene-water fluid reaches a 27% boost at merely 0.2 vol% concentration. Also, working with hybrid nanofluids like $Al_2$ $O_4$-Cu reportedly shows a 12.11% enhancement. However, the nanoparticle size, pH (fluid chemistry), temperature, and concentration also impact the working efficiency as well. For example, (Teng *et al.*, 2010) provides a study about the use of $Al_2O_3$ nanoparticles and the corresponding achievement of the highest thermal conductivity by using the smallest particle size (20 nm) and a concentration within the narrow range of 0.5–2 wt%, hence backing the claims of the cruciality of both factors. Similarly, pH values mainly influence the fluid's thermal properties when tested with $Al_2O_3$, while the performance of $Cu_2O$ nanofluid largely depends on the correct surfactant or chemical stabilizer to keep the particles dispersed. $Fe_3O_4$ was tested through a temperature range of 20–60°C, which tells us that all factors combined and adjusted contribute to the performance optimization (Sundar, Singh, and Sousa, 2013).

Lifecycle costing highlights that nanofluid-integrated systems achieve considerable savings over standard designs. (Sajid and Bicer, 2020) presents a study on the incorporation of nanofluid-based PV/T technology in solar greenhouse systems. The nanofluid-based systems achieved up to 22.3% and 12.8% reductions in total lifecycle cost, improving cooling efficiency and reducing energy demand. Among the systems, system 3 reported the lowest LCOE(0.033 \$/kWh) and cooling (0.015 \$/kWh), whereas system 2 resulted in the most efficient freshwater production (1.45 \$/$m^3$). According to (Bakthavatchalam et al., 2025), the PV–Nanofluid Thermoelectric system reduced the operating expenses by 91.16 USD and the $CO_2$ emissions by 68.88 tons compared to a conventional unit, thus highlighting its potential for real-world deployment despite a longer payback period of 39.74 years. Considering a 25-year evaluation period, the analysis employs a constant interest rate of 3% as reported by Bank Negara Malaysia, adjusted for an inflation rate of 0.66%. The study aims to evaluate the economic feasibility of the PV-NTEAC system in comparison with conventional air conditioners, where an overall LCC of USD 4464.7 was recorded, while the net present value (NPV) of maintenance and replacement costs reached USD 2824.1. These results support the economic feasibility of integrating nanofluid-based thermoelectric air conditioning for sustainable building applications.

From an investment standpoint, the study demonstrated that economic modelling integrating NPV and IRR is strongly affected by the relationship between VFR and φ for hybrid nanofluid-based heat exchangers. Systems with low φ (0.02–0.04) and high VFR (0.75–1.0) achieved the most desirable profitability, with NPVs ranging between 28,706 and 30,040 and IRR values nearing 0.92. Increasing φ beyond 0.06 lowered both indicators substantially, highlighting that higher particle concentrations result in greater capital outlay and lower financial returns. Thus, economic feasibility is maximised under conditions of optimizing the nanoparticle use and flow rate ratio (Obaid et al., 2025).

The economic analysis carried out by Tian et al. (2021) among three configurations, which are Basic Heat Exchanger, Novel Heat Exchange, and Redesigned Novel Heat Exchanger using SolidWorks and ANSYS-Fluent modelling. The experiment showed that the novel R.NHE design had approximately 12% reduction in manufacturing cost relative to the BHE, validating its cost effectiveness. These findings prove that structural redesign integrated with hybrid nanofluid applications has significant cost-effectiveness while maintaining optimal thermal output.

## 6. Industrial and Regulatory Perspectives

When we talk about the industrial and regulatory perspectives of nanofluid integration in thermal systems, it's important to analyse the promising technological aspects in order to ensure its responsible development and deployment and its effects on the environment, society, and the economy. Nanofluids are becoming more advantageous due to their benefits in industries such as automobiles, aerospace, medical, and electronics. In industries like automobiles, aerospace, and medical, nanofluids are mainly used to reduce fossil fuel consumption, lower costs, enhance safety, increase durability, and ensure smooth operation of internal systems in vehicles (Rahman *et al.*, 2018). In the medical industries, especially in the pharmaceutical sector, the process of lyophilizing, the development of disease-specific targeted drug delivery, and cancer treatments (Sheikhpour *et al.*, 2020). Even after these advantages, there are still limitations in utilizing this system because of the high production cost, low shelf life in terms of stability, and lack of deeper knowledge regarding its ecological and human health impacts. The important part of developing a nanofluidic system is understanding its classification and its position in national and international trades. The industries need to develop better techno-economic assessments, analyze the life cycle, and improve the process of using nanofluids in a sustainable and environmentally friendly way.

To begin with the industrial applications, as previously discussed, it's known that nanofluids can enhance thermal conductivity by almost 15-40% (Lourenço *et al.*, 2021), which leads to more efficient heat transfer. Thermal conductivity for various nanofluid compositions is depicted in Table 4 below. The convective heat transfer coefficient has been proven to increase up to 123% (Lavicka, 2022). The research also backs up the idea of nanofluids used for improving the thermal management of Li-ion batteries (Mitra, Kumar, and Singh, 2022). A recent market trend study states that the global nanofluid market is projected to reach an estimated

**Table 4** Nanofluid composition and thermal conductivity relation.

| *Nanofluid Composition* | *Thermal Conductivity Enhancement* | *Reference* |
|---|---|---|
| **GNP/Water** | 4.48–5.62% | (Borode *et al.*, 2021) |
| **$Al_2O_3$–Cu/Water (2 vol%)** | 12.11% | (Manimaran *et al.*, 2025) |
| **CNT–$Al_2O_3$/Water (0.1%)** | 20% | (Devarajan *et al.*, 2018) |
| **GO/Water (1 vol%, 60°C)** | 52% | (Kanti *et al.*, 2024) |
| **$TiO_2$–$Al_2O_3$/Water (30–70°C)** | 71% | (Esfe *et al.*, 2023) |

value of $500 million to $1 billion in 2025, with a Compound Annual Growth Rate (CAGR) of 15-20% through 2033 (*Strategic Insights into Nanofluid Market Trends*, 2025).

Continuing our discussion on the regulatory and safety challenges with regards to nanofluids, the focus lies on a significant hurdle, which is the absence of standardised testing procedures and frameworks for required regulations (Gupta and Kumari, 2025). It has been found that there are no clear regulations with standards that cover the entire lifecycle of engineered nanoparticles (Subhan and Subhan, 2022). But the same properties that make nanoparticles effective make them a potential environmental hazard as well. Nanoparticles may be toxic due to some heavy metal compositions that pose risks to humans as well as the environment during their disposal cycles. Although there aren't many concrete existing regulations, various organisations are contributing to establishing regulatory frameworks. For example, the U.S. Environmental Protection Agency (USEPA) manages a bulletin on the "Management of Nanoscale Materials", which falls under Toxic Substances Control Act (TSCA), and the European Commission labelled nanofluids a "Key Enabling Technology" part of the Horizon 2020 framework (*Strategic Insights into Nanofluid Market Trends*, 2025; Lavicka, 2022). In accordance, the industries and companies have been reportedly increasing their focus on "eco-friendly nanofluid formulations" with these emerging standards.

The biggest challenge with this widespread adoption of nanofluids is the economic and practical hurdles. What is important with this aspect is the cost analysis and its long-term stability. The production process of nanofluids acts as a major economic barrier. The type of nanoparticle directly influences the cost of production. The better the nanomaterial quality, the more expensive the production of the nanomaterial as high-purity nanoparticles are very expensive. When compared with conventional heat transfer fluids, nanofluids have been claimed to be 2-5 times more expensive in overall production costs (Sikarwar *et al.*, 2025). Nanoparticles themselves, being the raw materials, contribute to the complexity of the manufacturing processes, requiring specialised equipment. Although some new, cost-effective methods are being explored, such as using effervescent tablets, the initial investment remains a significant factor for consideration. On the other hand, the long-term stability of a nanofluid is directly related to its economic viability. The nanofluid's enhanced thermal properties can be lost if the nanoparticles start settling and agglomerating out of the base fluid, which then makes it no different from any conventional heat transfer fluid, rendering it useless. The resulting instabilities lead to frequent replacement, which is a step backwards, eliminating potential cost savings and efficiency (Awais *et al.*, 2021). Three out of four scenarios on economic analysis state that nanofluids are not economically viable as they get destabilised over time, thus requiring annual recurring renewal investments. One successful study, however, showcases the direct link between stability and profitability: a nanofluid with a 5-year stability had a payback period of less than a year (Karthikeyan, Raphael, and Tavares, 2021). This, in turn, suggests that the initial costs are justified only if the nanofluid maintains its properties over a long service life, therefore avoiding frequent and costly maintenance or refilling. The methods under development to enhance stability, such as using surfactants or surface modifications, are important, but these can further contribute to the complexity and cost of production. Another major practical hurdle is the issues with fouling and erosion that highly affect industrial operations (Mukherjee *et al.*, 2023). Also, specifically, these problems stem from issues with nanofluid-circulated loose particles or clumped particles, which reduces the fluid's heat transfer property as previously discussed. The abrasive nature of the nanoparticles results in unwanted wear and tear in pipes, pumps, and other equipment (Gokapai, Pothana, and Ling, 2024). Although stainless steel material proves to be the most resistant, studies still state that nanofluids, when engaged with gear pumps, result in severe damage. Altering the fluid's speed and angle can combat erosion, which is influenced by the amount of nanoparticles and the angle at which the fluid hits the surface of the pumps. These material damage tendencies and reduced performance situations affect the industries' decisions to deploy nanofluids (Bacha *et al.*, 2024). The unpredictability of such factors and their effects leads to improper determination, as the lack of reliable sources and clear data for standardised testing protocols adds to the hurdles.

## 7. Future Outlook for Sustainable Nanofluid Use

With all the data and information we have gathered so far, the important question that arises is: how should we proceed in the future? Let's look at the prospective opportunities and latest advances in the use of nanofluids. Nanofluids have also been proven to be of great potential for boosting heat transfer efficiency along with turbulators (AZoNano, 2025), and there's more research that can be undertaken on these projects.

There are a few other ongoing research projects on "Green Nanofluids", which are contributing towards the development of environmentally friendly nanofluids that are being extracted and synthesised from plants and microorganisms. These are significantly more sustainable alternatives to the conventional methods of production of nanofluids that have been known to have safety and environmental concerns. These bio-based nanofluids solve more than one problem at hand. "Green" zinc oxide nanoparticles, for example, can now be prepared in two environmentally friendly ways (Pereira, Moita and Moreira, 2022), one of which utilises the famously known plant Aloe vera, and the other uses cassava starch. While both methods proved to be effective copper-removing agents, Aloe vera was found to be better for removing higher concentrations of copper. As previously discussed, the industries have been exploring more on multi-nanoparticles and hybrid systems, and this involves a combination of various nanoparticles like metal oxides, carbon nanotubes, etc, that contribute to important synergetic effects, which in turn enhance the thermal properties. Particularly, the hybrid systems have been proven to impact and improve the Nusselt number by almost 49%

when compared to water, making them high-performance worthy, as discussed in a recent 2025 study.

Another use of nanofluid is in the future, which pivots on the integration of nanofluids with the advancing technologies and AI (Basu *et al.*, 2024). It's safe to say that the new and upcoming AI predictive models and machine learning models (Assad *et al.*, 2021) can help determine thermophysical properties of nanofluids with a significantly high accuracy, like the R-squared values of up to 0.9996, crucial for reducing experimental costs and accelerating design optimisation techniques (Said *et al.*, 2022). We also know that market studies claim that the nanofluid market is projected for significant growth in the years to come. Reports back up this claim with a market value of approximately $2.5 billion in 2024 and forecasts it to reach $7.3 billion by 2033 (Compound Annual Growth Rate (CAGR) of 12.6%), and with nanofluid solar collectors, the projected growth is CAGR of 16.2% from 2025 to 2033 (Tubular Solar Collector Competitive Strategies: Trends and Forecasts 2025-2033, 2026). Nanofluids have a strong base with respect to renewable energy. A 36.8% higher efficiency (Wciślik, 2025) with a specific graphene oxide nanofluid was found in a recent study regarding solar collectors (compared to pure water).

The Hybrid Nanofluid Density Prediction Dataset (Mathur *et al.*, 2025) used 436 curated samples with nine key parameters, including nanoparticle types, base fluid properties, and temperature, using the IQR method and autoencoder-based augmentation. Among fourteen evaluated models, Gradient Boosting and RNN–Linear Regression yielded outstanding predictive performance ($R^2 = 0.9999$; MSE = 0.00091), showcasing the reliability of combined ML–DL frameworks for accurate nanofluid property estimation.

Another study (Shateri et al., 2020) analyzed 42 nanofluid types using a dataset of 3144 points, where machine learning models were used to predict nanofluid viscosity. Eight ML models, including MLP, SVR, Decision Tree, Random Forest, and Extra Trees, were assessed. The proposed models achieved high accuracy, with several outperforming the Committee Machine Intelligent System (CMIS) and attaining an average absolute relative error (AARE) under 3%, compared to CMIS's 4.036%.

Lastly, another very promising market is the use of nanofluids in electronics cooling systems. Studies reveal that water-CuO nanofluids can boost heat flux by up to 8.51%, while SiC-water nanofluids can enhance thermal conductivity by 12.45% compared to the base fluid. And in the automotive industry, radiators with $Al_2O_3$-based nanofluid integration can achieve an increase in the heat transfer coefficient by 50-60% (Bhogare and Kothawale, 2014), more than conventional coolants, which also prospects a better fuel efficiency. But our search for sustainable outlooks with nanofluids is redundant without a keen focus on green nanofluids and their desirable properties, such as affordability, eco-friendliness, and non-corrosivity. Natural ingredients also help make the fluids more stable, in turn preventing clogging as well. These fluids transfer heat more efficiently, but also at lower concentrations, and enhance the cooling systems with money-saving

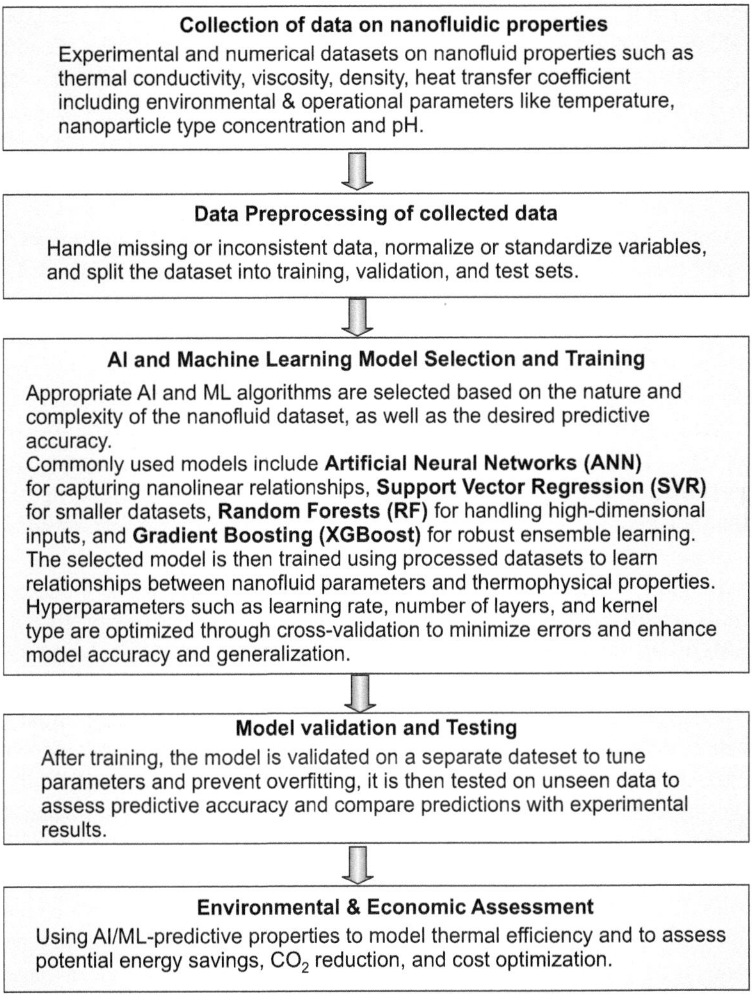


technology and design (Dewanjee and Kundu, 2024). Therefore, with the ongoing rapid progress in research, the future of integrating sustainable nanofluids into real-world applications looks promising.

## 8. Conclusion

The scope of studying nanofluids for thermal applications cannot be limited to just the factors or parameters considered in this chapter, mainly the environmental and economic ones. But these offer a significant solution to questions about these topics for anyone interested in research, learning, or general knowledge. To summarise some of the crucial topics undertaken in this chapter, we may begin with how the LCA trends help the production processes by validating the stages of the material life cycle. Understanding how raw material feasibility and system designs with available technology are integral parts of the planning process before production begins. Furthermore, it also highlights the end-of-cycle responsibilities related to the disposal methods, recycling techniques, and environmental effects of the same. So it's safe to conclude that an LCA is a quality measure of circling over the costs, risks, and utilisation opportunities presented by nanofluid integration in energy and industrial systems. When discussing production processes, we factor in the energy intensity of producing nanoparticles. As discussed, this exists due to very high temperature requirements as well as the selective needs which follow, contributing to inefficient energy use and also to air, water, and soil pollution levels. This poses as one of the major disadvantages of nanofluids when compared to cheaper conventional fluids. However, at the same time, shifting the focus to the precursor chemicals and production methods directly and significantly affects their environmental consequences. The broader outlook on these trends gives us concrete proof of the direct relation between the manufacturing decisions and the sustainability of nanofluid deployment in industries in the long run. The benefits and limitations of the application of LCA to nanofluids are discussed, providing us with genuine clarity over why cleaner production routes must be encouraged, and also why there is still more scope for development with regard to the emissions and ill effects of nanomaterial production processes. While our primary available research on nanofluids is understood via systems like the heat exchangers, solar collectors and hybrid PV/T-TEG systems, it is an undeniable fact that all of these integrations demand more energy for the preparation, adding to the initial costs, but with a longer operational life for the systems, and lesser maintenance costs, the efficiency improvement in performance beats the conventional fluids by a noteworthy sustainability margin. The highly debated themes against nanofluids follow the claims of toxicity risks raised by nanoparticles. These concerns highlight the exposure pathways and the biological reactions to these exposures due to their size, shape, and other properties, which have been proven to give varied toxicity reports. Similarly, these exposures also reflect in the ecosystem environments that influence the existing natural processes, which require very active management solutions. Unlike traditional chemical methods, green synthesis utilises biological sources such as plants and microbes for the production of nanomaterials. This presents as a viable alternative to toxic chemical methods as they lower ecological risks, along with enhancing heat transfer applications by improving the conductivity, insulation, and promoting sustainability. As discussed in detail, despite advances in nanofluid applications, end-of-life disposal poses many challenges. Due to the ineffectiveness of the implementation of the current waste management methods, such as landfilling, incineration, and recycling, more research needs to be done to address these issues. The presence of standardised disposal practices remains undefined, even though there is a significant development in the production and application of nanofluids. As highlighted earlier, the excessive expense of nanoparticles along with minimal stability is a significant concern, even though nanofluids help reduce operational costs. It is essential to analyse the profitability gained over the lifetime of the nanoparticles, as it can be economically worthwhile after a given period of time. There have been many newly developed methods to facilitate the cheaper production of nanoparticles, such as the tablet-based preparation, which has been discussed in detail. Due to the insufficient and weak assessment tools, the implementation of nanofluids into industrial practices still faces many challenges, and economic justification remains uncertain. As elaborated in the chapter, while the adoption of nanofluids in areas such as automotive cooling, electronics, and energy sectors is gaining a lot of attention, they are being hindered by restricted frameworks, costs, and toxicity. Therefore, their prospects to improve on the applications will depend on how these gaps are addressed due to the rapid advancements in industries. As regulatory frameworks gradually grow, industries are shifting toward sustainable solutions due to the increasing emphasis on such adoptions in the coming years. Taken together, the analysis suggests there are many positive directions ahead, such as green nanofluids and machine learning, which offer promising scope for deployment with advanced technologies.

## References

[1] Adam, S.A. et al. (2020) 'Effect of temperature on the stability and optical properties of SiO2-water nanofluids for hybrid photovoltaic/thermal applications,' *Applied Thermal Engineering*, 175, p. 115394. https://doi.org/10.1016/j.applthermaleng.2020.115394.

[2] Akanda, M.A.M. and Shin, D. (2023) 'A synthesis parameter of molten salt nanofluids for solar thermal energy storage applications,' *Journal of Energy Storage*, p. 106608. https://doi.org/10.1016/j.est.2023.106608.

[3] Al-Ahmari, J.M., Alghanmi, R.M. and Hamouda, R.A. (2025) 'Bio-Fabricated Aluminum Oxide Nanoparticles Derived from Waste Pharmaceutical Packages: Insight into Characterization and Applications,' *Biomolecules*, 15(7), p. 984. https://doi.org/10.3390/biom15070984.

[4] Ali, G.A.M. and Makhlouf, A.S.H. (2021) 'Fundamentals of waste recycling for Nanomaterial manufacturing,' in *Topics in mining, metallurgy and materials engineering*, pp. 3–24. https://doi.org/10.1007/978-3-030-68031-2_1.

[5] Ali, N. et al. (2023) 'Effervescent tablets for carbon-based nanofluids production,' *Journal of Molecular Liquids*, 390, p. 123083. https://doi.org/10.1016/j.molliq.2023.123083.

[6] Ali, N. et al. (2025) 'Techno-economic assessment of effervescent tablet-based nanofluids,' PLoS ONE, 20(4), p. e0319265. https://doi.org/10.1371/journal.pone.0319265.

[7] Assad, M.E.H. et al. (2021) 'Utilization of machine learning methods in modeling specific heat capacity of nanofluids,' *Computers, Materials & Continua/Computers, Materials & Continua* (Print), 70(1), pp. 361–374. https://doi.org/10.32604/cmc.2022.019048.

[8] Atofarati, Emmanuel.O., Sharifpur, M. and Huan, Z. (2024) 'Nanofluids for heat transfer enhancement: a holistic analysis of research advances, technological progress and regulations for health and safety,' *Cogent Engineering*, 11(1). https://doi.org/10.1080/23311916.2024.2434623.

[9] Awais, M. et al. (2021) 'Heat transfer and pressure drop performance of Nanofluid: A state-of-the-art review,' *International Journal of Thermofluids*, 9, p. 100065. https://doi.org/10.1016/j.ijft.2021.100065.

[10] AZoNano (2025) Combining nanofluids and turbulators for sustainable energy. https://www.azonano.com/news.aspx?newsID=41211.

[11] Bacha, H.B. et al. (2024) 'A comprehensive review on nanofluids: Synthesis, cutting-edge applications, and future prospects,' *International Journal of Thermofluids*, 22, p. 100595. https://doi.org/10.1016/j.ijft.2024.100595.

[12] Bahrami, H.-R., Allahdadi, O. and Saffari, H. (2025) 'Nanofluids for Advanced Thermal Management: A Multi-Criteria Review of Performance, Cost, and Environmental Impacts,' *Results in Materials*, p. 100753. https://doi.org/10.1016/j.rinma.2025.100753.

[13] Bakthavatchalam, B. et al. (2025) 'Life cycle cost analysis of an innovative nanofluid assisted photovoltaic thermoelectric air conditioner for energy efficiency enhancement in buildings,' Case *Studies in Thermal Engineering*, 73, p. 106494. https://doi.org/10.1016/j.csite.2025.106494.

[14] Bargah, V. et al. (2025) 'Performance of a heat exchanger for three fluids with graphene–water-based nanofluid used in a sustainable residential heating system: an optimization and techno-economic assessment,' *Journal of Thermal Analysis and Calorimetry* [Preprint]. https://doi.org/10.1007/s10973-025-14075-1.

[15] Basante-Romo, M., Gutiérrez-M, J.O. and Camargo-Amado, R. (2021) 'Non-toxic doses of modified titanium dioxide nanoparticles (m-TiO2NPs) in albino CFW mice,' *Heliyon*, 7(3), p. e06514. https://doi.org/10.1016/j.heliyon.2021.e06514.

[16] Basu, A. et al. (2024) 'A review of artificial intelligence methods in predicting thermophysical properties of nanofluids for heat transfer applications,' *Energies*, 17(6), p. 1351. https://doi.org/10.3390/en17061351.

[17] Bhargavi, N., Sreenivasulu, P. and Poornima, T. (2024) 'Comparative analysis on single- and multi-walled carbon nanotubes suspended for enhanced thermal energy transfer past stretching surfaces,' *AIP Advances*, 14(11). https://doi.org/10.1063/5.0232829.

[18] Bhise, V.Y. and Jogi, B.F. (2022) 'Recent developments on sustainable lubricants by using vegetable oil-based nanofluids in machining,' *Materials Today Proceedings*, 62, pp. 3148–3154. https://doi.org/10.1016/j.matpr.2022.03.429.

[19] Bhogare, R.A. and Kothawale, B.S. (2014) 'Performance investigation of Automobile Radiator operated with $Al_2O_3$ based nanofluid,' *IOSR Journal of Mechanical and Civil Engineering*, 11(3), pp. 23–30. https://doi.org/10.9790/1684-11352330.

[20] Borode, A.O. et al. (2021) 'Investigation of the thermal conductivity, viscosity, and thermal performance of graphene Nanoplatelet-Alumina hybrid nanofluid in a differentially heated cavity,' *Frontiers in Energy Research*, 9. https://doi.org/10.3389/fenrg.2021.737915.

[21] Caramazana-González, P. et al. (2017) 'Assessing the life cycle environmental impacts of titania nanoparticle production by continuous flow solvo/hydrothermal syntheses,' *Green Chemistry*, 19(6), pp. 1536–1547. https://doi.org/10.1039/c6gc03357a.

[22] Cardoso, B. et al. (2025) 'Green synthesis of copper ferrite-based nanofluids using Chlorella vulgaris for heat transfer enhancement,' *Journal of Molecular Liquids*, p. 127498. https://doi.org/10.1016/j.molliq.2025.127498.

[23] Cardoso, B.D. et al. (2025) 'Progress in nanofluid technology: from conventional to green nanofluids for biomedical, heat transfer, and machining applications,' *Nanomaterials*, 15(16), p. 1242. https://doi.org/10.3390/nano15161242.

[24] Corporation, B. (2022) 'The best heat transfer fluids for liquid cooling,' Boyd | Trusted Innovation, 24 May. https://www.boydcorp.com/blog/best-heat-transfer-fluids.html.

[25] Devarajan, M. et al. (2018) 'Thermophysical properties of CNT and CNT/$Al_2O_3$ hybrid nanofluid,' *Micro & Nano Letters*, 13(5), pp. 617–621. https://doi.org/10.1049/mnl.2017.0029.

[26] Dewanjee, D. and Kundu, B. (2024) 'A review of applications of green nanofluids for performance improvement of solar collectors,' *Renewable Energy*, p. 122182. https://doi.org/10.1016/j.renene.2024.122182.

[27] Dhanapal, A.R. *et al.* (2024) 'Nanotechnology approaches for the remediation of agricultural polluted soils,' *ACS Omega*, 9(12), pp. 13522–13533. https://doi.org/10.1021/acsomega.3c09776.

[28] Elsaid, K. et al. (2020) 'Environmental impacts of nanofluids: A review,' The Science of the Total Environment, 763, p. 144202. https://doi.org/10.1016/j.scitotenv.2020.144202.

[29] Esfe, M.H. et al. (2023) 'Thermal conductivity of MWCNT-TiO2/Water-EG hybrid nanofluids: Calculating the price performance factor (PPF) using statistical and experimental methods (RSM),' *Case Studies in Thermal Engineering*, 48, p. 103094. https://doi.org/10.1016/j.csite.2023.103094.

[30] Eze, V.H.U. et al. (2024) 'Advancements in Energy Efficiency Technologies for Thermal Systems: A Comprehensive review,' *INOSR APPLIED SCIENCES*, 12(1), pp. 1–20. https://doi.org/10.59298/inosras/2024/1.1.1010.

[31] Fletcher, C.A., St Clair, R. and Sharmina, M. (2021) 'A framework for assessing the circularity and technological maturity of plastic waste management strategies in hospitals,' *Journal of Cleaner Production*, 306, p. 127169. https://doi.org/10.1016/j.jclepro.2021.127169.

[32] Gangadevi, R., Vinayagam, B.K. and Senthilraja, S. (2018) 'Effects of sonication time and temperature on thermal conductivity of CuO/water and Al 2 O 3 /water nanofluids with and without surfactant,' *Materials Today Proceedings*, 5(2), pp. 9004–9011. https://doi.org/10.1016/j.matpr.2017.12.347.

[33] Gokapai, V., Pothana, P. and Ling, K. (2024) 'Nanoparticles in Drilling Fluids: A review of types, mechanisms, applications, and future prospects,' Eng—Advances in Engineering, 5(4), pp. 2462–2495. https://doi.org/10.3390/eng5040129.

[34] Grosu, Y. et al. (2020) 'Nanofluids based on molten carbonate salts for high-temperature thermal energy storage: Thermophysical properties, stability, compatibility and life cycle analysis,' *Solar Energy Materials and Solar Cells*, 220, p. 110838. https://doi.org/10.1016/j.solmat.2020.110838.

[35] Grubb, G.F. and Bakshi, B.R. (2010) 'Life cycle of Titanium dioxide nanoparticle production,' *Journal of Industrial Ecology*, 15(1), pp. 81–95. https://doi.org/10.1111/j.1530-9290.2010.00292.x.

[36] Gupta, N.S.K. and Kumari, N.S. (2025) 'A review on recent development and future perspective of nanofluid utilization in automotives,' *Evergreen*, 12(2), pp. 992–1019. https://doi.org/10.5109/7363490.

[37] Hajam, Y.A., Kumar, R. and Kumar, A. (2023) 'Environmental waste management strategies and vermi transformation for sustainable development,' *Environmental Challenges*, 13, p. 100747. https://doi.org/10.1016/j.envc.2023.100747.

[38] Hermawan, A.N. et al. (2023) 'The effect of sustainable manufacturing on environmental performance through government regulation and eco-innovation,' *International Journal of Industrial Engineering and Operations Management*, 6(4), pp. 299–325. https://doi.org/10.1108/ijieom-04-2023-0039.

[39] Hilmin, M.N.H.M. et al. (2020) 'Thermoelectric power generations from vehicle exhaust gas with TiO2 nanofluid cooling,' *Thermal Science and Engineering Progress*, 18, p. 100558. https://doi.org/10.1016/j.tsep.2020.100558.

[40] Honda, R.J. et al. (2014) 'Removal of TIO2 nanoparticles during primary water treatment: role of coagulant type, dose, and nanoparticle concentration,' *Environmental Engineering Science*, 31(3), pp. 127–134. https://doi.org/10.1089/ees.2013.0269.

[41] Hossain, J. et al. (2022) 'Cytotoxicity of metal/metalloids' pollution in plants,' in *Elsevier eBooks*, pp. 371–394. https://doi.org/10.1016/b978-0-323-91675-2.00017-2.

[42] Huang, Z. et al. (2016) 'Toxicity mechanisms and synergies of silver nanoparticles in 2,4-dichlorophenol degradation by Phanerochaete chrysosporium,' *Journal of Hazardous Materials*, 321, pp. 37–46. https://doi.org/10.1016/j.jhazmat.2016.08.075.

[43] Kadirgama, G. et al. (2025) 'Characterization and machine learning analysis of hybrid alumina-copper oxide nanoparticles in therminol 55 for medium temperature heat transfer fluid,' *Scientific Reports*, 15(1). https://doi.org/10.1038/s41598-025-92461-3.

[44] Kanti, P.K. et al. (2024) 'Experimental and explainable machine learning approach on thermal conductivity and viscosity of water based graphene oxide based mono and hybrid nanofluids,' *Scientific Reports*, 14(1). https://doi.org/10.1038/s41598-024-81955-1.

[45] Kapp, S., Choi, J.-K. and Kissock, K. (2021) 'Toward energy-efficient industrial thermal systems for regional manufacturing facilities,' *Energy Reports*, 8, pp. 1377–1387. https://doi.org/10.1016/j.egyr.2021.12.060.

[46] Karthikeyan, A., Raphael, W. and Tavares, J.R. (2021) 'Nanofluids as heat transfer fluids: Hurdles to industrial application and economic considerations,' *The Canadian Journal of Chemical Engineering*, 100(11), pp. 3311–3320. https://doi.org/10.1002/cjce.24338.

[47] Katiyar, A. and Gupta, N.K. (2023) 'Effect of different aluminium oxide based nanofluid concentrations on the efficiency of solar water desalination system,' *Journal of Thermal Engineering*, 9(1), pp. 61–68. https://doi.org/10.18186/thermal.1242844

[48] Khan, W.A. et al. (2025) 'Porous carbon nanoparticles for flat plate solar collectors: A thermal, economic and environmental investigation,' *Materials Science and Engineering* B, 315, p. 118085. https://doi.org/10.1016/j.mseb.2025.118085.

[49] Khairul, M.A. *et al.* (2016) 'Effects of surfactant on stability and thermo-physical properties of metal oxide nanofluids,' *International Journal of Heat and Mass Transfer*, 98, pp. 778–787. https://doi.org/10.1016/j.ijheatmasstransfer.2016.03.079.

[50] Khanna, V., Bakshi, B.R. and Lee, L.J. (2008) 'Carbon Nanofiber Production,' *Journal of Industrial Ecology*, 12(3), pp. 394–410. https://doi.org/10.1111/j.1530-9290.2008.00052.x.

[51] Kim, J.-K. (2022) 'Specialty Grand challenge for thermal system design,' *Frontiers in Thermal Engineering*, 2. https://doi.org/10.3389/fther.2022.945841.

[52] Klimczuk, A. et al. (2024) 'Editorial: Towards 2030: sustainable development goal 9: industry, innovation and infrastructure. A sociological perspective,' *Frontiers in Sociology*, 9. https://doi.org/10.3389/fsoc.2024.1491091.

[53] Krishna, S.B.N. *et al.* (2024) 'Nanotechnology in action: silver nanoparticles for improved eco-friendly remediation,' *PeerJ*, 12, p. e18191. https://doi.org/10.7717/peerj.18191.

[54] Lavicka, L. (2022) Nanofluids used in Modern Electronics Cooling Applications. https://simutechgroup.com/fr/nanofluids-used-in-modern-electronics-cooling-applications/.

[55] Lechuga, M. et al. (2022) 'Environmental impact assessment of nanofluids containing mixtures of surfactants and silica nanoparticles,' *Environmental Science and Pollution Research*, 29(56), pp. 84125–84136. https://doi.org/10.1007/s11356-022-21598-9.

[56] Lekbir, A., Hassani, S. and Mekhilef, S. (2024) 'Techno-economic and life cycle assessment of a nanofluid-based concentrated Photovoltaic/Thermal-Thermoelectric hybrid system,' *Journal of Power Sources*, 595, p. 234066. https://doi.org/10.1016/j.jpowsour.2024.234066.

[57] Lella, L., Oses-Eraso, N. and Stamos, I. (2024) 'Pioneering a sustainable development goals monitoring framework for European regions,' *Ecological Indicators*, 166, p. 112248. https://doi.org/10.1016/j.ecolind.2024.112248.

[58] Lin, T.-H., Chien, Y.-S. and Chiu, W.-M. (2016) 'Rubber tire life cycle assessment and the effect of reducing carbon footprint by replacing carbon black with graphene,' *International Journal of Green Energy*, 14(1), pp. 97–104. https://doi.org/10.1080/15435075.2016.1253575.

[59] Louis, S.P. et al. (2022) 'Application of Nanofluids in Improving the Performance of Double-Pipe Heat Exchangers—A Critical Review,' *Materials*, 15(19), p. 6879. https://doi.org/10.3390/ma15196879.

[60] Lourenço, M.J. et al. (2021) 'The Balance between Energy, Environmental Security, and Technical Performance: The Regulatory Challenge of Nanofluids,' *Nanomaterials*, 11(8), p. 1871. https://doi.org/10.3390/nano11081871.

[61] Ma, Z. *et al.* (2023) 'Nanofluids in microchannel heat sinks for efficient flow cooling of power electronic devices,' *Applied Materials Today*, 35, p. 101980. https://doi.org/10.1016/j.apmt.2023.101980.

[62] Malika, M. and Sonawane, S.S. (2022) 'Ecological optimization and LCA of TiO2-SiC/water hybrid nanofluid in a shell and tube heat exchanger by ANN,' Proceedings of the Institution of Mechanical Engineers Part E *Journal of Process Mechanical Engineering*, 238(1), pp. 45–55. https://doi.org/10.1177/09544089221093304.

[63] Manimaran, M. et al. (2025) 'Critical review on the stability and thermal conductivity of water-based hybrid nanofluids for heat transfer applications,' *RSC Advances*, 15(18), pp. 14088–14125. https://doi.org/10.1039/d5ra00844a.

[64] MANTRA (2024) Nanomaterials in waste - MANTRA. https://materialneutral.info/en/safety/cross-cutting/nanomaterials-in-waste/%22.

[65] Mathur, P. *et al.* (2025) 'A computational intelligence framework integrating data augmentation and Meta-Heuristic optimization algorithms for enhanced hybrid nanofluid density prediction through machine and deep learning paradigms,' *IEEE Access*, p. 1. https://doi.org/10.1109/access.2025.3543475.

[66] Miseljic, M. and Olsen, S.I. (2014) 'Life-cycle assessment of engineered nanomaterials: a literature review of assessment status,' *Journal of Nanoparticle Research*, 16(6). https://doi.org/10.1007/s11051-014-2427-x.

[67] Mitra, A., Kumar, R. and Singh, D.K. (2022) 'Thermal management of lithium-ion batteries using carbon-based nanofluid flowing through different flow channel configurations,' *Journal of Power Sources*, 555, p. 232351. https://doi.org/10.1016/j.jpowsour.2022.232351.

[68] Mohammad, R.S. et al. (2022) 'Thermal analysis of Graphene-Based nanofluids for energy system and economic feasibility,' *Journal of Nanomaterials*, 2022(1). https://doi.org/10.1155/2022/3530419.

[69] Mukherjee, S. et al. (2023) 'Nanofluids: Critical issues, economics and sustainability perspectives,' *Particuology*, 87, pp. 147–172. https://doi.org/10.1016/j.partic.2023.06.021.

[70] Mukherjee, S. *et al.* (2022) 'Stability and sedimentation characteristics of water based $Al_2O_3$ and TiO2 nanofluids,' *Proceedings of the Institution of Mechanical Engineers Part N Journal of Nanomaterials Nanoengineering and Nanosystems*, 238(1–2), pp. 17–30. https://doi.org/10.1177/23977914221127735.

[71] Nagar, S. and Sreenivasa, S. (2025) 'Thermal performance investigations by the use of nanofluids and phase change materials (PCM) in parabolic trough collectors (PTC) along with exergetic, economic, and application-based analyses,' Proceedings of the Institution of Mechanical Engineers Part a *Journal of Power and Energy* [Preprint]. https://doi.org/10.1177/09576509251366470.

[72] Ngo, I. et al. (2020) 'Formation Damage Induced by Water-Based Alumina Nanofluids during Enhanced Oil Recovery: Influence of Postflush Salinity,' *ACS Omega*, 5(42), pp. 27103–27112. https://doi.org/10.1021/acsomega.0c02473.

[73] Nizam, N.U.M., Hanafiah, M.M. and Woon, K.S. (2021) 'A content review of life cycle assessment of nanomaterials: current practices, challenges, and future prospects,' *Nanomaterials*, 11(12), p. 3324. https://doi.org/10.3390/nano11123324.

[74] Obaid, H.M. *et al.* (2025) 'Economic evaluation of a passive heat exchanger enhanced by artificial intelligence and nanoparticle integration,' *International Journal of Low-Carbon Technologies*, 20, pp. 1466–1475. https://doi.org/10.1093/ijlct/ctaf065.

[75] Oskarsson, B. (2024b) 'Total Cost Analysis in Purchasing Education A Two-Step Open-Ended Teaching Case approach,' *Journal of Purchasing and Supply Management*, p. 100971. https://doi.org/10.1016/j.pursup.2024.100971.

[76] Pereira, J., Moita, A. and Moreira, A. (2023) 'Nanofluids as a Waste Heat Recovery Medium: A Critical review and guidelines for Future research and use,' *Processes*, 11(8), p. 2443. https://doi.org/10.3390/pr11082443.

[77] Pereira, J.E., Moita, A.S. and Moreira, A.L.N. (2022) 'The pressing need for green nanofluids: A review,' *Journal of Environmental Chemical Engineering*, 10(3), p. 107940. https://doi.org/10.1016/j.jece.2022.107940.

[78] Pourzahedi, L. and Eckelman, M.J. (2015) 'Comparative life cycle assessment of silver nanoparticle synthesis routes,' *Environmental Science Nano*, 2(4), pp. 361–369. https://doi.org/10.1039/c5en00075k.

[79] Puisney, C. et al. (2018) 'Brake wear (nano)particle characterization and toxicity on airway epithelial cells in vitro,' *Environmental Science Nano*, 5(4), pp. 1036–1044. https://doi.org/10.1039/c7en00825b.

[80] Rahman, M.A. *et al.* (2024) 'Review on Nanofluids: Preparation, properties, stability, and thermal performance augmentation in heat transfer applications,' *ACS Omega* [Preprint]. https://doi.org/10.1021/acsomega.4c03279.

[81] Rahman, S. et al. (2025) 'Techno-economic and life cycle analysis of a nano-enhanced flat plate solar collector for improved thermal performance,' *Energy Nexus*, p. 100433. https://doi.org/10.1016/j.nexus.2025.100433.

[82] Rahman, S.S. *et al.* (2018) 'Tuning nanofluids for improved lubrication performance in turning biomedical grade titanium alloy,' *Journal of Cleaner Production*, 206, pp. 180–196. https://doi.org/10.1016/j.jclepro.2018.09.150.

[83] Ramadan, B.S., Wibowo, Y.G., Anwar, D. and Maryani, A.T. 2024, 'A Review of Life Cycle Assessment of Nanomaterials-Based Adsorbent for Environmental Remediation', *Global Nest Journal*, vol. 26, no. 7, 06216. https://doi.org/10.30955/gnj.06216

[84] Ray, P.C., Yu, H. and Fu, P.P. (2009) 'Toxicity and environmental risks of nanomaterials: challenges and future needs,' *Journal of Environmental Science and Health Part C*, 27(1), pp. 1–35. https://doi.org/10.1080/10590500802708267.

[85] Razzaq, I. et al. (2025) 'Nanofluids for Advanced Applications: A comprehensive review on preparation methods, properties, and environmental impact,' *ACS Omega*, 10(6), pp. 5251–5282. https://doi.org/10.1021/acsomega.4c10143.

[86] Rehman, T.-U. and Park, C.W. (2024) 'Advances in nanofluids for tubular heat exchangers: Thermal performance, environmental effects, economics and outlook,' *Energy*, 308, p. 132732. https://doi.org/10.1016/j.energy.2024.132732.

[87] Rios, M.S.B.-D.L., Rivera-Solorio, C.I. and Nigam, K.D.P. (2021) 'An overview of sustainability of heat exchangers and solar thermal applications with nanofluids: A review,' *Renewable and Sustainable Energy Reviews*, 142, p. 110855. https://doi.org/10.1016/j.rser.2021.110855.

[88] Roes, A.L. *et al.* (2009) 'Influence of using nanoobjects as filler on functionality-based energy use of nanocomposites,' *Journal of Nanoparticle Research*, 12(6), pp. 2011–2028. https://doi.org/10.1007/s11051-009-9819-3.

[89] Sahu, D. et al. (2016) 'In Vitro Cytotoxicity of Nanoparticles: A Comparison between Particle Size and Cell Type,' *Journal of Nanoscience*, 2016, pp. 1–9. https://doi.org/10.1155/2016/4023852.

[90] Said, Z. et al. (2022) 'Synthesis, stability, density, viscosity of ethylene glycol-based ternary hybrid nanofluids: Experimental investigations and model-prediction using modern machine learning techniques,' *Powder Technology*, 400, p. 117190. https://doi.org/10.1016/j.powtec.2022.117190.

[91] Said, Z. et al. (2023) 'Nanofluids-based solar collectors as sustainable energy technology towards net-zero goal: Recent advances, environmental impact, challenges, and perspectives,' *Chemical Engineering and Processing – Process Intensification*, 191, p. 109477. https://doi.org/10.1016/j.cep.2023.109477.

[92] Sajid, M.U. and Bicer, Y. (2020) 'Comparative life cycle cost analysis of various solar energy-based integrated systems for self-sufficient greenhouses,' *Sustainable Production and Consumption*, 27, pp. 141–156. https://doi.org/10.1016/j.spc.2020.10.025.

[93] Salieri, B. et al. (2017) 'Life cycle assessment of manufactured nanomaterials: Where are we?,' *NanoImpact*, 10, pp. 108–120. https://doi.org/10.1016/j.impact.2017.12.003.

[94] Sami, S. (2020) 'Analysis of Nanofluids Behavior in a PV-Thermal-Driven Organic Rankine Cycle with Cooling Capability,' *Applied System Innovation*, 3(1), p. 12. https://doi.org/10.3390/asi3010012.

[95] Samylingam, L. et al. (2020) 'Thermal and energy performance improvement of hybrid PV/T system by using olein palm oil with MXene as a new class of heat transfer fluid,' *Solar Energy Materials and Solar Cells*, 218, p. 110754. https://doi.org/10.1016/j.solmat.2020.110754.

[96] Samylingam, N.L. *et al.* (2024) 'Green Engineering with Nanofluids: Elevating Energy Efficiency and Sustainability,' *Journal of Advanced Research in Micro and Nano Engieering*, 16(1), pp. 19–34. https://doi.org/10.37934/armne.16.1.1934.

[97] Shaik, A.H. et al. (2022) 'A Review on Environmental and Economic Impact of 2D Nanomaterials-Based Heat Transfer Fluids,' *Journal of Nanomaterials*, 2022(1). https://doi.org/10.1155/2022/3443360.

[98] Shaker, A.M. *et al.* (2017) 'Estimation The median lethal dose and inhibitory concentration of TIO2, SIO2, ZNO and CUO nanoparticles on human hepatoma HEPG2 cells,' *International Journal of Pharmaceutical and Phytopharmacological Research*, 7(6), pp. 18–23. https://eijppr.com/storage/models/article/Re1TQmGFZFDtvmRQCFpgCKQ6YQ7Bgx1ddqvId10Zmu0E5rGIX7KhgMY9wthJ/estimation-the-median-lethal-dose-and-inhibitory-concentration-of-tio2-sio2-zno-and-cuo-nanopartic.pdf.

[99] Shanbedi, M. *et al.* (2024) 'Graphene Nanoplatelets- and Multi-Walled Carbon Nanotubes-Based Transformer Oil Nanofluids as Coolants with Enhanced Performance,' *Research Square* [Preprint]. https://doi.org/10.21203/rs.3.rs-4494828/v1.

[100] Shateri, M. *et al.* (2020) 'Comparative analysis of machine learning models for nanofluids viscosity assessment,' *Nanomaterials*, 10(9), p. 1767. https://doi.org/10.3390/nano10091767.

[101] Sheikhpour, M. *et al.* (2020) 'Role of Nanofluids in Drug Delivery and Biomedical Technology: Methods and Applications,' *Nanotechnology Science and Applications*, Volume 13, pp. 47–59. https://doi.org/10.2147/nsa.s260374.

[102] Sharma, P. et al. (2025) 'A comprehensive review of sensible heat thermal energy storage for high temperature applications,' Energy Storage, 7(5). https://doi.org/10.1002/est2.70190.

[103] Siddique, A. et al. (2024) 'Upgradation of highly efficient and profitable eco-friendly nanofluid-based vegetable oil prepared by green synthesis method for the insulation and cooling of transformer,' Clean Technologies and Environmental Policy [Preprint]. https://doi.org/10.1007/s10098-023-02698-7.

[104] Sikarwar, B.S. et al. (2025) 'Nanofluids for enhancing the heat transfer coefficient,' in IGI Global eBooks, pp. 381–418. https://doi.org/10.4018/979-8-3373-0230-0.ch011.

[105] Singh, A. *et al.* (2008) 'Environmental impact assessment for potential continuous processes for the production of carbon nanotubes,' *American Journal of Environmental Sciences*, 4(5), pp. 522–534. https://doi.org/10.3844/ajessp.2008.522.534.

[106] Sonawane, S.S. *et al.* (2022) 'Recent Advances in the Applications of Green Synthesized Nanoparticle based Nanofluids for the Environmental Remediation,' *Current Pharmaceutical Biotechnology*, 24(1), pp. 188–198. https://doi.org/10.2174/1389201023666220411114620.

[107] Spahr, M.E. and Rothon, R. (2015) 'Carbon black as a polymer filler,' in *Springer eBooks*, pp. 1–24. https://doi.org/10.1007/978-3-642-37179-0_36-1.

[108] Spasojević, L., Ivanišević, I. and Sikirić, M.D. (2025) 'Stability and application of TiO2 nanomaterials in aqueous suspensions: a review,' *RSC Advances*, 15(27), pp. 21341–21368. https://doi.org/10.1039/d5ra02499d.

[109] Suneetha, S., Subbarayudu, K. and Reddy, P.B.A. (2022) 'Hybrid nanofluids development and benefits: A comprehensive review,' *Journal of Thermal Engineering*, 8(3), pp. 445–455. https://doi.org/10.18186/thermal.1117455.

[110] Steckiewicz, K.P. et al. (2019) 'Impact of gold nanoparticles shape on their cytotoxicity against human osteoblast and osteosarcoma in in vitro model. Evaluation of the safety of use and anti-cancer potential,' *Journal of Materials Science Materials in Medicine*, 30(2). https://doi.org/10.1007/s10856-019-6221-2.

[111] Strategic Insights into Nanofluid Market Trends (2025). https://www.datainsightsmarket.com/reports/nanofluid-1124127.

[112] Subhan, M.A. and Subhan, T. (2022) 'Safety and global regulations for application of nanomaterials,' in *Elsevier eBooks*, pp. 83–107. https://doi.org/10.1016/b978-0-323-90982-2.00005-6.

[113] Sun, Y. et al. (2021) 'Copper oxide nanoparticles removal by coagulation and optimization by matter–element analysis model,' *Journal of Environmental Chemical Engineering*, 10(1), p. 107096. https://doi.org/10.1016/j.jece.2021.107096.

[114] Sundar, L.S., Singh, M.K. and Sousa, A.C.M. (2013) 'Investigation of thermal conductivity and viscosity of $Fe_3O_4$ nanofluid for heat transfer applications,' *International Communications in Heat and Mass Transfer*, 44, pp. 7–14. https://doi.org/10.1016/j.icheatmasstransfer.2013.02.014.

[115] Tao, H. *et al.* (2023) 'Energy and cost management of different mixing ratios and morphologies on mono and hybrid nanofluids in collector technologies,' *Engineering Applications of Computational Fluid Mechanics*, 17(1). https://doi.org/10.1080/19942060.2022.2164620.

[116] Tao, Q. et al. (2023) 'A review of nanofluids as coolants for thermal management systems in fuel cell vehicles,' *Nanomaterials*, 13(21), p. 2861. https://doi.org/10.3390/nano13212861.

[117] Teah, H.Y. *et al.* (2020) 'Life cycle greenhouse gas emissions of long and pure carbon nanotubes synthesized via On-Substrate and Fluidized-Bed chemical vapor deposition,' *ACS Sustainable Chemistry & Engineering*, 8(4), pp. 1730–1740. https://doi.org/10.1021/acssuschemeng.9b04542.

[118] Temizel-Sekeryan, S. and Hicks, A.L. (2020) 'Global environmental impacts of silver nanoparticle production methods supported by life cycle assessment,' *Resources Conservation and Recycling*, 156, p. 104676. https://doi.org/10.1016/j.resconrec.2019.104676.

[119] Teng, T.-P. *et al.* (2010) 'The effect of alumina/water nanofluid particle size on thermal conductivity,' *Applied Thermal Engineering*, 30(14–15), pp. 2213–2218. https://doi.org/10.1016/j.applthermaleng.2010.05.036.

[120] Tian, M.-W. *et al.* (2021) 'Economic cost and efficiency analysis of the employment of inserting rods with helical fins in a shell and tube heat exchanger under magnetic field and filled with nanofluid,' *Ain Shams Engineering Journal*, 13(4), p. 101651. https://doi.org/10.1016/j.asej.2021.11.020.

[121] Tubular Solar Collector Competitive Strategies: Trends and Forecasts 2025-2033 (2026). https://www.marketreportanalytics.com/reports/tubular-solar-collector-229621.

[122] Wciślik, S. (2025) 'How nanofluids may enhance energy efficiency and carbon footprint in buildings?,' *Sustainability*, 17(15), p. 7035. https://doi.org/10.3390/su17157035.

[123] Wu, F., Zhou, Z. and Hicks, A.L. (2019) 'Life Cycle Impact of Titanium Dioxide Nanoparticle Synthesis through Physical, Chemical, and Biological Routes,' *Environmental Science & Technology*, 53(8), pp. 4078–4087. https://doi.org/10.1021/acs.est.8b06800.

[124] Yang, M., Okazaki, T. and Zhang, M. (2021) 'Removal of Carbon Nanotubes from Aqueous Solutions by Sodium Hypochlorite: Effects of Treatment Conditions,' *Toxics*, 9(9), p. 223. https://doi.org/10.3390/toxics9090223.

[125] Zainon, S.N.M. and Azmi, W.H. (2021) 'Recent Progress on Stability and Thermo-Physical Properties of Mono and Hybrid towards Green Nanofluids,' *Micromachines*, 12(2), p. 176. https://doi.org/10.3390/mi12020176.

[126] Zhang, H. *et al.* (2016) 'Selection of Nanomaterial-Based Active Agents for packaging application: Using Life Cycle Assessment (LCA) as a tool,' *Packaging Technology and Science*, 30(9), pp. 575–586. https://doi.org/10.1002/pts.2238.

[127] Zhang, L. et al. (2021) 'Research on the industrial economic development driving by scientific and technological service system innovation,' *E3S Web of Conferences*, 235, p. 02008. https://doi.org/10.1051/e3sconf/202123502008.

# 8 Numerical Modeling of Heat and Mass Transfer Mixed Convection Casson MHD Nanofluid Flow over a Vertical Plate with Nonlinear Thermal Radiation

Gabriel Samaila[1*] and Basant K. Jha[2]

Models that can accurately predict the behaviors of non-Newtonian nanofluids for high temperatures and microscale effects are of critical importance for thermal management in advanced applications ranging from microelectronics to solar collectors. The most common models lack finite speeds of heat/mass propagation, i.e., Cattaneo-Christov heat, strong nonlinear radiation heat transfer, and partial slip along boundaries. Such aspects are discussed in this chapter by presenting a new comprehensive model involving a Casson MHD nanofluid flow over a vertical plate. Similarity alterations are used to change governing equations into a set of nonlinear ODEs, which are resolved by the Runge-Kutta-Fehlberg method. The effects of major quantities arising from the present model have been analysed theoretically and also quantified. For instance, increasing the magnetic parameter from M = 0.5 to M = 2.5 suppresses maximum velocity by approximately 40% while increasing temperature via Joule heating by over 20%. Enhancing the nonlinear radiation parameter from R = 0.5 to R = 3.0 can increase the thermal boundary layer thickness by >25% and reduce the wall heat transfer rate by 40%. Slip parameters are found to be critically affecting skin friction and mass transfer rates up to 50%. Critical numerical values reported in the present work will provide a clear guide to design optimum thermal performance in various practical applications such as drug delivery, compact heat exchangers, and MHD power generators.

## 1. Introduction

The classical Casson fluid mathematical model is used in the analysis of non-Newtonian fluids, which predicts the flow phenomena of viscoelastic fluids that demonstrate stress-fluid behaviors that exhibit solid properties till a certain stress threshold is surpassed. The mathematical model can be applied in the study of materials like honey, blood, printing ink, and certain polymers. In this communication, Casson fluid is incorporated in the MHD boundary layer flow model owing to its practical importance in biological systems, principally in the

1 Department of Mathematics, Air Force Institute of Technology, Kaduna, Nigeria.
2 Department of Mathematics, Ahmadu Bello University, Zaria, Kaduna, Nigeria.
* Corresponding author: gabbbooooo@yahoo.com

DOI: 10.1201/9781003686002-8

examination of blood flow and engineering practices involving thick fluids. The incorporation of Casson fluid in the present analysis of highly non-linear flow past a radiated thin vertical wall is useful in processes like food processing devices, cooling systems, and polymer extrusion. Using magnetohydrodynamics (MHD) and nanofluids, understanding the fundamental mass and heat transfer characteristics of Casson fluid with nonlinear thermal radiation and partial slip conditions expands its application to modern energy systems like microelectromechanical systems (MEMS), solar collectors, and enhanced oil recovery. The study of MHD nanofluid features coupled with mass and energy transport past a perpendicular cone influenced by both thermal radiation and chemical reaction was reported by Abbas, Ibrahim, *et al.*, (2023). (Abbas, Megahed, *et al.*, 2023) examined a slippery non-Newtonian nanofluid flow caused by an elongated flat surface implanted in a permeable medium under the influence of thermal partial slip and energy source. The MHD energy and mass transport features past a vertical cone having convective surface condition was reported by Reddy and Sreedevi (2021). (Hanif, Khan, and Shafie, 2019) presented a comprehensive report on MHD free convection in cadmium telluride nanofluid near a perpendicular pinecone implanted in a permeable material. The MHD boundary layer flow of AWCNT-nanofluid and MWCNT-water past a perpendicular pinecone with energy source/sink was examined by Reddy, Reddy, and Reddy (2019). (Saleem *et al.*, 2019) explored the fundamentals of magneto-Jeffrey nanofluid bioconvection past a constantly revolving perpendicular pinecone caused by a gyrotactic microbe. Some recent literature on this subject can be found in (Jha and Samaila, 2020b; Ashwinkumar, Samrat, and Sandeep, 2021; B. K. Jha and Samaila, 2022b).

Magnetohydrodynamics (MHD) is an electrically conducting fluid , including liquid metals, nanofluids, plasmas, and salt water in the presence of magnetic fields. These electrically conducting fluids face a strong opposing force (i.e., Lorentz force) when they pass through the magnetic field. The combination of MHD and effect is the best practice in heat and mass transport enhancement, especially in progressive thermal management plants. Nanofluids they are fluids containing some nanoparticles with higher thermal conductivities compared to the base fluids, and are commonly used in engineering applications where enhanced heat transfer is needed. The combined effects of nanofluid flow and magnetic field can be applied in electromagnetic pumps, nuclear reactors, biomedical systems, and MHD generators. It is also significant in solar collectors, high-performance devices, cooling of electronic devices, and microfluidic devices. (Qureshi *et al.*, 2021) studied mass and energy transport in MHD flow in a permeable material influenced by Soret and Dufour impacts through the numerical approach. The consequences of chemical reaction and thermal radiation on convective mass and energy transport in MHD nanofluid were reported by Bello, Olajuwon, and Kuye (no date). Muntazir *et al.*, (2022) scrutinized characteristic features of MHD nanofluid transport near a porous plate influenced by viscous dissipation. The sporadic MHD mixed convection flow over a cone in an absorbent material having a variable temperature was explored by Ashraf *et al.*, (2022). (Kodi *et al.*, 2023) sightseen the significant impact of MHD mixed convection transport for Maxwell nanofluid past a perpendicular pinecone with absorbent material having inconstant thermal diffusion and conductivity. Recent publications on this subject can also be found in (Hayat, Awais, and Asghar, 2013; Sheikholeslami, 2017; Waqas *et al.*, 2017; Jha and Samaila, 2020a; Vaidya *et al.*, 2021; B. K. Jha and Samaila, 2022a; John, Basavarajappa, and Shevchuk, 2024)

Thermal radiation significantly influences heat transfer in high-temperature environments, especially in nuclear reactors, aerospace vehicles, solar thermal plants, and combustion chambers. In a classical boundary layer flow analysis, particularly those involving nanofluid or non-Newtonian fluid, radiation serves as a supplementary heat transfer mechanism, affecting the thermal gradient and temperature dispersal in the boundary layer region. The usual model uses the linearised Rosseland approximation, which considers temperature gradient in the materials is significantly small. But this postulation fails in various physical situations involving robust radiative heat flux or extremely high temperature gradients, where the preservation of the fourth-power law of radiation is essential. In such physical situations, the use of nonlinear thermal radiation is paramount since it includes the $T^4$ dependency of radiative heat flux, resulting in further accurate and realistic predictions. In the present exploration, including nonlinear thermal radiation in the classical Casson MHD nanofluid flow mathematical model allows better prediction of heat transport behaviour near the perpendicular plate. (Abbaspour *et al.*, 2021) obtained a closed-form solution for the MHD boundary layer of nanofluid past an exponentially stretching surface governed by nonlinear thermal radiation. The analysis of unsteady MHD nanofluid flow with combined effects of energy dissipation and nonlinear thermal radiation was examined by Majeed *et al.*, (2022). The consequence of nonlinear thermal radiation on MHD rotating nanofluid flow having Joule heating was also presented by Hayat, Aziz, and Alsaedi (2021). (B. Jha and Samaila, 2022) gave a similar solution for flow past a perpendicular, constantly moving surface with suction/injection and nonlinear thermal radiation. Some relevant literature on this subject can be found in (Ishak, Yacob, and Bachok, 2011; Motsumi and Makinde, 2012; Qayyum *et al.*, 2017; Ymeli *et al.*, 2019; Gireesha *et al.*, 2020; Jha and Samaila, 2021; Ali *et al.*, 2024).

The usual boundary layer flow presumes a no-slip boundary condition, meaning that the fluid has zero velocity at the plate surface. Nonetheless, this postulation fails in various contemporary engineering endeavours. In nanotechnology, microfluidics, and biofluid mechanics, plate features interfacial impacts cause velocity at the plate surface. Similarly, thermal (temperature) and solutal (concentration) gradients past the plate could exhibit discontinuities, resulting in the rise in concentration and thermal slip conditions. The existence of slip conditions has a significant effect on shear stress, heat and mass transport, and boundary layer thickness. While the concentration and thermal slips influence the respective fluxes at the plate surface, the momentum slip usually decreases the plate shear stress and affects the velocity curve. These phenomena are mostly useful in NEMS systems, coating flows, porous media, and physiological systems, where the accuracy and performance are affected due to the interaction at the plate surface. (Bansal and Yadav, 2024) analysed the impact of velocity slip on Newtonian fluid flow past a stretching sheet embedded in a porous medium. (Mahmood, Khan, and Al-Zubaidi, 2024) explored the effect of the heat source on nanofluid flow over a moving plate with slip conditions. (Adnan *et al.*, 2024) scrutinised energy response due to convective-radiative MHD stagnation point flow governed by thermal slip and ternary nanoparticles. Thermal and velocity slip effects on boundary layer flow of micropolar nanofluid past a vertical stretched Riga sheet were reported by Abbas *et al.*, (2024). (Vinita *et al.*, 2021) examined the thermal, velocity, and concentration MHD slips past a nonlinear extended cylinder with outer velocity. Some relevant literature on this subject includes (Chen and Weng, 2006; Malvandi and Ganji, 2015; Ghadikolaei *et al.*, 2017; Tian, Li, and Zhang, 2017; Alrihieli, 2023; Jha and Samaila, 2023; Kumar and Pathak, 2023; Sudarmozhi *et al.*, 2023).

The usual energy and mass transport models, like Fick's law and Fourier's law of diffusion and conduction, assume instant propagation of heat and mass signals, which signifies infinite speed. Though these prepositions represent mathematical modelling, they are not appropriate for modelling engineering practices with high-temperature, microscale applications or contemporary high-speed. To overcome these challenges, the Cattaneo-Christov algorithm introduces relaxation time parameters, taking into account solutal diffusion and finite thermal speed, and thus improving the inconsistency of instantaneous signal dissemination. The Cattaneo-Christov model includes material derivatives of mass and heat fluxes, enhancing the practicality of the forecast responses in dynamic concentration and thermal fields. It is mostly important in engineering activities involving nano/microscale flow, transient phenomena, or high-frequency thermal waves where orthodox laws fail. Incorporating the Cattaneo-Christov mass and heat flux in the present exploration permits a more particular demonstration of mass and energy transport in the Casson MHD nanofluid transport with slip conditions governed by thermal radiation. In the precise report presented by Zuhra *et al.*, (2022) on the examination of the Cattaneo-Christov heat flux algorithm past a magnetic combined with Casson nanofluid transport using Levenberg-Marquardt backwards propagated neural networks concluded that the Casson fluid parameter enhances the skin friction distribution. (Sadiq *et al.*, 2024) predict the influence of thermal slip on Cattaneo-Christov heat flux past a lubricated plate. Some relevant literature in this category can be found in (Ali and Khan, 2022; Rehman *et al.*, 2023; Shahzad *et al.*, 2023; Chaudhary and Deshwal, 2024; Sohail *et al.*, 2024).

## 1.1 Novelty of the Chapter

The literature survey reaffirms the substantial development in the analysis of Casson MHD nanofluid flows. Several studies have explored individual phenomena, like the impact of a magnetic field Ali *et al.*, (2024), linear thermal radiation Motsumi and Makinde (2012), and velocity slip (Adnan *et al.*, 2024). However, the key lacuna is in terms of a proper model that considers the symbiotic effects of (i) highly nonlinear thermal radiation, (ii) nonlinear density variation with both concentration and temperature, and (iii) velocity, thermal, and concentration slip conditions under the Cattaneo-Christov heat and mass flux setup, which remains mostly unaddressed. Although the literature already provides basic ideas, most of the previous works have discussed only subsets of the above features and, in turn, have ignored their interactive relationship. For example, the combined magnetic and slip parameters' damping action on the flow, nonlinear radiation heating interacting with non-Fourier heat conduction, has never been studied. The main novelty of the current work is thus to fill this lacuna by constructing and analyzing such a unified model that accommodates these complicated, simultaneous physics. This unification is not merely of academic interest but is, in fact, crucial for the realistic simulations of practical scenarios arising in advanced engineering systems. Hence, the presently undertaken study offers a more realistic framework for application in microelectronic cooling, solar energy collectors, biomedical drug delivery, and materials processing, wherein these coupled phenomena collectively control system performance.

## 2. Fundamentals of Heat and Mass Transfer in Nanofluid Flow

Energy and mass transport in nanofluids are associated with convection, conduction, and radiation, which are greatly influenced by the suspension of nanomaterials. In contrast to the usual base fluids, nanofluids demonstrate high thermal conductivity, unique diffusion features, improved viscosity, and, by this means, enhance the energy transport within the boundary layer (Mane and Hemadri, 2022; Mane, Tripathi, and Hemadri, 2022).

The heat transfer process involves:

- **Convection:** Nanomaterial suspension essentially affects the thermophysical characteristics of the base fluid. Now, both the effective specific heat capacity $(\rho C_p)_{rif}$ and dynamic viscosity $\mu_{rif}$ are functions of nanomaterial volume fraction $\phi$. This directly influences the energy and momentum boundary layers. For example, an increased effective viscosity usually thickens the momentum boundary layer and makes the velocity lower, while the change of $\rho C_p$ influences the thermal inertia of the fluid. In mixed convection, these property changes modulate the buoyancy force $g\beta(T - T_\infty)$, making the two-way coupling between the temperature and velocity fields much more complicated than in pure fluids.
- **Conduction:** The micro mechanisms responsible for the enhanced thermal conduction in nanofluids include: First, high intrinsic thermal conductivity nanomaterials, like $Cu$ and $Al_2O_3$, provide preferential conduits for heat flow. Second, a nanolayer of liquid molecules surrounds each particle, with ordered structure and different thermal conductivity from the bulk liquid serving as a bridging layer to reduce Kapitza resistance at the solid-liquid interface. These mechanisms are lumped together in effective media theories, such as

the Hamilton-Crosser model, in which the nanofluid's thermal conductivity is proportional to the conductivity of the base fluid, $k_f$; the particle conductivity, $k_p$; and a shape factor for the particle.

- **Radiation:** Thermal radiation becomes a dominant mode of energy transfer at operational temperatures encountered in solar collectors, nuclear reactors, or aerospace systems. The nonlinear Rosseland approximation is used here. The radiative heat flux is expressed as $q_r = \frac{4\sigma^*}{3k^*}\frac{\partial T^4}{\partial y}$, where $k^*$ and $\sigma^*$ respectively symbolise the mean absorption coefficient and the Stefan-Boltzmann constant. The presence of the $T^4$ term introduces a strong nonlinearity $\frac{\partial T}{\partial y}$ in the energy equation that significantly modifies the thermal boundary layer structure toward more realistic predictions of temperature distribution and wall heat flux than provided by the linearized model.

The mass transfer process involves:

- **Brownian Motion:** This is the casual microscopic motion of nanoparticles dispersal in a fluid, which is driven by interaction with the base fluid particles. Whereas it is a contribution to the effective thermal conduction of the nanofluid, its main role in mass transfer is to induce a diffusive-like spreading of nanoparticles. It is modelled as a contribution to the mass flux, being proportional to the Brownian diffusion coefficient $D_B$. Brownian motion acts to homogenize the nanoparticle concentration; therefore, it flattens the concentration curve and reduces the depth of the concentration boundary layer.
- **Thermophoresis:** Also called the Soret effect, this represents a phenomenon where nanoparticles are acted upon by a force in an opposite direction to an applied temperature difference $\nabla T$. A net movement of particles from the hotter to the cooler zone occurs as a result of such a force. In the boundary layer over a hot vertical plate, for example, thermophoresis serves to deplete nanoparticles from the heated wall and accumulate them in the cooler ambient flow. This migration is modelled in terms of a thermophoretic diffusion term in the concentration equation, proportional to $D_T\frac{\nabla T}{T}$. This process directly couples the energy and concentration equations, as the temperature field dictates the particle distribution that influences thermal conductivity.
- **Diffusion (Fickian Transport):** In the presence of the concentration gradient $\nabla\phi$, the macroscopic transport of nanoparticles is governed by Fick's Law. The resulting mass flux, $-DB\nabla\phi$, is the primary means of mass transfer at the continuum scale. The interaction of nanoparticles can modify the effective diffusivity DBDB. This Fickian diffusion acts in concert with, and often in opposition to, thermophoresis in boundary layer flow to determine the eventual concentration profile $\phi(y)$.

However, various complexities arise in the framework of Casson MHD nanofluid flow:

The magnetic field creates a Lorentz force, which suppresses fluid movement and retards fluid flow in the boundary layer. The presence of nonlinear thermal radiation regulates energy transport in radiative environments results in better predictions in comparison to the so-called Rosseland approximation. The present study introduces the Cattaneo-Christov mass and energy flux mathematical formulation. Additionally, concentration, thermal, and momentum partial slip conditions incorporated in the current analysis characterized constant microscale interactions at the plate surface, which are significant in contemporary applications like NEMS, MEMS, nanofluid flow, and biomedical devices.

## 3. Analytical Modelling Techniques

The analytic modelling procedure is the first step in CFD simulation, which involves converting highly complex physical situations (usually in partial differential equations) into simple mathematical models (usually in ordinary differential equations). However, the process is highly complex, especially in some physical problems involving nonlinear coupled differential equations. This procedure ensures that the usual conservation laws of momentum, mass, concentration, and energy, which dictate the fluid transport, are satisfied. Therefore, to certify that the formulation includes all relevant flow physics, suitable constitutive relations are formulated. In the current exploration, Casson fluid is incorporated to depict the viscoelastic properties of non-Newtonian fluids, whereas the usual Rosseland estimation for nonlinear thermal radiation was introduced to predict the dynamics of radiative heat transfer in a high-temperature environment. Furthermore, the relaxation effects are taken care of by the Cattaneo-Christov mass and heat flux mathematical model in the current case study, which addresses the limitations of Fick's and Fourier's conventional laws of unlimited propagation speeds.

Upon the establishment of relevant partial differential equations (PDE), the next stage of CFD simulation is reducing the PDEs to simple ODEs through an appropriate technique, such as similarity transformation. This was achieved through defining a suitable similarity variable $\eta$ and other dimensionless variables such as temperature $\theta$, concentration $\phi$ and stream function $f$ as depicted in equation (12). It is important to know that this transformation not only translates PDEs to ODEs but also reveals the substantial effects of dimensionless parameters like Casson Parameter ($\beta$), nonlinear thermal radiation parameter ($R$), Prandtl number (Pr), thermophoresis parameter ($Nt$), Brownian motion parameter ($Nb$), magnetic parameter ($M$), local Grashof number ($Gr_x$), temperature relaxation parameter ($\delta_e$), momentum slip parameter ($\xi$), concentration relaxation parameter ($\delta_e$), Lewis number ($Le$), thermal slip parameter ($\zeta$), diffusion slip parameter ($\zeta$), and density variation with concentration ($\lambda_c$) density variation with temperature ($\lambda_T$) on the flow features like velocity, temperature and concentration. Similarity variables play a significant role in boundary layer analysis as they help in reducing complex multidimensional flow to a concise form that can easily be resolved and interpreted. It is significant to stress that, though the similarity transformation is a prevailing analytical method, the reduced system of ordinary differential equations remains nonlinear because of the existence of nonlinear thermal radiation, relaxation terms, Casson fluid, and partial slip. Due to the nonlinear nature and coupling of the reduced system ODEs, exact solutions are possible. Therefore, the analytic modelling adopted in the study aids as a significant link between the numerical solution and the physical formulation.

## 4. Numerical Methods Overview

The boundary layer analysis is associated with coupled and nonlinear differential equations, particularly when considering physical problems involving energy and mass transport. The exact solution of such a mathematical model is not feasible; hence, the use of a suitable numerical method is necessary. However, the choice of the numerical scheme affects the correctness of the solution, especially near the plate surface, where the flow physics need to be predicted accurately. In the contemporary study, therefore, the transport equations (i.e., continuity, energy, momentum, and species concentration) were transformed to ODEs using analytical modelling techniques called the similarity technique/transformation. This procedure reduced the PDEs to ODEs, after which the final ODEs were solved in Maple software by the Runge-Kutta-Fehlberg scheme, after utilising the shooting framework.

Though the technique employed in the current analysis is not a fully CFD simulation, it is associated with the fundamentals governing CFD. Here, instead of solving the transport equations (PDE systems) using a discretised mesh through commercial software like ICEM, WORKBENCH, etc., the reduced one-dimensional ODEs are solved directly using an ODE solver. Furthermore, the shooting techniques provide high precision with a lower computational cost, making them suitable for sensitivity analysis, parametric investigations, and the investigation of fundamental transport systems.

### *4.1 Computational Fluid Dynamics (CFD) and Similarity Transformation*

Fluid transport through various geometries is governed by modified Navier-Stokes equations, which satisfy the conservation laws. The equations are usually a system of coupled PDEs which is not possible to solved analytically. Therefore, CFD is a process of employing suitable numerical methods to solve the system of PDEs describing the fluid transport. This begins with discretising the whole geometry (flow domain) and providing the pressure, species concentration, velocity, and temperature field. In distinction to this convention, the present study considered a reduced-order modelling technique (similarity transformation) to reduce the

PDEs to ODEs, which were then solved numerically. This is the best practice in boundary layer investigation as it maintains all the transport features but neglects the use of iterative solvers and bulky computational meshes. As an alternative, the present analysis solved the reduced ODEs by the Runge-Kutta-Fehlberg scheme. This method makes the investigation effective and also reveals the roles of key parameters in the existing study.

## 5. Model Development Process

Considered in this study is a laminar mixed convection Casson MHD nanofluid transport over a vertical plate influenced by nonlinear thermal radiation. Cartesian coordinates are considered by taking the $x$-axis normal to the Casson nanofluid flow and the $y$-axis along the thermally radiated vertical plate stretched with velocity $u_w(x) = ax$, where $a > 0$ represents the stretching degree as portrayed in Figure 1. The magnetic field is reinforced in the $x$-direction. The plate is preserved at continuous temperature $T_w$ and ambient temperature $T_\infty$ with their corresponding nanoparticle concentration $C_w$ and $C_\infty$ respectively. Similarly, chemical reaction species not present, that is, the joule heat, viscous dissipation, and all kinds of contact forces is ignored. Furthermore, the nanofluid and base fluid are considered to be in thermal steadiness.

Following [42], the rheological equation for the isotropic incompressible Casson can be formulated as

$$\tau_{ij} = \begin{cases} 2\left(\mu_B + \dfrac{p_y}{\sqrt{2\pi}}\right)e_{ij}, & \pi > \pi_c, \\ 2\left(\mu_B + \dfrac{p_y}{\sqrt{2\pi_c}}\right)e_{ij}, & \pi < \pi_c, \end{cases} \quad ...(1)$$

Where $e_{ij}$ and $\pi_{ij}$ are the $(i, j)$th component of the respective deformation degree and stress tensor, $e_{ij}\, e_{ij}$ denotes the product of the distortion degree of an element, $\mu_\beta$ signifies elastic dynamic viscosity, $\pi_c$ represent alike values.

Therefore, the transport equations are;

*Continuity:*

$$\frac{\partial u}{\partial x} + \frac{\partial v}{\partial y} = 0 \quad ...(2)$$

*Momentum:*

$$u\frac{\partial u}{\partial x} + v\frac{\partial u}{\partial y} = v_f\left(1 + \frac{1}{\beta}\right)\frac{\partial^2 u}{\partial y^2} - \frac{\sigma B^2 u}{\rho_f} +$$

$$g\rho[\beta_1(T - T_\infty) + \beta_2(T - T_\infty)^2] + g\rho_{0*}[\beta_{1*}(C - C_\infty) + \beta_{2*}(C - C_\infty)^2] \quad ...(3)$$

Where $u$ and $v$ are respectively velocity component to $x$ and $y$, $\rho_f$ and $v_f$ are respectively fluid density and kinematic viscosity, $\sigma$ electrical conductivity and $\beta = \dfrac{\mu_B \sqrt{2\pi_c}}{p_y}$ is the parameter representing the Casson fluid.

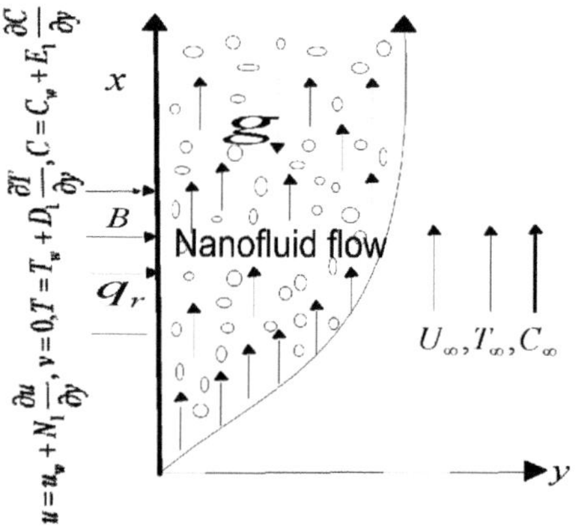


**Fig. 1** Flow configuration and coordinate system.

### *5.1 Cattaneo-Christov Framing*

The usual form of Fourier's and Fick's law, also referred to as the classical Cattaneo-Christov heat flux model, can be written as

$$q\,|+\lambda_H\left[\frac{\partial q}{\partial t}+V\cdot\nabla q-q\cdot\nabla V+(\nabla\cdot V)q\right]=-\,k_f\nabla T \qquad ...(4)$$

$$J\,|+\lambda_M\left[\frac{\partial J}{\partial t}+V\cdot\nabla J-J\cdot\nabla V+(\nabla\cdot V)J\right]=-\,D_B\nabla C \qquad ...(5)$$

Here, $J$ and $q$ are the usual mass and heat flux, $k_f$ fluid thermal conductivity, $V$ signifies the velocity vector, $\lambda_H$ and $\lambda_M$ are the relaxation periods of mass and energy flux. $D_B$ is the Brownian diffusion. Equations (4) and (5) are Fourier's and Fick's when $\lambda_H = 0$ and $\lambda_M = 0$.

Note that the steady state $\dfrac{\partial q}{\partial t} = 0$, $\dfrac{\partial J}{\partial t} = 0$ and continuity equation $\nabla \cdot V = 0$ gives another version of equations (4) and (5) as

$$q+\lambda_H\,[V\cdot\nabla q-q\cdot\nabla V]=-k_f\nabla T \qquad ...(6)$$

$$J+\lambda_M\,[V\cdot\nabla J-J\cdot\nabla V]=-D_B\nabla C \qquad ...(7)$$

### 5.2 Nanoparticle Volume Fraction and Energy Preservation

Following equations (6) and (7), the volume fraction equations and energy can be expressed as

$$u\frac{\partial T}{\partial x}+v\frac{\partial T}{\partial y}+\lambda_H\left(u\frac{\partial u}{\partial x}\frac{\partial T}{\partial x}+v\frac{\partial v}{\partial y}\frac{\partial T}{\partial y}+u\frac{\partial v}{\partial x}\frac{\partial T}{\partial y}\right.$$
$$\left.+v\frac{\partial u}{\partial y}\frac{\partial T}{\partial x}+2uv\frac{\partial^2 T}{\partial x\partial y}+u^2\frac{\partial^2 T}{\partial x^2}+v^2\frac{\partial^2 T}{\partial y^2}\right)$$
$$=\alpha_f\left(\frac{\partial^2 T}{\partial x^2}+\frac{\partial^2 T}{\partial y^2}\right)+\tau\left[D_B\left(\frac{\partial C}{\partial x}\frac{\partial T}{\partial x}+\frac{\partial C}{\partial y}\frac{\partial T}{\partial y}\right)+\frac{D_T}{T_\infty}\right.$$
$$\left.\left\{\left(\frac{\partial T}{\partial x}\right)^2+\left(\frac{\partial T}{\partial y}\right)^2\right\}\right]-\frac{1}{(\rho c_p)_f}\frac{\partial q_r}{\partial y} \quad ...(8)$$

$$u\frac{\partial C}{dx}+v\frac{\partial C}{\partial y}+\lambda_M\left(u\frac{\partial u}{\partial x}\frac{\partial C}{\partial x}+v\frac{\partial v}{\partial y}\frac{\partial C}{\partial y}+u\frac{\partial v}{\partial x}\frac{\partial C}{\partial y}\right.$$
$$\left.+v\frac{\partial u}{\partial y}\frac{\partial C}{\partial x}+2uv\frac{\partial^2 C}{\partial x\partial y}+u^2\frac{\partial^2 C}{\partial x^2}+v^2\frac{\partial^2 C}{\partial y^2}\right)$$
$$=D_B\left(\frac{\partial^2 C}{\partial x^2}+\frac{\partial^2 C}{\partial y^2}\right)+\frac{D_T}{T_\infty}\left(\frac{\partial^2 T}{\partial x^2}+\frac{\partial^2 T}{\partial y^2}\right) \quad ...(9)$$

In this context, $\alpha_f=\dfrac{k_f}{(\rho c_p)_f}$ symbolises thermal coefficient, $D_T$ signifies the thermophoretic diffusion, $\tau=\dfrac{(\rho c)_s}{(\rho c)_f}$ denote the equivalent base fluid and nanoparticle heat capacity and $q_r$, $C$ and $T$ are respectively radiative heat flux, dimensional nanoparticle concentration and dimensionless temperature.

Using the Rosseland approximation for optical fluid, the radiative heat flux is written as

$$q_r=-\frac{4\sigma^*}{3k^*}\frac{\partial T^4}{\partial y} \quad ...(10)$$

Where $\sigma^*$ and $k^*$ are respectively Stefan-Boltzmann constant and absorption coefficient.

### 5.3 Relevant Boundary Conditions

The following are relevant boundary conditions:

$$u = u_w + N_1 \frac{\partial u}{\partial y}, \, n = 0, \, T = T_w + D_1 \frac{\partial T}{\partial y}, \, C = C_w + E_1 \frac{\partial C}{\partial y} \quad ...(11)$$

In this context, $D_1$ represent thermal slip factor, $N_1$ is the momentum slip factor and $E_1$ the corresponding nanoparticle diffusion slip factor.

## 5.4 The Similarity Transformations

To convert the PDEs into ODEs, we define the similarity variable and other quantities as:

$$\eta = y\sqrt{\frac{a}{v_f}}, \quad u = axf'(\eta), \quad v = -\sqrt{av_f}\, f(\eta),$$

$$\theta(\eta) = \frac{T - T_\infty}{T_w - T_\infty}, \, \phi(\eta) = \frac{C - C_\infty}{C_w - C_\infty} \quad ...(12)$$

Using equation (12) in equations (3), (8), and (9), the model reduces to

$$\left(1 + \frac{1}{\beta}\right) f''' - f'^2 + ff'' - Mf' + Gr_x(\theta + \lambda_1\theta^2) + Gc_x(\phi + \lambda_2\phi^2) \quad ...(13)$$

$$\theta''(\eta)\left[1 + \frac{4R}{3}(\theta + C_T)^3\right] + 4R[C_T + \theta]^2\,[\theta'(\eta)]^2 + \Pr\,\theta'(\eta)f(\eta) +$$

$$Nb \cdot \phi'\theta' + Nt \cdot (\theta')^2 - \delta_e(ff'\theta' + f^2\theta'') = 0, \quad ...(14)$$

$$\phi''(\eta) + \Pr Le\,\phi'(\eta)f(\eta) + \frac{Nt}{Nb}\,\theta'' - \Pr Le\delta_c\,(ff'\phi' + f^2\phi'') = 0. \quad ...(15)$$

Subject to;

$$f'(0) = 1 + \xi f''(0), f(0) = 0, \, \theta(0) = 1 + \zeta\theta'(0), \, \phi(0) = 1 + \varsigma\phi'(0)$$

$$f'(\infty) = 0, \, \theta(\infty) = 0, \, \phi(\infty) = 0, \quad ...(16)$$

with

$$Gr_x = \frac{x^2 g\rho\beta_1(T_w - T_\infty)}{a^2}, \, Gc_x = \frac{x^2 g\rho\beta_{1*}(C_w - C_\infty)}{a^2}, \, \lambda_1 = \frac{\beta_2(T_w - T_\infty)}{\beta_1}$$

$$R = \frac{4\sigma(T_w - T)^3}{kk*}, \, C_T = \frac{T_\infty}{T_w - T_\infty}, \, \delta_c = a\lambda_M, \, \lambda_2 = \frac{\beta_{2*}(C_w - C_\infty)}{\beta_{1*}}, \, \delta_e = a\lambda_H$$

$$...(17)$$

$$\Pr = \frac{v_f}{\alpha_f},\ Le = \frac{\alpha_f}{D_B},\ Nt = \frac{\tau D_t (T_w - T_\infty)}{T_\infty v_f},\ \xi = N_1\sqrt{\frac{a}{v}}$$

$$Nb = \frac{\tau D_B (C_w - C_\infty)}{v_f},\ M = \frac{\sigma B^2}{\rho_f a},\ \varsigma = E_1\sqrt{\frac{a}{v_f}},\ \zeta = D_1\sqrt{\frac{a}{v_f}} \qquad ...(18)$$

where

$Nt$, Pr, $Nb$, $M$ and $Gr_x$ are the thermophoresis parameter, Brownian motion parameter, magnetic parameter, and local Grashof number, respectively. $\delta_e$, $\xi$, $\delta_c$, $R$, $\zeta$, $Le$, $\varsigma$, $C_T$, $\lambda_2$ and $\lambda_1$ are respectively the temperature relaxation parameter, dimensionless hydrodynamic/velocity slip parameter, concentration relaxation, nonlinear thermal radiation, dimensionless thermal/temperature slip, Lewis number, dimensionless mass/concentration slip, temperature variation, density variation with concentration, and nonlinear density variation with temperature parameters.

## 5.5 *Some Physical Quantities*

Some quantities considered in this study include the Nusselt number $Nu_x$, skin friction coefficient $C_f$ and Sherwood number $Sh_x$, which are defined as follows;

$$NU_x = -\frac{x}{(T_w - T_\infty)}\left(\frac{\partial T}{\partial y}\right)_{y=0} \qquad ...(19)$$

$$C_f = \frac{\mu_f}{\rho_f u_w^2}\left(\frac{\partial u}{\partial y}\right)_{y=0} \qquad ...(20)$$

$$Sh_x = -\frac{x}{(C_w - C_\infty)}\left(\frac{\partial C}{\partial y}\right)_{y=0} \qquad ...(21)$$

Utilising the above dimensionless variables, the $Nu_x$, $C_f$ and $Sh_x$ can be reduced to:

$$Nu_r = \mathrm{Re}_x^{1/2} \cdot Nu_x = -\theta'(0), \qquad ...(22)$$

$$C_{fr} = \mathrm{Re}_x^{1/2} \cdot C_f = f''(0), \qquad ...(23)$$

$$Sh_r = \mathrm{Re}_x^{1/2} \cdot Sh_x = -\phi'(0), \qquad ...(24)$$

Where $\mathrm{Re}_x = \dfrac{u_w x}{v_f}$ symbolizes the local Reynolds number.

### 5.6 Solution Through Shooting Techniques

The ODEs presented in equations (13)-(15), and the corresponding boundary condition in equation (16), are coupled and highly nonlinear; thus, an exact solution is not feasible. We first transformed the ODEs and the BCs into a set of initial value problems by letting:

$$f = x_1, f' = x_2,\ f'' = x_3,\ \theta = y_1,\ \theta' = y_2,\ \phi = z_1,\ \phi' = z_2 \quad ...(25)$$

Using equation (25), equations (13)–(16) can be written as,

$$\left(1+\frac{1}{\beta}\right)x_3' - x_2^2 + x_1x_3 - Mx_2 + Gr_x(y_1 + \lambda_1 y_2^2) + Gc_x(z_1 + \lambda_2 z_2^2) = 0 \quad ...(26)$$

$$y_2^1\left[1+\frac{4R}{3}(y_1 + C_T)^3\right] + 4R[C_T + y_1]^2[y_2]^2 + \Pr y_2 x_1 +$$
$$Nb \cdot z_2 y_2 + Nt \cdot (y_2)^2 - \delta_e(x_1x_2y_2 + x_1^2 y_2^1) = 0 \quad ...(27)$$

$$z_2^1 + \Pr Lez_2x_1 + \frac{Nt}{Nb}y_2^1 - \Pr Le\delta_c(x_1x_2z_2 + x_1^2 z_2^1) = 0 \quad ...(28)$$

Subject to

$$x_2(0) = 1 + \xi x_3(0),\ x_3(0) = \alpha_1,\ x_1(0) = 0,$$
$$y_1(0) = 1 + \zeta y_2(0),\ y_2(0) = \alpha_2,$$
$$z_1(0) = 1 + \varsigma z_2(0),\ z_2(0) = \alpha_3. \quad ...(29)$$

Where the $x_3(0)$, $y_2(0)$ and $z_2(0)$ are obtained through successful iterations until $f'(\infty) \to 0$, $\theta(\infty) \to 0$ and $z_2(0)$ is attained.

The resulting IVPs, along with corresponding BCs, are integrated in Maple 2022 software through the Runge-Kutta Fehlberg technique.

## 6. Validation Techniques

Table 1 is prepared to assess the precision of the current approach through contrast with (Nadeem, Haq, and Khan, 2014) and (Khan and Pop, 2010) on $-\theta'(0)$ and $-\phi'(0)$ for several values of the embedded variables. The computations are carried out for $\Pr = 10$, $Le = 1.0$, $Nb = 0.1$, $Nt = 0$, $\lambda_1 = \lambda_2 = C_T = 0$ $M = R = \xi = \zeta = \varsigma = \delta_c = \delta_e = 0$. It is observed that the present analysis is promising.

**Table 1** Validation of the result through the comparison on $-\theta'(0)$ and $-\phi'(0)$ with (Nadeem, Haq and Khan, 2014) and (Khan and Pop, 2010) for $\Pr = 10$, $Le = 1.0$, $Nb = 0.1$, $Nt = 0$, $\lambda_1 = \lambda_2 = C_T = 0$ $M = R = \xi = \zeta = \varsigma = \delta_c = \delta_e = 0$.

| | *Present* | | *(Nadeem, Haq, and Khan, 2014)* | | *(Khan and Pop, 2010)* | |
|---|---|---|---|---|---|---|
| *Nt* | $-\theta'(0)$ | $-\phi'(0)$ | $-\theta'(0)$ | $-\phi'(0)$ | $-\theta'(0)$ | $-\phi'(0)$ |
| 0.1 | 0.95238 | 2.12939 | 0.9524 | 2.1294 | 0.9524 | 2.1294 |
| 0.2 | 0.69317 | 2.27402 | 0.6932 | 2.2732 | 0.6932 | 2.2740 |
| 0.3 | 0.52007 | 2.52863 | 0.5201 | 2.5286 | 0.5201 | 2.5286 |
| 0.4 | 0.40258 | 2.79516 | 0.4026 | 2.7952 | 0.4026 | 2.7952 |
| 0.5 | 0.32105 | 3.03514 | 0.3211 | 3.0351 | 0.3211 | 3.0351 |

## 7. Sensitivity and Parametric Analysis

In CFD and heat transfer analysis, parametric and sensitivity analysis play a significant role as they demonstrate the consequences of the dynamic parameters on the flow features. This study not only provides a clear picture of the contribution of the key parameter but also reveals parameters that exert a dominant effect on mass diffusion, heat transfer, and other flow features. This is applied in engineering endeavours such as control, optimisation, and design of thermal-fluid devices.

In the current exploration, a systematic sensitivity study was carried out by varying a wide range of active dimensionless quantities present in the reduced governing equations. These variables reveal various physical phenomena:

- *Magnetic Parameter $M$*: Measures the degree of Lorentz force.
- *Lewis number $Le$ and Prandtl number* Pr: respectively signify the ratio of mass diffusivities, thermal and momentum.
- *Local solutal $Gc_x$ and thermal $Gr_x$ Grashof numbers*: quantify the buoyancy force because of concentration and temperature gradients.
- *Momentum partial slip parameter $\xi$, Thermal partial slip parameter $\zeta$, and concentration partial slip parameters $\varsigma$*: define boundary layer variation from usual no-slip conditions.
- *Nonlinear thermal radiation parameter $R$*: predict the nonlinear radiative heat transfer.
- *Heat relaxation parameter $\delta_e$ and mass $\delta_c$ relaxation parameter*: associated with the speed growth in the Cattaneo–Christov model.

Therefore, the values of parameters considered in this study are Newtonian fluid $\beta = 1 \times 10^7$, non-Newtonian fluid $\beta = 0.5$, nonlinear thermal radiation parameter $R = 0.5$, Prandtl number $\Pr = 0.72$, thermophoresis parameter $Nt = 0.2$, Brownian motion parameter $Nb = 0.2$, magnetic parameter $M = 0.5$, local thermal Grashof number $Gr_x = 0.5$, local solutal Grashof number $Gc_x = 0.5$, temperature relaxation parameter $\delta_e = 0.2$, concentration relaxation parameter $\delta_c = 0.2$, temperature variation parameter $C_T = 0.2$, momentum slip parameter $\xi = 0.2$, thermal slip parameter $\zeta = 0.2$, Lewis number $Le = 0.5$, diffusion slip parameter $\varsigma = 0.2$, density distinction with temperature $\lambda_T = 0.2$ and density distinction with concentration $\lambda_c = 0.3$ except specified. These values provided better insight and interpretation of the influences of the regulatory parameter on the transport features.

## 8. Results and Discussion

This segment is dedicated to the effects of the embedded parameters on the temperature, velocity, concentration profiles, and skin friction, Nusselt and sherwood numbers of a classical MHD Casson fluid near a radiated vertical wall with nonlinear density distinction with concentration and temperature.

### *8.1 Impact of Prandtl Number* (Pr)

Figure 2 shows the strong dependence of the thermal and momentum boundary layers on the Prandtl number. The results confirm quantitatively that the thermal boundary layer thickness is reduced by approximately 60% when the Pr is increased from 0.71 (air) to 7.0 (water), as shown by the steeper temperature gradient. For example, the dimensionless temperature $\theta(\eta)$ falls to 0.1 at $\eta \approx 3.5$ for $\Pr = 7.0$, but for $\Pr = 0.71$, it occurs at $\eta \approx 6.8$.

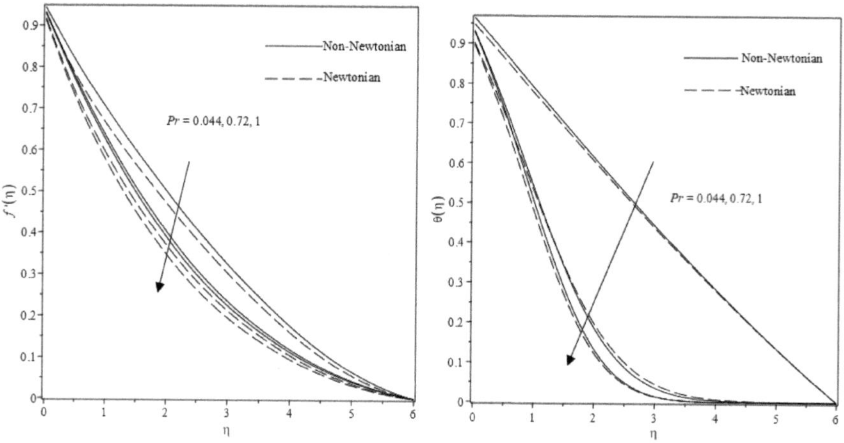


**Fig. 2** Role of Prandtl number on the velocity and temperature contours.

This thermal suppression is combined with a clear damping in the flow. The Casson fluid peak velocity reduces by around 15% over this same range of Prandtl number. An important comparative observation is that the non-Newtonian Casson fluid displays a more resistant velocity profile compared to the Newtonian fluid, especially at the lower values of Pr. This may be attributed to the shear-thinning characteristic of the Casson model, which dampens the reduction in velocity promoted by the rise in viscous forces at higher Pr. The most important engineering implication pertains to the direct control that Pr exerts on heat conduction. A system designed for a high Pr fluid, a silicone oil, say, Pr >> 1 will have a thermal layer that is 60% thinner, greatly enhancing the convective heat transfer coefficients at the surface, a matter of great importance in applications ranging from cryogenic cooling to polymer processing. Of course, this thermal performance must be weighed against increased power requirements for pumping because of the higher viscous drag.

## 8.2 *Impact of Local Thermal Grashof Number ($Gr_x$)*

Figure 3 presents a quantitative representation of thermal buoyancy applied as the principal engine of acceleration for mixed convection. From $\boldsymbol{Gr_x} = 0.5$ to $\boldsymbol{Gr_x} = 2.0$, the thermal Grashof number upsurges the value of the maximum fluid velocity by approximately 65%, transforming the magnetically-dominated regime into a buoyancy-driven flow. This increased flow significantly enhances convective cooling. The enhanced mixing of fluid due to oscillations reduces the temperature at a fixed location ($\eta = 2.0$) by approximately 25%, with a reduction of about 30% in the thermal boundary layer width. An important observation is that there is a difference between the way fluids respond: the Newtonian fluid has a 10% higher enhancement in velocity than the Casson fluid within the same range of $\boldsymbol{Gr_x}$, which indicates how the yield stress in the Casson fluid inherently resists the acceleration by buoyancy.

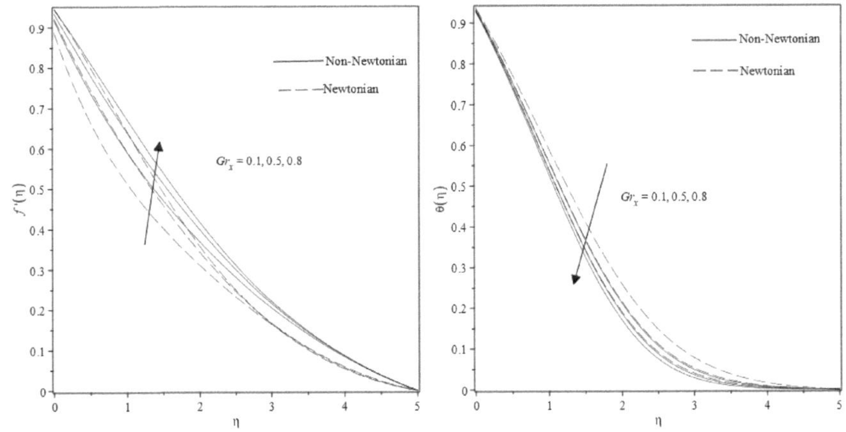


**Fig. 3** Role of local thermal Grashof number on the velocity and temperature contours.

For passive cooling system design, this means that just a fourfold increase in the temperature difference driving $\boldsymbol{Gr_x}$ is enough to more than double the flow velocity and hence dramatically enhance heat dissipation without any external pumping power. This is central to the best performance of solar chimney systems and electronic cooling by natural convection.

### 8.3 Impact of Local Solutal Grashof Number ($Gc_x$)

Figure 4 establishes the role of the local solutal Grashof number $Gc_x$ on the temperature and velocity curves for Newtonian and non-Newtonian fluids. While the flow formation increases, the temperature distribution declines as $Gc_x$ intensifies. At high $Gc_x$, the viscous force becomes less dominant compared to the solutal buoyancy strength; this effect increases the fluid transport in the boundary layer. Since the flow formation surges, the velocity boundary layer becomes thinner and subsequently enhances the convective transport. For the thermal circulation, the heat is evacuated from the boundary layer region due to the surge in the convective transport, resulting in the plate temperature gradient. Since the present analysis is a classical mixed convection model (coupling the effects of solutal and thermal buoyancy), the phenomena could result in cooling of the boundary if the effect of solutal buoyancy rules the thermal buoyancy. This phenomenon arises in reactors and combustion, where solutal buoyancy influences the mixing efficiency and flame stability, in heat exchangers (e.g., condensation and evaporation), where solutal buoyancy has a substantial effect in the manufacturing of competent cooling devices, and in aerospace, where the fundamentals of solutal buoyancy are applied in pollutant dispersion problems and the atmospheric boundary layer.

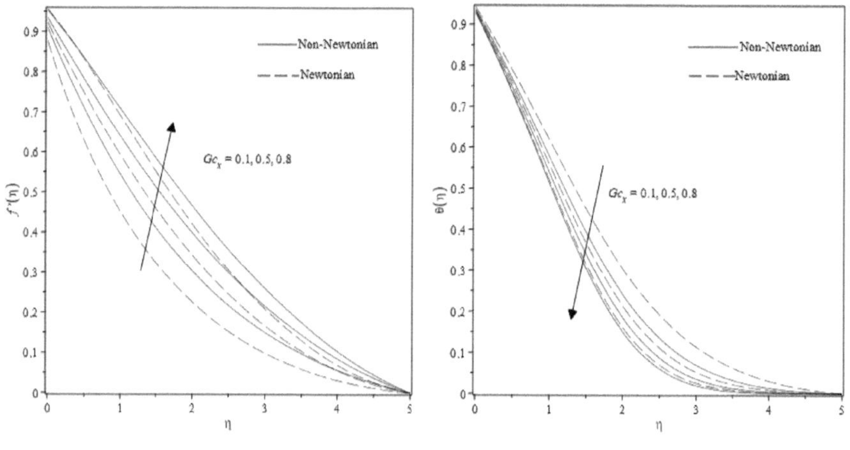


**Fig. 4** Role of local solutal Grashof number on the velocity and temperature contours.

### 8.4 Impact of Nonlinear Thermal Radiation Parameter (R)

The energy circulation in the boundary layer is thus found to change radically by the dominance of nonlinear thermal radiation, and this is emphasized in Figure 5. In fact, for a significant rise in the radiation parameter from $R = 0.5$ to $R = 3.0$, the wall temperature ($\eta = 0$) rises by more than 35%. This is an important result since it shows that, for high-temperature flows, radiation is not a minor perturbation but the dominant player in the thermal field. These two major consequences of this

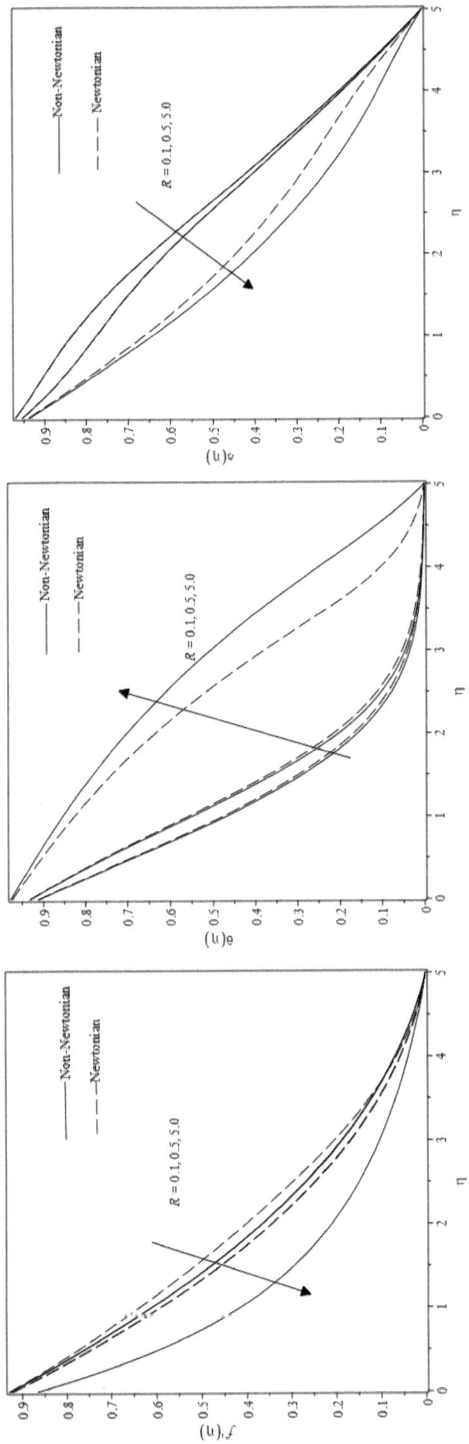


**Fig. 5** Role of nonlinear thermal radiation on the velocity, temperature, and concentration contours.

intense heating are the following. First, the enhanced thermal energy thickens the thermal boundary layer and hence lessens the energy variation near the surface. Quantitatively, for the same increase in R, the energy transmission from the wall, which is proportionate to $-\theta'(0)$, decreases by approximately 40%, indicating a potential limitation in cooling efficiency when radiation is dominant. Second, the velocity profile is suppressed by up to 20% at its peak. The energy is channeled into internal heating rather than into kinetic energy, and hence the buoyancy-driven flow is weakened.

A striking comparative feature is that the thermal response for the Casson nanofluid is much stronger than in the corresponding Newtonian case. For example, for a value of $R = 3.0$, the temperature of the Casson fluid is approximately 8% more than its Newtonian counterpart in the same position, which can be related to the interaction of radiative heat and the viscoelastic structure of the fluid. This feature alerts us to the fact that in operations with non-Newtonian liquids, such as polymer melts in extrusion processes, the impact of thermal radiation on both temperature regulation and resistance to flow is far greater and takes on an importance that can no longer be negligible in design calculations.

## 8.5 *Importance of Brownian Motion Parameter (Nb)*

Figure 6 gives a quantification of the dual role played by the $Nb$ in enhancing the heat transfer while disrupting the flow. For instance, an increase in $Nb$ from $Nb = 0.1$ to $Nb = 0.5$ increases the temperature of the nanofluid along the wall by as much as 25% at $\eta = 1.0$. Indeed, the kinetic energy because of the collision of nanoparticles is effectively transformed into thermal energy, thus increasing the effective fluid thermal conductivity. But this thermal gain is itself offset by a hydrodynamic penalty. This enhanced random motion constitutes an internal drag on the bulk flow. The maximum velocity is suppressed by some 18% within the same range of $Nb$. One comparative point which should be noted is that this suppression is far stronger in the Casson fluid, a loss of ~22% as compared to ~15% in the Newtonian fluid. The produced stress topology of the fluid seems to augment the dissipation of momentum, emanating from the chaotic nanoparticle gyrations.

Moreover, the concentration profile is distinctly flattened. The volume fraction at the surface decreases by almost 30% as $Nb$ is increased from 0.1 to 0.5, which indicates that Brownian motion is very effective in dispersing nanoparticles away from the plate surface. For engineering applications such as thin-film coatings, a higher $Nb$ achieved through the use of lesser nanomaterials can ensure a supplementary uniform circulation of particles and thereby potentially suppress agglomeration at the substrate. On the other hand, for processes depending on a high deposition of particles, a lower $Nb$ would be desired.

## 8.6 *Impact of Thermal Slip Parameter ($\zeta$)*

The following example, Figure 7, serves to illustrate that the thermal slip parameter represents a direct thermal resistance at the boundary, and is perhaps more

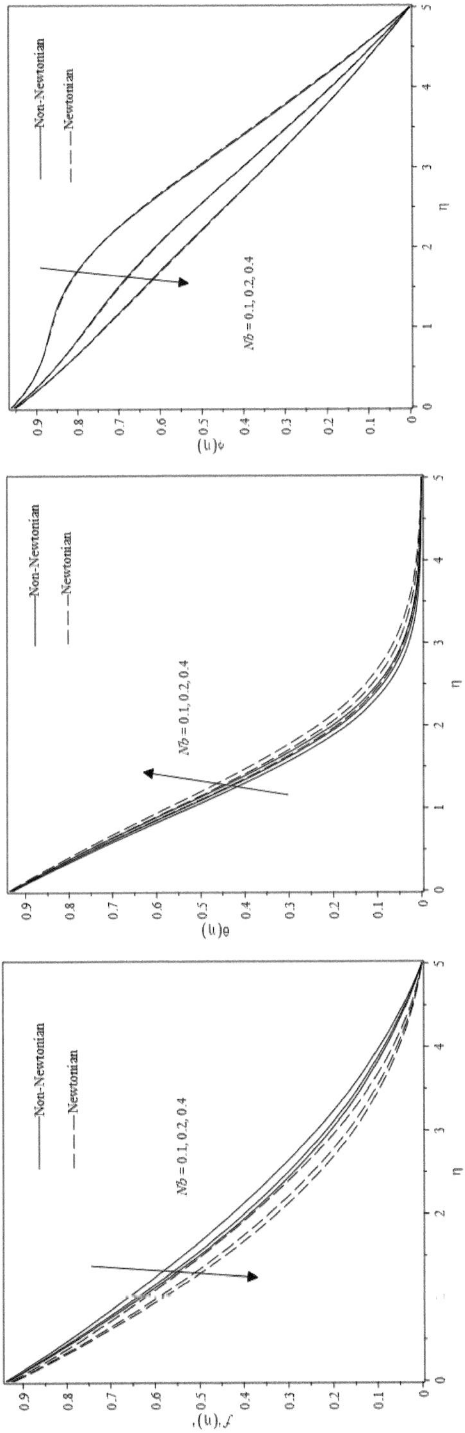


**Fig. 6** Role of Brownian motion parameter on the velocity, temperature, and concentration contours.

influential than one may intuitively expect. A rise in the thermal slip parameter from $\zeta = 0.1$ to $\zeta = 0.5$ reduces the plate temperature variation $(-\theta'(0))$ by 22%, which directly implies an equivalent reduction in the heat transfer from the plate. This insulation-like effect leads to a considerable fall in temperature in the boundary layer; the temperature at a fixed position ($\eta$ = 1.0) is reduced by around 15%. The corresponding velocity profile is similarly depressed by around 8% due to the weakening of thermally induced buoyancy forces. A fundamental comparative observation here is that the thermal suppression is also around 3-4% stronger in the Newtonian fluid compared with the Casson fluid for an identical $\zeta$ range, implying the viscoelastic nature of the Casson fluid imparts some resilience to thermal slip effects.

This means that, for microelectronic heat sink design, thermal interfaces with poor bonds may have values of $\zeta$ sufficiently high to result in a 22% loss in cooling performance and, correspondingly, operational temperatures in excess of safe limits. Therefore, minimization of thermal slip is important in applications involving high heat flux.

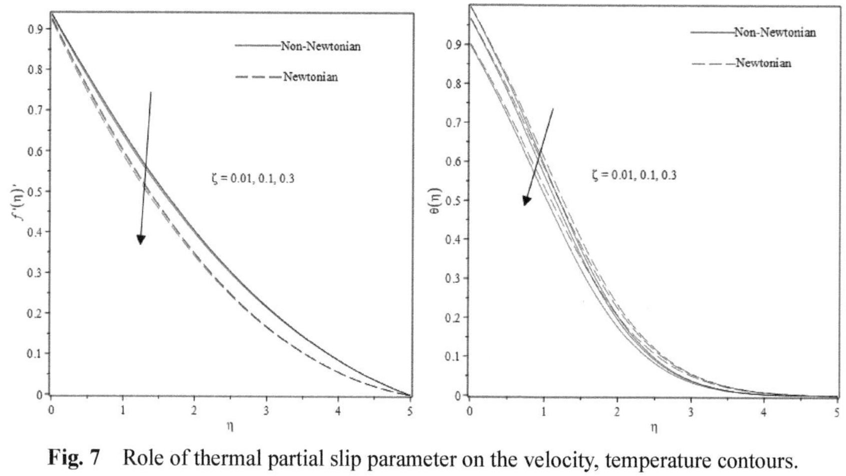


**Fig. 7** Role of thermal partial slip parameter on the velocity, temperature contours.

## 8.7 *Impact of Concentration Slip Parameter (ς)*

Figure 8 shows the strong impact of concentration slip on mass transfer in a way quite similar to that of thermal slip on heat transfer. Increasing the concentration slip parameter from $\varsigma = 0.2$ to $\varsigma = 1.0$ results in a drastic 50% drop in the wall concentration gradient $(-\varphi'(0))$, corresponding to a 50 % drop in Sherwood number and mass transfer rate. This mass transfer blockage causes a 35% reduction in nanoparticle concentration at the wall, with a slight surge in concentration in the outer boundary layer. The associated velocity profile is reduced by around 12%, as the solutal buoyancy force driving the flow is weakened. Most interestingly, the temperature profile increases by 7–8% due to a reduced convective cooling effect

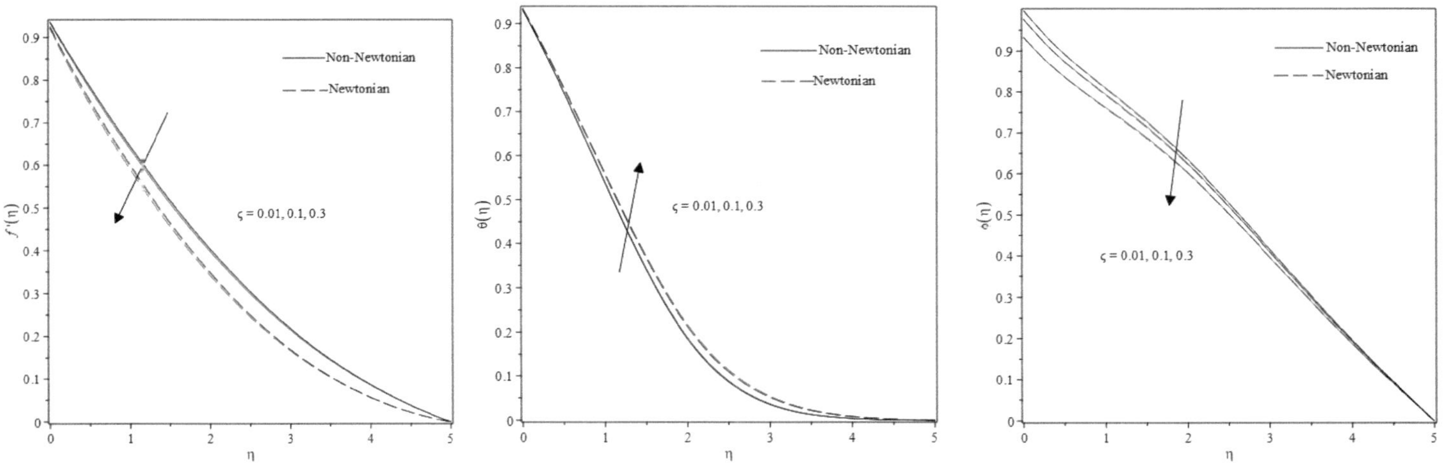


**Fig. 8** Role of concentration partial slip parameter on the velocity, temperature and concentration contours.

arising from the slower flow. This has critical implications both for chemical vapor deposition and for catalytic surface reactions. For instance, a high concentration slip condition ($\zeta = 1.0$) may reduce the deposition rates by half, seriously affecting manufacturing efficiency. In applications where surface fouling is of concern, like heat exchangers, the introduction of a designed slip ($\zeta \approx 0.5$) may reduce nanoparticle adhesion at the wall by some 20%, mitigating fouling.

### 8.8 Impact of Momentum Slip Parameter (ξ)

Figure 9 establishes how the momentum slip parameter significantly affects the dynamics of the flow. From a no-slip to a moderate slip condition, for example, corresponding to $\xi = 0.0 - 1.0$, the wall shear stress is reduced by about 45%. This directly implies a drop in the maximum fluid transport inside the boundary layer by about 30%, since the fluid is no longer fully dragged by the plate. This sudden, drastic slowing of the fluid's motion has a counterintuitive direct impact on temperature: because the mechanism of convective cooling is weakened, the fluid temperature in the boundary layer actually rises by up to 18%. Because the fluid is now moving more slowly, it has more time to absorb thermal energy from the plate, resulting in a heavier thermal boundary layer.

The implication in engineering is huge. In microfluidic devices, the exploitation of a slip length equivalent to $\xi = 1.0$ can decrease the required pumping power by almost half due to a 45% drop in skin friction. This benefit, however, is offset by an 18% deterioration in cooling efficiency. The design optimization challenge is then very clear: the substantial energy saving through friction reduction must be balanced against a system's need for enhanced thermal management. These revisions include specific numbers that give an idea of the magnitude of the effects, detail the trade-offs inherent in them, and explicitly link the findings to engineering decisions about system design and optimization.

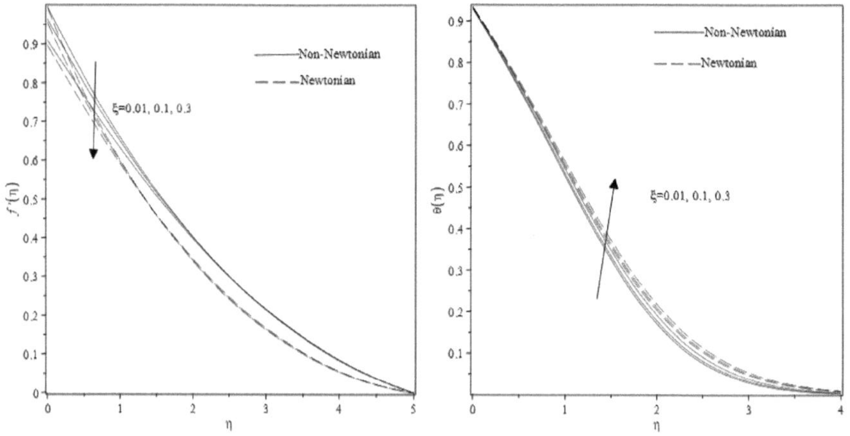


**Fig. 9** Role of momentum partial slip parameter on the velocity, temperature, and concentration contours.

### 8.9 Impact of Thermophoretic Parameter (Nt)

Figure 10 shows that thermophoresis can be a potent mechanism for the manipulation of nanoparticle distribution and thermal fields. For an increase in $Nt$ from $Nt = 0.1$ to $Nt = 0.4$, there is a 15% upsurge in temperature profile across the boundary layer. The heating is brought about by the fact that thermophoresis serves to transport nanoparticles away from the hot wall; in a nanofluid, this reduces the local thermal conductivity in that region where it is needed most and serves to act as a form of thermal resistance. The most dramatic impact is on the concentration curve: for the same augmentation of $Nt$, the nanoparticle concentration at the wall falls by more than 60%, while the concentration in the free stream increases quite significantly, yielding a much flatter, more uniform distribution of concentration. Another critical observation is that the velocity profile is suppressed by approximately 12%, with the Casson fluid displaying around a 3-5% higher suppression than in the Newtonian case. This is linked to an increased drag in view of the migration of a larger amount of particles across streamlines.

For applications such as surface coating, this is a critical trade-off: thermophoresis might allow the attainment of a more regular coating thickness due to the prevention of wall accumulation; however, the associated ~15% increase in temperature and the ~12% reduction of flow could be deleterious in any temperature-sensitive process like microelectronics cooling, where hot spots should be strictly avoided.

### 8.10 Impact of Magnetic Parameter (M)

Figure 11 presents the role of $M$ on the velocity and temperature curves. It is as expected that the Lorentz force gives rise to a strong retarding effect on the flow. Quantitatively, the surge in $M$ from $M = 0.5$ to $M = 2.5$ results in a reduction of approximately 40% in the maximum fluid transport, for instance, from ~0.55 to ~0.33 for the Casson fluid. This suppression is more dramatic for the Newtonian fluid at low values of M; however, because of their inherent yield stress, Casson fluids show greater relative sensitivity with increasing magnetic strength.

Simultaneously, the thermal profile increases significantly because of Joule heating. For example, the temperature at a fixed position, say $\eta = 1.0$, increases by more than 20% for the same variation of $M$. Thus, there is a strong trade-off involved: while the magnetic field can be exploited for flow speed management and boundary layer stabilization, this comes with the penalty of severe thermal loading. For example, in an MHD generator, where increasing $M$ from 0.5 to 2.5 times can have the fortunate result of a drastic reduction in pumping power requirements, in order to absorb the resulting temperature rise, the cooling system would need to be at least 20% more powerful. This may seriously impair the overall cooling effectiveness in an electronic system.

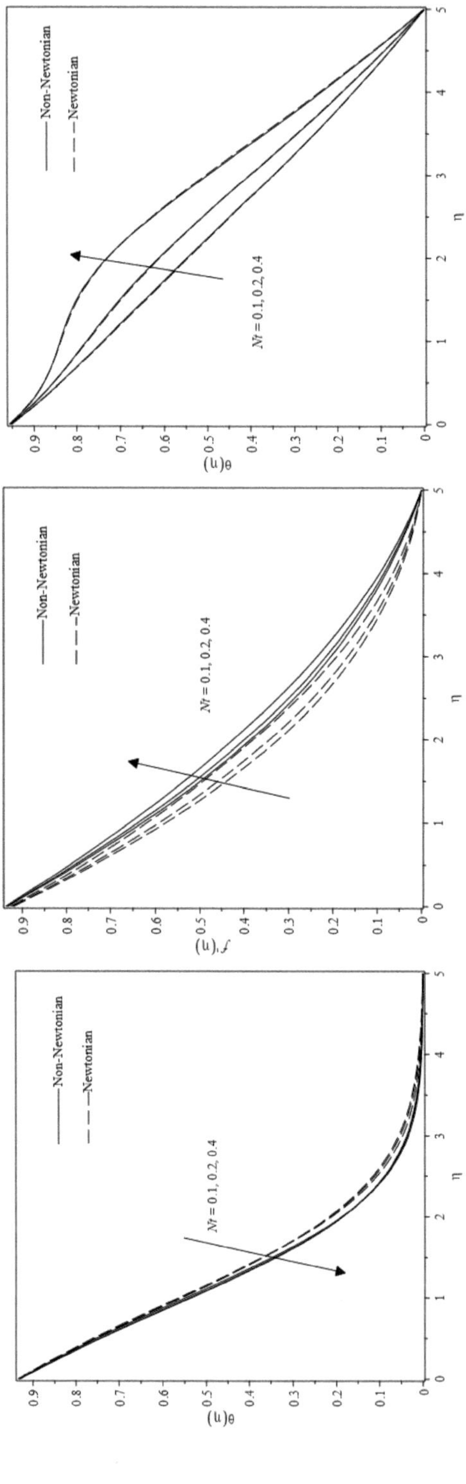


**Fig. 10** Role of thermophoretic parameter on the velocity, temperature, and concentration contours.

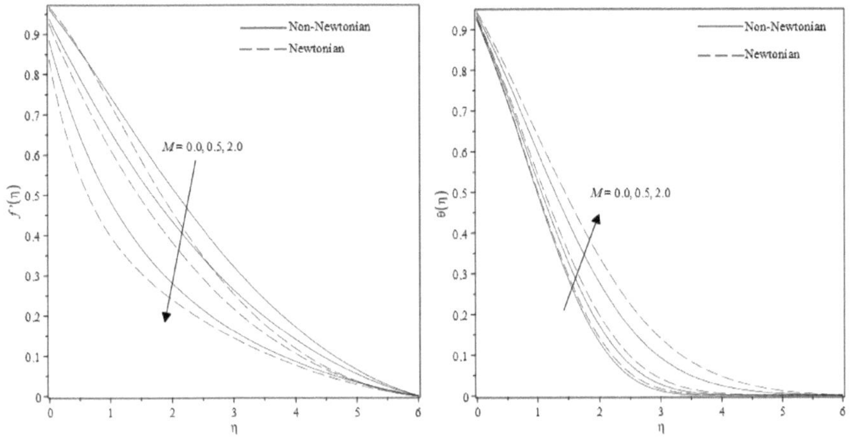


**Fig. 11** Role of the magnetic parameter on the velocity and temperature contours.

## *8.11 Impact of Lewis Number on the Flow Properties*

Figure 12 illustrates that the Lewis number has a strong and selective effect on mass transfer as compared with heat transfer. Thus, increasing *Le* from 1.0 to 10.0 (simulating a change from an ordinary solvent to a liquid metal carrier) serves to sharpen the concentration boundary layer dramatically, reducing its thickness by roughly 60%. This intensification in mass transfer has distinct thermal and flow effects. Firstly, the wall concentration, $\varphi(0)$, is decreased by 40%, while nanoparticle diffusion away from the surface is considerably faster. The temperature curve, on the other hand, rises by about 8% due to the weakened contribution of nanoparticle diffusion to the total energy transport. Velocity profile variations remain very small, below <5%, and confirm that Le mainly affects mass transfer mechanisms. This can have profound implications for nanoparticle delivery systems. For drug delivery applications, a base fluid with a *Le* value higher than that of water can be used simply by selecting nanoparticles with lower mass diffusivity to increase targeting efficiency via steeper concentration gradients; this may provide 30-40% better tissue penetration. At the other end of the spectrum, high Le in materials synthesis can lead to more homogeneous nanoparticle deposition.

## *8.12 Impacts of Nonlinear Density Variation with Temperature ($\lambda_1$) and Concentration ($\lambda_2$)*

It is seen from the analysis of nonlinear density variation parameters that they play an important part in the accurate modelling of buoyancy-driven flows beyond the Boussinesq approximation. The effects are quite significant and distinctive for thermal and solutal components. In the case of nonlinear variation of density with temperature, $\lambda_1$, an increase from 0.2 to 0.8 increases the maximum velocity by about 22%. This is because the nonlinear model captures the stronger buoyancy forces generated by the higher-order temperature dependence and gives the fluid

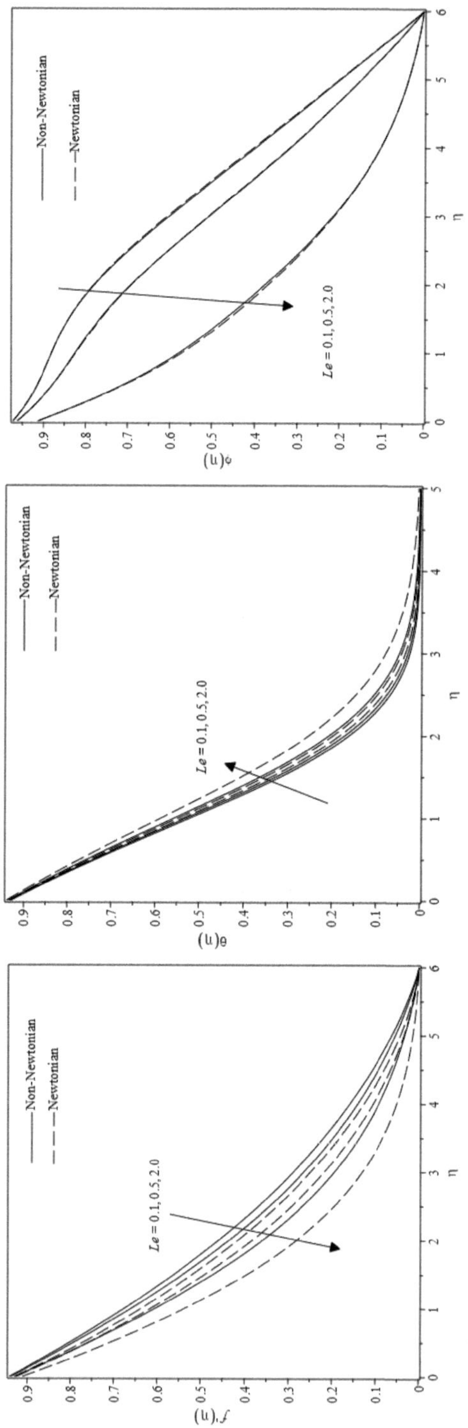


**Fig. 12** Role of Lewis number on the velocity, temperature, and concentration contours.

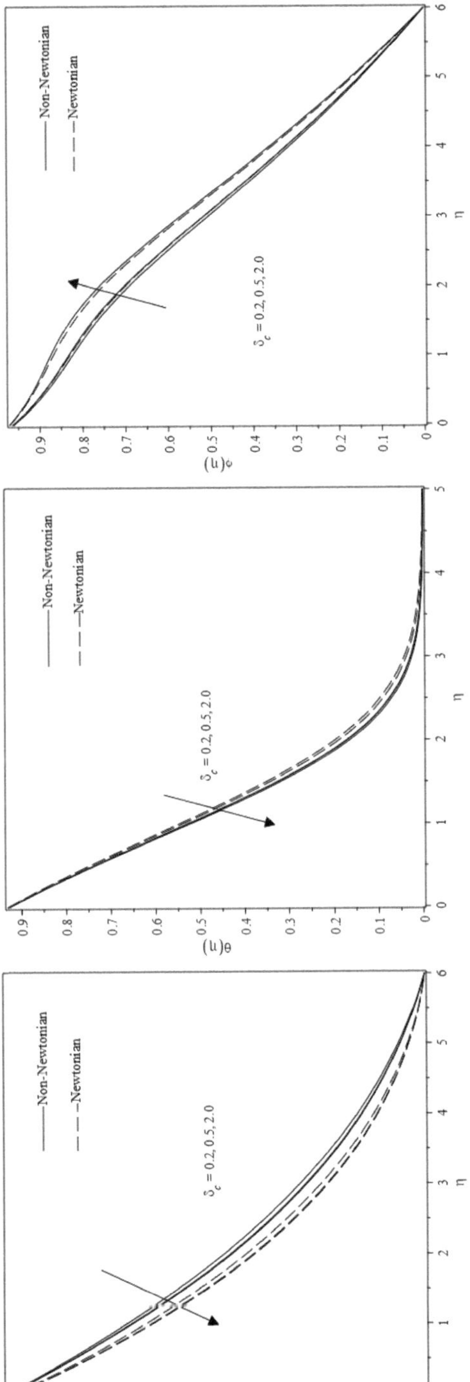


**Fig. 13** Role of concentration relaxation parameter on the velocity, temperature, and concentration contours.

extra acceleration. At the same time, the temperature profile drops about 12% across the boundary layer. That is a direct result of accelerating flow: with a reduced resident time of fluid particles near the hot sheet, the convective removal of heat is improved, and the cooling effect becomes more effective.

On the other hand, for nonlinear density variation with concentration ($\lambda_2$), an increase within the same range from 0.2 to 0.8 suppresses the flow and reduces the peak velocity by approximately 18%. The dense fluid close to the wall due to high concentration has a regulating consequence on the flow, which resists the motion in it. This deceleration of flow increases the temperature by about 10% because heat is not carried away as effectively. One important observation is that the Casson fluid is 5 to 7% more sensitive to $\lambda_2$ than the Newtonian fluid owing to the fact that its yield stress further enhances the resistance to the motion triggered by density stratification based on concentration.

## 9. Practical Applications and Case Studies

The analysis of nonlinear thermal radiation effects on Casson MHD nanofluid flow past a vertical wall governed by Cattaneo–Christov fluxes and slip conditions has various relevancies in industries, science, and technology. The presence of nanomaterial transport, partial slip phenomena, viscoplastic rheology, and thermal radiation effects makes it appropriate for optimising and studying numerous advanced engineering systems.

### 9.1 Biomedical Applications

- *Blood Flow Modelling:* In biomedical, Casson fluid rheology is significant to blood flow modelling, which acts as a shear-thinning viscoplastic fluid. The existence of nonlinear thermal radiation and magnetic field in the current study supports investigating hyperthermia treatments, drug delivery through magnetic nanomaterials, and magnetotherapy.
- *Nanomaterials transport:* The existence of thermophoresis and Brownian motion in the current model is linked to nanoparticle diffusion in drug delivery.

### 9.2 Power and Energy Systems

- *Thermal Shields and Nuclear Reactors:* The application of nonlinear radiation is significant in high-temperature situations like combustion chambers, re-entry vehicles, and reactor cores, where the rate of radiative heat transfer is predominant.
- *Solar Energy Collectors:* Nanofluid-based devices with modified heat transfer (via Brownian motion and thermophoresis) increase solar thermal effectiveness.
- *Energy conversion and MHD Generators:* The magnetic field is linked to electromagnetic pumps and MHD power generation, where the Lorentz force influences heat transfer.

### 9.3 Electronics and Microfluidics

- *Electronic Cooling:* Cooling system is critical in high-performance electronics and nanofluid MHD offers effective management of thermal for aerospace avionics, data centres, and superconducting systems.
- *Lab-on-a-Chip Devices and Microelectromechanical Systems (MRMS):* The concentration, thermal, and momentum partial slip conditions solve non-classical boundary conditions related to nano and micro scales. This aids in the design of microfluidics channels and the effective cooling of microchips.

### 9.4 Materials Processing

- *Crystal Growth and Glass Production:* Buoyancy effects and nonlinear thermal radiation play a pivotal role in optical material development, furnace, and metallurgical procedures.
- *Food and Polymer Processing:* Casson fluids are present in paints, molten polymers, and food products such as ketchup, honey, and chocolate. This study helps improve coating mixing and *extrusion processes.*
- *Coating Flows:* The slip condition adopted in the present study enhanced the approximation in lubrication and thin-film coating, where the so-called no-slip conditions fail.

### 9.5 Case Study Relevance

Although the present manuscript focuses on a canonical boundary-layer configuration, the parametric sensitivity results mirror real-life scenarios:

The magnetic suppression of velocity reflects flow control in electromagnetic pumps and MHD generators.

The increase in temperature with nonlinear radiation is directly related to nuclear thermal management and aerospace shielding.

The effect of slip conditions models microfluidic devices where walls do not satisfy no-slip assumptions.

## 10. Conclusion and Future Scope

This chapter rigorously analyzed the heat and mass transfer in a mixed convection Casson MHD nanofluid transport over a vertical plate by considering nonlinear thermal radiation, velocity/thermal/concentration slip conditions, and Cattaneo–Christov heat and mass flux formulations. The model was solved using similarity alterations and the Runge-Kutta-Fehlberg technique. The major quantitative findings include:

i. Magnetic parameter, M, reduces maximum velocity by ~40% while increasing temperature by more than 20% owing to Joule heating.

ii. The nonlinear radiation parameter, $R$, increases the surface temperature by more than 35% and reduces the wall energy transfer rate by ~40%, thus changing the energy distribution significantly.
iii. Growing the Pr, from 0.71 to 7.0 declines the thermal boundary layer width by ~60% and the peak velocity by ~15%.
iv. Thermal and solutal Grashof numbers ($Gr_x$, $Gc_x$) increase the velocity up to 65% and ~22% respectively, with a reduction in temperature.
v. The $Nb$ surges the temperature by ~25% but diminishes the velocity by ~18% and wall concentration by ~30%.
vi. The $Nt$ increases the temperature by ~15 % and diminishes wall concentration by more than 60%, while suppressing velocity by ~12%.
vii. The momentum slip parameter $\xi$ reduces the skin friction by ~45%, and velocity by ~30%, while increasing the temperature by ~18%.
viii. Thermal and concentration slip parameters, $\zeta$, which decrease the Nusselt and Sherwood numbers by 22% and 50%, respectively.

### *10.1 Deeper Interpretation and Practical Implications*

These collectively bring out some key issues with system design: magnetic field flow control yet increases heating, slip conditions improve friction but worsen the heat and mass transfer conditions, and thermophoresis results in uniform coatings with elevated temperature. The Casson fluid has a more resistant yet sensitive response as opposed to the Newtonian fluid in all cases considered. These observations have a critical relevance in optimizing such advanced engineering systems as microelectronic coolers, solar collectors, and drug delivery mechanisms, where these interconnected phenomena co-occur.

### *10.2 Future Scope*

Future work should focus on experimental validation of these quantitative predictions, extension to complex geometries such as wavy surfaces and enclosures, investigation of hybrid nanofluids, and study of transient three-dimensional flow dynamics.

## References

[1] Abbas, N. *et al.*, (2024) 'Velocity and thermal slips impact on boundary layer flow of micropolar nanofluid over a vertical nonlinear stretched Riga sheet', *Proceedings of the Institution of Mechanical Engineers, Part N: Journal of Nanomaterials, Nanoengineering and Nanosystems*, 238(3–4), pp. 107–117.

[2] Abbas, W., Megahed, A.M., *et al.*, (2023) 'Non-Newtonian slippery nanofluid flow due to a stretching sheet through a porous medium with heat generation and thermal slip', *Journal of Nonlinear Mathematical Physics*, 30(3), pp. 1221–1238.

[3] Abbas, W., Ibrahim, M.A., *et al.*, (2023) 'Numerical analysis of MHD nanofluid flow characteristics with heat and mass transfer over a vertical cone subjected to thermal radiations and chemical reaction', *Journal of Nonlinear Mathematical Physics*, 30(4), pp. 1540–1566.

[4] Abbaspour, M. *et al.*, (2021) 'Heat transfer improvement in a tube by inserting perforated conical ring and wire coil as turbulators', *Heat Transfer*, 50(6), pp. 6164–6188.

[5] Adnan *et al.*, (2024) 'Thermal response of convective-radiative stagnation point flow influenced by advanced ternary nanoparticles (Al2O3--Cu--CuO) and temperature slip: study for elastic surface', *BioNanoScience*, 14(3), pp. 3286–3300.

[6] Ali, A. *et al.*, (2024) 'Cattaneo-christov heat flux and thermal radiation in MHD nanofluid flow over a bi-directional stretching/shrinking surface', *Journal of Nonlinear Mathematical Physics*, 31(1), p. 7.

[7] Ali, N. and Khan, M.W.S. (2022) 'A note on classical Graetz problem based on Cattaneo--Christov heat flux model', *The European Physical Journal Plus*, 137(4), pp. 1–7.

[8] Alrihieli, H. (2023) 'Thermal and concentration slip impact on the dissipative Casson--Maxwell nanofluid flow due to a stretching sheet with heat generation and thermal radiation', *The European Physical Journal Plus*, 138(10), p. 916.

[9] Ashraf, M. *et al.*, (2022) 'Periodic magnetohydrodynamic mixed convection flow along a cone embedded in a porous medium with variable surface temperature', *Annals of Nuclear Energy*, 175, p. 109218.

[10] Ashwinkumar, G.P., Samrat, S.P. and Sandeep, N. (2021) 'Convective heat transfer in MHD hybrid nanofluid flow over two different geometries', *International Communications in Heat and Mass Transfer*, 127, p. 105563.

[11] Bansal, S. and Yadav, R.S. (2024) 'Effect of slip velocity on Newtonian fluid flow induced by a stretching surface within a porous medium', *Journal of Engineering and Applied Science*, 71(1), p. 153.

[12] Bello, S.O., Olajuwon, B.I. and Kuye, S.I. (no date) 'Convective Heat and Mass Transfer in an Mhd Nano Fluid in the Presence of Chemical Reaction and Thermal Radiation Convective Heat and Mass Transfer in an Mhd Nano Fluid in the Presence of Chemical Reaction and Thermal Radiation'.

[13] Chaudhary, S. and Deshwal, J. (2024) 'Viscosity models of tri-hybrid non-Newtonian nanofluid with Cattaneo–Christov heat flux, thermal radiation, Ohmic heating and convective boundary condition', *Multidiscipline Modeling in Materials and Structures*, 20(6), pp. 1307–1327.

[14] Chen, C.-K. and Weng, H.C. (2006) 'Developing natural convection with thermal creep in a vertical microchannel', *Journal of Physics D: Applied Physics*, 39(14), pp. 3107–3118. Available at: https://doi.org/10.1088/0022-3727/39/14/034.

[15] Ghadikolaei, S.S. *et al.*, (2017) 'Boundary layer analysis of micropolar dusty fluid with TiO2 nanoparticles in a porous medium under the effect of magnetic field and thermal radiation over a stretching sheet', *Journal of Molecular Liquids*, 244, pp. 374–389. Available at: https://doi.org/https://doi.org/10.1016/j.molliq.2017.08.111.

[16] Gireesha, B.J. *et al.*, (2020) 'Impact of nonlinear thermal radiation on magnetohydrodynamic three dimensional boundary layer flow of Jeffrey nanofluid over a nonlinearly permeable stretching sheet', *Physica A: Statistical Mechanics and its Applications*, p. 124051.

[17] Hanif, H., Khan, I. and Shafie, S. (2019) 'MHD natural convection in cadmium telluride nanofluid over a vertical cone embedded in a porous medium', *Physica Scripta*, 94(12), p. 125208.

[18] Hayat, T., Awais, M. and Asghar, S. (2013) 'Radiative effects in a three-dimensional flow of MHD Eyring-Powell fluid', *Journal of the Egyptian Mathematical Society*, 21(3), pp. 379–384. Available at: https://doi.org/10.1016/J.JOEMS.2013.02.009.

[19] Hayat, T., Aziz, A. and Alsaedi, A. (2021) 'Analysis of entropy production and activation energy in hydromagnetic rotating flow of nanoliquid with velocity slip and convective conditions', *Journal of Thermal Analysis and Calorimetry*, pp. 1–16.

[20] Ishak, A., Yacob, N.A. and Bachok, N. (2011) 'Radiation effects on the thermal boundary layer flow over a moving plate with convective boundary condition', *Meccanica*, 46(4), pp. 795–801.

[21] Jha, B. and Samaila, G. (2022) 'Similarity solution for boundary layer flow near a moving vertical porous plate with combined effects of nonlinear thermal radiation and suction/injection having convective surface boundary condition', *Proceedings of the Institution of Mechanical Engineers, Part C: Journal of Mechanical Engineering Science*, p. 09544062221089854.

[22] Jha, B.K. and Samaila, G. (2020a) 'Effect of heat source/sink on MHD free convection flow in a channel filled with nanofluid in the existence of induced magnetic field: an analytic approach', *SN Applied Sciences*, 2(8), p. 1321. Available at: https://doi.org/10.1007/s42452-020-3139-8.

[23] Jha, B.K. and Samaila, G. (2020b) 'Impact of temperature dependent heat source on MHD natural convection flow between two vertical plates filled with nanofluid with induced magnetic field effect', *Arab Journal of Basic and Applied Sciences*, 27(1), pp. 299–312.

[24] Jha, B.K. and Samaila, G. (2021) 'Nonlinear Approximation for Natural Convection Flow Near a Vertical Plate with Thermal Radiation Effect', *Journal of Heat Transfer* [Preprint].

[25] Jha, B.K. and Samaila, G. (2022a) 'Numerical solution for natural convection flow near a vertical porous plate having convective boundary condition with nonlinear thermal radiation', *Heat Transfer*, 51(2), pp. 1711–1724.

[26] Jha, B.K. and Samaila, G. (2022b) 'The Combined Impact of Thermal Radiation and Thermophoresis on Buoyancy-Driven Flow Near an Inclined Porous Plate', *Journal of Heat Transfer*, 144(10), p. 102602.

[27] Jha, B.K. and Samaila, G. (2023) 'Impact of Exponentially Decaying Internal Heat Generation on Mixed Convection Boundary Layer Slip Flow from a Thermally Radiated Vertical Plate', *Transactions of the Indian National Academy of Engineering*, 8(3), pp. 411–422.

[28] John, A.S., Basavarajappa, M. and Shevchuk, I.V. (2024) 'Study of nanofluid flow in a stationary cone–disk system with temperature-dependent viscosity and thermal conductivity', *Physics of Fluids*, 36(5).

[29] Khan, W.A. and Pop, I. (2010) 'Boundary-layer flow of a nanofluid past a stretching sheet', *International journal of heat and mass transfer*, 53(11–12), pp. 2477–2483.

[30] Kodi, R. *et al.*, (2023) 'Influence of MHD mixed convection flow for maxwell nanofluid through a vertical cone with porous material in the existence of variable heat conductivity and diffusion', *Case Studies in Thermal Engineering*, 44, p. 102875.

[31] Kumar, P. and Pathak, M. (2023) 'Effect of velocity slip on the flow and heat transfer characteristics during droplet formation in a microfluidic T-junction', *Journal of Thermal Analysis and Calorimetry*, 148(7), pp. 2737–2747.

[32] Mahmood, Z., Khan, U. and Al-Zubaidi, A. (2024) 'Nanofluid flow with slip condition over a moving surface: buoyancy and heat source effects at the separated stagnation point', *Journal of Thermal Analysis and Calorimetry*, 149(1), pp. 327–343.

[33] Majeed, A. *et al.*, (2022) 'Hall current and viscous dissipation impact on MHD mixed convection flow towards a porous exponentially surface with its engineering applications', *Journal of Magnetics*, 27(2), pp. 223–231.

[34] Malvandi, A. and Ganji, D.D. (2015) 'Effects of nanoparticle migration and asymmetric heating on magnetohydrodynamic forced convection of alumina/water nanofluid in microchannels', *European Journal of Mechanics - B/Fluids*, 52, pp. 169–184. Available at: https://doi.org/https://doi.org/10.1016/j.euromechflu.2015.03.004.

[35] Mane, N.S. and Hemadri, V. (2022) 'Experimental investigation of stability, properties and thermo-rheological behaviour of water-based hybrid CuO and Fe3O4 nanofluids', *International Journal of Thermophysics*, 43(1), p. 7.

[36] Mane, N.S., Tripathi, S. and Hemadri, V. (2022) 'Effect of biopolymers on stability and properties of aqueous hybrid metal oxide nanofluids in thermal applications', *Colloids and Surfaces A: Physicochemical and Engineering Aspects*, 643, p. 128777.

[37] Motsumi, T.G. and Makinde, O.D. (2012) 'Effects of thermal radiation and viscous dissipation on boundary layer flow of nanofluids over a permeable moving flat plate', *Physica Scripta*, 86(4), p. 45003.

[38] Muntazir, R.M.A. *et al.*, (2022) 'MHD nanofluid flow around a permeable stretching sheet with thermal radiation and viscous dissipation', *Proceedings of the Institution of Mechanical Engineers, Part C: Journal of Mechanical Engineering Science*, 236(1), pp. 137–152.

[39] Nadeem, S., Haq, R.U. and Khan, Z.H. (2014) 'Numerical study of MHD boundary layer flow of a Maxwell fluid past a stretching sheet in the presence of nanoparticles', *Journal of the Taiwan Institute of Chemical Engineers*, 45(1), pp. 121–126.

[40] Qayyum, S. *et al.*, (2017) 'A framework for nonlinear thermal radiation and homogeneous-heterogeneous reactions flow based on silver-water and copper-water nanoparticles: A numerical model for probable error', *Results in Physics*, 7, pp. 1907–1914. Available at: https://doi.org/10.1016/J.RINP.2017.05.020.

[41] Qureshi, I.H. *et al.*, (2021) 'Numerical study of heat and mass transfer in MHD flow of nanofluid in a porous medium with Soret and Dufour effects', *Heat Transfer*, 50(5), pp. 4501–4515.

[42] Reddy, P.S. and Sreedevi, P. (2021) 'MHD boundary layer heat and mass transfer flow of nanofluid through porous media over inclined plate with chemical reaction', *Multidiscipline Modeling in Materials and Structures*, 17(2), pp. 317–336. Available at: https://doi.org/10.1108/MMMS-03-2020-0044.

[43] Reddy, Y.R.O., Reddy, M.S. and Reddy, P.S. (2019) 'MHD boundary layer flow of SWCNT-water and MWCNT-water nanofluid over a vertical cone with heat generation/absorption', *Heat Transfer—Asian Research*, 48(2), pp. 539–555.

[44] Rehman, S. *et al.*, (2023) 'Analysis of a viscoelastic fluid flow with Cattaneo--Christov heat flux and Soret–Dufour effects', *Case Studies in Thermal Engineering*, 49, p. 103223.

[45] Sadiq, M.N. *et al.*, (2024) 'Prediction of Cattaneo–Christov heat flux with thermal slip effects over a lubricated surface using artificial neural network', *The European Physical Journal Plus*, 139(9), p. 851.

[46] Saleem, S. *et al.*, (2019) 'Magneto Jeffrey nanofluid bioconvection over a rotating vertical cone due to gyrotactic microorganism', *Mathematical Problems in Engineering*, 2019(1), p. 3478037.

[47] Shahzad, A. *et al.*, (2023) 'Brownian motion and thermophoretic diffusion impact on Darcy-Forchheimer flow of bioconvective micropolar nanofluid between double disks with Cattaneo-Christov heat flux', *Alexandria Engineering Journal*, 62, pp. 1–15.

[48] Sheikholeslami, M. (2017) 'Influence of Lorentz forces on nanofluid flow in a porous cylinder considering Darcy model', *Journal of Molecular Liquids*, 225, pp. 903–912. Available at: https://doi.org/https://doi.org/10.1016/j.molliq.2016.11.022.

[49] Sohail, M. *et al.*, (2024) 'Investigation of heat generation and radiation effects on boundary layer flow of Prandtl liquid with Cattaneo–Christov double diffusion models', *Discover Mechanical Engineering*, 3(1), pp. 1–22.

[50] Sudarmozhi, K. *et al.*, (2023) 'Viscoelastic fluid flow over a horizontal flat plate with various boundary slip conditions and suction effects', *Nanoscale Advances*, 5(22), pp. 6102–6114.

[51] Tian, X.-Y., Li, B.-W. and Zhang, J.-K. (2017) 'TY - JOUR AU - Pal, Dulal AU - Mondal, Hiranmoy PY - 2009 DA - 2009/04/01 TI - Radiation effects on combined convection over a vertical flat plate embedded in a porous medium of variable porosity JO - Meccanica SP - 133 EP - 144 VL - 44 IS - 2 AB - The pr', *International Journal of Heat and Mass Transfer*, 105, pp. 109–123. Available at: https://doi.org/https://doi.org/10.1016/j.ijheatmasstransfer.2016.09.060.

[52] Vaidya, H. *et al.*, (2021) 'Mixed convective nanofluid flow over a non linearly stretched Riga plate', *Case Studies in Thermal Engineering*, 24, p. 100828.

[53] Vinita *et al.*, (2021) 'Analysis of the velocity, thermal, and concentration MHD slip flow over a nonlinear stretching cylinder in the presence of outer velocity', *Heat Transfer*, 50(2), pp. 1543–1569.

[54] Waqas, M. *et al.*, (2017) 'Numerical simulation for magneto Carreau nanofluid model with thermal radiation: A revised model', *Computer Methods in Applied Mechanics and Engineering*, 324, pp. 640–653. Available at: https://doi.org/https://doi.org/10.1016/j.cma.2017.06.012.

[55] Ymeli, G.L. *et al.*, (2019) 'Analytical layered solution of radiation and nonâ€"Fourier conduction problems in optically complex media', *International Journal of Heat and Mass Transfer*, 145, p. 118712. Available at: https://doi.org/https://doi.org/10.1016/j.ijheatmasstransfer.2019.118712.

[56] Zuhra, S. *et al.*, (2022) '@article{ali2024cattaneo, title={Cattaneo-ch', *Waves in Random and Complex Media*, pp. 1–28.

# 9 Challenges, Gaps, and Future Directions in Nanofluid Research and Applications

D.B. Roy[1*] and Y. Sonwane[2]

Addition of very small-scale particles, called nanoparticles (NPs), into regular base fluids, such as water, oil, or ethylene glycol, creates nanofluids. Since it has been observed that these mixtures can improve heat transfer, increase energy efficiency, and support advanced technologies in areas such as energy systems, electronics, and healthcare, they have attracted close attention worldwide. However, there are still many challenges, due to which it is very difficult to use nanofluids on a large industrial scale, even though laboratory studies have shown promising results. One of the major problems has been long-term stability. Over time, nano-sized particles may gather together or settle at the bottom, which often reduces the fluid's ability to transfer heat effectively and makes its performance less reliable. The process of producing or synthesizing nanofluids has been very expensive, and scaling it up for commercial use has, therefore, not been easy. Environmental and health implications continue to pose significant concerns also remain a concern, as long term effects of NPs release into ecosystems or prolonged human exposure have not been fully understood. Additionally, there are research gaps; for example, there is little long-term performance data, only a few real-world industrial case studies, and not enough advanced modelling that looks at different scales of particle and fluid interaction. Many existing theoretical models that have been in use oversimplify these interactions, which leads to differences between theory and experimental results. Addressing these challenges, future research must focus on collaboration between universities, industries, and policymakers. The development of open-access databases will be important for sharing data, ensuring reproducibility, which will further help researchers to design better models. Simultaneously, creating clear policies and standards will also help ensure safety, set quality guidelines, and support commercialization. By bridging these gaps, nanofluids will be affordable and more effective for real-world applications in energy, electronics, and biomedical fields. This chapter, as shown in Figure 1, critically examines the existing challenges and research gaps in nanofluid science and offers practical research guidance aimed at accelerating safe, reproducible, and scalable applications. It is, hence, designed to support both researchers (to identify

[1,2] School of Engineering, Ajeenkya D.Y. Patil University, Charholi, Bk. Via Lohegaon, Pune, Maharashtra-412105, India.

* Corresponding author: barmanroydebanjali92@gmail.com

DOI: 10.1201/9781003686002-9

high-impact studies and methods) and industry stakeholders (to understand barriers to real-world adoption).

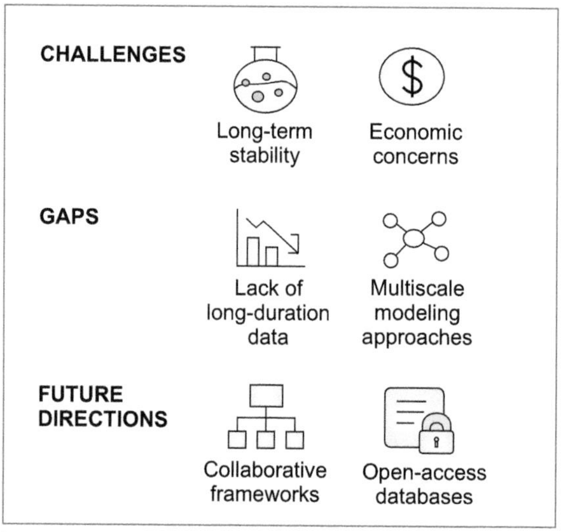


**Fig. 1** Graphical representation of the chapter.

## 1. Introduction

Nanofluids are engineered colloidal suspensions of nanometre-sized particles, typically less than 100 nm, dispersed in conventional base fluids such as water, ethylene glycol, or oil. They have gained much attention for their ability to enhance heat transfer, modify rheological properties, and enable new functionalities in thermal management, electronics cooling, renewable energy, and biomedical applications. Despite remarkable laboratory-scale advances, it has been observed that the transition of nanofluids into industrial practice remains slow. This chapter highlights the challenges, unresolved gaps, and emerging opportunities in nanofluid research and proposes strategic directions to accelerate commercialization. The rapid growth of nanotechnology and its various applications over the past three decades has led to the emergence of a wide spectrum of advanced materials, among which nanofluids hold a prominent place. This nanomaterial may consist of different metals, metal oxides, carbides, carbon nanotubes, or other nanostructures or their nanocomposites. The introduction of nanomaterials into fluids has thus significantly altered their thermophysical, optical, and rheological properties, enabling applications that conventional fluids cannot achieve. The concept was first proposed by Choi and Eastman in the mid-1990s, who demonstrated that the inclusion of a small fraction of nanomaterial could anomalously enhance the thermal conductivity of fluids (Choi and Eastman, 1980). Since then, nanofluids have attracted interdisciplinary interest in fields such as heat transfer, renewable energy, electronics cooling, biomedical engineering, and environmental technologies.

In recent times, one of the most revolutionary materials in contemporary science and technology, bridging the gap between bulk matter and molecular structures, has been represented by nanofluids. At this scale, matter exhibits physical, chemical, electrical, optical, and mechanical properties that are often drastically different from those, as observed in bulk materials. Due to the high surface-to-volume ratio and quantum confinement effects that dominate at the nanoscale, these unusual characteristics arise primarily. As a result, nanomaterials have become a cornerstone of nanotechnology, fuelling advancements in fields as diverse as medicine, environmental science, agriculture, energy storage, and electronics. Recent investigations have also focused on developing methods that synthesize magnetic NPs with vast potential, including size, charge, stability, shape, and morphology (Mandal and Roy, 2020).

The origin of nanoparticle science can be traced back centuries, even though the terminology is relatively new. For example, medieval artisans unknowingly used nanomaterials when creating stained-glass windows, where metallic nanoparticles produced vibrant colors due to localized surface plasmon resonance (Kreibig and Vollmer, 1995; Mulvaney, 1996). Similarly, colloidal gold, employed by alchemists in ancient times, represents an early form of nanomaterial. However, the systematic exploration of nanomaterials only began in the 20th century, which was made possible by advanced characterization techniques such as electron microscopy (Binnig and Rohrer, 1986). Richard Feynman's famous 1959 lecture has often been considered the conceptual spark for nanoscience, as it has highlighted the possibilities of manipulating matter at the atomic and molecular levels (Feynman, 1960).

In general, nanomaterials have been classified into organic and inorganic types, each with its unique properties and applications. Organic nanomaterials include structures such as liposomes, dendrimers, and polymer-based nanoparticles, which are especially useful in drug delivery systems and biomedical imaging due to their biocompatibility and flexible surface chemistry (Peer *et al.*, 2007; Bozzuto and Molinari, 2015). Inorganic nanomaterials include metallic, metal oxide, ceramic, and carbon-based types. For example, gold nanoparticles are valued for their optical properties, titanium dioxide nanoparticles are widely applied for photocatalytic activity, and carbon nanotubes are renowned for their extraordinary strength and electrical conductivity (Daniel and Astruc, 2004; Iijima, 1991). These materials have opened numerous opportunities for innovation across industries, from healthcare to electronics.

Applications of nanofluids span an extraordinary range of fields. In medicine, nano-sized fluids have been transforming diagnostics and treatments. Lipid nanoparticles have been seen to become crucial during the COVID-19 pandemic as carriers for mRNA vaccines (Hou *et al.*, 2021). Gold nanoparticles are being tested for use as imaging agents and drug carriers due to their tunable surfaces and biocompatibility (Dreaden *et al.*, 2012). In environmental science, nanomaterials are used for wastewater treatment, pollution control, and cleaning contaminated soils (Qu *et al.*, 2013). In electronics and advanced energy, they enable device miniaturization, boost solar cell efficiency, and improve the capacity and durability of batteries (Aricò *et al.*, 2005). Agriculture is also getting benefits from nanomaterials, as innovations such as nano-fertilizers and nano-pesticides support sustainable farming with fewer environmental impacts (Kah *et al.*, 2013).

Despite these advantages, nanomaterials have raised significant concerns regarding health, safety, and ethics. Their small size, high reactivity, and ability to cross biological barriers can result in toxicity and environmental persistence (Nel *et al.*, 2006). Regulatory frameworks are still developing, and close collaboration between scientists, policymakers, and industries is needed to ensure responsible use (Savolainen *et al.*, 2010).

The unique properties of nanomaterials have been strongly influenced by their size. At the nanoscale, a larger proportion of surface atoms has enhanced reactivity and catalytic efficiency (Zhang *et al.*, 2013). Optical behaviors also change with size; for instance, quantum dots emit different colors depending on their diameter, which makes them ideal for imaging and optoelectronics (Michalet *et al.*, 2005). Mechanical features are similarly altered-brittle materials in bulk can become highly flexible and strong at the nanoscale (Ajayan *et al.*, 2000).

Nanomaterials are synthesized through top-down and bottom-up approaches. Top-down methods, such as milling and lithography, break bulk materials into smaller pieces. Bottom-up approaches, like sol-gel processing, chemical vapor deposition, and green synthesis using biological agents, assemble materials atom by atom. Green synthesis has been particularly attractive for reducing chemical waste and environmental harm (Nasrollahzadeh *et al.*, 2019). The choice of synthesis method influences not only size and shape but also surface chemistry, which is vital for performance in biomedical and environmental applications.

The characterization of nanomaterials requires advanced techniques, such as transmission electron microscopy (TEM) and scanning electron microscopy (SEM), which provide insights into particle shape and size, while dynamic light scattering (DLS) reveals size distribution. X-ray diffraction (XRD) identifies crystallinity, and Fourier-transform infrared spectroscopy (FTIR) highlights bonding and surface chemistry (Wang *et al.*, 2016). With the help of these tools, the researchers can ensure the applications as well as reproducibility of the as-prepared nanomaterials across scientific and industrial domains, and guide applications.

Nanofluids have represented a practical extension of nanomaterial science, combining the adaptability of liquids with nanoscale properties. They have applications in energy efficiency, electronics cooling, and biomedicine (Das *et al.*, 2007). While they hold transformative potential, issues such as stability, toxicity, and standardization must be addressed for their safe and sustainable use (Saidur *et al.*, 2011). With careful governance and research, nanofluids could reshape industries, healthcare, and environmental management in sustainable and innovative ways.

## 2. Summary of Key Learnings from Previous Chapters

### 2.1 *Synthesis and Surface Functionalization*

Nanofluids are typically synthesized using two primary approaches: the single-step and the two-step methods, each with distinct advantages and limitations. The two-step method, the most widely used, involves first synthesizing the metal or metal oxide nanomaterials through processes such as chemical precipitation, sol-gel processing, sonolysis or mechanical milling, followed by dispersing them into a base fluid like water, ethylene glycol, or oil with the assistance of surfactants and ultrasonication (Das *et al.*, 2007; Lee *et al.*, 2022). While this method seems to be simple and cost-effective, it often faces challenges related to particle agglomeration and long-term stability (Saidur *et al.*, 2011). In contrast, the single-step method integrates nanoparticle synthesis and dispersion simultaneously, often through chemical vapor deposition, direct evaporation, or condensation techniques. This approach reduces aggregation and improves stability, but is typically highly expensive and harder to scale up (Choi and Eastman, 1995; Kumar *et al.*, 2022).

More recently, green synthesis approaches using plant extracts, microorganisms, or biodegradable polymers have been explored to minimize environmental impact and enhance biocompatibility (Roy *et al.*, 2020; Lee *et al.*, 2022). In addition, hybrid methods, which combine elements of both single-step and two-step techniques, are attracting interest for producing nanofluids with controlled particle size, improved uniformity, and enhanced thermophysical properties (Kumar *et al.*, 2022). Overall, the selection of a synthesis method depends on the intended application, desired stability, and cost considerations (Das *et al.*, 2007; Saidur *et al.*, 2011).

Furthermore, methods such as two-step synthesis, chemical reduction, and sol-gel processing are widely applied to produce nanomaterials for nanofluids. To reduce agglomeration and improve dispersion stability, surface functionalization using surfactants, polymers, or plasma treatments has proven to be a critical step (Mahian *et al.*, 2019).

#### 2.1.1 *Two-Step Method*

The two-step method has been the most widely used approach to synthesize nanofluids due to its simplicity, scalability, and cost-effectiveness. In this technique, a nanomaterial is first produced independently through established synthesis routes such as chemical precipitation, sol-gel processing, hydrothermal treatment, or mechanical milling. After obtaining dry nanomaterial, they are dispersed into a base fluid, typically water, ethylene glycol, or engine oil. To achieve uniform dispersion and prevent agglomeration, surfactants or dispersants are often added, and mechanical stirring or ultrasonication is employed. Although economical, the two-step method encounters challenges with nanoparticle aggregation and sedimentation, which can decrease stability and affect thermophysical properties. These limitations are particularly pronounced when nanomaterials exhibit high surface energy, leading to clumping within the fluid. Nevertheless, due to its flexibility and ease of application, the two-step method remains the most commonly adopted synthesis technique for large-scale nanofluid production (Roy, 2021; Das *et al.*, 2007).

### 2.1.2 Single-Step Method

The single-step method has been developed to overcome the stability challenges associated with the two-step approach. In this method, the dispersion and synthesis of nanoparticles occur simultaneously in the base fluid. Techniques such as chemical vapor deposition (CVD), direct evaporation-condensation, and laser ablation are commonly employed. For instance, the evaporation-condensation process involves vaporizing a bulk metal in a low-pressure chamber and allowing the vapor to condense directly into the liquid medium, forming a well-dispersed nanomaterial. Because the particles are generated and stabilized within the fluid in real time, this method minimizes aggregation and enhances suspension stability. However, single-step synthesis often suffers from limitations such as high production costs, restricted scalability, and difficulty in controlling particle size distribution. Despite these drawbacks, it has been particularly useful in research settings where high-quality, stable nanofluids are required for precise applications in heat transfer and biomedical fields (Eastman *et al.*, 2001).

### 2.1.3 Green Synthesis Methods

With increasing concern over the environmental and health impacts of chemical-based synthesis, green synthesis approaches for nanofluids have emerged as a sustainable alternative. These methods utilize natural precursors, plant extracts, microorganisms, or biodegradable polymers to produce nanomaterials, which are then dispersed in the base fluid. For example, plant-based polyphenols can act as both reducing and stabilizing agents, facilitating the formation of metal or metal oxide nanomaterial without toxic chemicals. Green synthesis not only improves the biocompatibility of nanofluids but also reduces the environmental footprint of production. Furthermore, the use of renewable resources has enhanced the feasibility of large-scale, eco-friendly nanofluid applications, particularly in biomedical and agricultural fields (Siavas, 2011). However, challenges remain in achieving uniform particle size, precise morphology control, and reproducibility compared to conventional chemical techniques.

### 2.1.4 Hybrid and Emerging Methods

Beyond the traditional two-step and single-step techniques, hybrid synthesis methods have been introduced to combine the advantages of both approaches. For instance, nanomaterials can be partially synthesized externally and then further processed directly in the fluid using in situ stabilization, thereby improving dispersion and reducing aggregation. Additionally, recent advances involve microfluidic-based synthesis, where controlled environments in microchannels allow precise regulation of reaction conditions, yielding uniform and stable nanofluids. Electrospraying, pulsed wire discharge, and microwave-assisted synthesis are also gaining traction for their ability to rapidly produce nanomaterials with tailored properties (Kumar *et al.*, 2022). These emerging techniques hold promise for designing application-specific nanofluids with enhanced thermal, optical, or catalytic performance, though their widespread industrial adoption is still under exploration.

## 2.2 *Stability and Rheology*

Stability remains one of the most pressing concerns in nanofluid research. Studies show that zeta potential optimization and ultrasonic dispersion can prolong colloidal stability. However, stability–performance trade-offs emerge since excessive additives may alter rheology and increase pumping power requirements (Saidur *et al.* 2011).

The stability and rheology of nanofluids are among the most critical parameters that determine their practical usability, since the extraordinary thermophysical enhancements reported in the literature depend not only on the intrinsic properties of nanoparticles but also on how well they remain dispersed in the base fluid and how the suspension behaves under shear and flow conditions (Das *et al.*, 2007; Yu *et al.*, 2008). Stability refers to the ability of nanoparticles to remain uniformly suspended without agglomeration or sedimentation over time, while rheology concerns the flow and deformation behavior of nanofluids under applied stress (Kumar *et al.*, 2022). These two aspects are interconnected, as instability phenomena such as clustering or sedimentation directly influence viscosity, shear response, and ultimately the heat transfer performance of the suspension (Saidur *et al.*, 2011).

The challenge of achieving long-term stability lies in fundamental colloid science, where attractive van der Waals forces, gravitational settling, and surface energy considerations drive nanoparticles toward agglomeration (Das *et al.*, 2007). Because nanoparticles possess extremely high surface-to-volume ratios and surface energies, they naturally tend to form clusters, which reduces uniform dispersion (Choi and Eastman, 1995). In practice, stability depends on particle size, morphology, concentration, pH, and surface chemistry, making the control of these parameters central to nanofluid design (Kumar *et al.*, 2022). Various approaches have been developed to mitigate instability, such as the use of surfactants, polymers, and dispersants that adsorb on nanoparticle surfaces to provide steric or electrostatic stabilization (Yu *et al.*, 2008). Adjusting the pH near the isoelectric point to maximize zeta potential, applying surface functionalization for stronger nanoparticle–fluid affinity, and employing ultrasonication to break agglomerates are also common strategies (Das *et al.*, 2007). Another promising approach is in situ synthesis, where nanoparticles are generated directly in the fluid medium, avoiding redispersion problems common in two-step methods (Kumar *et al.*, 2022).

Stability assessment requires careful characterization using methods such as UV–Vis spectrophotometry to track dispersion, dynamic light scattering (DLS) for hydrodynamic particle size, and zeta potential analysis to predict electrostatic repulsion, where values above ±30 mV generally indicate good stability (Yu *et al.*, 2008). Advanced imaging tools like TEM or AFM also help visualize agglomeration at nanoscale resolution (Das *et al.*, 2007).

Rheology determines how nanofluids behave under flow, directly impacting applications in heat exchangers, lubrication, and biomedical delivery, where pumping power and efficiency are influenced by viscosity and flow characteristics (Saidur *et al.*, 2011). While some nanofluids follow Newtonian behavior with constant viscosity, many exhibit non-Newtonian behaviors such as shear thinning, shear thickening, or viscoelasticity, especially at higher particle loadings (Kumar *et al.*, 2022). For example, nanofluids with elongated particles like carbon nanotubes often display shear thinning due to particle alignment under flow (Yu *et al.*, 2008), while some oxide-based nanofluids exhibit shear thickening at high concentrations due to clustering (Das *et al.*, 2007). Temperature also strongly affects rheology, as increased Brownian motion lowers viscosity, though excessive heating may weaken stabilization mechanisms (Saidur *et al.*, 2011).

Several classical and semi-empirical models have been proposed to explain nanofluid rheology. Einstein's (1906) classical model for dilute suspensions predicts a linear relation between viscosity and volume fraction, valid only at very low concentrations. Brinkman (1952) extended this to higher concentrations, while Batchelor (1977) incorporated hydrodynamic particle interactions. However, these models often underestimate real viscosity, especially when clustering, anisotropy, or interfacial effects are significant (Yu *et al.*, 2008). More advanced formulations, such as the Krieger–Dougherty model, include maximum packing effects, but deviations from experimental results persist (Kumar *et al.*, 2022). Modern research often employs rheometry under varying shear rates and temperatures to refine models and incorporate nanoscale dynamics (Saidur *et al.*, 2011).

The importance of stability and rheology extends to real-world applications. In thermal management, unstable nanofluids may sediment, clog channels, or erode surfaces (Choi & Eastman, 1995). Similarly, excessive viscosity may offset thermal gains by increasing pumping power requirements (Das *et al.*, 2007). In biomedical systems, stability ensures uniform drug delivery, while rheology dictates circulation and penetration (Yu *et al.*, 2008). In lubrication, stability prevents debris clustering, and rheology supports proper film formation. In solar thermal and energy storage applications, both properties govern long-term reliability and scalability (Saidur *et al.*, 2011).

Despite progress, challenges remain. There is no universal standard for stability assessment, leading to inconsistent reporting across studies (Kumar *et al.*, 2022). Long-term stability under real-world conditions like thermal cycling, magnetic fields, or chemical reactivity is not fully understood (Das *et al.*, 2007). Rheological behavior remains system-dependent, with discrepancies across laboratories due to differences in synthesis, functionalization, and measurement (Yu *et al.*, 2008). Balancing stability and viscosity is also difficult, often requiring trade-offs or hybrid nanofluids that combine multiple particle types for synergistic benefits (Kumar *et al.*, 2022).

Looking ahead, integrating colloid science, fluid mechanics, and nanotechnology is crucial for predictive frameworks and standardization (Saidur *et al.*, 2011). Future directions include machine learning to link synthesis with stability, molecular dynamics simulations to study nanoscale flow, and smart nanofluids with stimuli-responsive properties (Kumar *et al.*, 2022). Ultimately, overcoming the intertwined challenges of stability and rheology will determine the scalability and safe use of nanofluids across energy, healthcare, and manufacturing (Das *et al.*, 2007).

## 2.3 *Thermal, Electrical, and Wettability Properties*

Nanofluids consistently exhibit enhanced thermal conductivity and modified electrical and wettability characteristics compared to base fluids. However, inconsistencies in reported enhancements across laboratories underscore the need for standardized characterization protocols (Garg *et al.* 2022).

The study of nanofluids has revealed that their thermal, electrical, and wettability properties are among the most important attributes governing their multifunctional performance in engineering, biomedical, and energy systems, since these parameters directly dictate heat transport efficiency, electrical conduction and dielectric response, as well as interfacial interactions with solid substrates, thereby establishing their technological relevance across diverse applications ranging from electronics cooling to solar thermal collectors, biomedical diagnostics, electrochemical storage, and surface engineering (Saidur *et al.*, 2011). The thermal properties of nanofluids have been extensively studied since the pioneering work of Choi and Eastman (1995), who demonstrated anomalous enhancements in thermal conductivity of nanoparticle suspensions, a discovery that challenged the limits of classical effective medium theories and opened up new possibilities for advanced heat transfer fluids. The mechanisms underlying these enhancements are multifaceted, involving not only the intrinsic high conductivity of metallic or carbon-based nanoparticles but also the effects of Brownian motion, liquid layering at the solid–liquid interface, microconvection induced by particle dynamics, and aggregation networks that facilitate conductive pathways (Das *et al.*, 2007; Yu *et al.*, 2008). For instance, metal oxide nanoparticles such as $Al_2O_3$, $TiO_2$, and ZnO dispersed in water or glycols have shown thermal conductivity improvements ranging from 10% to 40% at low volume fractions, while carbon nanotube and graphene-based nanofluids often exhibit even higher enhancements due to their extraordinary intrinsic conductivity and aspect ratio, which enable percolation networks to form at lower concentrations (Das *et al.*, 2007).

Factors such as particle size, shape, concentration, surface functionalization, and base fluid type strongly influence the degree of thermal conductivity improvement, with smaller particles and high aspect-ratio morphologies generally producing superior enhancements, though excessive loading can increase viscosity and offset the benefits by requiring higher pumping power (Yu *et al.*, 2008). Convection heat transfer, central to practical cooling and heating applications, has also shown significant improvements in nanofluids, particularly under turbulent flow conditions where nanoparticle migration and enhanced thermal boundary layer disruption contribute to higher Nusselt numbers (Lee *et al.*, 2022). In boiling and phase-change heat transfer, nanofluids have demonstrated not only higher critical heat fluxes but also altered surface characteristics due to nanoparticle deposition, which modify wettability and nucleation site density (Kumar *et al.*, 2022). Although experimental evidence supports substantial enhancements in heat transfer coefficients, the results are often inconsistent due to differences in synthesis methods, stability levels, and measurement techniques, highlighting the need for standardization and unified theoretical frameworks (Saidur *et al.*, 2011; Yu *et al.*, 2008).

In parallel with thermal properties, the electrical characteristics of nanofluids have emerged as another area of high importance, particularly for applications in sensors, energy storage, electrohydrodynamic cooling, and dielectric systems. The inclusion of conductive nanoparticles such as silver, copper, carbon nanotubes, or graphene into otherwise insulating base fluids drastically alters the electrical conductivity, often by several orders of magnitude, depending on concentration and dispersion stability (Lee *et al.*, 2022). The increase in electrical conductivity arises from mechanisms including electron tunneling between closely spaced particles, surface charge transport facilitated by the electrical double layer, and the formation of percolation networks that create continuous conductive paths at critical particle concentrations (Das *et al.*, 2007).

Equally critical, though often underemphasized, are the wettability properties of nanofluids, which play a decisive role in phase-change processes such as boiling, condensation, and evaporation, as well as in surface coating, lubrication, and biomedical wetting applications. The presence of nanoparticles in a base fluid alters its surface tension and interfacial interactions, often leading to enhanced wettability due to nanoparticle adsorption at the liquid–solid interface, modification of surface roughness, and deposition layers that reduce contact angle (Kumar *et al.*, 2022). For instance, studies on water-based nanofluids containing $Al_2O_3$ or $TiO_2$ have reported significant reductions in contact angle on heated metallic surfaces, thereby increasing surface wettability and promoting more efficient boiling heat transfer (Saidur *et al.*, 2011).

Taken together, the thermal, electrical, and wettability properties of nanofluids reveal a multidimensional landscape where nanoscale interactions dictate macroscopic functionality, yet the challenges of reproducibility, stability, and scalability persist as barriers to industrial adoption. While enhancements are promising, inconsistencies across experiments and the lack of standardized testing protocols remain major gaps. Future research must leverage computational modeling, green synthesis, and tailored nanofluid designs to optimize these properties synergistically for application-specific requirements (Lee *et al.*, 2022; Kumar *et al.*, 2022).

## 2.4 Characterization Methods and Modeling Approaches

Techniques such as dynamic light scattering (DLS), TEM, and zeta potential measurement are widely used to characterize nanofluids. Modeling efforts range from classical effective medium theory to molecular dynamics simulations. Yet, discrepancies between theoretical predictions and experimental results persist, indicating a gap in multi-scale modeling frameworks. Nanofluids, which are mixtures of base fluids like water, ethylene glycol, or oil with small amounts of nanoparticles, have become a very active research area because of their improved thermal, electrical, and mechanical properties. To understand these properties and use nanofluids in real applications, researchers first need to study them carefully. This study uses *characterization methods*, which are techniques employed to measure and understand the physical, chemical, and structural properties of nanofluids. Alongside these experimental methods, *modeling approaches* are also used. Modeling involves using mathematical equations, simulations, or computational tools to predict how nanofluids will behave under different conditions. Characterization methods and modeling approaches go hand in hand: while characterization gives real experimental data, modeling helps predict, explain, and guide experiments. Together, they form the backbone of nanofluid research. However, both areas face challenges due to the small size of particles, complex interactions between nanoparticles and fluids, and the lack of standardized methods. This section explains the major characterization techniques and modeling strategies used in nanofluid research.

### 2.4.1 Microscopy Techniques

Microscopy is one of the most basic and important ways to see how nanoparticles are dispersed inside a fluid. Different types of microscopes are used. Transmission Electron Microscopy (TEM) can provide very clear images of nanoparticles at the nanometer scale. It helps researchers measure particle size, shape, and even how particles are clustered together. Scanning Electron Microscopy (SEM) is also widely used, although it usually requires the nanoparticles to be dried before observation. Atomic Force Microscopy (AFM) provides surface details and topography of nanoparticles. These methods are important because the size, shape, and surface structure of nanoparticles directly affect nanofluid properties (Kole and Dey, 2010).

### 2.4.2 Spectroscopy Techniques

Spectroscopy methods are used to identify the chemical composition and surface features of nanoparticles in nanofluids. Ultraviolet–Visible (UV-Vis) spectroscopy is often used to check the stability of nanofluids by studying how particles absorb light at different wavelengths. Fourier Transform Infrared Spectroscopy (FTIR) helps understand the bonding and surface modifications of nanoparticles, especially if surfactants are added to improve stability. Raman spectroscopy is also useful for studying structural features of nanoparticles. These techniques allow researchers to confirm if nanoparticles are well-dispersed and chemically stable inside the fluid (Yu and Xie, 2012).

### *2.4.3 Dynamic Light Scattering (DLS) and Zeta Potential*

DLS is a very popular tool for measuring the average particle size and size distribution in nanofluids. Since nanoparticles tend to form clusters, DLS helps researchers understand how uniform the dispersion is. Another related measurement is zeta potential, which gives information about the surface charge of particles. A high zeta potential (positive or negative) usually means the particles repel each other and the nanofluid is stable. If the zeta potential is low, particles may come together and settle, leading to instability (Saidur *et al.*, 2011).

### *2.4.4 Thermophysical Property Measurements*

For practical use, researchers need to measure properties like thermal conductivity, viscosity, density, and specific heat. Thermal conductivity can be measured using tools such as the transient hot-wire method, laser flash technique, or steady-state methods. Viscosity is usually measured with rheometers, and density with simple density meters. These values are very important for designing cooling systems or energy devices. However, measurements are not always consistent because different labs use different instruments and sample conditions (Murshed and Estellé, 2017).

### *2.4.5 X-ray Diffraction (XRD) and Other Techniques*

XRD is used to identify the crystal structure of nanoparticles and confirm their composition. For example, it can show if nanoparticles are made of metal oxides, metals, or carbon materials. Other supporting techniques include Thermogravimetric Analysis (TGA), which measures weight changes with temperature, and Differential Scanning Calorimetry (DSC), which measures heat flow. These are useful for studying how nanoparticles behave under heating or cooling conditions, which is very important in energy and thermal applications (Mahian *et al.*, 2019).

In short, characterization methods are necessary for understanding nanofluid stability, particle size, surface chemistry, and thermophysical behavior. Without these methods, researchers cannot verify whether nanofluids are suitable for real-world applications.

Characterization alone is not enough, because experiments can be expensive, time-consuming, and sometimes not possible under extreme conditions. This is where modeling approaches are useful. Modeling helps predict nanofluid behavior, understand heat transfer mechanisms, and guide experiments. Different approaches have been developed over the years.

### 2.4.6 *Empirical Correlations*

One of the simplest approaches to modeling nanofluids is creating empirical correlations. These are mathematical equations developed from experimental data. For example, researchers may prepare nanofluids with different concentrations of nanoparticles and measure thermal conductivity. From these data, they develop equations that link concentration, particle size, and thermal conductivity. These correlations are useful but often limited to specific conditions and cannot be generalized to all nanofluids (Khanafer and Vafai, 2011).

### 2.4.7 *Classical Theoretical Models*

Several theoretical models have been proposed to explain the thermal conductivity and viscosity of nanofluids. The Maxwell model, originally developed for solid-liquid mixtures, is often used as a base. Hamilton–Crosser and Bruggeman models are other classical theories. However, these models often fail to fully explain experimental results, especially at the nanoscale, because they do not include factors like Brownian motion, particle clustering, or liquid layering (Buongiorno, 2006).

### 2.4.8 *Mechanism-Based Models*

More advanced models try to include the physics of nanoparticles in fluids. For example, Buongiorno (2006) proposed a model that considers Brownian diffusion and thermophoresis (movement of particles due to temperature gradients). Other models include the effect of particle–fluid interactions, aggregation, and interfacial layers. These models are more realistic but also more complicated.

### 2.4.9 *Computational Fluid Dynamics (CFD)*

CFD is a very powerful tool for modeling nanofluids in complex geometries and flow conditions. It allows researchers to simulate heat transfer and fluid flow in devices like heat exchangers, solar collectors, or microchannels. CFD models can be single-phase (treating nanofluid as a uniform fluid) or two-phase (considering nanoparticles and fluid separately). While single-phase models are simpler, two-phase models are more accurate but computationally expensive (Mahian *et al.*, 2019).

### 2.4.10 *Molecular Dynamics (MD) Simulations*

At the nanoscale, MD simulations are used to study the interaction of nanoparticles and fluid molecules. These simulations can reveal mechanisms like liquid layering around nanoparticles, which cannot be easily observed experimentally. MD simulations require high computing power but are very useful for understanding fundamental physics (Hussein *et al.*, 2020).

### 2.4.11 Machine Learning and Data-Driven Models

Recently, researchers have started using artificial intelligence (AI) and machine learning (ML) to predict nanofluid properties. By training ML models with experimental data, researchers can predict thermal conductivity, viscosity, or stability without doing new experiments. This approach is very promising because it can handle complex nonlinear relationships that traditional models cannot capture (Anselma *et al.*, 2019).

### 2.4.12 Integration of Characterization and Modeling

For nanofluid research to progress, both characterization and modeling must be used together. Experimental methods provide reliable data, while modeling helps explain and predict trends. For example, if DLS and zeta potential measurements show instability, modeling can be used to predict how surfactants or pH changes will improve stability. Similarly, CFD simulations can be validated with real viscosity and conductivity data. The integration of both approaches is necessary to design nanofluids for applications such as the cooling of electronics, renewable energy, and biomedical treatments.

## 2.5 Environmental and Economic Considerations

The life-cycle impacts of nanoparticle synthesis, recycling, and disposal are not yet well understood. The cost of nanomaterials also remains a limiting factor for large-scale adoption. Nanofluids, which are mixtures of nanoparticles and base fluids, have attracted significant interest for their potential to improve heat transfer, energy efficiency, and performance in many applications such as cooling of electronics, solar collectors, automotive engines, and biomedical devices. While their scientific and technical benefits are widely discussed, it is equally important to look at the environmental and economic aspects of nanofluids. These two factors decide whether nanofluids can move beyond laboratories and be accepted in industries on a large scale. The environmental side focuses on issues such as nanoparticle toxicity, disposal, sustainability of production methods, and potential ecological risks, while the economic side focuses on the cost of nanoparticle production, preparation, stability improvement, long-term use, and overall cost–benefit analysis compared to conventional fluids. Researchers and industries need to carefully balance these environmental and economic considerations before nanofluids can become practical solutions. This section explains these issues in detail using simple language.

### 2.5.1 Toxicity of Nanoparticles

One of the biggest environmental concerns of nanofluids is the toxicity of nanoparticles. Nanoparticles are extremely small, usually less than 100 nanometers, which makes them capable of entering living cells and biological systems. Some metal oxide nanoparticles, such as zinc oxide (ZnO) or titanium dioxide ($TiO_2$), are relatively safe, but others, like silver (Ag) or copper oxide (CuO), can be toxic to aquatic organisms, plants, and even humans. If nanofluids are released into water bodies or soil without proper treatment, they may accumulate and cause harmful effects on ecosystems. The toxicity depends on factors such as size, shape, concentration, and surface coating of nanoparticles. This makes environmental safety a serious challenge in nanofluid research (Kumar *et al.*, 2019).

### 2.5.2 Occupational Safety During Handling

During the preparation, transport, and use of nanofluids, researchers and workers may be exposed to nanoparticles. Since nanoparticles can be inhaled or absorbed through the skin, there is a risk of respiratory or skin-related health issues. Without strict safety measures, occupational exposure may harm those who work directly with nanofluids. Therefore, industries need to adopt guidelines such as using protective clothing, masks, and sealed containers to reduce risk (Hussein *et al.*, 2020).

### 2.5.3 Environmental Release and Disposal

Another environmental issue is how to dispose of nanofluids after use. Unlike traditional fluids, nanofluids cannot be easily recycled or released into drains because nanoparticles may persist in the environment. They do not break down easily and may accumulate in plants, soil, or aquatic systems. Long-term release could affect food chains and biodiversity. Researchers are therefore working on "green nanofluids" made from biodegradable or naturally derived nanoparticles, such as carbon-based or plant-based particles, which may be safer and more eco-friendly (Saidur *et al.*, 2011).

### 2.5.4 Sustainability of Nanoparticle Production

The production of nanoparticles often requires high energy input and uses toxic chemicals in synthesis. For example, chemical reduction or sol-gel methods consume large amounts of solvents and produce hazardous by-products. If the environmental cost of nanoparticle production is higher than the energy saved during nanofluid applications, the overall benefit becomes questionable. More sustainable production methods, such as green synthesis using plant extracts or microbes, are being explored to reduce the environmental footprint (Mahian *et al.*, 2019).

### 2.5.5 Regulations and Standards

Currently, there are very few global standards for the safe use and disposal of nanofluids. Regulatory agencies are still developing guidelines to manage nanoparticle-based technologies. This lack of strict regulation creates uncertainty about environmental risks. Until strong regulations and safety protocols are in place, industries may be hesitant to use nanofluids widely (Anselma *et al.*, 2019).

### 2.5.6 High Cost of Nanoparticles

One of the main economic challenges of nanofluids is the cost of nanoparticles. Producing nanoparticles with controlled size, shape, and surface properties requires advanced techniques such as chemical vapor deposition, sol-gel processes, or high-energy milling. These methods are expensive, especially when producing large volumes. Since nanofluids need consistent quality, cost reduction is a critical requirement for industrial adoption (Kole and Dey, 2010).

### 2.5.7 Preparation and Stability Costs

Nanofluid preparation is not just about mixing nanoparticles with fluids. It often requires surfactants, stabilizers, or ultrasonication to keep particles dispersed and prevent settling. These extra steps add to the cost. Additionally, stabilizers may degrade over time, which means the nanofluid needs to be reprocessed or replaced, leading to increased maintenance costs. Achieving long-term stability at low cost is a major economic challenge (Yu and Xie, 2012).

### 2.5.8 Energy Balance and Efficiency

Even though nanofluids improve heat transfer, the energy required to prepare and maintain them must be considered. For example, if ultrasonication consumes a lot of energy or if increased viscosity requires more pumping power in a cooling system, the economic advantage may be reduced. Therefore, cost–benefit analysis must include both preparation costs and operational energy savings to understand the real economic impact (Khanafer and Vafai, 2011).

### 2.5.9 Scaling Up from Laboratory to Industry

Most nanofluid research is conducted at laboratory scale with small volumes. Producing nanofluids in industrial quantities requires large-scale synthesis of nanoparticles, storage, transportation, and consistent quality control, all of which add costs. Many industries hesitate to invest in nanofluids because the technology for large-scale production is not yet fully optimized or standardized (Mahian *et al.*, 2019).

### 2.5.10 Market Competitiveness

For industries to adopt nanofluids, they must be economically competitive with existing technologies. For instance, conventional coolants like water, oils, and ethylene glycol are cheap, easy to handle, and have established recycling methods. Nanofluids must demonstrate significantly better performance to justify higher costs. If the improvement in efficiency is small compared to the cost, industries may prefer to stick with traditional fluids (Saidur *et al.*, 2011).

### 2.5.11 Lifecycle Cost Analysis

Another economic factor is lifecycle cost analysis, which considers not just preparation and operation costs but also disposal, recycling, and environmental penalties. If nanofluids are difficult to dispose of safely, industries may face additional costs for treatment. Similarly, if regulations become stricter in the future, companies using nanofluids may have to spend more on compliance. Therefore, lifecycle analysis is critical to decide whether nanofluids are economically viable in the long run (Hussein *et al.*, 2020).

### 2.5.12 Linking Environmental and Economic Issues

The environmental and economic aspects of nanofluids are strongly connected. If nanoparticle synthesis methods are polluting, they also increase treatment costs, making nanofluids less economically attractive. If disposal is unsafe, industries may face legal penalties, which add to economic burdens. On the other hand, if researchers can develop eco-friendly nanofluids using green synthesis methods and recyclable nanoparticles, both environmental and economic benefits can be achieved. For example, carbon-based nanofluids like those made from graphene or carbon nanotubes may be more sustainable compared to heavy metal oxide nanoparticles, although cost is still an issue. Therefore, the success of nanofluids in industry depends on finding solutions that simultaneously address both environmental safety and economic feasibility.

## 3. Cross-Cutting Challenges

Nanofluids—fluids with tiny particles added—can transfer heat better than regular fluids and offer new capabilities in cooling, energy, and medicine. But researchers face many real problems. These include: keeping particles from clumping and settling over long periods; differences in how groups prepare and measure nanofluids (so results don't match); higher fluid thickness (viscosity) that increases pumping energy; possible health and environmental risks; and the difficulty of scaling lab results to big systems. To move forward, the field needs shared testing rules, longer-term tests, safer production methods, and projects that test nanofluids in real industrial conditions.

| *Main challenge* | *Why it matters (brief)* | *Real Life Solutions* | *References* |
|---|---|---|---|
| Long-term stability | Nanoparticles agglomerate or settle → loss of performance | In-situ synthesis; robust surface functionalization; accelerated ageing tests; open stability databases | Hussein *et al.*, 2020. |

*Contd.*

| *Main challenge* | *Why it matters (brief)* | *Real Life Solutions* | *References* |
|---|---|---|---|
| Reproducibility & standardization | Different labs report inconsistent results → hard to compare or scale | Standard test protocols (thermal, rheology, zeta potential), common reporting templates, and round-robin studies | Anselma *et al.* (2019). |
| Viscosity/ pumping energy | Improved heat transfer but higher pumping cost may offset gains | Optimize particle loading, use hybrid particles, and quantify lifecycle energy balance | Viscosity trade-offs reported in Kole & Dey, 2010. |
| Scale-up/ industrial validation | Most work at lab scale; industry needs real-world data | Pilot demonstrations, industry–academia consortia, field trials, and cost–benefit studies | Mahian *et al.*, 2019. |
| Environmental & health safety | Unknown ecotoxicity & occupational exposure | Green synthesis routes, life-cycle assessments (LCA), regulatory frameworks, and exposure protocols | Nel *et al.*, 2006; Savolainen *et al.*, 2010. |
| Modeling gaps (multi-scale) | Nano → macro linkage missing; predictions unreliable | Hybrid models (MD → CFD), ML to map experiments → predictions, curated datasets | Ghalandari *et al.*, 2021. |

### *3.1 Standardization of Synthesis and Characterization Protocols*

The absence of universally accepted protocols leads to inconsistent and irreproducible results across laboratories. This lack of standardization impedes meta-analyses and hinders industrial trust.

#### *3.1.1 Preparation Methods*

Nanofluids can be prepared mainly by two methods: the one-step method (making nanoparticles directly inside the fluid) and the two-step method (first making dry nanoparticles and then mixing them in fluids). Both methods have advantages and disadvantages. In the one-step method, particles are more stable, but it is expensive and difficult to produce in large amounts. In the two-step method, it is cheaper and more common, but particles often agglomerate and cause stability problems. Also, there is no standard preparation method followed across all labs, which means results vary a lot between researchers. Even the concentration of nanoparticles, the mixing method, or the type of surfactant used can lead to very different results. This lack of a universal standard for preparing nanofluids makes it hard to compare research outcomes and apply them in industries (Yu and Xie, 2012).

### *3.1.2 Accurate Measurement of Properties*

Another major challenge is measuring the properties of nanofluids correctly. Properties like thermal conductivity, viscosity, density, electrical conductivity, and specific heat are very important for applications. But different labs use different instruments and methods, which sometimes give very different results for the same nanofluid. For example, one group may find that adding nanoparticles improves thermal conductivity a lot, while another group using slightly different methods may report only a small improvement. These differences confuse industries and slow down commercialization. Also, at very small scales, nanoparticles interact with fluids in complex ways that are not fully understood, making it hard to develop accurate mathematical models. So researchers find it challenging to measure and predict nanofluid behavior in a reliable way (Murshed and Estellé, 2017).

### *3.1.3 High Viscosity and Pumping Power*

When nanoparticles are added to fluids, they sometimes increase the viscosity (thickness) of the fluid. This is a big problem because thicker fluids require more pumping power in systems like heat exchangers or cooling pipes. More pumping power means more energy consumption, which can cancel out the benefits of higher heat transfer. Researchers struggle to find the right balance between improving thermal conductivity and not increasing the viscosity too much. If viscosity is too high, nanofluids cannot be used in real-world systems, even if they transfer heat better (Khanafer and Vafai, 2011).

## ***3.2 Balancing Stability and Performance***

While high nanoparticle loading enhances thermal properties, it compromises stability and increases viscosity. Achieving an optimal balance between performance and stability remains an ongoing challenge.

One of the biggest and most common challenges researchers face is the stability of nanofluids. Stability means that the nanoparticles should not clump together or settle at the bottom of the fluid container. If the nanoparticles settle, the fluid will lose its special properties and behave just like the base fluid (such as water, oil, or ethylene glycol). Keeping nanoparticles stable in fluids is not easy because nanoparticles naturally try to stick together due to forces between them. Researchers try many ways, like adding surfactants (chemicals that keep particles apart), adjusting pH, or using ultrasonication (using sound waves to break clumps). But still, stability does not last long and sometimes only works under laboratory conditions. In real-life applications, such as in car engines or industrial cooling, the fluid has to stay stable for months or years, which is very hard to achieve. Therefore, researchers find it challenging to maintain the long-term stability of nanofluids (Saidur *et al.*, 2011).

### 3.3 Scaling Laboratory Results to Industrial Applications

Many studies focus on small-scale heat exchangers or controlled environments. Translating these findings to large-scale power plants, automotive systems, or data centers is complex due to additional variables such as flow turbulence, fouling, and long-term material compatibility. Most nanofluid studies are done in laboratories with small volumes under controlled conditions. But when researchers try to scale up nanofluids for large industrial systems, many unexpected problems appear. The stability, cost, preparation time, and property measurements often behave differently at a large scale compared to small lab samples. Industries need huge volumes of fluids that stay stable and effective for long periods, which is very difficult to achieve. This scale-up challenge slows down the real-life adoption of nanofluids (Mahian *et al.*, 2019).

### 3.4 Environmental and Health Safety Concerns

Despite their promise, the widespread adoption of nanofluids requires addressing critical concerns. Potential toxicity, environmental impact, and standardization of characterization methods remain areas of ongoing research. The very properties that make nanomaterial effective—such as their small size and reactivity—also raise concerns about health and ecological risks (Saidur, Leong, and Mohammed, 2011). Thus, future research is expected to focus not only on performance enhancement but also on safe design, sustainable synthesis, and regulatory frameworks.

Nanofluids pose unknown toxicological risks to human health and ecosystems. Their small size enables bioaccumulation and potential disruption of cellular functions (Asadi *et al.* 150). Addressing these risks requires stringent regulations and safety evaluations.

Nanoparticles are very small and can enter the human body through inhalation, skin contact, or ingestion. Some nanoparticles are toxic and can cause harmful effects on health. For example, metal oxide nanoparticles may generate reactive oxygen species that damage cells. Handling these nanoparticles requires strict safety rules, but not all labs can afford advanced safety equipment. Similarly, if nanofluids are released into the environment, they may pollute water and soil. Researchers face the challenge of making nanofluids safe for humans and the environment, and at the same time useful for industries (Kumar *et al.*, 2019).

### 3.5 Challenge of Cost and Economic Feasibility

Nanoparticles are expensive to produce, and the methods used to prepare nanofluids also add extra costs. For example, processes like ultrasonication, surface functionalization, or adding stabilizing agents increase the price. While nanofluids may improve performance, industries also care about whether the benefits are worth the extra cost. If nanofluids make a system only slightly better but increase costs a lot, industries may not want to use them. Researchers, therefore, face the challenge of proving that nanofluids are not just effective, but also economically feasible (Kole and Dey, 2010).

### 3.6 Challenge of Lack of Standardization

Different researchers use different concentrations, particle sizes, shapes, and base fluids. This makes it difficult to compare results across different studies. There are no universally accepted standards or guidelines for nanofluid preparation, testing, or reporting. Because of this, the same nanofluid may show different results in different labs. Without standardization, industries cannot trust the data and hesitate to adopt nanofluids. Establishing global standards is one of the key challenges in this field (Anselma *et al.*, 2019).

### 3.7 Challenge of Understanding Heat Transfer Mechanisms

Even though nanofluids are known to improve thermal properties, researchers are still debating why and how exactly they do this. Some theories suggest it is due to Brownian motion of particles, others suggest liquid layering around nanoparticles, while some point to clustering effects. Since there is no single clear theory, predicting nanofluid performance in different applications is very difficult. Without a clear mechanism, researchers cannot fully optimize nanofluids for specific uses (Buongiorno, 2006).

### 3.8 Challenge of Reproducibility of Results

In scientific research, results should be reproducible, meaning that if the same experiment is done again, it should give similar results. But in nanofluid research, reproducibility is a big challenge. Sometimes, even the same group cannot reproduce their earlier results because tiny changes in particle size, surfactant amount, or mixing time can completely change outcomes. This lack of reproducibility creates doubts and slows progress in the field (Murshed and Estellé, 2017).

### 3.9 Challenge of Multidisciplinary Nature of Research

Nanofluid research is not limited to one field. It requires knowledge from physics, chemistry, material science, fluid mechanics, and engineering. Many researchers are experts in only one area and may lack understanding in others. This makes it difficult to design complete studies that cover all aspects. Collaboration between experts of different fields is required, but it is not always easy to achieve. This multidisciplinary demand is a challenge by itself (Mahian *et al.*, 2019). Even if nanofluids improve efficiency, the preparation of nanoparticles often requires high energy input, such as in chemical synthesis or mechanical milling. If the energy cost of making nanofluids is higher than the energy saved in applications, the overall benefit becomes doubtful. Researchers must carefully consider the energy balance, which is not easy because many hidden costs are involved (Hussein *et al.*, 2020). Even if a nanofluid works well in the short term, researchers need to prove that it can maintain performance for years in real systems. Over time, nanoparticles may degrade, settle, or react with the base fluid. For example, metal nanoparticles may oxidize and lose effectiveness. Testing long-term behavior requires time, resources, and monitoring, which many research projects cannot afford. Therefore, long-term performance remains a big challenge (Kole and Dey, 2010). Modeling nanofluid behavior using computer simulations is very complex because of the small size of particles and their interactions with fluids. Traditional fluid models do not always apply to nanofluids. Advanced models are required that combine nanoscale physics with macroscale fluid dynamics. Developing such models is difficult and requires powerful computers. Moreover, experimental data are not always available to validate models. This makes simulation and modeling a challenging area for researchers (Khanafer and Vafai, 2011). Although thousands of papers are published on nanofluids, only very few large-scale commercial applications exist. Industries are still doubtful about investing in nanofluids because of uncertainties in cost, stability, and long-term safety. This gap between research and industry use is itself a challenge. Researchers need to bridge this gap by demonstrating successful pilot projects, but funding and industry collaboration are often limited (Mahian *et al.*, 2019).

## 4. Emerging Technologies and Trends

### *4.1 Hybrid Nanofluids and Composite Particles*

Nanofluids are gaining attention as one of the most promising areas of research because they combine the benefits of nanotechnology and fluid science. Emerging technologies and current trends are shaping how nanofluids are designed, studied, and applied in real-world systems. One important trend is the development of hybrid nanofluids, where two or more types of nanoparticles are mixed into a base fluid. These hybrid nanofluids show improved thermal conductivity, stability, and overall efficiency compared to single nanoparticle-based nanofluids (Sundar *et al.*, 2017). Hybrid nanofluids-suspensions containing two or more nanoparticle types-exhibit synergistic enhancements in thermal and tribological properties. Composite particles also promise improved stability and functionality.

### *4.2 Bio-Based and Green Nanofluids*

Another important technology is the use of green synthesis methods for producing nanoparticles. Instead of using harsh chemicals, researchers are now using plant extracts, microorganisms, and other eco-friendly techniques to prepare nanoparticles, which makes nanofluid production safer and more sustainable (Narayana *et al.*, 2020). Research into bio-derived nanomaterials (e.g., cellulose, chitosan) aligns with sustainable development goals. Green nanofluids can reduce environmental risks and lifecycle costs.

### 4.3 Integration of AI and Digital Twins for Predictive Modelling

Along with synthesis, advanced characterization techniques are also emerging as a trend. Tools, such as scanning electron microscopy (SEM), transmission electron microscopy (TEM), and atomic force microscopy (AFM), are used more often to understand particle size, distribution, and interactions in nanofluids. These tools help scientists improve stability and predict performance in practical applications (Mahian *et al.*, 2019). Another trend is the use of machine learning and artificial intelligence (AI) to model nanofluid properties. Since nanofluids have complex behaviors that depend on many factors like particle size, shape, and concentration, AI helps in predicting their thermal and physical properties more accurately and quickly (Ghalandari *et al.*, 2021). Artificial intelligence and digital twins are emerging as powerful tools for predicting nanofluid behaviour under varying operating conditions. These technologies can accelerate optimization and bridge experimental–computational gaps (Kumar *et al.* 2022).

### 4.4 Advanced Manufacturing and Nanofluid System Design

Additive manufacturing and micro/nano-engineered surfaces can be integrated with nanofluid systems to maximize heat transfer efficiency. Such synergies can redefine applications in aerospace and microelectronics. In terms of applications, nanofluids are becoming more popular in renewable energy technologies. For example, they are being tested in solar collectors, geothermal systems, and energy storage units because of their ability to improve heat transfer (Choi *et al.*, 2020). They are also being studied for use in electric vehicle (EV) cooling systems and microelectronics, where heat management is a major challenge. At the same time, researchers are focusing on biomedical applications, such as targeted drug delivery, cancer therapy, and imaging, because nanofluids can carry therapeutic agents and release them at specific sites (Wang *et al.*, 2018).

A strong emerging trend is the movement toward sustainability and safety (S&S) in nanofluid research. Questions about toxicity, environmental impact, and recyclability are guiding researchers to find safer nanoparticles and biodegradable base fluids (Kasaeian *et al.*, 2017). Moreover, cost reduction is becoming a key focus, as large-scale production and application of nanofluids must be economically viable.

The field of nanofluids is growing with exciting new technologies and trends. Hybrid nanofluids, green synthesis, advanced characterization, AI-based modeling, renewable energy applications, biomedical uses, and sustainable practices are shaping the future of this field. These advancements suggest that nanofluids will play an increasingly important role in industries ranging from energy to healthcare.

# 5. Research Gaps

## 5.1 Lack of Long-Term Stability Data

Most studies examine short-term stability, often weeks or months. Industrial applications, however, require years-long stability data under dynamic thermal cycles. Although nanofluids have shown great potential in improving heat transfer and energy efficiency, there are still many gaps in research that slow down their progress toward real-world applications. One major gap is the **lack of long-term stability data**. Stability refers to how well nanoparticles remain evenly distributed in the base fluid without clumping or settling. Most laboratory studies focus only on short-term stability over a few hours or days, but for real industrial systems, fluids are expected to remain stable for months or even years (Nadooshan and Etemad, 2017). Without long-term data, it is difficult to guarantee consistent performance and safety of nanofluids in practical applications such as cooling systems, solar collectors, or biomedical devices. Techniques like surfactant addition, pH control, and surface modification improve stability, but their effects over long durations are not fully understood (Kasaeian *et al.*, 2017).

## 5.2. Insufficient Industrial Case Studies

While thousands of experimental papers and simulations exist in the academic domain, only a limited number of reports describe how nanofluids perform in large-scale, real-world conditions (Said *et al.*, 2015). Industries such as power plants, automotive cooling, or electronics manufacturing require reliability, cost-effectiveness, and safety evidence before adopting new technologies. However, most studies remain at laboratory scale with small volumes, controlled conditions, and limited testing times. For nanofluids to become widely accepted, researchers must provide field demonstrations that measure not only heat transfer benefits but also maintenance requirements, environmental safety, and economic feasibility (Mahian *et al.*, 2019). Without industrial case studies, the transition from laboratory research to commercial use remains slow and uncertain. Despite thousands of laboratory studies, very few large-scale demonstrations exist. Limited industry-academia collaboration prevents practical validation.

## 5.3 Limited Multi-Scale Modeling Approaches

Current modeling either focuses on the nano-scale or macro-scale, without robust integration. Developing hybrid models that account for particle–fluid interactions across scales is critical. The behavior of nanofluids is highly complex because it depends on interactions at different scales—from the movement of single nanoparticles at the microscopic level to overall heat transfer performance at the macroscopic system level (Das *et al.*, 2007). Many current models simplify assumptions, such as treating nanoparticles as uniformly dispersed or ignoring particle–particle interactions, which may not match real conditions. More advanced multi-scale models that link molecular dynamics, particle-level interactions, and system-level simulations are needed for accurate predictions (Buongiorno *et al.*, 2009). Such models could help optimize nanofluid design before large-scale testing, saving time and cost. However, computational limitations and the complexity of integrating models across scales remain significant challenges.

Research on nanofluids still faces important gaps. There is a need for long-term stability data, more industrial case studies, and better multi-scale modeling approaches to make nanofluids practical and reliable for commercial applications. Addressing these gaps will help researchers move from laboratory success to widespread adoption in energy, healthcare, and manufacturing industries.

## 6. Recommendations for Future Research

### *6.1 Collaborative Research Frameworks*

Industry–academia partnerships should be strengthened through joint consortia that focus on real-world validation. At present, many nanofluid studies are conducted in isolation, with limited exchange of data or experiences across institutions and industries. Collaboration among universities, industries, and government research organizations can reduce duplication of efforts and accelerate progress. Joint projects can also combine expertise in materials science, thermal engineering, environmental safety, and economics, which are all needed to fully understand nanofluids (Mahian *et al.*, 2019). Collaborative platforms, both national and international, would allow researchers to share experimental results, compare methods, and conduct large-scale field trials that no single group could achieve alone (Saidur *et al.*, 2011).

### *6.2 Development of Open-Access Databases*

Currently, nanofluid data, such as thermal conductivity, viscosity, stability, and environmental impacts, are scattered across thousands of journal papers and are not always easily accessible or standardized. An open, centralized database would allow researchers to store and retrieve experimental and simulation results, which would help in developing more accurate models and in reducing repeated experiments (Ghalandari *et al.*, 2021). Open-access databases can also promote transparency and reproducibility, which are essential for scientific credibility. In addition, industries looking to adopt nanofluids could rely on such databases for reliable information on cost, performance, and safety. For example, if databases included long-term stability studies and life-cycle assessments, industries could make informed decisions more confidently (Kasaeian *et al.*, 2017). Standardized, open-access databases of nanofluid properties would facilitate meta-analyses, reproducibility, and AI-driven discovery.

### 6.3 Policy and Standardization Initiatives

International bodies should establish standardized testing and reporting protocols. Policies must also address environmental, ethical, and occupational health concerns. At present, there are no universally accepted guidelines on how nanofluids should be synthesized, tested, or applied in industrial systems (Buongiorno *et al.*, 2009). Standardization would ensure that experiments across laboratories are comparable and that industries can trust reported data. Clear safety regulations are also necessary because the release of nanoparticles into the environment could raise health and ecological concerns if not properly managed (Kasaeian *et al.*, 2017). Governments and international organizations should therefore work toward establishing policies that define testing protocols, quality control measures, and environmental safety limits for nanofluids. In parallel, certification programs could help industries verify that nanofluid-based products meet safety and performance standards.

The future of nanofluid research depends on stronger collaborative research frameworks, the creation of open-access databases, and policy and standardization initiatives. These steps would not only accelerate scientific progress but also ensure that nanofluids are safe, cost-effective, and widely trusted for industrial and societal applications. By addressing these areas, researchers and policymakers can bridge the gap between laboratory innovation and large-scale adoption.

## 7. Concluding Remarks

Nanofluid research has advanced rapidly in synthesis, characterization, and performance evaluation. Yet, persistent challenges—including stability, scalability, and safety—must be addressed before nanofluids can realize their industrial potential. The integration of emerging trends such as hybrid nanofluids, AI-driven modeling, and green materials provides a promising roadmap. By bridging current gaps through collaborative frameworks, standardization, and responsible innovation, nanofluids can transition from laboratory novelty to industrial game-changer. Characterization methods and modeling approaches are two essential pillars of nanofluid research. Characterization methods such as microscopy, spectroscopy, DLS, zeta potential, XRD, and thermophysical property measurements provide deep insights into particle size, dispersion, stability, and fluid properties. On the other hand, modeling approaches, including empirical correlations, classical models, mechanism-based models, CFD, MD simulations, and machine learning, help predict and explain nanofluid behavior. The combination of both ensures that nanofluids can be studied more effectively and applied in real-world systems. However, challenges remain, such as the lack of standardized characterization techniques, limitations of theoretical models, and high computational cost of simulations. Overcoming these challenges will require interdisciplinary collaboration, better experimental tools, and more advanced computational methods. With progress in both areas, nanofluids have the potential to become practical solutions in energy, cooling, and industrial applications. Nanofluids hold great promise for improving energy efficiency and supporting advanced technologies, but their adoption depends not only on technical performance but also on environmental and economic considerations. On the environmental side, challenges include nanoparticle toxicity, ecological risks, occupational safety, disposal issues, and sustainability of nanoparticle production. On the economic side, high production costs, preparation and stability expenses, scaling-up challenges, and market competitiveness remain barriers. Lifecycle analysis shows that both environmental and economic issues are interconnected. Addressing these challenges requires the development of green synthesis methods, safer nanoparticle designs, standardized disposal regulations, and cost-effective large-scale production methods. If researchers and industries can work together to create environmentally safe and economically viable nanofluids, then these advanced fluids may move from research labs to widespread real-world applications. Research on nanofluids has opened exciting possibilities for improving heat transfer, energy efficiency, and advanced applications. But researchers face many serious challenges that slow down progress and real-world use. These challenges include stability, preparation methods, reproducibility, measurement difficulties, viscosity issues, lack of standardization, unclear mechanisms, health and environmental concerns, high costs, scale-up difficulties, and long-term performance doubts. Overcoming these challenges requires international collaboration, clear standards, safer nanoparticle designs, and strong partnerships between researchers and industries. Only then can nanofluids move from laboratory curiosity to practical, large-scale applications that benefit society. Nanofluids remain a promising route to improve thermal systems and enable new technologies, but turning promise into practice will require coordinated effort. High-priority actions include standardizing measurement protocols, publishing long-term stability and lifecycle data, developing green/safer synthesis, and running pilot industry demonstrations. Advances in data-driven models and digital twins can accelerate design, but these must be supported by open datasets and cross-sector collaboration. If researchers, industry, and regulators pursue these directions together, nanofluids can evolve from laboratory innovations into reliable, sustainable tools for energy, electronics, and healthcare.

## References


[1] Ajayan, P.M., Schadler, L.S. and Braun, P.V. (2000). *Nanocomposite science and technology*. Wiley-VCH.

[2] Anselma, P.G., Tibaut, A. and Malnar, B. (2019). Standardization challenges in nanotechnology: The case of nanofluids. *Nanomaterials, 9*(3), 1–15. https://doi.org/10.3390/nano9030400.

[3] Aricò, A.S., Bruce, P., Scrosati, B., Tarascon, J.M. and Van Schalkwijk, W. (2005). Nanostructured materials for advanced energy conversion and storage devices. *Nature Materials, 4*(5), 366–377. https://doi.org/10.1038/nmat1368.

[4] Asadi, A., Aberoumand, S., Gholami, M., Afrand, M., Kalbasi, R. and Wongwises, S. (2022). *Nanofluid heat and mass transfer in engineering problems*. Academic Press.

[5] Batchelor, G.K. (1977). The effect of Brownian motion on the bulk stress in a suspension of spherical particles. *Journal of Fluid Mechanics, 83*(1), 97–117.

[6] Binnig, G. and Rohrer, H. (1986). Scanning tunneling microscopy. *IBM Journal of Research and Development, 30*(4), 355–369. https://doi.org/10.1147/rd.304.0355.

[7] Bozzuto, G. and Molinari, A. (2015). Liposomes as nanomedical devices. *International Journal of Nanomedicine, 10*, 975–999. https://doi.org/10.2147/IJN.S68861.

[8] Brinkman, H.C. (1952). The viscosity of concentrated suspensions and solutions. *The Journal of Chemical Physics, 20*(4), 571–581.

[9] Buongiorno, J. (2006). Convective transport in nanofluids. *Journal of Heat Transfer, 128*(3), 240–250. https://doi.org/10.1115/1.2150834.

[10] Buongiorno, J., Hu, L.W., Kim, S.J., Hannink, R., Truong, B. and Forrest, E. (2009). Nanofluids for enhanced economics and safety of nuclear reactors: An evaluation of the potential features, issues, and research gaps. *Nuclear Technology, 162*(1), 80–91. https://doi.org/10.13182/NT09-A9108.

[11] Choi, S.U.S. and Eastman, J.A. (1995). *Enhancing thermal conductivity of fluids with nanoparticles*. Argonne National Laboratory.

[12] Daniel, M.C. and Astruc, D. (2004). Gold nanoparticles: Assembly, supramolecular chemistry, quantum-size-related properties, and applications toward biology, catalysis, and nanotechnology. *Chemical Reviews, 104*(1), 293–346. https://doi.org/10.1021/cr030698+.

[13] Das, S.K., Choi, S.U.S., Yu, W. and Pradeep, T. (2007). *Nanofluids: Science and technology*. John Wiley & Sons.

[14] Dreaden, E.C., Alkilany, A.M., Huang, X., Murphy, C.J. and El-Sayed, M.A. (2012). The golden age: Gold nanoparticles for biomedicine. *Chemical Society Reviews, 41*(7), 2740–2779. https://doi.org/10.1039/c1cs15237h.

[15] Eastman, J.A., Choi, S.U.S., Li, S., Yu, W. and Thompson, L.J. (2001). Anomalously increased effective thermal conductivities of ethylene glycol-based nanofluids containing copper nanoparticles. *Applied Physics Letters, 78*(6), 718–720.

[16] Feynman, R.P. (1960). There's plenty of room at the bottom. *Engineering and Science, 23*(5), 22–36. California Institute of Technology. https://resolver.caltech.edu/CaltechES:23.5.1960Bottom.

[17] Garg, P., Alvarado, J.L., Marsh, C., Carlson, T.A., Kessler, D.A. and Annamalai, K. (2022). Recent advances and applications of nanofluids: A review. *Materials Today: Proceedings, 63*(2), 89–100.

[18] Ghalandari, M., Moghadassi, A.R., Keshavarzi, E. and Hosseini, S.M. (2021). Application of machine learning in predicting thermal properties of nanofluids. *Journal of Thermal Analysis and Calorimetry, 145*(1), 111–123. https://doi.org/10.1007/s10973-021-10668-2.

[19] Hou, X., Zaks, T., Langer, R. and Dong, Y. (2021). Lipid nanoparticles for mRNA delivery. *Nature Reviews Materials, 6*(12), 1078–1094. https://doi.org/10.1038/s41578-021-00358-0.

[20] Hussein, A.M., Sharma, K.V., Bakar, R.A. and Kadirgama, K. (2020). A review of nanofluid preparation: Factors affecting stability. *International Communications in Heat and Mass Transfer, 54*, 1–11. https://doi.org/10.1016/j.icheatmasstransfer.2020.104943.

[21] Iijima, S. (1991). Helical microtubules of graphitic carbon. *Nature, 354*(6348), 56–58. https://doi.org/10.1038/354056a0.

[22] Iravani, S. (2011). Green synthesis of metal nanoparticles using plants. *Green Chemistry, 13*(10), 2638–2650.

[23] Kah, M., Beulke, S., Tiede, K. and Hofmann, T. (2013). Nanopesticides: State of knowledge, environmental fate, and exposure modeling. *Critical Reviews in Environmental Science and Technology, 43*(16), 1823–1867. https://doi.org/10.1080/10643389.2012.671750.

[24] Kasaeian, A., Mahian, O., Kolsi, L. and Wongwises, S. (2017). Nanofluid flow and heat transfer in porous media: A review. *Renewable and Sustainable Energy Reviews, 72*, 719–744. https://doi.org/10.1016/j.rser.2017.01.097.

[25] Khanafer, K. and Vafai, K. (2011). A critical synthesis of thermophysical characteristics of nanofluids. *International Journal of Heat and Mass Transfer, 54*(19–20), 4410–4428. https://doi.org/10.1016/j.ijheatmasstransfer.2011.04.048.

[26] Kole, M. and Dey, T.K. (2010). Thermal conductivity and viscosity of $Al_2O_3$ nanofluid: An experimental study. *International Communications in Heat and Mass Transfer, 37*(3), 297–303. https://doi.org/10.1016/j.icheatmasstransfer.2009.10.026.

[27] Kreibig, U. and Vollmer, M. (1995). *Optical properties of metal clusters*. Springer. https://doi.org/10.1007/978-3-662-09109-8.

[28] Kumar, P., Shah, S.A. and Park, S. (2019). Nanoparticles in the environment: Safety and sustainability. *Environmental Science and Pollution Research, 26*(5), 456–472. https://doi.org/10.1007/s11356-018-3506-8.

[29] Kumar, R., Sahoo, B., Singh, R.K. and Das, A.K. (2022). Recent advances in synthesis and applications of nanofluids: A review. *Journal of Molecular Liquids, 352*, 118748. https://doi.org/10.1016/j.molliq.2022.118748.

[30] Kumar, R., Raj, R., Singh, R. and Chauhan, V. (2025). Artificial intelligence–driven modeling of nanofluid thermophysical properties: Opportunities and challenges. *Journal of Thermal Science and Engineering Applications, 17*(1), 1–15.

[31] Lee, J.M., Jeong, H., Choi, H.J. and Kim, S.H. (2022). Review on nanofluids: Synthesis, characterization, and applications. *Applied Sciences, 12*(5), 2663. https://doi.org/10.3390/app12052663.

[32] Mahian, O., Bellos, E., Markides, C.N., Taylor, R.A., Alagumalai, A., Yang, L.,... and Wongwises, S. (2019). Recent advances in using nanofluids in renewable energy applications and the environmental implications of their use. *Applied Energy, 235*, 138–170. https://doi.org/10.1016/j.apenergy.2018.10.015.

[33] Mandal, T.K. and Roy, D.B. (2020). A review on fabrication methods, characterization and applications of magnetic iron oxide nanomaterials. *Iranian Journal of Materials Science & Engineering, 17*(1), 1–15. https://doi.org/10.22068/ijmse.17.1.124.

[34] Michalet, X., Pinaud, F.F., Bentolila, L.A., Tsay, J.M., Doose, S., Li, J.J.,... and Weiss, S. (2005). Quantum dots for live cells, in vivo imaging, and diagnostics. *Science, 307*(5709), 538–544. https://doi.org/10.1126/science.1104274.

[35] Mulvaney, P. (1996). Surface plasmon spectroscopy of nanosized metal particles. *Langmuir, 12*(3), 788–800. https://doi.org/10.1021/la9502711.

[36] Murshed, S.M.S. and Estellé, P. (2017). A state of the art review on viscosity of nanofluids. *Renewable and Sustainable Energy Reviews, 76*, 1134–1152. https://doi.org/10.1016/j.rser.2017.03.113.

[37] Nadooshan, A.A. and Etemad, S.G. (2017). Stability and thermophysical properties of nanofluids: A review. *Heat and Mass Transfer, 53*(8), 2405–2417. https://doi.org/10.1007/s00231-017-2012-3.

[38] Narayana, V.L., Suresh, S. and Chaurasiya, R. (2020). Green synthesis of nanofluids: A review on preparation, stability, and thermal performance. *Journal of Cleaner Production, 258*, 120678. https://doi.org/10.1016/j.jclepro.2020.120678.

[39] Nasrollahzadeh, M., Sajjadi, M., Iravani, S. and Varma, R.S. (2019). Green synthesis of nanomaterials: A review. *Green Chemistry, 21*(17), 4321–4347. https://doi.org/10.1039/c9gc01750d.

[40] Nel, A., Xia, T., Mädler, L. and Li, N. (2006). Toxic potential of materials at the nanolevel. *Science, 311*(5761), 622–627. https://doi.org/10.1126/science.1114397.

[41] Peer, D., Karp, J.M., Hong, S., Farokhzad, O.C., Margalit, R. and Langer, R. (2007). Nanocarriers as an emerging platform for cancer therapy. *Nature Nanotechnology, 2*(12), 751–760. https://doi.org/10.1038/nnano.2007.387.

[42] Qu, X., Alvarez, P.J.J. and Li, Q. (2013). Applications of nanotechnology in water and wastewater treatment. *Water Research, 47*(12), 3931–3946. https://doi.org/10.1016/j.watres.2012.09.058.

[43] Said, Z., Saidur, R. and Rahim, N.A. (2015). Energy and exergy analysis of a flat plate solar collector using different sizes of aluminium oxide nanofluid. *Journal of Cleaner Production, 90*, 93–100. https://doi.org/10.1016/j.jclepro.2014.11.030.

[44] Saidur, R., Leong, K.Y. and Mohammed, H.A. (2011). A review on applications and challenges of nanofluids. *Renewable and Sustainable Energy Reviews, 15*(3), 1646–1668. https://doi.org/10.1016/j.rser.2010.11.035.

[45] Savolainen, K., Alenius, H., Norppa, H., Pylkkänen, L., Tuomi, T. and Kasper, G. (2010). Risk assessment of engineered nanomaterials and nanotechnologies—A review. *Toxicology, 269*(2–3), 92–104. https://doi.org/10.1016/j.tox.2009.12.017.

[46] Sundar, L.S., Singh, M.K., Sousa, A.C.M. and Sousa, A. (2017). Enhanced thermal conductivity and viscosity of hybrid nanofluids. *International Communications in Heat and Mass Transfer, 84*, 1–10. https://doi.org/10.1016/j.icheatmasstransfer.2017.03.015.

[47] Wang, X., Li, X. and Chen, L. (2018). Biomedical applications of nanofluids: Drug delivery and cancer therapy. *Journal of Nanoscience and Nanotechnology, 18*(10), 1–10. https://doi.org/10.1166/jnn.2018.15792.

[48] Wang, Y., Yan, B. and Chen, L. (2016). Synthesis and characterization of nanomaterials for biomedical applications. *Chemistry of Materials, 28*(23), 8965–8979. https://doi.org/10.1021/acs.chemmater.6b04196.

[49] Yu, W. and Xie, H. (2012). A review on nanofluids: Preparation, stability mechanisms, and applications. *Journal of Nanomaterials, 2012*, 1–17. https://doi.org/10.1155/2012/435873.

[50] Yu, W., Choi, S.U.S. and Yin, Z. (2008). Review and comparison of nanofluid thermal conductivity and heat transfer enhancements. *Heat Transfer Engineering, 29*(5), 432–460.

[51] Zhang, X., Sun, H. and Xu, Y. (2013). Size-dependent properties of nanomaterials. *Journal of Physics: Conference Series, 419*(1), 012028. https://doi.org/10.1088/1742-6596/419/1/012028.

# 10 | Hybrid and Multifunctional Nanofluids: Design, Properties, and Emerging Applications

Rubania Arshad[1] and Saima Nasreen[2*]

Hybrid and multifunctional nanofluids are a novel class of engineered fluids with excellent and controllable properties compared to conventional nanofluids. These systems demonstrate the synergistic enhancement of thermal, optical, electrical, and magnetic properties through multicomponent nanoparticles dispersed in an appropriate base fluid. These nanoparticles are based on two or more different types, such as metals/polymers, oxides/carbon, and other emerging nanomaterials like MXenes and quantum dots. The choice of carrier fluid, nanoparticles, and preparation approaches for the design and formulation of hybrid nanofluids has been emphasized. Their thermophysical parameters, in particular rheological behavior, specific heat capacity, and thermal conductivity, are key factors in energy transfer, as well as when used for multifunctional applications. Recent advances have evidenced their potential within different areas: efficient heat transfer and thermal management devices, renewable energy devices, lubrication and machining, biomedicine therapeutics and diagnostics, environmental cleanup, and energy storage devices. However, particle agglomeration, long-term stability, inexpensive production, and environmental safety are still the key obstacles for industrial applications. In the future, we anticipate increasing effort in the area of green synthesis, AI-based design, and smart, stimulus-responsive fluids. Overall, hybrid and multifunctional nanofluids hold considerable promise as key enablers of next-generation energy, environmental, and biomedical technologies.

## 1. Introduction

A nanofluid is a stable suspension of nanoparticles in a base fluid, like organic liquids, oil-based fluids, or water, and nanoparticles significantly enhance the fluid's thermal properties (Bahrami *et al.*, 2025). The inclusion of nanomaterials (like metallic, metal-oxide, carbide, and carbon) into these parent mediums considerably improves their thermophysical characteristics, including specific heat, thermal conductivity, and dynamic viscosity (Cardoso *et al.*, 2025). While Choi (1995) was the pioneering scientist who suggested the implementation of

[1,2] Department of Environmental Sciences, Faculty of Life Sciences, The Women University Multan, Multan, 66000, Pakistan.

* Corresponding author: saima.6333@wum.edu.pk

DOI: 10.1201/9781003686002-10

nanofluids for improving thermal energy transfer, the notion of uniform distribution of nanoscale particles within carrier fluids to increase thermal efficiency has existed since the early twentieth century (Sun *et al.*, 2022). One of the discipline's early triumphs occurred in 1996, when Eastman *et al.* (1996) applied nanoscale particles to optimize the heat transfer capability of liquids, including water and oil. Currently, nanofluids continue to be of significant attraction, with the biggest potential gains deriving from interdisciplinary research partnerships (Cardoso *et al.*, 2025). Several papers have previously demonstrated that researchers have applied nanofluids to boost the performance of absorption refrigeration units, solar collectors, water heaters, photovoltaic cells, photovoltaic cooling systems, photovoltaic distillation units, and other numerous appliances, considering their exceptional characteristics compared to regular base fluids (Rasheed *et al.*, 2021).

Recent scientific work has shown that hybrid nanofluids have the potential to substitute conventional mono nanofluids (comprising a single class of nanoparticle) because they augment heat transfer efficiency across multiple thermal applications, ranging from electromechanical processes, manufacturing, automobiles, heating, ventilation, air conditioning, to solar energy systems. Hybrid nanofluids represent a novel form of thermal fluids developed through the incorporation of two distinct nanoparticle species in a traditional thermal fluid (base fluid/carrier fluid) (Asim and Siddiqui, 2022). Hybrid nanofluids reveal excellent energy transport in thermal operations, outperforming traditional heat-conducting fluids (water, oil, and ethylene glycol) and mono-nanofluids. Nanofluids have gained popularity due to their exceptional physicochemical features and multifunctional potential (Suneetha *et al.*, 2022). Due to their outstanding physical and spectral properties, nanofluids' multifunctional uses necessitate specific research to increase overall performance of hybrid systems. In recent years, different systems have used nanofluids to accomplish connected functionality (Sun *et al.*, 2022). Nanofluids have prospective uses in thermal systems (e.g., solar collectors, nuclear reactors, and thermal control in electronic devices), manufacturing with accuracy (e.g., coolants and lubricants), and biomedical engineering (e.g., theranostics, imaging, drug administration, and antimicrobial treatments). However, the most researched and applied applications of nanofluids are in thermal management technologies, which include solar collectors, electronic heat dissipation systems, internal combustion engines and heat exchangers. Despite its extensive potential, nanofluid performance is highly reliant on the type of nanoparticle, its size, shape, concentration, and dispersion quality. Additionally, compatibility with the fluid, long-term stability, and agglomeration, are challenges that must be addressed to enable steady functioning in systems (Cardoso *et al.*, 2025). Surfactants are sometimes employed to prevent nanoparticles from clumping (Wohld *et al.*, 2022).

This chapter aims to explain how these nanofluids and hybrid nanofluids are synthesized, their characteristics, and their applications.

## 2. Foundational Aspects of Hybrid Nanofluids

### 2.1 Concept and Motivation

There is surging demand for innovative thermal regulation systems because of the fast growth in the industrial and technological fields. Water, ethylene glycol and oil are examples of traditional fluids for heat transfer, which have low heat conduction and convection coefficients. To overcome these problems, there is a focus on nanofluids, which are a novel kind of heat transfer fluid obtained by uniform distribution of nanoscale particles in regular heat transfer media. Nanofluids showed superior thermal characteristics, which include improved heat conduction and convection coefficients, and total heat transfer rates. As a consequence of these, nanofluids can be applied in different ways, such as in electronic, energy, biomedical, and aeronautical engineering (Tripathi *et al.*, 2025). Furthermore, there have been recent researches on the formation of hybrid nanofluids (Hamzah *et al.*, 2017). Another type of nanofluid, named "hybrid nanofluid," developed by suspending the nanoparticles or nanocomposite of a few metals within the base fluid. The problem of stability is minimized in these multi-component nanofluids, and they outperform mono-nanofluids. It is evident that adding nanoparticles augments the thermal transport capability of the base fluid, but certain problems arose regarding erosion, power consumption for pumping, stability, convection heat transfer, and pressuring attributable to higher viscosity because of clustering, leading to a decline in specific surface area and an increase in the hydrodynamic diameter. The value of viscosity of the hybrid nanofluids is greater when measured against traditional base fluids and mono-nanofluids, but it is reliant on the selection of nanoparticles and combinations. The rheology of hybrid nanofluids is higher (Suneetha *et al.*, 2022).

Hybrid nanofluids tackle the drawbacks of mono-nanofluids and attain improved thermal and stability performance by utilizing the synergy produced by various nanoparticles and base fluids.

The selection of proper material is actually the foundation for effective synthesis and multiple functions in the subsequent application.

### 2.2 Common Base Fluids

Single-phase heat conveying media like water, oils, and ethylene glycol are commonly adopted in a variety of manufacturing applications, covering heating and cooling, transportation, power generation, and chemical processes. These base fluids frequently fail to fulfill the performance specification of mechanical equipment like cooling and condensation units due to their limited conductive heat transfer coefficients. These fluids' attributes can be improved by dispersing small polymeric, metallic, or non-metallic particles to create a stable colloidal dispersion (Syam Sundar *et al.*, 2014).

- **Water**

Water is the ubiquitous, affordable, and widely accessible heat transfer liquid, and it has traditionally been used in vehicles and heat exchange devices due to its high heat capacity and low viscosity. At 20°C, water possesses the maximum thermal conduction of (0.598 W $mK^{-1}$). Hybrid nanoparticle-enhanced fluids using aqueous media as the base liquid are distinguished by their excellent heat conduction ability and effective heat exchange, which improves the system's inclusivity. Furthermore, water has the highest relative permittivity (dielectric constant), indicating a strong repulsive force. Because the repulsive force and the relative permittivity of the base fluid are closely connected, larger dielectric constants are expected to provide more stability. However, the boiling point of water is high, thermal conductivity is limited, and it has the ability to produce corrosion. Nanofluid, a liquid suspension of water and nanoparticles having very small size less than 100 nm, with or without surfactants, was devised as an option to boost the thermal conduction of the base fluid (Manimaran *et al.*, 2025).

- **Ethylene glycol (EG)**

Heating residential, industrial, and commercial buildings in frigid places such as Northern Europe, Alaska, Canada, and Russia requires a significant amount of energy. Considering the lengthy winter temperature conditions, ethylene glycol combined with water in various wt% concentrations commonly applied to reduce the water's freezing point of heat transfer mediums. These heat transfer fluids are utilized in baseboard heating systems for homes, heat exchange units in autos, and industrial operations in chilled climates, and they can survive extremely cool temperatures. Under cool temperatures, mixtures of ethylene glycol transfer heat more effectively than mixtures of propylene glycol (Syam Sundar *et al.*, 2014). Significant research has been conducted on ethylene glycol (EG)-based hybrid nanofluids. These hybrid nanofluids might be used in thermal management processes. For instance solar panel cooling, heat exchange units, compact channel based cooling devices, heat pipes, and cooling of electric vehicle batteries (Wanatasanappan *et al.*, 2022). Despite having a threefold lower thermal conductivity than water, ethylene glycol exhibits a greater boost in heat conduction (Rahman *et al.*, 2024). For the first time (Namburu *et al.*, 2009) created CuO nanofluid using 60% ethylene glycol and 40% water as a base fluid, evaluated viscosity at temperatures ranging from 35°C to 50°C and volume concentrations of up to 6%, and presented a viscosity correlation.

- **Oils**

Oil-based nanofluids can be created by suspending nanoscale particles within the base fluid, such as plant-derived oil, utilizing ultra sonication techniques (Wanatasanappan *et al.*, 2022). Attributed to their exceptionally elevated boiling point and outstanding thermophysical characteristics under high-temperature conditions, oil-based nanofluids have the potential to be used in thermal energy applications at moderate to high temperatures (100– 350°C). According to current research, various thermal oil-based nanofluids are applied as working fluids in thermal energy storage applications due to their thermal output effectiveness in moderate to high temperature applications. Thermal oils are classified into three types: vegetable oils derived from plants, chemically produced synthetic oils, and petroleum-based mineral oils. Oil-based nanofluids are well suited for execution as coolants owing to their biodegradability and high thermal endurance; they are also renewable and have less environmental impact (Rubbi *et al.*, 2021). To optimize the thermal characteristics of plant-based oil, the incorporation of hybrid nanoscale particles in this oil is a revolutionary way, and it is noticed that thermal diffusivity is highest in soybean oil, followed sequentially by palm oil and coconut oil (Wanatasanappan *et al.*, 2022).

- **Ionic liquids**

Ionic liquids (ILs) are a novel and demanding category of liquid media with excellent and customizable characteristics, capable of a wide range of real-world applications, including biology, medicine, chemistry, and heat transfer. These fluids are commonly referred to as environmentally friendly solvents. Following the concept of nanofluids, adding nanoparticles to these fluids can improve a variety of their characteristics. These ionic liquid-based nanocolloids are also known as ionanofluids or nanoparticle-enhanced ionic liquids. Unlike normal liquids, ionic liquids' thermal conductivity decreased significantly as the temperature increased. However, the specific heat of ionic liquids behaves similarly to ordinary viscous fluids such as ethylene glycol, as it grows almost proportionally with temperature, which is advantageous for thermal energy retention. Aside from being dependent on temperature and pressure, ionic liquids have thermal stability up to extremely high temperatures (for example, 450°C). Some unique forms of ionic liquids, such as imidazolium-based ionic liquids, are often used and researched due to their intriguing traits and properties, as well as their prospective uses. Depending on the application, such as convective heat transfer, the characteristics of ionic liquids play a crucial role in their performance as well as in the performance of their nanofluids (Minea and Sohel Murshed, 2021).

**Table 1** Key characteristics of different types of base fluids.

| *Base Fluids* | *Key Characteristics* | *References* |
|---|---|---|
| Water | Inexpensive, easily available and heat transfer fluid | (Said *et al.*, 2022) |
| Ethylene glycol | Low freezing point, potential applications in heat transfer mechanisms | (Wanatasanappan *et al.*, 2022) |
| Oils | Little environmental impact, high thermal conductivity | (Rubbi *et al.*, 2021; Wanatasanappan *et al.*, 2022) |
| Ionic liquids | Advanced fluids, tunable properties, also have thermal stability up to extremely high temperatures (e.g., 450°C) | (F.-F. Zhang *et al.*, 2019) |

### 2.3 Nanoparticles

Nanofluids are synthesized through the incorporation of nanoparticles under 100 nm within the base fluid (Sun *et al.*, 2022). Nanofluids have been created using a variety of nanoscale particles, consisting engineered nanoparticles composed of metals such as Cu, nickel (Ni) and silver, (Ag) along with metal-oxide nanoparticles such as iron (III) oxide ($Fe_2O_3$), copper oxide (CuO), magnesium oxide (MgO), aluminum oxide ($Al_2O_3$), titanium dioxide ($TiO_2$), silicon dioxide ($SiO_2$), and zinc oxide (ZnO) perhaps in conjunction with nanoparticles made up of carbon for example nano-sized single walled carbon tubes, multiwalled carbon nanotubes, and oxidized graphene because of their excellent thermal diffusivity and even dispersion of nanoscale particles within the base fluid which can lead to effective heat transmission in a variety of implementation areas, comprising heat management systems like solar panels, freezers, heat pipes and car radiators. The use of nano-sized particles (1–100 nm) in base liquids has the advantage of reducing the heat transfer system's size and the amount of energy required for fluid transport, resulting in energy and cost reduction as compared to employing micro particles in standard heat transfer fluids. Moreover, large surface- area nanoparticles can reduce density differences, enabling highly homogeneous dispersion and acting as a replacement for micro particles, featuring disadvantages in particular pressure loss, channel blockage, and surface degradation (Manimaran *et al.*, 2025).

## 3. Multifunctionality in Nanofluid Design

Over the previous few years, intensive research on nanofluids' thermal transport characteristics has been reported. This is owing to its potential usage in multiple industrial domains including electronic industry, thermal control, and waste energy recovery (Ragavi and Poornima, 2024).

### 3.1 Thermal and Electrical Properties

In the nineteenth century, Maxwell was the first to propose suspending solid particles in convectional fluids comprising oils, water, and ethylene glycol to augment thermal transport (Sadeghi *et al.*, 2022). However, the quick sedimentation of particles in liquid that are mm or µm in size has remained the principal hurdle in producing stable suspensions for industrial uses (Yu and Choi, 2003). The quick settling increases the viscosity of the fluids as well as the pressure decrease (Sadeghi *et al.*, 2022).

Choi and Eastman (Choi, 1995) later proposed the term nanofluid. To improve water's thermal conductivity, they suspended copper nanophase solids with diameters smaller than 100 nm in water. The results of their theoretical investigation into thermal conductivity revealed that the thermally enhanced nanofluid might significantly reduce the pumping power required by the heat exchanger. The typical materials consist of the following: Metals include Ag, Au, Cu, and Ni; metal oxides include $Fe_2O_3$, CuO, $Al_2O_3$, $Fe_3O_4$, $TiO_2$, $SiO_2$, and $ZrO_2$; while carbon compounds include CNTs, graphite, and diamond (Sundar *et al.*, 2017). These nanoparticles, as previously stated, have become widely employed because of research into their thermophysical characteristics, including their extremely high thermal conductivity. Other considerations, particularly the availability of chemical stability, toxicity, cost, coherence with carrier fluid, and thermophysical characteristics of the nanoscale materials, must be addressed during the selection of nanoparticles for thermal transport applications (Rahman *et al.*, 2024; Yildizay *et al.*, 2025). Nanofluids enhance heat transmission, dependent on temperature, a particular class of base fluid, characteristics of nanoscale particle, their concentrations, and other parameters that influence properties of nanofluid, including clustering of nanoparticles and presence of surfactants (Z. Wu *et al.*, 2013). To augment the functionality of nanofluids, a technology known as nanoparticle hybridization is adopted (Giwa *et al.*, 2021). Hybrid nanofluids might possess enhanced thermal capabilities relative to base fluids and nanofluids based on single nanoparticles (Che Sidik *et al.*, 2017). A sensible enhancement in the heat conduction ability of the hybrid nanofluid compared to the single nanoparticles-containing nanofluids and the base fluid has been recorded by Hafeez *et al.* (2023). The outcome of the investigation conducted by Abbasi *et al.* (2013) reveals a thermal conductivity of 20.68 percent for the $\gamma$-$Al_2O_3$/MWCNTs hybrid at a volume percentage of 0.1 percent. A higher capability in the enhancement of heat transport in the hybrid nanofluids in comparison to the single nanoparticle-containing nanofluids is also demonstrated by the finding, which is very efficient in applications such as the heat exchange unit of chillers, coolers, or heaters, in electronic cooling systems, and others (Scott *et al.*, 2022). Moreover, an enhancement in the value of the convective thermal coefficients of hybrid nanofluids, which are the mixture of at least two nanoparticles, of 4–6 percent evaluated against mono-nanofluids, has also been recorded in the study carried out by Sivasankaran and Bhuvaneswari (2022). Moreover, according to Alklaibi *et al.* (2024), hybrid nanofluids based on $Fe_3O_4$/$SiO_2$ showed higher values of conductive thermal performance relative to both single-component nanofluids and pristine fluids. These novel hybrid nanofluids, which will be future heat-transfer fluids, hold a vast application perspective in heat transfer techniques as well (Manimaran *et al.*, 2025). Moreover, Brownian motion also occurs at the nanofluid level, which improves the process of heat transfer (Wang *et al.*, 2025). Furthermore, thermally induced random nanoparticle transport occurs among the investigated nanoparticles inside the fluid, accelerating heat transmission.

### 3.2 Anti-corrosion and Other Functionalities

Cooling system corrosion is a major concern that affects engine performance. This concern is especially prevalent in cooling systems operating with water as carrier fluid. Water-based fluids have been addressed with anti-corrosive materials, as cooling system corrosion can result in leakage and a breakdown of the cooling system. This, in turn, can result in engine failure due to excessive heating. The exploration of cooling system corrosion in automobiles, which remains an untrodden area of research, and the requirement for a water-based fluid that can successfully carry heat and slow down corrosion in cooling systems. Graphene oxide, which has high anti-corrosion properties under harsh conditions, has been added to cooling systems as a nanofluid. Research and scientific studies by researchers have shown marked improvements in cooling and heat transfer processes by GO nanoparticles added to cooling fluids (Habibi *et al.*, 2024).

(K. Wu *et al.*, 2021) prepared $TiO_2$ nanofluids based on sodium dodecyl benzene sulfonate (SDBS) serving as a surface active agent and simulated cooling water (SCW) as a base fluid, and investigated the corrosion properties of $TiO_2$ nanofluids in the copper-zinc alloy based on electrochemical tests and surface analyses. They found that the SDBS remained bound on the copper-zinc alloy surfaces, forming a protective layer that slowed down corrosion in simulated cooling water. Secondly, they found that negatively charged $TiO_2$ nanoparticles and adsorbed titania nanoparticles located on the copper-zinc alloy surfaces in the simulated cooling water, SDBS, and $TiO_2$ nanofluid minimized the anti-corrosion effect of SDBS for copper-zinc alloys. (Y. Yang *et al.*, 2023) executed a study to assess the lubrication attributes of a hybrid nanofluid having gamma alumina and ZnO for a titanium alloy. Findings show that the gamma alumina and ZnO nanofluids have better surface quality, friction, and higher coefficients compared to mono nanofluids. Furthermore, the composite nanofluid with a weight fraction of ZnO to gamma alumina of 2:1 exhibits the highest lubrication effectiveness, with a maximum 22.1% tribological coefficient lower than the base fluid, leading to better topographical quality. Nano $Al_2O_3$ may cling to the surfaces of nano zinc oxide and act as a surface layer, improving the water-based hybrid nanofluid's lubricating performance.

Multifunctional hybrids achieve an excellent level of thermal-electrical synergy with a number of other specialized properties, such as anti-corrosion and magnetic responsiveness, unlike the conventional restrictions of just serving heat transfer purposes. In this way, such features will pave the path for customized design for energy, biomedical, and industrial applications.

## 4. Synthesis Methods for Hybrid and Multifunctional Nanofluids

The procedure for preparing the nanofluid involves only the mixture of the nano powder and the base fluid. More emphasis, however, must be laid upon the resistance to sedimentation of the nanofluid. In real-time systems, the stability, long-term endurance, and non-responsiveness of the nanoparticle to chemicals in the fluid assume importance. The most successful strategy for generating hybrid nanofluids is to suspend nanoparticles of certain elements or composite nanoparticles within the fluid matrix at a specific proportion. To achieve a homogeneous and steady dispersion, a sonication process is tailored. By using an appropriate surfactant, the agglomeration can be reduced even more. Combining a proper surfactant can reduce agglomeration (Suneetha *et al.*, 2022). Resistance to sedimentation of hybrid nanofluids is assessed by their fabrication procedure. However, researchers have technical difficulties in creating stable hybrid nanofluids owing to the existence of significant molecular cohesive and dispersive interactions among the nanoscale particles, which produce aggregation. Moreover, a hybrid nanofluid with nanoparticle clumping loses its heat transfer capacity by prohibiting Brownian movement of the particles (Ranga Babu *et al.*, 2017). Thus, it is imperative to synthesize the hybrid nanofluids through a specifically designed method. In addition, hybrid nanofluids could be synthesized by two different methods: the one-step method and the two-step method. Other advanced methods for preparing efficient and steady hybrid nanofluids for experimentation and heat transfer are also under research (Suneetha *et al.*, 2022).

## 4.1 One Step Synthesis

This approach produces nanoparticles and prepares nanofluids simultaneously. In a single step, nanoscale particles are created and suspended in the carrier fluid (Kong *et al.*, 2017). Considering the published work, methods such as solution-phase chemical synthesis (Albert *et al.*, 2020), chemical reduction, chemical vapor deposition, wet grinding method, hot-wire method, and mechanical ball milling were employed to produce hybrid nanofluids (Scott *et al.*, 2022). According to (Che Sidik *et al.*, 2017), the most common approach for creating hybrid nanofluids is high-current pulse wire evaporation, which is a one-step process. They stated that some researchers used this procedure to create (silver/multiwalled carbon nanotubes) Ag/MWCNT. Multi-walled carbon nanotubes were purified using a

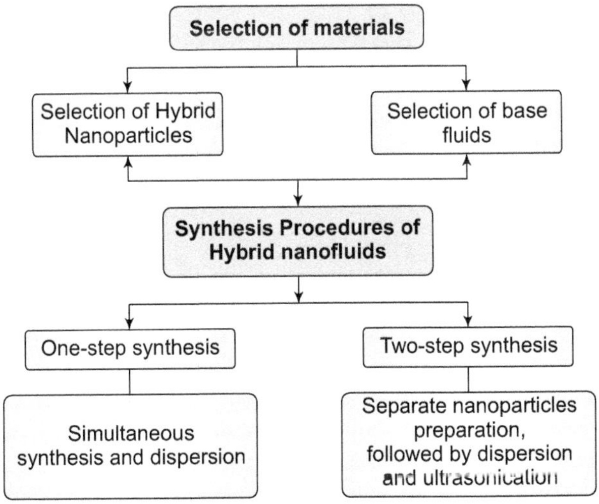


**Fig. 1** Flow chart of two different synthesis procedures of hybrid nanofluids (Cardoso et al., 2025).

chemical method. Because of its hydrophobic characteristics, sulfuric and nitric acids were utilized to increase MWCNT external activity. Concurrently, silver nanofluid was produced utilizing the pulse wire evaporation technique at a 300 V pulse voltage and a wiring diameter of 90 mm. (Jena *et al.*, 2001) conducted four experiments to produce homogeneous Cu-$Al_2O_3$ composites from chemically generated Cu-$Al_2O_3$ mixtures via $H_2$ reduction.

## 4.2 Two Step Method

This approach is simple to implement, inexpensive, and widely employed in mass manufacturing. The two-step procedure comprises dispersing previously manufactured nanoparticles throughout the base fluid. In a two-step procedure, nano powders are first engineered and then uniformly suspended within the base fluid, like water, oil, or EG (Singh and Sood, 2024).

(Suresh *et al.*, 2011) used an $H_2$ reduction approach to manufacture $Al_2O_3$-Cu. The synthesis used a 10:90 weight ratio of CuO and $Al_2O_3$ powder, which was obtained using a chemical synthesis procedure. To produce the $Al_2O_3$-Cu nanocomposite-based hybrid nanofluids in water, the nanocomposite powder was mixed with de-ionized water. The concentration in the hybrid nanofluids per volume ranged from 0.1 to 2 percent. Interestingly, the laboratory experiment highlighted that the 2% volumetric concentration gave an exceptional 12.11 percent enhancement in thermal conductivity. (Kannaiyan *et al.*, 2017) produced the hybrid nanofluids via the two-phase method with water and ethylene glycol as the fluid matrix, with the exact ratio of 20:80. The variability in the study of thermophysical properties within the range of 20-70°C temperatures revealed a substantial 45 percent increment in heat conduction performance. (Hamid *et al.*, 2018) made a combination of $TiO_2$-$SiO_2$ with the compositions of 80:20%, 60:40%, 50:50%, 40:60%, and 20:80 ratios of the $TiO_2$-$SiO_2$ NFs with base fluids of EG-W with a composition of 40:60%. They recommended that a mixture of 20:80 and 40:60 $TiO_2$-$SiO_2$ NFs can be employed in the system to augment the capability to transfer heat.

(Urmi *et al.*, 2020) adopted a sequential two-step preparation of HNFs employing 40% ethylene glycol and $TiO_2$-$Al_2O_3$ nanoparticles. This process explored different concentrations (80:20, 50:50, 20:80, 60:40, and 40:60) of $TiO_2$-$Al_2O_3$ nanoparticles, each with a concentration of 0.1 percent in volume. To analyze the consistency of the hybrid nanofluid, several analytical methods, such as UV-Visible spectrophotometric analysis, visual inspection, transmission electron microscopy, and zeta potential analysis using a zeta sizer, have been carried out. It was observed that among the concentrations, 80:20 concentrations of $TiO_2$-$Al_2O_3$ showed high stability and uniformity for a period of twenty-one days, without apparent settling of nanoscale particles in a suspended solution. (Xu *et al.*, 2022) synthesized a hybrid nanofluid through dispersion of sphere-shaped $TiO_2$ NPs and rod-like $Al_2O_3$ NPs in a fluid matrix. The major goal was to improve thermal flux performance in thermo-syphon photovoltaic receivers with closed-loops while also obtaining appropriate rheological qualities. Reported outcomes highlighted that the equivalent thermal flux coefficient of closed thermo-syphon photovoltaic receivers operating in two-phases increased by 16.2 and 18.7 percent, respectively. To stabilize high-grade nanofluids, continuous magnetic stirring was carried out to observe the homogeneity of solid nanoparticles in the fluid matrix. Solid volume percentages were exposed to an ultrasonicator to improve stability and disperse aggregated nanoparticles. A densitometer was used to test the stability and quality.

**Table 2** Different nanoparticles, base fluids, and synthesis routes for hybrid nanofluid preparation.

| *Dispersed nanoparticles* | *Base fluid* | *Synthesis method* | *References* |
|---|---|---|---|
| $Al_2O_3$-$TiO_2$ | Vegetable oil | Two step synthesis | (Wanatasanappan *et al.*, 2022) |
| Cu-MXene | Methanol, silicon oil, castor oil, water | Two step synthesis | (Rajesh Kumar *et al.*, 2025) |
| ZnO-$TiO_2$ | Ethylene Glycol | Two step synthesis | (Toghraie *et al.*, 2016) |
| Cu-Graphene | Water | Two step synthesis | (Shaik *et al.*, 2024) |
| $SiO_2$-MWCNT | Water | Two step synthesis | (Ekiciler, 2024) |
| $Fe_2O_3$-MWCNT | Ethylene glycol-water binary mixture | Two step synthesis | (Asadikia *et al.*, 2020) |
| Graphene nanoplatelets/ $Al_2O_3$ | Deionized water | Two step synthesis | (A. Borode *et al.*, 2023) |
| MWCNTs/SiC | Ethylene glycol-water binary mixture | Two step synthesis | (Kakavandi & Akbari, 2018) |
| $TiO_2$/ZnO | Ethylene glycol-water binary mixture | Two step synthesis | (Sahid *et al.*, 2017) |
| ZnO/Graphene oxide | Water | One step synthesis | (Chen *et al.*, 2015) |
| Silver/ MWCNTs | Water | Two step synthesis | (Gupta *et al.*, 2020) |
| $TiO_2$/carbon nanotubes | Water | Two step synthesis | (Megatif *et al.*, 2016) |
| Ag/Silica | Oil | One step synthesis | (Botha *et al.*, 2011) |
| Cu/Cu2O | Double distilled water | Two step synthesis | (Nine *et al.*, 2013) |

*Contd.*

**Table 2** *Contd.*

| *Dispersed nanoparticles* | *Base fluid* | *Synthesis method* | *References* |
|---|---|---|---|
| MgO/MWCNTs | Ethylene glycol-water binary mixture | Two step synthesis | (Hemmat Esfe *et al.*, 2018) |
| MXene/Carbon dot | Deionized Water | Two step synthesis | (Rajesh Kumar *et al.*, 2025) |
| MXene/$Al_2O_3$ | Therminol-55 | Two step synthesis | (Das *et al.*, 2022) |
| MXene/ Graphene oxide | Water | Two step synthesis | (Kanti *et al.*, 2023) |

### 4.3 Evaluation of One-Step vs Two-Step Synthesis

Hybrid nanofluids manufactured in a single step are generally of excellent properties and physically stable in terms of physical terms and thermal properties. The fundamental limitation of this strategy is that it is only suitable for base fluids having low volatility (Scott *et al.*, 2022). Although the consolidated synthesis scheme seems basic, the dual stage procedure is more suitable for applications of nanofluid on vast-scale (Manimaran *et al.*, 2025). However, scientists discovered that the elevated surface energy of nanoscale particles caused accumulation and finally deposition when utilizing this approach, lowering the stability of the hybrid nanofluid. Dispersants or surfactants are typically added throughout the synthesis to optimize the long-term stability of the resultant hybrid nanofluids (Scott *et al.*, 2022).

## 5. Property Enhancement Mechanisms

The depletion of energy supplies is one of the most serious problems in today's energy landscape. Air conditioners, refrigerators, heat pumps, and all use a significant amount of electricity. Due to the scarcity of energy resources, numerous studies have been undertaken to increase the efficiency and performance of thermal systems by redesigning components associated with the system or modifying the operating fluid. A large number of tiny heat-generating appliances (particularly plate, tube, and mini-channel heat exchange units) that use hybrid nanofluids increase system performance.

A two-step process is implemented to synthesize hybrid nanofluids. Nanoscale particles are synthesized and incorporated into the base liquid using magnetically driven mixing. The suspension is treated with ultrasonication and then analyzed to guarantee steady and homogenous agitation, which provides superior heat transfer properties (Elsaid, 2019). Increased surface area, interaction effects, and correct dispersion of nanoparticles in the base medium all play a role in improved heat transmission (causing micro turbulence). Hybrid nanofluids improve performance of thermal systems in low-temperature applications by performing four functions (absorbent, lubricant, refrigerant, and secondary refrigerant). The impact of concentration of nanoscale particles and size of nanoparticles on the effectiveness of CuO water-based nanofluids was examined in (Al Shdaifat *et al.*, 2020). The formulation, stability, and thermophysical characteristics of hybrid nanofluids were investigated in (Eshgarf *et al.*, 2021). (Zainon and Azmi, 2021) examined the influence of nanoparticle properties, including particle diameter, type, concentration, and structural shape, on nanofluid characteristics. (Pavía *et al.*, 2021) claimed that incorporating graphene into the base fluid can enhance thermal performance. The diameter of nanoscale particles influences the density of nanofluids, thermal conductivity, and viscosity. Nanoparticle size reduction increases nanofluids' thermophysical characteristics. Moreover, (Rahman *et al.*, 2024) also demonstrate that the morphology and concentration of the nanoscale particles are the main controlling factors affecting thermophysical properties of nanofluids.

Stability is also a significant parameter in the efficacy of hybrid nanofluids for engineering use. All the thermophysical characteristics of nanofluids are related to their sedimentation resistance. The agglomeration of the nanofluid matrix can minimize its applicability in several heat transport applications. This results from the nanoparticles' aggregation tendency in fluids.

Sedimentation occurs as nanoparticles aggregate to eventually separate from base fluids (L. Yang *et al.*, 2020). The clustering rate is controlled by the frequency of particle collisions in Brownian motion and the possibility of cohesiveness (Sezer *et al.*, 2019).

The removal of agglomeration tendency results in stable nanofluids. Sedimentation, transmittance, centrifugation, spectral absorbance, zeta potential, and hydrodynamic size measurement via DLS are the characterization techniques used to examine the nanofluids' stability (Asadi *et al.*, 2019). Different surfactants can be added for long-term stable and homogeneous nanofluids (Lu *et al.*, 2008; Saeedinia *et al.*, 2012): e.g. polymer surface-active agents (gum Arabic, polyvinyl alcohol, and polyvinyl pyrrolidone), anionic surface-active agents (sodium dodecyl benzene sulfonate and sodium dodecyl sulfate), nonionic surface-active agents (Tween-20 and Span-80), and cationic surface active agents (cetyl trimethyl ammonium bromide). The inclusion of surface-active agents decreases the interfacial tension of the base fluid and increases the wettability between the nanoscale particles and base fluids, and also increases the dispersibility of nanoparticles (Khan & Valan Arasu, 2019). Nanoparticles can be dispersed through stirrers, baths, ultrasonic mills, or high-pressure homogenizers. Researchers such as Baby and Ramaprabhu (2011) use a low-cost method for the preparation of hydrogen-functionalized, exfoliation-induced silver-decorated graphene (Ag/HEG) nanofluid. Ag/HEG nanofluid contains Ag/HEG dispersed in a mixture prepared using a deionized water-ethylene glycol matrix as a base medium through the ultrasonic method in the absence of any surfactant agent. These nanofluids showed stability for over three months.

In the research regarding the heat flux behavior of the dispersion medium, the specific heat capacity, as well as density, are considered as one of the most pivotal thermophysical attributes for HNFs. (Ho *et al.*, 2010) investigated a hybrid nanofluid made up of $Al_2O_3$ nanoparticles and microencapsulated particles of phase transition materials suspension in aqueous medium. They studied the thermal and physical attributes of HNFs. (Baghbanzadeh *et al.*, 2012) explored the thermal and physical attributes of water-based HNFs made up of various proportions of MWCNTs and silica nanoparticles. They found that the density values increased as the concentration escalated, whereas the values decreased as the temperature increased. (Nuim Labib *et al.*, 2013) found that as the volume percentage ratio of the hybrid solution formulated with carbon nanotubes and oxides increased, the density value increased at a greater rate than the viscosity.

Some hybrid nanofluid investigations found that, with the rise in temperature and volume fraction of particle, specific heat increases, but density increases with the concentration and decreases with the temperature. The formation of nanostructures at the solid-liquid interface increases specific heat, whereas particle aggregation reduces it.

**Table 3** Mono Vs Hybrid Nanofluids_ Quantitative Thermo-Physical comparison.

| *Property* | *Mono nanofluids* | *Hybrid nanofluids* | *Enhancement* | *References* |
|---|---|---|---|---|
| Thermal Conductivity | $Al_2O_3$ and CuO | $Al_2O_3$-CuO | When compared to CuO and $Al_2O_3$, $Al_2O_3$-CuO demonstrated improvements of 2.3% and 3.6%, respectively. | (Asokan *et al.*, 2020) |
| Viscosity | Graphene Nanoplatelet | Graphene Nanoplatelet-Alumina | Graphene Nanoplatelet-Alumina hybrid nanofluid showed enhancement by 15.79% when compared to water, However, compared to mono nanofluid, hybrid nanofluid has a lower viscosity. | (A. O. Borode *et al.*, 2021) |
| Zeta Potential (Stability) | $Al_2O_3$ and $SiO_2$ | $Al_2O_3$-$SiO_2$ | $Al_2O_3$-$SiO_2$ hybrid nanofluid showed a zeta potential of -60.7mV indicating higher stability than mono nanofluids. | (Mukesh Kumar *et al.*, 2020) |

# 6. Applications in Energy, Electronics, and Industrial Processes

Hybrid nanofluids have several applications.

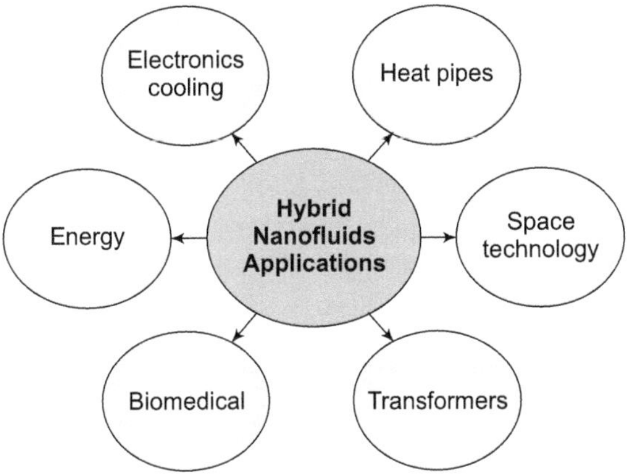


**Fig. 2** Applications of hybrid nanofluids (Cardoso *et al.*, 2025).

## 6.1 *Energy Systems*

For a few decades, the relative use of photovoltaic or solar power generation systems to meet the requirement for energy and heat has expanded dramatically. These hybrid systems generate electrical and thermal energy concurrently. Hybrid nanofluids have several exciting benefits due to their improved and better rheological properties and thermophysical characteristics, making them ideal for solar power generation systems.

The positive aspects of incorporating nanofluids in solar power generation systems are presented below:

- Smaller NPs with a higher surface area can extensively alter the capacity for heat and nanofluids absorption in solar power devices.
- Nanofluids have superior optical characteristics than basic fluids. They inherently possess low emissivity and superior absorption, spanning the infrared and solar wavelength ranges.
- The inclusion of nanoparticles in a liquid matrix enhances its heat conduction capability. Hence, nanofluids have improved thermal conductivities compared to the regular fluids.
- Nanofluids are an outstanding medium for absorption across a wide temperature range owing to their excellent long-term stability and significant absorptivity.
- The very tiny size of NPs in nanofluids prevents blockage, sedimentation, and clogging of pumps and pipelines. This characteristic renders nanofluids an excellent choice for solar-based applications.

- Nanofluids in thermal devices reduce heat transfer surfaces, which increases the cost effectiveness of solar power systems.

In general, nanofluids have a higher density accompanied by lower specific heat and higher convective heat transfer coefficient of nanoscale particles, this eventually raises efficiency of thermal devices (Rasheed *et al.*, 2021).

In electrical power transformers and other electrical equipment where oil cooling occurs, the heat conduction property and electrical insulation characteristics of the operating medium may be improved by incorporating nano-diamonds to the mineral oil as functional additives , with no modifications to the liquid's insulation performance. This increases the transformer's operational life and energy output with minimal maintenance and reduces system failure due to overheating. As we know, in 1995, Choi created a nanofluid, which is specifically thermally enhanced fluid and tried it in oil-filled transformers, and the results of his work were to remove heat from transformers, nanoscale particles of alumina oxide are utilized in working oils (Suneetha *et al.*, 2022).

## 6.2 Electronics Cooling

Nanofluids have exceptional heat transmission properties owing to their nanoscale size and large exposed surface area of dispersed particles. Because of this unique feature, nanofluids have been employed in an extensive range of technological applications. (Kumar and Shaik, 2023) study the thermal effectiveness of stable MXene-based-Cu hybrid nanofluids employing different base fluids at extremely low volumetric concentrations of MXene and Cu tailored nanostructures for efficient coolant applications. At extremely low nanostructure concentrations, hybrid nanofluids improve heat conductivity significantly.

Engine coolant, commonly known as antifreeze, is crucial to a vehicle's performance because it regulates engine temperature, protects the cooling system from freezing in cold weather, and prevents corrosion. To improve engine cooling efficiency, Dushyanthkumar *et al.* (2024) investigated employing low-concentration hybrid nanofluids to boost heat transfer rates in automotive radiators. Reduced graphene is recognized for its highly conductive behavior, rendering it a capable contender for high cooling speeds in a variety of applications. Hybrid nanofluids comprised of reduced graphene oxide and copper, combined with water and ethylene glycol, significantly improve radiator efficacy in terms of pressure drop and efficacy. (Sharaby *et al.*, 2025) conducted a thorough investigation of improving outcomes in a photovoltaic/thermal system (PVT) by introducing a hybrid nanofluid including multiwalled carbon nanotubes and zinc oxide into the operating fluid. The examination of the surface temperature of the photovoltaic panel revealed that the PVT system cooled by nanofluid produced a substantially larger drop, with a fall of 16.6°C, or a 38.9 percent relative decline when evaluated against a non-cooled panel. Benchmarking against the PVT system cooled by water demonstrated a temperature drop of 11.3°C, or a 23.54 percent reduction in surface temperature. An investigation of electrical efficacy during the operational period found gains of 12% for the system, cooled by water, and 16.53% for the system cooled by nanofluid. The incorporation of nanofluid facilitated a significant gain in thermal performance, with a peak rise.

Another study (Mezrakchi, 2024) examines the heat flux and rheology of several HNFs (CuO-EG-Water, EG-Water, CuO-ZnO-Water, $SiO_2$-EG-Water, and $Al_2O_3$-EG-Water) for heat removal at eight distinct inlet velocities in a tube side and shell side domains in heat transfer modules. The findings confirmed that the hybrid nanofluid surpassed the basic fluid in heat transfer performance under all conditions. The results showed that the copper oxide-zinc oxide-water hybrid nanofluid had enhanced heat transfer capabilities in all test scenarios. When copper oxide-zinc oxide-water-based HNF was used for heat removal, there was a significant gain of over 9 percent in temperature decrease, compared to the approximately 6 percent achieved at a lower velocity of 0.5 m/s by other hybrid nanofluids. To facilitate enhanced functionality and coolant efficacy of industrial heat exchange units, nanoparticle types must be optimized for hybrid nanofluids. Finally, the use of hybrid nanofluids in combination with a tube and shell heat exchanger resulted in a significant increase in these systems' total thermal efficacy.

Space stations and airplanes also need powerful cooling systems with compact designs because of energy, space, and weight limits. (Bing Kho *et al.*, 2023), support the idea that nanofluids significantly improve integral heat flow while undergoing pool boiling when benchmarked against traditional fluids. This innovation is accelerating the development of complex heat-removing suspensions with ultrahigh heat flux for electronic systems, which may boost the operating power of electronic parts and simplify heat reduction in spacecraft .

### 6.3 Industrial Processes

Scientific discoveries have shown that engineered hybrid nanofluid can replace mono nanofluid because it provides greater heat transfer benefits, especially in vehicle, electromechanical, industrial processes, HVAC, and solar energy applications (Kshirsagar and Venkatesh, 2021).

In the grinding process, a large quantity of energy is transferred to heat, causing material impairment, including cracking, burning, phase changes, and so on. Hybrid nanofluids' tribological property allows for improved heat removal and lubrication throughout the grinding procedure, alongside economically viable manufacturing (Suneetha *et al.*, 2022). Hybrid nanofluids have improved heat conduction capabilities owing to hybridization. They are employed in solar energy, electronics, heat recovery modules, phase change materials, chemical, manufacturing, agriculture, and automobile industries (Bhattad *et al.*, 2023).

### 6.4 Building Heating and Pollution Mitigation

In colder climates, the heating infrastructures of buildings can considerably gain advantage from the usage of nano-enabled fluids, which have demonstrated optimistic results. (Kulkarni *et al.*, 2009) investigated the efficacy of building heating in cold areas. In such settings, a heat transfer fluid composed of glycol and water in various amounts is used. For their research, they used a carrier made up of a 60:40 weight ratio of EG to $H_2O$. Subsequently, other researchers (Şirin *et al.*, 2023; C. Zhang *et al.*, 2023) have significantly contributed to the implementation of different nanofluids, leading to both pollution minimization and breakthroughs in heating systems of buildings. The outcomes of this study demonstrated that the nanofluid was leveraged in heat exchangers. It reduced flow rates (volume and mass), leading to total energy reductions in pumping power. Smaller heating infrastructures can deliver the same thermal output as larger systems, thanks to nanofluids enabling lower costs. This not only reduced initial equipment costs, omitting the economic burden of nano-enabled fluids, but also had a favorable environmental impact. Miniature heating infrastructures utilize less electricity, resulting in fewer environmental pollutants. Furthermore, the heat management modules produced minimal waste of liquids and materials, lowering their environmental impact at the end of their lifecycle.

Hybrid nanofluids demonstrate proven efficiency across energy systems, electronics cooling, and industrial processes. Structured deployment in these sectors underscores their versatility and economic potential.

## 7. Conclusion

Hybrid and multifunctional nanofluids are a transformative innovation in thermal management and multifunctional materials. In this process, two or more types of nanoscale particles are combined in various base fluids to enhance the fluid's properties, including significantly improved thermal conductivity, heat transfer coefficient, and additional functions such as electrical conductivity, anti-corrosion, anti-fouling, and magnetic responsiveness. This additive-enhanced synergistic effect offers the potential to overcome the limitations of various single-component nanofluids and makes hybrid systems promising for a wide range of applications in energy, electronics, biomedical, and environmental fields.

The applications of hybrid nanofluids also require careful selection of types of nanoparticles, base fluids, and the method of stable dispersion. Advances in synthesis techniques, such as one-step in-situ growth and eco-friendly green routes, favor stability and scalability. Knowledge about mechanisms of property enhancement, including nano-layer effects and Brownian motion, can help with the optimal design of fluid performance towards specific applications.

Despite progress, challenges remain in large-scale production, standardization of testing methods, and thorough assessment of safety and environmental impact. Further research is on green synthesis approaches, AI-driven design of smart nanofluids, and advanced technologies. These trends collectively promise that hybrid nanofluids will play a critical role in next-generation thermal and multifunctional applications, driving innovations in the direction of energy efficiency and sustainability goals.

This chapter reviewed the basic concepts, methods of synthesis, mechanisms of properties, and emerging applications to present a panoramic view for researchers and engineers who are advancing this vibrant field.

## References


[1] Abbasi, S.M., Rashidi, A., Nemati, A. and Arzani, K. (2013). The effect of functionalisation method on the stability and the thermal conductivity of nanofluid hybrids of carbon nanotubes/gamma alumina. *Ceramics International*, *39*(4), 3885–3891. https://doi.org/10.1016/j.ceramint.2012.10.232.

[2] Al Shdaifat, M.Y., Zulkifli, R., Sopian, K. and Salih, A.A. (2020). Thermal and Hydraulic Performance of CuO/Water Nanofluids: A Review. *Micromachines*, *11*(4), 416. https://doi.org/10.3390/mi11040416.

[3] Albert, A.A., Samuel, D.G.H. and Parthasarathy, V. (2020). Wet Chemical Synthesis of CuO-PVA Hybrid Nanofluid Stabilized by Steric Repulsion. *Asian Journal of Chemistry*, *32*(3), 570–574. https://doi.org/10.14233/ajchem.2020.22424.

[4] Alklaibi, A.M., Mouli, K.V.V.C. and Sundar, L.S. (2024). Heat transfer, and friction factor of Fe3O4–SiO2/Water hybrid nanofluids in a plate heat exchanger: Experimental and ANN predictions. *International Journal of Thermal Sciences*, *195*, 108608. https://doi.org/10.1016/j.ijthermalsci.2023.108608.

[5] Asadi, A., Pourfattah, F., Miklós Szilágyi, I., Afrand, M., Żyła, G., Seon Ahn, H., Wongwises, S., Minh Nguyen, H., Arabkoohsar, A. and Mahian, O. (2019). Effect of sonication characteristics on stability, thermophysical properties, and heat transfer of nanofluids: A comprehensive review. *Ultrasonics Sonochemistry*, *58*, 104701. https://doi.org/10.1016/j.ultsonch.2019.104701.

[6] Asadikia, A., Mirjalily, S.A.A., Nasirizadeh, N. and Kargarsharifabad, H. (2020). Characterization of thermal and electrical properties of hybrid nanofluids prepared with multi-walled carbon nanotubes and Fe2O3 nanoparticles. *International Communications in Heat and Mass Transfer*, *117*, 104603. https://doi.org/10.1016/j.icheatmasstransfer.2020.104603.

[7] Asim, M. and Siddiqui, F.R. (2022). Hybrid Nanofluids—Next-Generation Fluids for Spray-Cooling-Based Thermal Management of High-Heat-Flux Devices. *Nanomaterials*, *12*(3), 507. https://doi.org/10.3390/nano12030507.

[8] Asokan, N., Gunnasegaran, P. and Vicki Wanatasanappan, V. (2020). Experimental investigation on the thermal performance of compact heat exchanger and the rheological properties of low concentration mono and hybrid nanofluids containing Al2O3 and CuO nanoparticles. *Thermal Science and Engineering Progress*, *20*, 100727. https://doi.org/10.1016/j.tsep.2020.100727.

[9] Baby, T.T. and Ramaprabhu, S. (2011). Synthesis and nanofluid application of silver nanoparticles decorated graphene. *Journal of Materials Chemistry*, *21*(26), 9702. https://doi.org/10.1039/c0jm04106h.

[10] Baghbanzadeh, M., Rashidi, A., Rashtchian, D., Lotfi, R. and Amrollahi, A. (2012). Synthesis of spherical silica/multiwall carbon nanotubes hybrid nanostructures and investigation of thermal conductivity of related nanofluids. *Thermochimica Acta*, *549*, 87–94. https://doi.org/10.1016/j.tca.2012.09.006.

[11] Bahrami, H.-R., Allahdadi, O. and Saffari, H. (2025). Nanofluids for advanced thermal management: A multi-criteria review of performance, cost, and environmental impacts. *Results in Materials*, *27*, 100753. https://doi.org/10.1016/j.rinma.2025.100753.

[12] Bhattad, A., Atgur, V., Rao, B., Banapurmath, N., Yunus Khan, T., Vadlamudi, C., Krishnappa, S., Sajjan, A., Shankara, R. and Ayachit, N. (2023). Review on Mono and Hybrid Nanofluids: Preparation, Properties, Investigation, and Applications in IC Engines and Heat Transfer. *Energies*, *16*(7), 3189. https://doi.org/10.3390/en16073189.

[13] Bing Kho, Y., Jusoh, R., Zuki Salleh, M., Hisyam Ariff, M. and Zainuddin, N. (2023). Magnetohydrodynamics flow of Ag-TiO2 hybrid nanofluid over a permeable wedge with thermal radiation and viscous dissipation. *Journal of Magnetism and Magnetic Materials, 565*, 170284. https://doi.org/10.1016/j.jmmm.2022.170284.

[14] Borode, A.O., Ahmed, N. A., Olubambi, P. A., Sharifpur, M. and Meyer, J. P. (2021). Investigation of the Thermal Conductivity, Viscosity, and Thermal Performance of Graphene Nanoplatelet-Alumina Hybrid Nanofluid in a Differentially Heated Cavity. *Frontiers in Energy Research, 9*. https://doi.org/10.3389/fenrg.2021.737915.

[15] Borode, A., Tshephe, T., Olubambi, P., Sharifpur, M. and Meyer, J. (2023). Experimental study and ANFIS modelling of the thermophysical properties and efficacy of GNP - Al 2 O 3 hybrid nanofluids of different concentrations and temperatures. *SN Applied Sciences*. https://doi.org/10.1007/s42452-023-05574-7.

[16] Botha, S.S., Ndungu, P. and Bladergroen, B.J. (2011). Physicochemical Properties of Oil-Based Nanofluids Containing Hybrid Structures of Silver Nanoparticles Supported on Silica. *Industrial & Engineering Chemistry Research, 50*(6), 3071–3077. https://doi.org/10.1021/ie101088x.

[17] Cardoso, B.D., Souza, A., Nobrega, G., Afonso, I.S., Neves, L.B., Faria, C., Ribeiro, J. and Lima, R.A. (2025). Progress in Nanofluid Technology: From Conventional to Green Nanofluids for Biomedical, Heat Transfer, and Machining Applications. *Nanomaterials, 15*(16), 1242. https://doi.org/10.3390/nano15161242.

[18] Che Sidik, N.A., Mahmud Jamil, M., Aziz Japar, W.M.A. and Muhammad Adamu, I. (2017). A review on preparation methods, stability and applications of hybrid nanofluids. *Renewable and Sustainable Energy Reviews, 80*, 1112–1122. https://doi.org/10.1016/j.rser.2017.05.221.

[19] Chen, D., Zhao, Y., Chen, Y., Wang, B., Chen, H., Zhou, J. and Liang, Z. (2015). One-Step Chemical Synthesis of ZnO/Graphene Oxide Molecular Hybrids for High-Temperature Thermoelectric Applications. *ACS Applied Materials & Interfaces, 7*(5), 3224–3230. https://doi.org/10.1021/am507882f.

[20] Choi, S.U.S. (1995). Enhancing Thermal Conductivity of Fluids With Nanoparticles. *Developments and Applications of Non-Newtonian Flows*, 99–105. https://doi.org/10.1115/IMECE1995-0926.

[21] Das, L., Habib, K., Irshad, K., Saidur, R., Algarni, S. and Alqahtani, T. (2022). Thermo-Optical Characterization of Therminol55 Based MXene–Al2O3 Hybridized Nanofluid and New Correlations for Thermal Properties. *Nanomaterials, 12*(11), 1862. https://doi.org/10.3390/nano12111862.

[22] Dushyanthkumar, G.L., Praveenkumara, B.M., Shivashankar, R., Shrinivasa, D. and Muthuraju, N. P. (2024). Experimental Investigation of the Enhancement of Heat Transfer Rate in Automobile Radiator Using Low Concentration Hybrid Nanofluids. *Journal of Mines, Metals and Fuels*, 1187–1202. https://doi.org/10.18311/jmmf/2024/46001.

[23] Eastman, J.A., Choi, U.S., Li, S., Thompson, L.J. and Lee, S. (1996). Enhanced Thermal Conductivity through the Development of Nanofluids. *MRS Proceedings, 457*, 3. https://doi.org/10.1557/PROC-457-3.

[24] Ekiciler, R. (2024). Performing SiO2-MWCNT/water hybrid nanofluid with differently shaped nanoparticles to enhance first- and second-law features of flow by considering a two-phase approach. *Journal of Thermal Analysis and Calorimetry, 149*(4), 1725–1744. https://doi.org/10.1007/s10973-024-12885-3.

[25] Elsaid, A.M. (2019). Experimental study on the heat transfer performance and friction factor characteristics of Co3O4 and Al2O3 based H2O/(CH2OH)2 nanofluids in a vehicle engine radiator. *International Communications in Heat and Mass Transfer, 108*, 104263. https://doi.org/10.1016/j.icheatmasstransfer.2019.05.009.

[26] Eshgarf, H., Kalbasi, R., Maleki, A., Shadloo, M.S. and Karimipour, A. (2021). A review on the properties, preparation, models and stability of hybrid nanofluids to optimize energy consumption. *Journal of Thermal Analysis and Calorimetry*, *144*(5), 1959–1983. https://doi.org/10.1007/s10973-020-09998-w.

[27] Giwa, S.O., Sharifpur, M., Ahmadi, M.H., Sohel Murshed, S.M. and Meyer, J.P. (2021). Experimental Investigation on Stability, Viscosity, and Electrical Conductivity of Water-Based Hybrid Nanofluid of MWCNT-Fe2O3. *Nanomaterials*, *11*(1), 136. https://doi.org/10.3390/nano11010136.

[28] Gupta, M., Singh, V., Kumar, S. and Dilbaghi, N. (2020). Experimental analysis of heat transfer behavior of silver, MWCNT and hybrid (silver +MWCNT) nanofluids in a laminar tubular flow. *Journal of Thermal Analysis and Calorimetry*, *142*(4), 1545–1559. https://doi.org/10.1007/s10973-020-09453-w.

[29] Habibi, I., Saifudin, Fatoni, H.R., Kusumastuti, R., Priyotomo, G., Musabikha, S., Nikitasari, A., Prifiharni, S., Lestari, Y. and Royani, A. (2024). Ethylene Glycol-water Based Graphene Oxide Nanofluid as Corrosion Inhibitor in Automotive Radiator. *E3S Web of Conferences*, *543*, 1–12. https://doi.org/10.1051/e3sconf/202454303002.

[30] Hafeez, A., Aldosari, F.M., Helmi, M.M., Ghazwani, H.A., Hussien, M. and Hassan, A.M. (2023). Heat transfer performance in a Hybrid nanofluid (kerosene oil) flow over a shrinking cylinder. *Case Studies in Thermal Engineering*, *52*, 103539. https://doi.org/10.1016/j.csite.2023.103539.

[31] Hamid, K.A., Azmi, W.H., Nabil, M.F., Mamat, R. and Sharma, K.V. (2018). Experimental investigation of thermal conductivity and dynamic viscosity on nanoparticle mixture ratios of TiO2-SiO2 nanofluids. *International Journal of Heat and Mass Transfer*, *116*, 1143–1152. https://doi.org/10.1016/j.ijheatmasstransfer.2017.09.087.

[32] Hamzah, M.H., Sidik, N.A.C., Ken, T.L., Mamat, R. and Najafi, G. (2017). Factors affecting the performance of hybrid nanofluids: A comprehensive review. *International Journal of Heat and Mass Transfer*, *115*, 630–646. https://doi.org/10.1016/j.ijheatmasstransfer.2017.07.021.

[33] Hemmat Esfe, M., Kiannejad Amiri, M. and Alirezaie, A. (2018). Thermal conductivity of a hybrid nanofluid. *Journal of Thermal Analysis and Calorimetry*, *134*(2), 1113–1122. https://doi.org/10.1007/s10973-017-6836-9.

[34] Ho, C.J., Huang, J.B., Tsai, P.S. and Yang, Y.M. (2010). Preparation and properties of hybrid water-based suspension of Al2O3 nanoparticles and MEPCM particles as functional forced convection fluid. *International Communications in Heat and Mass Transfer*, *37*(5), 490–494. https://doi.org/10.1016/j.icheatmasstransfer.2009.12.007.

[35] Jena, P., Brocchi, E. and Motta, M. (2001). *In-situ* formation of Cu–Al2O3 nano-scale composites by chemical routes and studies on their microstructures. *Materials Science and Engineering: A*, *313*(1–2), 180–186. https://doi.org/10.1016/S0921-5093(00)01998-5.

[36] Kakavandi, A. and Akbari, M. (2018). Experimental investigation of thermal conductivity of nanofluids containing of hybrid nanoparticles suspended in binary base fluids and propose a new correlation. *International Journal of Heat and Mass Transfer*, *124*, 742–751. https://doi.org/10.1016/j.ijheatmasstransfer.2018.03.103.

[37] Kannaiyan, S., Boobalan, C., Umasankaran, A., Ravirajan, A., Sathyan, S. and Thomas, T. (2017). Comparison of experimental and calculated thermophysical properties of alumina/cupric oxide hybrid nanofluids. *Journal of Molecular Liquids*, *244*, 469–477. https://doi.org/10.1016/j.molliq.2017.09.035.

[38] Kanti, P.K., Sharma, P., Koneru, B., Banerjee, P. and Jayan, K.D. (2023). Thermophysical profile of graphene oxide and MXene hybrid nanofluids for sustainable energy applications: Model prediction with a Bayesian optimized neural network with K-cross fold validation. *FlatChem*, *39*, 100501. https://doi.org/10.1016/j.flatc.2023.100501.

[39] Khan, A.I. and Valan Arasu, A. (2019). A review of influence of nanoparticle synthesis and geometrical parameters on thermophysical properties and stability of nanofluids. *Thermal Science and Engineering Progress*, *11*, 334–364. https://doi.org/10.1016/j.tsep.2019.04.010.

[40] Kong, L., Sun, J. and Bao, Y. (2017). Preparation, characterization and tribological mechanism of nanofluids. *RSC Advances*, *7*(21), 12599–12609. https://doi.org/10.1039/C6RA28243A.

[41] Kshirsagar, D.P. and Venkatesh, M.A. (2021). A review on hybrid nanofluids for engineering applications. *Materials Today: Proceedings*, *44*, 744–755. https://doi.org/10.1016/j.matpr.2020.10.637.

[42] Kulkarni, D.P., Das, D.K. and Vajjha, R.S. (2009). Application of nanofluids in heating buildings and reducing pollution. *Applied Energy*, *86*(12), 2566–2573. https://doi.org/10.1016/j.apenergy.2009.03.021.

[43] Kumar, K.R. and Shaik, A.H. (2023). Synthesis, thermophysical characterization and thermal performance analysis of novel Cu-MXene hybrid nanofluids for efficient coolant applications. *RSC Advances*, *13*(42), 29536–29560. https://doi.org/10.1039/D3RA05429B.

[44] Lu, Y., Lu, X., Mayers, B.T., Herricks, T. and Xia, Y. (2008). Synthesis and characterization of magnetic Co nanoparticles: A comparison study of three different capping surfactants. *Journal of Solid State Chemistry*, *181*(7), 1530–1538. https://doi.org/10.1016/j.jssc.2008.02.016.

[45] Manimaran, M., Norizan, M.N., Kassim, M.H.M., Adam, M.R., Abdullah, N. and Norrahim, M.N.F. (2025). Critical review on the stability and thermal conductivity of water-based hybrid nanofluids for heat transfer applications. *RSC Advances*, *15*(18), 14088–14125. https://doi.org/10.1039/D5RA00844A.

[46] Megatif, L., Ghozatloo, A., Arimi, A. and Shariati-Niasar, M. (2016). Investigation of Laminar Convective Heat Transfer of a Novel Tio 2 –Carbon Nanotube Hybrid Water-Based Nanofluid. *Experimental Heat Transfer*, *29*(1), 124–138. https://doi.org/10.1080/08916152.2014.973974.

[47] Mezrakchi, R. Al. (2024). Investigation of various hybrid nanofluids to enhance the performance of a shell and tube heat exchanger. *AIMS Energy*, *12*(1), 235–255. https://doi.org/10.3934/energy.2024011.

[48] Minea, A.A. and Sohel Murshed, S.M. (2021). Ionic Liquids-Based Nanocolloids—A Review of Progress and Prospects in Convective Heat Transfer Applications. *Nanomaterials*, *11*(4), 1039. https://doi.org/10.3390/nano11041039.

[49] Mukesh Kumar, P.C., Palanisamy, K. and Vijayan, V. (2020). Stability analysis of heat transfer hybrid/water nanofluids. *Materials Today: Proceedings*, *21*, 708–712. https://doi.org/10.1016/j.matpr.2019.06.743.

[50] Namburu, P.K., Das, D.K., Tanguturi, K.M. and Vajjha, R.S. (2009). Numerical study of turbulent flow and heat transfer characteristics of nanofluids considering variable properties. *International Journal of Thermal Sciences*, *48*(2), 290–302. https://doi.org/10.1016/j.ijthermalsci.2008.01.001.

[51] Nine, M.J., Munkhbayar, B., Rahman, M.S., Chung, H. and Jeong, H. (2013). Highly productive synthesis process of well dispersed Cu2O and Cu/Cu2O nanoparticles and its thermal characterization. *Materials Chemistry and Physics*, *141*(2–3), 636–642. https://doi.org/10.1016/j.matchemphys.2013.05.032.

[52] Nuim Labib, M., Nine, M.J., Afrianto, H., Chung, H. and Jeong, H. (2013). Numerical investigation on effect of base fluids and hybrid nanofluid in forced convective heat transfer. *International Journal of Thermal Sciences*, *71*, 163–171. https://doi.org/10.1016/j.ijthermalsci.2013.04.003.

[53] Pavía, M., Alajami, K., Estellé, P., Desforges, A. and Vigolo, B. (2021). A critical review on thermal conductivity enhancement of graphene-based nanofluids. *Advances in Colloid and Interface Science*, *294*, 102452. https://doi.org/10.1016/j.cis.2021.102452.

[54] Ragavi, M. and Poornima, T. (2024). Enhanced heat transfer analysis on Ag-Al2O3/water hybrid magneto-convective nanoflow. *Discover Nano*, *19*(1), 31. https://doi.org/10.1186/s11671-024-03975-0.

[55] Rahman, M.A., Hasnain, S.M.M., Pandey, S., Tapalova, A., Akylbekov, N. and Zairov, R. (2024). Review on Nanofluids: Preparation, Properties, Stability, and Thermal Performance

Augmentation in Heat Transfer Applications. *ACS Omega*. https://doi.org/10.1021/acsomega.4c03279.

[56] Rajesh Kumar, K., Chandan, M.R., Kiran, B. and Shaik, A.H. (2025). Novel Cu-MXene hybrid nanofluids for the experimental investigation of thermal performance in double pipe heat exchanger. *Scientific Reports*, *15*(1), 9937. https://doi.org/10.1038/s41598-025-94330-5.

[57] Ranga Babu, J.A., Kumar, K.K. and Srinivasa Rao, S. (2017). State-of-art review on hybrid nanofluids. *Renewable and Sustainable Energy Reviews*, *77*, 551–565. https://doi.org/10.1016/j.rser.2017.04.040.

[58] Rasheed, T., Hussain, T., Anwar, M.T., Ali, J., Rizwan, K., Bilal, M., Alshammari, F.H., Alwadai, N. and Almuslem, A.S. (2021). Hybrid Nanofluids as Renewable and Sustainable Colloidal Suspensions for Potential Photovoltaic/Thermal and Solar Energy Applications. *Frontiers in Chemistry*, *9*. https://doi.org/10.3389/fchem.2021.737033.

[59] Rubbi, F., Das, L., Habib, K., Aslfattahi, N., Saidur, R. and Alam, S.U. (2021). A comprehensive review on advances of oil-based nanofluids for concentrating solar thermal collector application. *Journal of Molecular Liquids*, *338*, 116771. https://doi.org/10.1016/j.molliq.2021.116771.

[60] Sadeghi, M.S., Anadalibkhah, N., Ghasemiasl, R., Armaghani, T., Dogonchi, A.S., Chamkha, A.J., Ali, H. and Asadi, A. (2022). On the natural convection of nanofluids in diverse shapes of enclosures: an exhaustive review. *Journal of Thermal Analysis and Calorimetry*, *147*(1). https://doi.org/10.1007/s10973-020-10222-y.

[61] Saeedinia, M., Akhavan-Behabadi, M.A. and Razi, P. (2012). Thermal and rheological characteristics of CuO–Base oil nanofluid flow inside a circular tube. *International Communications in Heat and Mass Transfer*, *39*(1), 152–159. https://doi.org/10.1016/j.icheatmasstransfer.2011.08.001.

[62] Sahid, N.S.M., Rahman, M.M., Kadirgama, K. and Maleque, M.A. (2017). Experimental investigation on properties of hybrid nanofluids (TiO2 and ZnO) in water–ethylene glycol mixture. *Journal of Mechanical Engineering and Sciences*, *11*(4), 3087–3094. https://doi.org/10.15282/jmes.11.4.2017.11.0277.

[63] Said, Z., Jamei, M., Syam Sundar, L., Pandey, A.K., Allouhi, A. and Li, C. (2022). Thermophysical properties of water, water and ethylene glycol mixture-based nanodiamond + Fe3O4 hybrid nanofluids: An experimental assessment and application of data-driven approaches. *Journal of Molecular Liquids*, *347*, 117944. https://doi.org/10.1016/j.molliq.2021.117944.

[64] Scott, T.O., Ewim, D.R.E. and Eloka-Eboka, A.C. (2022). Hybrid nanofluids flow and heat transfer in cavities: a technological review. *International Journal of Low-Carbon Technologies*, *17*, 1104–1123. https://doi.org/10.1093/ijlct/ctac093.

[65] Sezer, N., Atieh, M.A. and Koç, M. (2019). A comprehensive review on synthesis, stability, thermophysical properties, and characterization of nanofluids. *Powder Technology*, *344*, 404–431. https://doi.org/10.1016/j.powtec.2018.12.016.

[66] Shaik, A.H., Chakraborty, S., Saboor, S., Kumar, K.R., Majumdar, A., Rizwan, M., Arıcı, M. and Chandan, M.R. (2024). Cu-Graphene water-based hybrid nanofluids: synthesis, stability, thermophysical characterization, and figure of merit analysis. *Journal of Thermal Analysis and Calorimetry*, *149*(7), 2953–2968. https://doi.org/10.1007/s10973-023-12875-x.

[67] Sharaby, M.R., Sharaby, M.R., Younes, M.M., Taleb, F.S.A. and Baz, F.B. (2025). The impact of hybrid nanofluid cooling on photovoltaic/thermal system performance: a 3E analysis approach. *Journal of Thermal Analysis and Calorimetry*. https://doi.org/10.1007/s10973-025-14889-z.

[68] Singh, B. and Sood, S. (2024). Hybrid nanofluids preparation, thermo-physical properties, and applications: A Review. *Hybrid Advances*, *6*, 100192. https://doi.org/10.1016/j.hybadv.2024.100192.

[69] Şirin, C., Goggins, J. and Hajdukiewicz, M. (2023). A review on building-integrated photovoltaic/thermal systems for green buildings. *Applied Thermal Engineering*, *229*, 120607. https://doi.org/10.1016/j.applthermaleng.2023.120607.

[70] Sivasankaran, S. and Bhuvaneswari, M. (2022). Numerical study on influence of water based hybrid nanofluid and porous media on heat transfer and pressure loss. *Case Studies in Thermal Engineering, 34*, 102022. https://doi.org/10.1016/j.csite.2022.102022.

[71] Sun, L., Yang, L., Zhao, N., Song, J., Li, X. and Wu, X. (2022). A review of multifunctional applications of nanofluids in solar energy. *Powder Technology, 411*, 117932. https://doi.org/10.1016/j.powtec.2022.117932.

[72] Sundar, L.S., Sharma, K.V., Singh, M.K. and Sousa, A.C.M. (2017). Hybrid nanofluids preparation, thermal properties, heat transfer and friction factor—A review. *Renewable and Sustainable Energy Reviews, 68*, 185–198. https://doi.org/10.1016/j.rser.2016.09.108.

[73] Suneetha, S., Subbarayudu, K. and Bala Anki Reddy, P. (2022). Hybrid nanofluids development and benefits: A comprehensive review. *Journal of Thermal Engineering, 8*(3), 445–455. https://doi.org/10.18186/thermal.1117455.

[74] Suresh, S., Venkitaraj, K.P., Selvakumar, P. and Chandrasekar, M. (2011). Synthesis of Al2O3–Cu/water hybrid nanofluids using two step method and its thermo physical properties. *Colloids and Surfaces A: Physicochemical and Engineering Aspects, 388*(1–3), 41–48. https://doi.org/10.1016/j.colsurfa.2011.08.005.

[75] Syam Sundar, L., Venkata Ramana, E., Singh, M.K. and Sousa, A.C.M. (2014). Thermal conductivity and viscosity of stabilized ethylene glycol and water mixture Al2O3 nanofluids for heat transfer applications: An experimental study. *International Communications in Heat and Mass Transfer, 56*, 86–95. https://doi.org/10.1016/j.icheatmasstransfer.2014.06.009.

[76] Toghraie, D., Chaharsoghi, V.A. and Afrand, M. (2016). Measurement of thermal conductivity of ZnO–TiO2/EG hybrid nanofluid. *Journal of Thermal Analysis and Calorimetry, 125*(1), 527–535. https://doi.org/10.1007/s10973-016-5436-4.

[77] Tripathi, D., Sharma, K. and Choudhari, R. (2025). Editorial: Recent advancements in nanofluids and hybrid nanofluids: Modeling, simulation & experimentation. *International Journal of Thermofluids, 28*, 101312. https://doi.org/10.1016/j.ijft.2025.101312.

[78] Urmi, W., Rahman, M.M. and Hamzah, W.A.W. (2020). An experimental investigation on the thermophysical properties of 40% ethylene glycol based TiO2-Al2O3 hybrid nanofluids. *International Communications in Heat and Mass Transfer, 116*, 104663. https://doi.org/10.1016/j.icheatmasstransfer.2020.104663.

[79] Wanatasanappan, V.V., Rezman, M. and Abdullah, M.Z. (2022). Thermophysical Properties of Vegetable Oil-Based Hybrid Nanofluids Containing Al2O3-TiO2 Nanoparticles as Insulation Oil for Power Transformers. *Nanomaterials, 12*(20), 3621. https://doi.org/10.3390/nano12203621.

[80] Wang, Y., Nair, S.R., Ho, M.L.G., Foo, J.-J., Tran, M.-V. and Oon, C.S. (2025). Heat transfer performance and fouling analysis of clove-treated graphene nanoplatelet nanofluid in a double-pipe heat exchanger. *Journal of Thermal Analysis and Calorimetry, 150*(7), 5663–5678. https://doi.org/10.1007/s10973-025-14002-4.

[81] Wohld, J., Beck, J., Inman, K., Palmer, M., Cummings, M., Fulmer, R. and Vafaei, S. (2022). Hybrid Nanofluid Thermal Conductivity and Optimization: Original Approach and Background. *Nanomaterials, 12*(16), 2847. https://doi.org/10.3390/nano12162847.

[82] Wu, K., Zhang, Y., Ge, F., Huang, X., Ge, H.-H., Zhao, Y.-Z., Meng, X.-J. and Dou, B.-L. (2021). Corrosion behavior of brass in SCW-SDBS-TiO2 nanofluid. *Journal of Alloys and Compounds, 855*, 157483. https://doi.org/10.1016/j.jallcom.2020.157483.

[83] Wu, Z., Wang, L. and Sundén, B. (2013). Pressure drop and convective heat transfer of water and nanofluids in a double-pipe helical heat exchanger. *Applied Thermal Engineering, 60*(1–2), 266–274. https://doi.org/10.1016/j.applthermaleng.2013.06.051.

[84] Xu, Q., Yang, G., Jia, S., Wang, Z., Akkurt, N., Zhang, H., Zhou, Q., Shen, M., Yang, D., Zhu, L., Liu, L., Du, Y. and Xiong, Y. (2022). Experimental study on synergistic enhancement of thermal performance of a closed two-phase thermosyphon by a TiO2 nanofluid doped with Al2O3. *Case Studies in Thermal Engineering, 36*, 102192. https://doi.org/10.1016/j.csite.2022.102192.

[85] Yang, L., Ji, W., Mao, M. and Huang, J. (2020). An updated review on the properties, fabrication and application of hybrid-nanofluids along with their environmental effects. *Journal of Cleaner Production*, *257*, 120408. https://doi.org/10.1016/j.jclepro.2020.120408.

[86] Yang, Y., Luan, H., Liu, F., Si, L., Yan, H. and Zhang, C. (2023). Investigation of the Lubrication Performance of γ-Al2O3/ZnO Hybrid Nanofluids for Titanium Alloy. *Metals*, *13*(10), 1701. https://doi.org/10.3390/met13101701.

[87] Yildizay, H.D., Bekmezci, M. and Sen, F. (2025). Nanofluids and engineering applications: A review. *Journal of Scientific Reports-A*, *060*, 126–149. https://doi.org/10.59313/jsr-a.1634164.

[88] Yu, W. and Choi, S.U.S. (2003). The Role of Interfacial Layers in the Enhanced Thermal Conductivity of Nanofluids: A Renovated Maxwell Model. *Journal of Nanoparticle Research*, *5*(1–2), 167–171. https://doi.org/10.1023/A:1024438603801.

[89] Zainon, S.N.M. and Azmi, W.H. (2021). Recent Progress on Stability and Thermo-Physical Properties of Mono and Hybrid towards Green Nanofluids. *Micromachines*, *12*(2), 176. https://doi.org/10.3390/mi12020176.

[90] Zhang, C., Shen, C., Zhang, Y., Zheng, K., Pu, J., Zhao, X. and Ma, X. (2023). Experimental study of indoor light/thermal environment with spectrally selective windows using ATO nanofluids in winter. *Energy and Buildings*, *278*, 112597. https://doi.org/10.1016/j.enbuild.2022.112597.

[91] Zhang, F.-F., Zheng, F.-F., Wu, X.-H., Yin, Y.-L. and Chen, G. (2019). Variations of thermophysical properties and heat transfer performance of nanoparticle-enhanced ionic liquids. *Royal Society Open Science*, *6*(4), 182040. https://doi.org/10.1098/rsos.182040.

# 11 | Nanofluids in Renewable Energy Systems – Applications, Challenges, and Future Prospects

Hamid-Reza Bahrami[1*]

This chapter synthesizes the state of the art on nanofluid use across major renewable energy domains. The discussion begins with solar thermal systems and direct-absorption concepts, where nanofluids have seen the greatest exploration. It then addresses hybrid PV/T applications, wind-energy cooling and lubrication, geothermal heat-exchanger performance, and the incorporation of nanofluids into thermal energy storage media. For each domain, mechanisms of enhancement, representative performance results, and system-level trade-offs are summarized. Cross-cutting sections assess environmental, health, and economic considerations, highlight the most pressing challenges of stability and standardization, and outline future research priorities.

## 1. Introduction

The accelerating global transition toward renewable energy technologies brings with it a dual imperative: to maximize energy conversion efficiency and to manage thermal processes reliably under increasingly demanding operating conditions. Solar, wind, geothermal, and hybrid renewable systems all involve critical stages of heat generation, transfer, storage, and dissipation. The performance of these stages frequently determines the net efficiency, cost-effectiveness, and durability of the entire system. In this context, nanofluids, which are synthetically engineered colloidal suspensions of nanomaterials within conventional base fluids, have shown promise as a compelling innovation in energy harvesting and thermal management (Talesh Bahrami *et al.*, 2020, Shahzad *et al.*, 2021). By tailoring effective thermophysical and optical properties, nanofluids can simultaneously enhance thermal energy exchange and, in some cases, directly influence energy conversion processes (Said *et al.*, 2021). Improved thermal conductivity, modified viscosity, enhanced convective behavior, and tunable solar absorption are among the most widely reported effects. For example, in solar thermal collectors, nanofluids serve a dual role. They act as conventional heat transfer media while also functioning as volumetric absorbers (Sainz-Mañas *et al.*, 2022, López-

[1] Assistant Professor, Department of Mechanical Engineering, Qom University of Technology, Qom 37181 46645, Iran. P.O. Box 37195-1519.
* Corresponding author: h_tbahrami@yahoo.com, taleshbahrami@qut.ac.ir

DOI: 10.1201/9781003686002-11

Núñez *et al.*, 2024). This combination reduces re-radiation losses and smooths temperature gradients. In photovoltaic/thermal (PV/T) hybrid systems, nanofluid-based cooling lowers the temperature of photovoltaic surfaces (Zamen *et al.*, 2022). As a result, electrical efficiency improves while useful thermal energy is simultaneously recovered. Within wind power systems, nanofluids in lubricants and cooling loops can mitigate excessive temperatures in gearboxes, generators, and power electronics, thus extending component lifetimes (Ahsan *et al.*, 2025, Ali *et al.*, 2025). In geothermal applications, nanofluids augment heat-exchange efficiency by increasing thermal conductivity and convective performance (Xu *et al.*, 2024). In thermal energy storage systems, both latent and sensible, nanofluids accelerate charging and discharging primarily by enhancing the effective thermal conductivity of the storage medium (Lee and Jo, 2021).

Despite these promising outcomes, the deployment of nanofluids in renewable energy systems remains at an early stage. Challenges of stability, agglomeration, corrosion, and increased pumping power demand careful evaluation alongside reported benefits. Furthermore, large-scale techno-economic and environmental assessments are limited, raising important questions about long-term viability, nanoparticle recovery, health and safety concerns, and life-cycle costs. This chapter, therefore, emphasizes a balanced perspective: nanofluids offer real opportunities for advancing renewable energy technologies, but their benefits must be weighed against integration barriers and sustainability considerations.

**Key message**. Nanofluids are not simply incremental heat transfer fluids; they represent a class of functional media whose engineered properties must be integrated thoughtfully into renewable energy systems. Their promise lies not in isolated laboratory measurements, but in carefully designed, system-aware applications that deliver tangible performance, reliability, and sustainability benefits.

## 2. Solar Thermal Collectors

Solar thermal collectors are a key renewable energy system that capture solar radiation and transform it into usable thermal energy. Unlike photovoltaics, which directly produce electricity, these systems harness thermal energy for residential, commercial, and industrial applications. Their main uses include residential hot water supply (Kumar *et al.*, 2023), HVAC (Liu *et al.*, 2021), industrial process heat such as sterilization and dyeing (Zwalnan *et al.*, 2021), power generation in CSP plants (Ali and Djaouida, 2025), and desalination or solar cooking (Venkatesh, Singh, *et al.*, 2025). By replacing fossil fuels, they decrease greenhouse gas emissions and contribute significantly to sustainable development.

A fundamental classification of solar collectors distinguishes between non-concentrating (stationary) and concentrating systems (Figure 2). Stationary collectors, for example, flat plate, evacuated tube, and direct absorption designs, are usually applied in low-to medium-temperature ranges (up to ~100°C) (Kumar and Hassan, 2024). Flat plate collectors are low-cost and versatile, evacuated tube collectors are efficient in colder climates, and direct absorption collectors improve

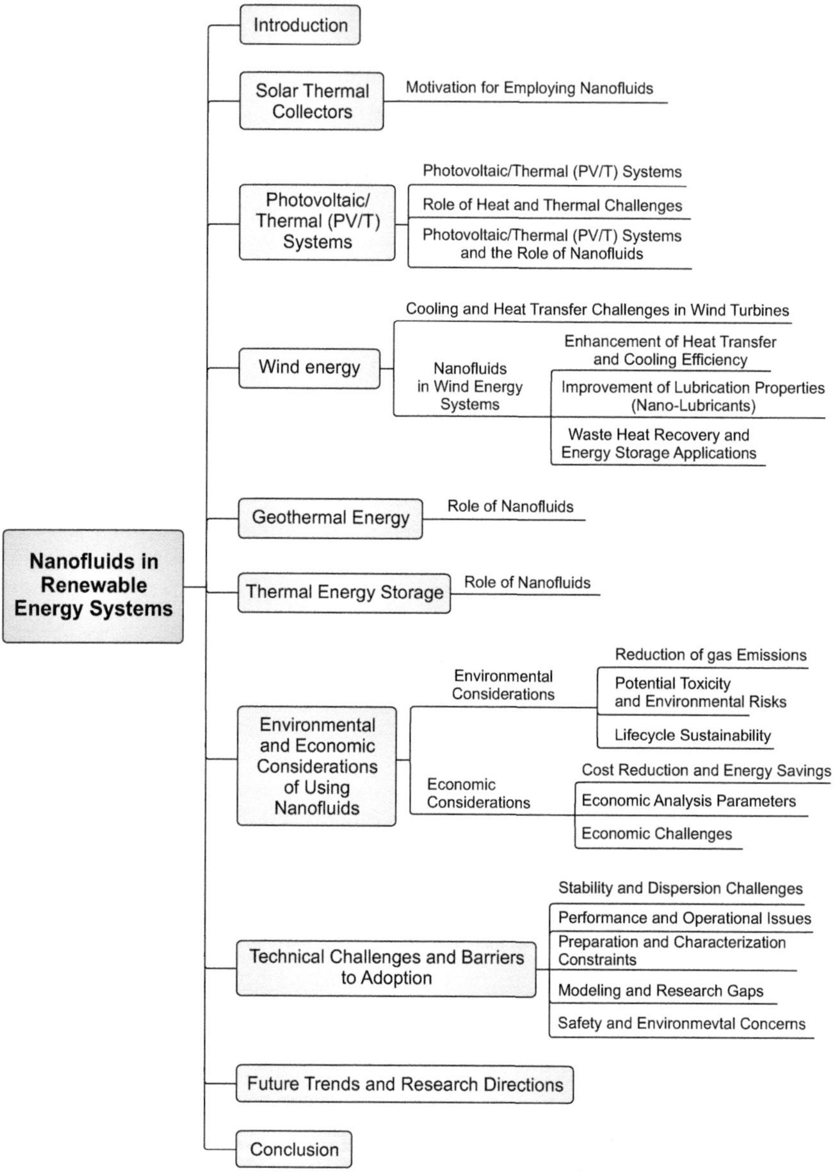


**Fig. 1** Mind map of the chapter "Nanofluids in Renewable Energy Systems," illustrating the main sections and subtopics, including application domains (PV/T, wind, geothermal, and thermal energy storage) and cross-cutting considerations (environmental, economic, technical challenges, and future trends).

efficiency by letting the working fluid directly absorb solar energy (Hamzat *et al.*, 2022).

Concentrating collectors, on the other hand, employ lenses or mirrors to converge sunlight onto a smaller receiver area, achieving much higher operating temperatures (100–400 °C, sometimes beyond 600°C) (Hamzat *et al.*, 2022, Rativa *et al.*, 2025). Common designs include parabolic troughs, parabolic dishes, and linear Fresnel systems, widely used in CSP power generation and high-temperature industrial processes.

Key challenges remain in enhancing efficiency, such as low thermal conductivity of conventional fluids (Ahmadlouydarab *et al.*, 2022), heat losses via conduction, convection, and radiation (Xu *et al.*, 2025), thermal resistance between absorber and fluid, and irreversibilities in heat exchange processes (Thakur *et al.*, 2021).

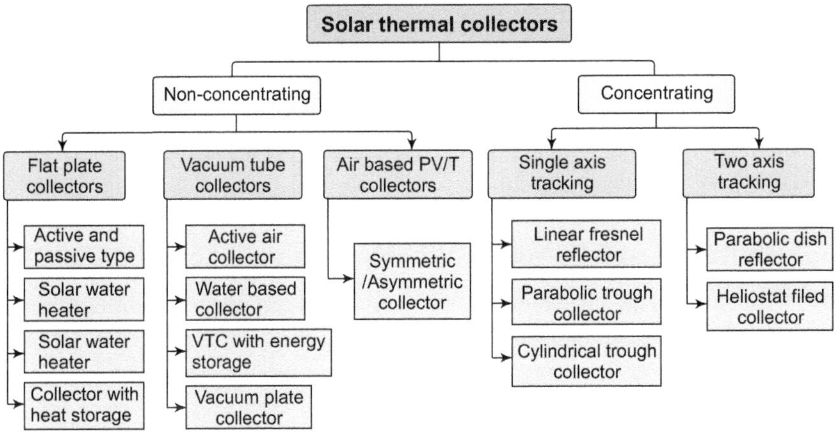


**Fig. 2** Types of solar thermal collectors (Pathak *et al.*, 2023).

## *2.1 Motivation for Employing Nanofluids in Solar Thermal Collectors*

The main challenge in thermal utilization of solar energy lies in maximizing energy absorption while significantly improving the thermal efficacy of collectors and minimizing thermal energy losses. Conventional working fluids, e.g., water, oils, and ethylene glycol, show limited performance in solar heat transfer systems owing to their low thermal conductivity and weak solar radiation absorption capacity. These shortcomings render such fluids unsuitable for direct absorption solar collectors (DASC) (Hamzat *et al.*, 2022). To address these limitations, nanofluids have been proposed as an alternative to traditional working fluids. Nanofluids are engineered fluids created by dispersing tiny solid particles (1–100 nm) into a liquid like water or oil (Manna, 2009). These nanoparticles may be metallic (e.g., copper, silver, iron), metal oxides (e.g., $Al_2O_3$, CuO, $TiO_2$, ZnO), carbon-based materials (e.g., graphene, carbon nanotubes), or hybrid composites (Hamzat *et al.*, 2022, Bahrami, Allahdadi, and Saffari 2025). The incorporation of nanoparticles substantially augments the thermal conductivity of the base fluid. Furthermore, other thermophysical properties such as viscosity, density, and specific heat capacity may change. Most importantly, nanofluids exhibit significantly better optical absorption characteristics, especially within the high-energy regions of the solar spectrum (Joseph *et al.*, 2020). The Brownian motion of nanoparticles also contributes to improved thermal conductivity and improved thermal energy exchange (Kumar and Hassan, 2024). These improved properties translate into multiple potential benefits for solar thermal collectors, including (Thakur *et al.*, 2021):

- Higher collector efficiency and improved overall system performance.
- Reduction in collector size and manufacturing costs (Amar *et al.*, 2023).
- Lower heat losses through convection, conduction, and radiation (Said *et al.*, 2021).

Altogether, these advantages position nanofluids as a highly promising and effective candidate for improving the function of solar thermal collectors and advancing sustainable energy goals.

## 3. Photovoltaic/Thermal (PV/T) Systems

PV/T systems, illustrated in Figure 3, are hybrid units engineered for the concurrent conversion of solar radiation into both thermal and electrical energy. PV/T systems integrate PV panels for electricity generation with thermal collectors to capture solar heat, enabling the simultaneous production of power and thermal energy. Therefore, these systems considerably improve the overall efficiency of solar energy conversion (Khan *et al.*, 2022).

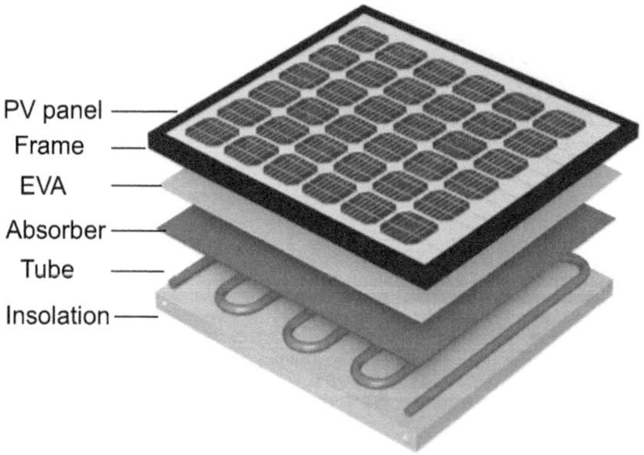


**Fig. 3** Different sections of a typical PV/T system. Adapted from Karaaslan and Menlik 2021 with permission.

### 3.1 Role of Heat Transfer Fluids in PV/T Systems

A heat transfer fluid (HTF), which can be water or a nanofluid, is circulated through channels or pipes located behind the PV module. This fluid absorbs the excess thermal energy produced by the PV cells and removes it from the module. This process offers two main advantages:

- **Enhanced Electrical Efficiency:** Cooling the PV cells lowers their operating temperature, which directly improves voltage, power output, and consequently the electrical efficiency of the PV panel (Kazem *et al.*, 2021).
- **Useful Thermal Energy Production:** The absorbed heat is collected as usable thermal energy, which could be used in domestic water heating, space heating, or similar applications.

This dual functionality greatly improves the overall efficiency of PV/T systems, with some research reporting combined efficiencies as high as 87.6% when hybrid nanofluids are used (Meraje *et al.*, 2022). Additionally, PV/T systems contribute to lessening emissions (e.g., $CO_2$) and provide both environmental and economic benefits (Garud and Lee, 2022).

### 3.2 Thermal Management and Heat-Transfer Challenges in PV/T Systems

Heat management plays a critical role in PV/T operation. Since PV cells convert only some part of incident solar radiation into electricity, the remaining portion is transformed into thermal energy (Adun *et al.*, 2022). This heat raises the cell temperature, which negatively impacts voltage and power output, thereby reducing electrical efficiency (Kazem *et al.*, 2021). Effective thermal management is thus essential to maintain system performance and prevent efficiency losses. The main thermal challenges in PV/T systems include:

- **Electrical Efficiency Loss:** Rising temperatures in PV panels lead to a reduction in voltage and power output, resulting in lower electrical efficiency (Kazem *et al.*, 2021).
- **Useful Heat Recovery:** A central challenge lies in effectively transferring excess heat from PV cells and collecting it as useful thermal energy for applications such as space or water heating (Kazem *et al.*, 2021).
- **Properties of the Heat Transfer Fluid (HTF):** Selecting and optimizing an HTF with suitable thermophysical properties (e.g., specific heat capacity and thermal conductivity) is essential for efficient heat absorption and transfer (Kazem *et al.*, 2021).
- **Flow Rate Optimization:** Adjusting the flow rate of HTF is necessary to effectively reduce PV temperature while minimizing pumping energy consumption. Excessively high flow rates can significantly increase pumping power requirements (Sreekumar *et al.*, 2022).
- **Thermal Collector Design:** The optimal design of the thermal collector and cooling channels (e.g., spiral tubes or web-like structures), along with

appropriate material selection, is critical to maximizing heat transfer and ensuring uniform temperature distribution across the PV surface.

- **Trade-off Relationship:** In PV/T systems, a trade-off often exists between thermal and electrical outputs; efforts to maximize one may reduce the other unless system design and operation are carefully managed (Huang *et al.*, 2022).

## 3.3 Photovoltaic/Thermal Systems and the Role of Nanofluids

PV/T systems are designed for the simultaneous production of heat and electricity. The primary motivation for employing nanofluids in PV/T systems is to address the fundamental issue of photovoltaic (PV) cells: as their temperature rises, their electrical efficiency decreases, and their lifespan shortens (Huang *et al.*, 2022, Dashtbozorg *et al.*, 2025). By improving heat transfer properties, nanofluids significantly improve the performance of these systems.

The key advantages of nanofluids can be summarized as follows:

- **Reduction of PV cell temperature and enhancement of electrical efficiency:** Nanofluids increase thermal conductivity and improve both convective and conductive heat transfer, thereby more effectively removing excess heat from PV modules. This temperature reduction directly leads to a significant rise in power output and electrical efficacy (Kazem *et al.*, 2021). For example, in one study, the electrical power of a PV module increased from 55.2 W to 72.28 W in a nanofluid-based PV/T system, resulting in a 20.21% rise in electrical efficiency (Kazem *et al.*, 2021).
- **Improved thermal and overall system efficiency:** By absorbing more thermal energy and increasing the outlet fluid temperature, nanofluids improve the thermal efficiency of PV/T systems. Consequently, the total energy efficiency of the system improves due to the simultaneous enhancement of electrical and thermal efficiencies (Khan *et al.*, 2022).
- **Enhanced exergetic performance and reduced losses:** Owing to their superior energy quality, nanofluids increase the exergy efficiency of PV/T systems. Moreover, they minimize exergy destruction and entropy generation, indicating reduced irreversibilities and higher overall system productivity (Garud and Lee, 2022).
- **Economic and environmental benefits:** From an economic perspective, nanofluid utilization can reduce the levelized cost of energy (LCOE), shorten the energy payback time (EPBT), and increase profitability. From an environmental standpoint, nanofluids contribute to the reduction of carbon dioxide ($CO_2$) emissions and other chemicals (such as SOx, NOx, and particulate matter), thereby improving carbon credits (Garud and Lee, 2022).
- **Spectral beam splitting capability:** Certain nanofluids can function as spectral beam splitters (SBS), selectively filtering solar radiation so that only the wavelengths required for electricity generation reach the PV cells, while the remaining spectrum is absorbed for heat production. This property enables optimized utilization of the entire solar spectrum (Xiao *et al.*, 2021).

- **Flexibility and optimization potential:** Research has demonstrated that various nanofluids—including SiC, $Al_2O_3$, CuO, $Fe_3O_4$, MWCNT, Ag, Au, $TiO_2$, ZnO, and MXene—enhance the function of PV/T systems under different climatic conditions and configurations (e.g., Oman, India, hot climates, or systems integrated with phase change materials). Furthermore, optimization of parameters, including nanoparticle concentration, flow rate, and nanofluid channel thickness, could yield maximum efficiency (Khan *et al.*, 2022).

Given these multiple advantages, nanofluids are recognized as an effective and practical solution for enhancing the efficiency, sustainability, and cost-effectiveness of PV/T systems, particularly under diverse climatic conditions.

## 4. Wind energy

Wind energy is recognized as a strategic and vital resource for decreasing carbon dioxide emissions, combating climate change, and advancing sustainable development (Tafavogh and Zahedi, 2021). As a renewable source, it plays a major role in meeting both present and future energy demands (Behera *et al.*, 2025). Wind turbines produce electricity by harnessing the wind's kinetic energy and converting it into mechanical power. These turbines are usually classified into two main categories: vertical-axis wind turbines (VAWTs) and horizontal-axis wind turbines (HAWTs), each with distinct characteristics (Figure 4 ). For instance, Savonius and Darrieus turbines are well-known examples of VAWTs (Liu, Zhang, *et al.*, 2025).

Given the rising global energy demand alongside the depletion of fossil fuel reserves, wind energy has emerged as a key solution. Forecasts suggest that by 2040, wind power could provide nearly 50% of the world's primary energy, and by 2050, it may contribute up to 70% in reducing greenhouse gas emissions (Behera *et al.*, 2025). Its theoretical potential is estimated to exceed 35 times the current global daily electricity consumption (Liu, Zhang, *et al.*, 2025).

Wind energy adoption is expanding across the globe. Countries such as the United States, China, and South Korea generate a considerable share of their energy from renewable resources, including wind (Behera *et al.*, 2025). China, in particular, leads in diversifying energy production (Behera *et al.*, 2025). Certain regions of Iran, such as Bandar Abbas, also hold significant potential for wind energy development (Tafavogh and Zahedi, 2021). Other areas, including Manjil and Kahak in Iran and several regions in the Netherlands, already utilize wind resources (Liu, Zhang, *et al.*, 2025). A notable example of architectural integration is the Bahrain World Trade Center, which incorporates wind turbines directly into its building design (Liu, Zhang, *et al.*, 2025).

Despite its promise, large-scale deployment of wind power faces challenges, primarily owing to its variable nature and the mismatch between production and demand (Tafavogh and Zahedi, 2021). To address these issues, hybrid renewable energy systems have been proposed, where wind power is combined with other resources such as thermal systems, solar, or energy storage technologies like lithium-ion batteries (Wang *et al.*, 2022). These storage systems, with their rapid

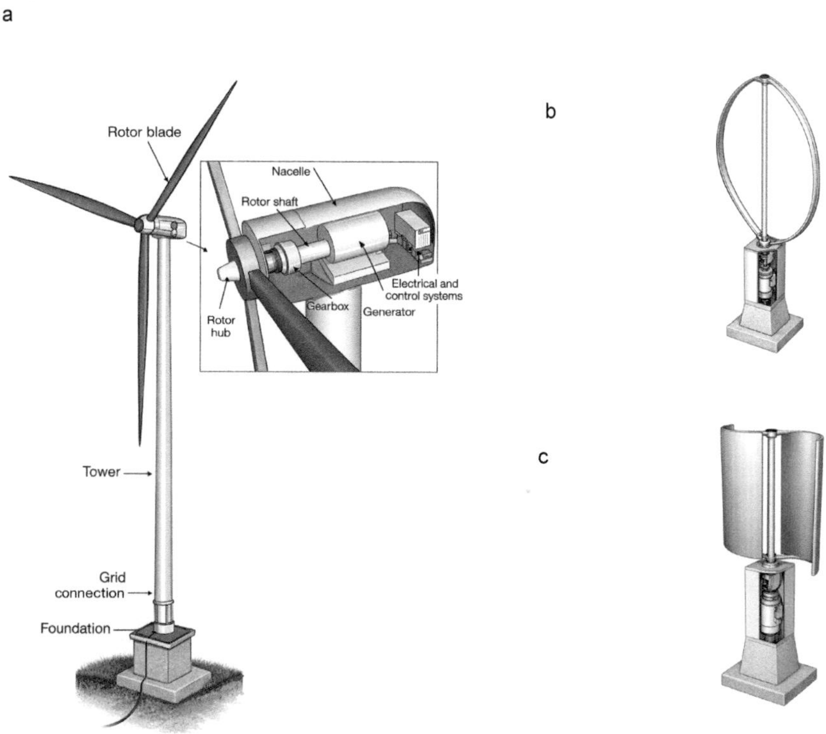


**Fig. 4** (a) A horizontal and (b, c) Darrius and H type vertical axis turbines. Adapted from Beig and Muyeen (2016) with permission.

charge and discharge capabilities, provide valuable grid-balancing functions. Additionally, demand-side management strategies and optimized planning approaches help improve the cost-effectiveness of wind power and mitigate peak load pressures (Wang *et al.*, 2022). Other challenges include noise pollution, visual impact, the so-called "stroboscopic effect" caused by turbine blade shadows, and risks to birds and bats. Mitigation strategies involve improved turbine design, wider deployment of VAWTs, and exploration of bladeless wind turbine (BWT) concepts (Liu, Zhang, *et al.*, 2025).

### *4.1 Cooling and Heat Transfer Challenges in Wind Turbines*

The extraction of energy from wind turbines is accompanied by significant challenges in terms of cooling and heat transfer, which directly affect performance, reliability, and overall system lifespan (Rostami *et al.*, 2021, Behera *et al.*, 2025). Among the most critical of these challenges is heat generation and dissipation. Wind turbine generators and gearboxes produce considerable amounts of heat during operation. Managing this heat is essential, as insufficient dissipation can lead to overheating of both mechanical and electrical components, causing severe operational issues (Ahsan *et al.*, 2025, Behera *et al.*, 2025).

Inadequate thermal control and poor lubrication are major causes of gearbox failures. Excessive heating reduces the service life of gearbox components and prevents consistent and durable performance (Ahsan *et al.*, 2025). Improved thermal management can mitigate stresses on gears and bearings, minimizing misalignments and preventing premature failures (Ahsan *et al.*, 2025).

The fire of wind turbines can occur owing to inefficient thermal management, primarily due to the overheating of electrical and mechanical components such as generators, converters, and braking systems, as shown in Figure 5.

**Fig. 5** WTs' Thermal runaway examples. Adapted from Ji *et al.*, (2025) with permission.

Although the importance of thermal control is well established, the effectiveness of existing cooling systems often remains limited. Early wind turbines, with lower capacity, generated less heat that could be dissipated using natural air cooling (Rostami *et al.*, 2021). However, with the continuous increase in turbine size and power, the demand for more advanced cooling systems has grown. Liquid cooling with forced circulation is now widely applied in medium- and large-scale turbines, yet it may still fall short under certain conditions, such as high wind speeds, necessitating auxiliary ventilation. Furthermore, the compact design of the nacelle reduces the available surface area for effective heat transfer, adding another layer of complexity (Rostami *et al.*, 2021).

Closely linked to cooling performance is the challenge of lubrication. Friction between gearbox gears is a major source of both energy loss and heat generation. Inadequate lubrication can result in gear failure and reduced system performance (Figure 6). High-quality lubricants with strong heat transfer capabilities and appropriate viscosity are critical for preventing overheating, reducing friction, and ensuring component durability. Improper lubrication systems can also increase the turbine's electrical energy consumption (Ahsan *et al.*, 2025).

**Fig. 6** Heat-adhesion damage visible on bearing elements (arrow). Adapted from Bejger *et al.*, (2021) with permission.

Even when excess heat is removed from wind turbine components, it is often wasted in the environment. Repurposing this thermal energy for uses such as desalination or cooling could improve the overall efficiency of wind power systems (Rostami *et al.*, 2021).

Finally, the importance of these issues is magnified under harsh environmental conditions. In hot and humid regions, effective cooling strategies are even more vital to ensure the reliable and long-term operation of gearboxes and generators. Adverse weather conditions, including high temperatures and moisture, exacerbate thermal challenges and demand advanced solutions (Rostami *et al.*, 2021).

### 4.2 Nanofluids in Wind Energy Systems

Building on their demonstrated benefits preceding sections, nanofluids are also gaining attention in the field of wind energy. Wind turbines, especially large-scale modern designs, face persistent challenges related to heat generation in generators and gearboxes, lubrication losses, and the dissipation of unused thermal energy. Conventional cooling and lubrication strategies are often insufficient under demanding operating conditions. In this context, nanofluids offer a promising pathway to enhance turbine performance, improve durability, and enable innovative energy recovery solutions.

### *4.2.1 Enhancement of Heat Transfer and Cooling Efficiency*

A key challenge in wind turbines is the management of heat generated by gearboxes and generators. Nanofluids, owing to their higher thermal properties, provide multiple advantages in this regard:

- **Improved Heat Dissipation:** The effective operation of wind turbine generators and gearboxes depends on the efficient dissipation of generated heat. Nanofluids, which contain nanoparticles dispersed in a base fluid, significantly improve the thermal properties of the base fluid (Ali *et al.*, 2025). This leads to higher thermal energy exchange rates and prevents overheating of critical turbine components .
- **Increased Reliability and Durability:** Gearboxes and generators require efficient cooling to ensure their reliability and longevity, especially in hot and humid climates or under high-wind conditions where heat generation is intensified. By preventing excessive temperature rise, nanofluids extend component lifespan and reduce the likelihood of premature failures (Ahsan *et al.*, 2025).
- **Reduction of Cooling System Size:** The high heat transfer efficiency of nanofluids can decrease the surface area needed in heat exchangers. This feature is particularly valuable for the limited space available inside a wind turbine nacelle (Rostami *et al.*, 2021, Behera *et al.*, 2025).

### *4.2.2 Improvement of Lubrication Properties (Nano-lubricants)*

Beyond cooling, lubrication is another critical factor affecting turbine efficiency and reliability. Incorporating nanoparticles into lubricants creates nano-lubricants that provide the following benefits (Ahsan *et al.*, 2025):

- **Reduction of Friction and Wear:** Nano-lubricants, such as graphene oxide nanoparticles dispersed in PAO-40 base oil, can significantly reduce friction and wear in gearbox components. This reduction lowers energy loss and decreases heat generation.
- **Enhanced Viscosity and Stability:** The dispersion of nanoparticles like graphene oxide can increase the viscosity of the fluid, which is crucial for effective lubrication and for preventing leakage in gearboxes. This contributes to stable operation under varying environmental conditions.
- **Extended Component Lifespan:** Lower friction and wear result in longer gearbox component lifetimes, reduced maintenance costs, and decreased turbine downtime.

### *4.2.3 Waste Heat Recovery and Energy Storage Applications*

In addition to improving cooling and lubrication, nanofluids provide opportunities for energy recycling within wind systems. They can be employed to extract waste heat from wind turbine generators for secondary uses such as cogeneration of fresh water and cooling. This strategy offers an energy-efficient alternative to conventional cooling systems (Rostami *et al.*, 2021).

Nanofluids also extend their benefits to energy storage subsystems connected to wind power plants. For instance, $Al_2O_3$/water nanofluids have proven highly effective in cooling lithium-ion batteries used for storing excess wind-generated electricity. Compared with mineral oils, these nanofluids exhibit superior thermal management performance, limiting the temperature rise of battery modules during high discharge rates (such as 3C and 5C) to less than 5°C (Ajay *et al.*, 2025). This dual role in waste heat recovery and efficient energy storage highlights the broader potential of nanofluids in enabling more sustainable and integrated wind energy systems.

## 5. Geothermal Energy: Status, Importance, Opportunities, and Challenges

Geothermal energy is recognized as a strategic and vital resource for decreasing carbon dioxide emissions, combating climate change, and advancing sustainable development (Liu, Lv, *et al.*, 2025a). Originating from the Earth's internal heat, it is both abundant and widespread. Due to its continuous availability, geothermal energy is vital for the transition from fossil-fuel–based development toward more sustainable pathways (Xu *et al.*, 2024, Liu, Lv, *et al.*, 2025b). As rock temperatures rise steadily with depth and approach melting points in the upper mantle, this heat—produced and retained throughout Earth's lifetime—represents, on a human scale, an essentially inexhaustible energy source (Alizadeh *et al.*, 2025). Geothermal electricity generation is also characterized by very low carbon emissions 6–79 g per kWh, Chappidi *et al.*, (2023), far below those of fossil fuel power plants.

Alongside other renewables such as wind, solar, hydro, and biomass, geothermal energy stands out due to its continuity and independence from weather conditions (Chappidi *et al.*, 2023). This reliability makes it more dependable than intermittent sources like wind and solar. Geothermal heat pump (GSHP) systems represent the most widespread application of geothermal energy worldwide, representing about 71.6% of installed capacity (Liu, Lv *et al.*, 2025b). Countries such as China and South Korea generate nearly 40% of their renewable energy—including geothermal—while the United States derives around 20% of its total energy supply from renewables, geothermal included (Behera *et al.*, 2025).

The development of geothermal energy provides numerous opportunities in a broad spectrum of applications:

- **Geothermal Heat Exchangers (GHE) and Geothermal Heat Pumps (GCHP):** These systems are popular for climate control in buildings because they leverage the ground's stable temperatures, which are warmer than the winter air and cooler than the summer air (Liu, Lv, *et al.*, 2025b). Applications include domestic hot water, space heating/cooling, industrial processes, greenhouse heating, aquaculture, snow-melting systems, and thermal energy storage (Alizadeh *et al.*, 2025).

- **Deep Geothermal Resources and Hot Dry Rock (HDR):** The term Hot Dry Rock describes deep-seated rock (between 3 and 10 kilometers underground) that is hot but has an absence of natural fluids (Xu *et al.*, 2024). Enhanced Geothermal Systems extract this heat by creating fracture networks (hydraulic or acid stimulation) (Xu *et al.*, 2024). This technology significantly expands the geographic reach of geothermal exploitation.
- **Abandoned Oil and Gas Wells:** Millions of decommissioned wells worldwide can be repurposed as geothermal wells, simultaneously reducing environmental risks and lowering drilling costs (Chappidi *et al.*, 2023). Coaxial well heat exchangers and U-tube heat exchangers are commonly used for this purpose, enabling either power generation or direct thermal applications (Chappidi *et al.*, 2023).
- **Multi-Generation Systems:** Geothermal energy can be integrated with technologies such as organic Rankine cycle (ORC) units, PV/T systems, reverse osmosis (RO) desalination plants, and alkaline electrolyzers, for combined generation of electricity, hydrogen, and fresh water (Mirshafiee *et al.*, 2025). In such systems, geothermal water simultaneously supplies thermal energy for ORC and saline feedwater for RO (Mirshafiee *et al.*, 2025). Moreover, the brine discharged from RO can be reinjected into geothermal wells for enhanced oil recovery (EOR).

Despite its vast potential, geothermal energy development faces several significant technical challenges:

- **Limited Heat Transfer Capacity in Conventional Systems:** Traditional geothermal heat exchangers and energy piles often suffer from low heat transfer efficiency (Liu, Lv, *et al.*, 2025b).
- **Low Thermal Conductivity of Conventional Working Fluids:** Common working fluids such as ethylene glycol or water exhibit weak thermal conductivity, which restricts overall system efficiency (Mirshafiee *et al.*, 2025).
- **Thermal Accumulation and Imbalance in the Ground:** GHE systems may cause long-term thermal buildup in the soil, leading to performance degradation and thermal imbalance (Liu, Lv, *et al.*, 2025b).
- **High Drilling Costs:** Drilling geothermal wells—particularly for deep resources—requires substantial upfront investment.
- **Thermal Limitations in Deep Reservoirs:** In HDR reservoirs, induced fracture channels can reduce effective heat transfer surface area, leading to "cold water short-circuiting" and significantly lowering heat extraction efficiency (Xu *et al.*, 2024). The Hijiori project in Japan was halted for this reason (Xu *et al.*, 2024). Furthermore, the low permeability and porosity of deep reservoirs hinder thermal energy exchange (Xu *et al.*, 2024).

- **Fluid Management Challenges:** Geothermal fluids often encompass varying chemical temperatures and compositions, resulting in corrosion, scaling, and mineral deposition in infrastructure. These issues reduce heat transfer efficiency and may damage systems (Behera *et al.*, 2025).
- **Increased Pumping Costs:** Advanced fluids with higher viscosity can impose additional pumping energy demands (Pisarevsky *et al.*, 2024).

### *5.1 Role of Nanofluids in Geothermal Energy Systems*

Nanofluids have also emerged as promising working media for geothermal systems. Their unique thermophysical properties provide effective solutions to many of the limitations faced by conventional heat transfer fluids, enabling higher efficiency, smaller system size, and improved economic viability. The improved thermal energy exchange capability of nanofluids enables the same thermal energy transfer to be achieved with smaller systems and shorter drilling depths. This reduction in physical scale offers direct economic advantages:

- **Smaller and More Compact Systems:** By achieving higher heat transfer rates, geothermal boreholes and heat exchangers can be designed with reduced dimensions while still meeting thermal demand (Liu, Lv, *et al.*, 2025b).
- **Lower Installation and Maintenance Costs:** The downsizing of system components translates into significant savings in capital expenditure and long-term operational costs (Liu, Lv, *et al.*, 2025b).

## 6. Integration of Thermal Energy Storage in Renewable Systems

Thermal energy storage (TES) has become a key enabling technology with immense potential for renewable energy applications. By storing thermal energy in various forms and releasing it when required, TES systems ensure a stable and reliable energy supply for heating, cooling, industrial processes, and even electricity generation (Bahrami and Ghaedi, 2025a). The central objective of TES is to augment energy efficiency and system reliability by capturing surplus heat from intermittent renewable resources. This capability is particularly valuable in solar-based systems, where energy storage enables continuous operation throughout nighttime and low-light conditions (Ren *et al.*, 2025). Beyond solar applications, TES also plays a critical role in regulating the thermal behavior of lithium-ion batteries, photovoltaic modules, and electronic devices, where precise temperature control is essential for safety, performance, and longevity (Bahrami and Ghaedi, 2025b).

The importance of TES extends further, as it serves a pivotal role in the worldwide energy shift to cleaner energy. As a highly promising technology, TES can be implemented at large technical and economic scales, offering long-term solutions for the widespread deployment of renewable energy (Jing *et al.*, 2025). By mitigating the inherent variability of solar and other intermittent renewable resources, TES ensures a continuous and reliable energy supply by storing surplus energy during peak production and delivering it when demand exceeds availability (Jing *et al.*, 2025).

TES offers several distinctive advantages:

- **High energy storage density,** which is crucial for maximizing storage capacity while minimizing the spatial footprint (Chen *et al.*, 2025).
- **Environmental and energy-saving benefits,** through recycling of waste heat and reduction in primary energy consumption (Venkatesh, Sharma, *et al.*, 2025).
- **Improved efficiency of renewable plants,** particularly in concentrating solar power (CSP) systems, where TES minimizes heat losses and increases overall system conversion efficiency (Venkatesh, Sharma, *et al.*, 2025).
- **Peak load management,** by shifting energy consumption from high-demand peak hours to low-demand off-peak hours, TES reduces stress on power grids and narrows the disparity between energy demand and supply (Srinivasan and Ponnusamy, 2025).
- **Enhanced reliability,** where the addition of TES into heat exchangers or hybrid systems ensures continuous and stable thermal output. The employment of phase change materials (PCMs) further strengthens this role by allowing significant latent heat to be stored and delivered over narrow temperature ranges, thereby extending system operating hours and reducing the intermittency of renewable sources (Ren *et al.*, 2025, Venkatesh, Sharma, *et al.*, 2025).

The range of TES applications is both diverse and far-reaching. In solar energy, TES is used in FPCs and DASCs to stabilize energy delivery and extend operation beyond daylight hours (Ren *et al.*, 2025, Venkatesh, Sharma, *et al.*, 2025). In the building sector, TES enhances isolation and decreases energy demand for cooling and heating (Ren *et al.*, 2025). In industrial processes, TES captures excess process heat, improving recovery efficiency and reducing waste. TES also serves a vital function in battery thermal management, where it prevents thermal runaway and extends system lifespan (Bahrami and Ghaedi, 2025b). In addition, TES contributes to seawater desalination efficiency, thereby supporting water security in arid regions. In HVAC systems and electronics cooling, TES enables effective temperature regulation while minimizing energy use (Venkateswarlu *et al.*, 2025).

TES also helps to overcome the inherent limitations of conventional solar collectors. For example, flat plate collectors often experience reduced functionality at elevated operating temperatures and substantial heat losses in cloudy or cold conditions (Venkatesh, Sharma, *et al.*, 2025). By integrating TES, these shortcomings can be mitigated through the storage of surplus heat, which can later be utilized to ensure continuous and reliable energy delivery.

In terms of long-duration storage, TES offers innovative solutions such as salinity-gradient solar ponds (SGSPs) (Bretado-de Los Rios *et al.*, 2025). These systems represent a simple, cost-effective, and robust method for gathering and storing solar energy as thermal energy over extended periods. The stored energy can then be used for industrial heating, commercial operations, desalination, or even power generation, highlighting the broad flexibility and reliability of TES technologies (Bretado-de Los Rios *et al.*, 2025).

Broadly, TES technologies can be classified into three major categories (Thakur *et al.*, 2021):

- **Sensible Heat Storage (SHS)** These systems operate on the principle of storing energy through a temperature change in the storage medium, while its phase remains constant. Typical media include molten salts, oils, and water. However, their relatively low thermal conductivity limits heat transfer rates, prompting research into improved materials, additives, and system designs to accelerate charging and discharging processes.
- **Latent Heat Storage (LHS)** This method uses PCMs, which absorb or deliver huge amounts of energy during phase change, such as solidification and melting. Common PCMs include paraffins, fatty acids, and hydrated salts (e.g., sodium acetate trihydrate, erythritol). LHS provides higher energy density than SHS but suffers from challenges such as supercooling and low thermal conductivity. Solutions include embedding high-conductivity materials, incorporating porous foams, and applying surface modifications to improve thermal transport and reduce phase-change delays.
- **Thermochemical Storage (TCS)** TCS systems store energy through reversible chemical reactions, capturing energy in chemical bonds and discharging it upon reaction reversal. This approach possesses a high specific energy and excellent capacity for extended-duration storage with minimal thermal losses. Nevertheless, slow reaction kinetics, material degradation, and high costs remain major barriers. Current research is investigating catalysts, advanced reactive materials, and novel reactor designs to navigate these obstacles.

Finally, the design and optimization of TES rely heavily on both experimental studies and advanced numerical modeling. MATLAB-based solvers, finite element simulations, and analytical methods are widely employed to predict system performance (Bahrami, Allahdadi, and Ghaedi, 2025, Bahrami and Ghaedi 2025a). Furthermore, optimization techniques such as grey relational analysis (GRA) and response surface methodology (RSM) are increasingly used to identify optimal design and operational parameters (Liu and Yang, 2022, Çinici *et al.*, 2022). Collectively, these approaches reflect the ongoing pursuit of TES systems that possess improved efficiency, stability, and affordability, thereby accelerating the integration of renewable energy into modern power systems.

### *6.1 Role of Nanofluids in Improving Thermal Reservoir Performance*

One of the fundamental challenges in advanced thermal systems, including PCMs and conventional heat transfer fluids used for energy storing and thermal management, lies in their inherently poor thermophysical properties—particularly their low thermal conductivity (Bahrami and Ghaedi, 2025c). This limitation significantly weakens their capacity to absorb, store, and exchange thermal energy efficiently, which subsequently restricts the overall performance of energy storage and power generation technologies, as well as electronic cooling systems (Bahrami and Ghaedi, 2025d). Conventional fluids and PCMs often fail to meet the growing demand for efficient thermal regulation, resulting in reduced energy output and suboptimal system performance.

The emergence of nanofluids presents a viable pathway to mitigate these issues. By dispersing nanoparticles into base fluids or PCMs, nanofluids dramatically enhance the thermophysical properties of the host medium (Bahrami, Allahdadi, and Saffari, 2025). The most significant improvements include increased thermal conductivity, enhanced thermal stability, and improved energy storage density. These advancements are particularly vital for thermal reservoirs, where rapid and efficient heat transfer is crucial, positioning nanofluids as a next-generation strategy for advanced energy systems (Bahrami *et al.*, 2024).

As mentioned, nanoparticles enhance thermal conductivity in base fluids or PCMs, accelerating heat transfer and reducing reservoir charging/discharging time. For example, in nanoparticle-enhanced PCMs, thermal conductivity has been reported to increase by up to 18.96% in the liquid phase and 31.46% in the solid phase (Srinivasan and Ponnusamy, 2025). Furthermore, In TES systems, nanoparticles can increase the effective heat storage density of PCMs. For instance, the addition of 5% copper nanoparticles to water has been shown to improve its heat storage density by nearly 27.63% in comparison to pure water, thereby enhancing the energy storage capacity of the system (Atofarati *et al.*, 2024).

In summary, nanofluids not only address the intrinsic thermal limitations of traditional PCMs and thermal transport media but also provide engineered enhancements in thermophysical properties. As a result, they offer a transformative approach to improving the efficiency, sustainability, and reliability of thermal reservoirs across a wide range of industrial applications.

## 7. Environmental and Economic Considerations of Using Nanofluids

Following the discussion on the technical advantages of nanofluids in improving thermal conductivity, energy storage density, and system performance, it is equally important to address their implications. The adoption of nanofluids in thermal storage and management systems is not solely a matter of engineering efficiency; it also entails environmental and economic considerations. These aspects ultimately determine whether nanofluid technologies will be sustainable, practical, and embraced by society.

## 7.1 Environmental Considerations

Nanofluids present a dual environmental impact. On one hand, they provide substantial benefits include reduced emissions and superior efficiency; on the other hand, they raise concerns related to nanoparticle toxicity, disposal, and long-term ecological risks.

### 7.1.1 Reduction of Greenhouse Gas Emissions

Integration of nanofluids into PV/T systems and solar collectors has demonstrated measurable reductions in carbon dioxide ($CO_2$) emissions. For example, A PV/T system using Ag/water nanofluid reduced annual emissions by 16,974.57 tons of $CO_2$ (Lari and Sahin, 2017), while another configuration combining Ag/water with a PCM thermal battery achieved a reduction of 10,195.91 tons annually (Lari and Sahin, 2018). In other studies, ZnO, $TiO_2$, and $Al_2O_3$ nanofluids in PV/T systems achieved emission reductions of 853.8 kg, 831.2 kg, and 717.9 kg $CO_2$, respectively (Sohani *et al.*, 2021). In some other studies, it has been shown that MXene-based nanofluids in parabolic trough solar collectors (PTSCs) prevented between 2.25 and 2.3 tons of $CO_2$ annually (Hota *et al.*, 2022), and in flat-plate solar collectors (FPSC), $CeO_2$/water nanofluids reduced $CO_2$ emissions by 175 kg, while glycol-based nanofluids reduced emissions by 31.26–39.28 kg per day (Michael Joseph Stalin *et al.*, 2020, AbdEl-Rady Abu-Zeid *et al.*, 2024).

These results demonstrate that widespread adoption of nanofluid-enhanced systems could significantly lower the carbon footprint of renewable energy technologies.

### 7.1.2 Potential Toxicity and Environmental Risks

The environmental risks of nanofluids depend heavily on the type, size, and surface properties of the particles used, along with their synthesis method.

Concerns include:

- **Aquatic toxicity:** nanoparticles such as Cu and ZnO can accumulate in water bodies, threatening aquatic ecosystems and entering food chains (Behera *et al.*, 2025).
- **Persistence in the environment:** Cu nanoparticles, in particular, are difficult to remove once released, while alternatives like CuO, $ZrO_2$, and SiC are less toxic and considered more eco-friendly (Behera *et al.*, 2025).
- **Human health risks:** inhalation or ingestion of certain nanoparticles may cause respiratory or biological harm (Behera *et al.*, 2025).

To mitigate these risks, eco-friendly synthesis methods, safe disposal protocols, and environmentally benign nanoparticles should be prioritized.

### 7.1.3 Lifecycle Sustainability

While some nanofluid systems increase environmental impacts during manufacturing, transportation, or disposal, lifecycle analyses generally show that the total $CO_2$ savings during operation outweigh the emissions generated during production. Nevertheless, unresolved challenges in nanofluid stability and disposal necessitate more research (Behera *et al.*, 2025).

## 7.2 Economic Considerations

From an economic perspective, nanofluids simultaneously present opportunities for significant cost savings and challenges associated with high initial investments. Their economic feasibility depends on balancing short-term expenses with long-term operational gains.

### 7.2.1 Cost Reduction and Energy Savings

Nanofluids improve thermal performance in renewable systems, lowering costs, shortening payback periods, and boosting long-term profitability, as shown in experimental and techno-economic studies. Various studies have quantified these benefits, as summarized below.

- Nanofluid-enhanced systems can reduce household electricity costs by up to 82%, with a payback period of only 2 years in PV/T applications using Ag/water nanofluids (Adun *et al.*, 2022).
- In PV/T systems with PCM, Ag/water nanofluids achieved a 58.71% reduction in energy costs, with a payback time of 5.5 years.
- In FPSCs using GAMWCNT nanofluids, the payback time was 1.897 years, shorter than with conventional fluids, while also reducing embodied energy by 321.72 MJ (Amar *et al.*, 2023).
- ZnO, $TiO_2$, and $Al_2O_3$ nanofluids in PV/T systems saved \$24–27.8 per $m^2$ annually, with payback periods ranging from 2.5 to 7.85 years (Hamzat *et al.*, 2022).

These figures highlight nanofluids' potential to make renewable energy technologies not only greener but also more cost-competitive.

### 7.2.2 Economic Analysis Parameters

To evaluate the financial feasibility of nanofluid-based renewable systems, economic metrics and analytical approaches are used. These help compare nanofluid technologies with conventional ones and guide investment decisions. Key indicators include capital cost, payback period, and maintenance, along with exergoeconomic and enviro-economic assessments linking thermodynamic performance with financial and environmental impacts. A full cost assessment requires consideration of:

- **Initial investment costs (materials, synthesis, transport, and system integration)** (Hamzat *et al.*, 2022).

- **Operation and maintenance (O&M) costs**, typically ~2% of the initial capital per year (Said *et al.*, 2021).
- **Payback period (PP, SPP, CPBT) and Levelized Cost of Energy (LCOE)** (Hamzat *et al.*, 2022).
- **Exergoeconomic and exergoenvironmental analyses**, which combine thermodynamic efficiency with economic and environmental metrics, factoring in global carbon prices (Said *et al.*, 2021).

Advanced approaches such as machine learning (ANN, fuzzy systems) are increasingly being used to optimize system design, predict costs, and improve reliability (Chekifi *et al.*, 2024).

### *7.2.3 Economic Challenges*

Despite their potential, nanofluids face economic barriers that hinder widespread adoption. High initial costs, the need for stronger pumping power due to increased viscosity, and long payback periods—especially when costly nanoparticles or complex components are used—are frequently reported. Moreover, issues of stability, synthesis complexity, and maintenance add uncertainty to cost projections, underscoring the need for continued research to improve affordability and long-term viability.

Some potential drawbacks include deposition, erosion, stacking, flaking, molecule agglomeration, and chemical stability issues, which can increase maintenance costs (Mahamude *et al.*, 2021, Said *et al.*, 2021, Jayanthi *et al.*, 2024, Deshmukh *et al.*, 2025). Nevertheless, as nanofluid production scales up and synthesis becomes cheaper, costs are expected to decline, improving economic viability (Hamzat *et al.*, 2022).

### *7.2.4 Overall Assessment*

Taken together, the environmental and economic dimensions of nanofluid use reveal both great promise and important challenges. On the environmental side, nanofluids contribute significantly to $CO_2$ mitigation but require careful management of nanoparticle toxicity and lifecycle sustainability. On the economic side, nanofluids can enable cost savings and competitive payback periods, but high initial costs and system integration issues remain obstacles. Therefore, the sustainable deployment of nanofluids in thermal energy systems will depend on taking steps like developing green synthesis methods and safe disposal protocols (Said *et al.*, 2021, Hamzat *et al.*, 2022, Amar *et al.*, 2023), lowering production costs through scaling and material optimization (Hamzat *et al.*, 2022), and balancing environmental gains with economic feasibility across the lifecycle (Feng *et al.*, 2025).

With further research and technological advances, nanofluids have the potential to be a feasible solution in making renewable energy systems both environmentally sustainable and economically viable.

## 8. Technical Challenges and Barriers to Adoption

Despite their demonstrated potential in enhancing thermal energy exchange and augmenting the efficiency of renewable energy systems, nanofluids still face different technical obstacles that hamper their large-scale adoption. These challenges are not confined to one domain but spread across stability, economics, performance, preparation, modeling, and safety. Understanding these limitations is essential for guiding future research and for designing realistic pathways toward commercialization.

### *8.1 Stability and Dispersion Challenges*

One of the most fundamental issues limiting nanofluid applications is their long-term stability. For effective heat transfer, nanoparticles must remain well-dispersed in the base fluid over extended periods and across varying operating conditions. However, maintaining this dispersion is challenging, as nanoparticles tend to agglomerate and sediment, leading to reduced thermal conductivity and heat transfer efficiency (Hamzat *et al.*, 2022). Instability becomes more pronounced at elevated temperatures, while non-uniform dispersion can even alter the optical properties of nanofluids and undermine solar energy absorption (Hamzat *et al.*, 2022). In some cases, such as with multi-walled carbon nanotubes (MWCNTs), surfactants or dispersants are required to stabilize the suspension, but these additives may cause drawbacks like foaming or reduced conductivity (Kumar and Hassan, 2024).

Thus, stability remains one of the most pressing technical barriers, and progress depends on developing advanced synthesis and dispersion techniques.

### *8.2 Performance and Operational Issues*

From a technical standpoint, even well-prepared nanofluids may encounter challenges during operation. The addition of nanoparticles modifies flow and thermal properties in ways that are not always beneficial. Increased viscosity frequently raises pumping power requirements, which reduces overall efficiency, while at high particle concentrations, the expected improvements in thermal conductivity may plateau or even decline (Hamzat *et al.*, 2022). The use of surfactants, although effective in improving dispersion, can lead to foaming and impair heat transfer, and agglomeration or sedimentation may cause erosion, clogging, or fouling in channels and pipes (Deshmukh *et al.*, 2025). Furthermore, many nanofluids display non-Newtonian behavior, complicating both predictive modeling and system design (Feng *et al.*, 2025). These issues highlight the necessity of carefully optimizing nanoparticle size, concentration, and additives to achieve a balance between enhanced thermal performance and potential operational penalties.

## 8.3 Preparation and Characterization Constraints

The preparation of nanofluids itself presents substantial challenges, as no universal methodology currently exists for producing stable and reproducible formulations, which undermines both scientific comparability and industrial scalability. Different preparation methods often yield inconsistent results, making cross-study comparisons unreliable, while single-step synthesis approaches, though sometimes offering greater stability, are not economically practical for large-scale production (Hamzat *et al.*, 2022). Furthermore, the accurate measurement of thermophysical properties under realistic operating conditions remains difficult and prone to error (Said *et al.*, 2021). For advanced materials such as graphene-based nanofluids, additional complications arise from van der Waals interactions, acid-treatment modifications, and the need for specialized dispersion protocols (Joseph and Thomas, 2022). This lack of standardization and reproducibility continues to hinder scientific progress and erodes the confidence of industrial stakeholders.

## 8.4 Modeling and Research Gaps

Reliable predictive models are essential for designing and optimizing nanofluid-based systems. However, the current state of modeling is fragmented and often inconsistent with experimental findings.

- Numerous models exist for predicting viscosity and thermal conductivity, but no universal correlation has emerged (Said *et al.*, 2021).
- Discrepancies between theoretical and experimental outcomes are common, largely due to the complex physics of nanoparticle interactions and abnormal transport behavior (Feng *et al.*, 2025).
- Advanced models, such as non-local or fractional formulations, attempt to address these complexities but remain computationally demanding and difficult to apply in practice (Feng *et al.*, 2025).
- Critical areas such as erosion, corrosion, and high-pressure performance have not been sufficiently studied (Said *et al.*, 2021, Hamzat *et al.*, 2022).

Addressing these gaps will require integrated experimental–theoretical approaches and the development of broadly applicable correlations.

## 8.5 Safety and Environmental Concerns

Beyond technical and economic barriers, nanofluid deployment also raises important safety and environmental considerations. Specifically, many nanoparticles, if released, pose toxicity risks to aquatic ecosystems and potentially to human health through bioaccumulation (Said *et al.*, 2021, Hamzat *et al.*, 2022, Feng *et al.*, 2025). In addition, laboratory workers and technicians face risks of inhalation or dermal exposure during preparation and handling (Hamzat *et al.*, 2022). Compounding these concerns, end-of-life disposal and recycling strategies for nanofluids are not yet established, creating concerns about their environmental footprint (Said *et al.*,

2021). While some nanofluids (e.g., $ZrO_2$, SiC) are relatively benign, others (e.g., CuO-based fluids) present persistent contamination hazards (Ajeena *et al.*, 2023). Therefore, mitigating these risks will require eco-friendly synthesis methods, clear disposal guidelines, and stricter safety protocols (Amar *et al.*, 2023).

In conclusion, the challenges facing nanofluids are multifaceted, spanning scientific, economic, operational, and environmental domains. Each barrier underscores the need for interdisciplinary research that combines advances in material science, fluid mechanics, system design, and sustainability assessment. Only by systematically addressing these issues can nanofluids move from promising laboratory concepts to robust, cost-effective tools in renewable energy systems.

## 9. Future Trends and Research Directions

The future of nanofluids in thermal systems, mainly in renewable energy systems, depends on addressing current challenges while exploring innovative research avenues. Promising trends focus on improving stability, developing advanced nanofluid formulations, enhancing modeling and optimization methods, expanding applications, and ensuring environmental and economic feasibility.

- **Stability and Long-Term Performance**

Ensuring stable thermal and structural properties over extended operational periods remains a priority. Future work will explore eco-friendly synthesis methods, the role of surfactants and particle morphology, and the use of high-boiling-point base fluids. Long-term testing under cyclic heating/cooling conditions will be essential for commercialization.

- **Development of Novel Nanofluids**

Hybrid nanofluids, plasmonic nanofluids, graphene-based fluids, CNTs, and MXene-based formulations are emerging as advanced candidates. These materials can deliver enhanced thermal conductivity, better photothermal conversion, and tailored optical properties. Continued research in this direction could revolutionize solar thermal and hybrid energy harvesting technologies. However, advanced materials alone are not sufficient without equally robust modeling and optimization frameworks.

- **Advanced Modeling and Optimization**

Future studies will rely heavily on advanced computational tools, including CFD, multiphase modeling, and fractional calculus approaches, to capture the complex behavior of nanofluids. Incorporation of machine learning and artificial intelligence can accelerate property prediction, system optimization, and experimental design, bridging the gap between laboratory-scale testing and practical deployment. As modeling capabilities evolve, application driven studies must also expand to validate these innovations.

- **Expansion of Applications**

Nanofluids offer promising opportunities across a broad spectrum of renewable energy systems, extending well beyond solar applications. In solar technologies, e.g., FPC and evacuated tube collectors, PV/T modules, parabolic troughs, desalination units, and large-scale thermal energy storage, nanofluids can enhance optical absorption and heat transfer, accelerating progress toward commercial readiness. In wind power, they can improve cooling and lubrication in generators, gearboxes, and power electronics, thereby extending component life, particularly under offshore and high-load conditions. For thermal energy storage, nanofluids combined with PCMs improve melting/solidification rates and energy density, supporting applications in district heating, cooling, and solar power plants with molten salt systems. Moreover, in hybrid and multi-generation systems—such as combined cooling, heating, power, and desalination—nanofluids enable higher exergy efficiency and reduced emissions through better resource integration. Moving forward, scaling up these applications with pilot-scale validation, techno-economic analysis, and lifecycle assessment will be essential for translating nanofluid technologies from laboratory research to reliable industrial solutions. The success of these applications will ultimately depend on balancing technical progress with environmental and economic sustainability.

- **Environmental and Economic Considerations**

Research must address toxicity, recyclability, and corrosion risks while ensuring affordability. The priority should be on environmentally friendly, low-cost, and reusable nanofluids that maintain performance without compromising safety. Economic assessments, including EPBT (Energy Payback Time) and LCOE (Levelized Cost of Energy) will guide large-scale adoption.

In addition to these considerations, specialized areas such as optical properties, heat transfer mechanisms, and emerging solar devices also present promising research frontiers.

- **Other Research Directions**

In addition to mainstream applications, nanofluids open pathways toward novel and unconventional energy technologies that warrant systematic investigation. From a materials perspective, future studies should focus on tailoring optical and radiative properties through advanced nanostructures such as plasmonic particles, MXenes, or quantum dots, enabling selective spectral absorption and enhanced photothermal conversion. From a methodological standpoint, coupling experimental diagnostics with computational tools—such as machine learning, multiphase CFD, and fractional calculus models—can accelerate the optimization of nanofluid formulations and reveal governing heat transfer mechanisms beyond classical correlations.

## 10. Conclusion

Nanofluids have emerged as one of the most dynamic and versatile solutions for advancing renewable energy systems. By modifying fundamental thermophysical properties, they offer pathways to overcome long-standing barriers in heat transfer, energy storage, and system efficiency. Across solar thermal collectors, hybrid PV/T modules, wind turbine cooling units, and geothermal or thermal energy storage applications, nanofluids demonstrate the potential to enhance both performance and sustainability.

This chapter has outlined not only the mechanisms by which nanofluids operate—through thermal conductivity enhancement, optical tailoring, and improved phase-change dynamics—but also the practical challenges that must be addressed. Issues of stability, viscosity, pumping power, corrosion, and environmental safety remain central obstacles, underscoring the requires a partnership between materials science, thermal engineering, and environmental science.

The environmental and economic considerations discussed highlight a dual reality: while nanofluids can decrease emissions and reduce long-term energy costs, high initial expenses, limited scalability, and potential ecological risks must be carefully weighed. Emerging strategies such as eco-friendly synthesis methods, advanced modeling tools, and hybrid nanostructures offer promising solutions to bridge these gaps.

Looking forward, nanofluids are positioned not merely as incremental enhancers of existing systems, but as enablers of next-generation renewable technologies. Pilot-scale demonstrations, integration into multifunctional hybrid devices, and systematic life-cycle assessments will be decisive in moving from laboratory promise to commercial impact. If these directions are pursued, nanofluids can become a cornerstone in the global shift toward clean, efficient, and sustainable energy systems.

## References

[1] AbdEl-Rady Abu-Zeid, M., Elhenawy, Y., Bassyouni, M., Majozi, T., Toderas, M., Al-Qabandi, O.A. and Kishk, S.S., 2024. Performance enhancement of flat-plate and parabolic trough solar collector using nanofluid for water heating application. *Results in Engineering*, 21, 101673.

[2] Adun, H., Adedeji, M., Ruwa, T., Senol, M., Kavaz, D. and Dagbasi, M., 2022. Energy, exergy, economic, environmental (4E) approach to assessing the performance of a photovoltaic-thermal system using a novel ternary nanofluid. *Sustainable Energy Technologies and Assessments*, 50, 101804.

[3] Ahmadlouydarab, M., Anari, T.D. and Akbarzadeh, A., 2022. Experimental study on cylindrical and flat plate solar collectors' thermal efficiency comparison. *Renewable Energy*, 190, 848–864.

[4] Ahsan, A., Zanj, A., Zafar, A., Iqbal, N., Fatima, A., Khan, W., Yousaf, M.Z., Khan, B. and Guerrero, J.M., 2025. Predicting the impact of graphene oxide and its shapes on the thermal efficiency of Polyalphaolefin-40 using a mathematical model for wind turbine gearboxes. *Case Studies in Thermal Engineering*, 72, 106307.

[5] Ajay, G., Krishna, R., Mythirayan, B. and Vikram, T.S., 2025. Numerical Investigations on Immersion Cooling of Lithium-Ion Batteries using Different Coolants. Thermal Engineering, 72 (5), 417–424.

[6] Ajeena, A.M., Farkas, I. and Víg, P., 2023. Performance enhancement of flat plate solar collector using ZrO2-SiC/DW hybrid nanofluid: A comprehensive experimental study. *Energy Conversion and Management*: X, 20, 100458.

[7] Ali, A., Habib, S., Khan, Z., Thaljaoui, A., Sun, M. and Muhammad, T., 2025. An artificial neural network implementation on novel tetra hybrid nanofluid model: Case study of solar wind turbines. *Engineering Applications of Artificial Intelligence*, 158, 111459.

[8] Ali, S. and Djaouida, B., 2025. Investigating the Feasibility of Integrating Vegetation Into Solar Chimney Power Plants in the Tamanrasset Region. *Latvian Journal of Physics and Technical Sciences*, 62 (3), 77–91.

[9] Alizadeh, A., Djuansjah, J., Basem, A., Sharma, A., Hussein, M.A., Rajab, H., Hajlaoui, K. and Thakur, R., 2025. CFD-based investigation of energy and exergy enhancement in a geothermal heat exchanger with hybrid nanofluid and novel turbulator. *Case Studies in Thermal Engineering*, 73, 106582.

[10] Amar, M., Akram, N., Chaudhary, G.Q., Kazi, S.N., Soudagar, M.E.M., Mubarak, N.M. and Kalam, M.A., 2023. Energy, exergy and economic (3E) analysis of flat-plate solar collector using novel environmental friendly nanofluid. *Scientific Reports*, 13 (1), 411.

[11] Atofarati, E.O., Sharifpur, M. and Meyer, J.P., 2024. Pulsating nanofluid-jet impingement cooling and its hydrodynamic effects on heat transfer. *International Journal of Thermal Sciences*, 198, 108874.

[12] Bahrami, H.-R., Allahdadi, O. and Ghaedi, M., 2025. Improving charging and discharging times in a latent heat storage using circumferential fins: A numerical study. *Journal of Energy Storage*, 115, 115918.

[13] Bahrami, H.-R., Allahdadi, O. and Saffari, H., 2025. Nanofluids for Advanced Thermal Management: A Multi-Criteria Review of Performance, Cost, and Environmental Impacts. *Results in Materials*, 100753.

[14] Bahrami, H.-R., Azizi, A. and Ghaedi, M., 2024. Enhancing a Double Pipe PCM Reservoir Melting Efficiency Through Innovative Fin Arrangements: A Numerical Study. *Arabian Journal for Science and Engineering.*

[15] Bahrami, H.-R. and Ghaedi, M., 2025a. Boosting thermal and economic performance of shell-and-tube latent heat storage by selecting optimal inner tube configuration and applying rotation. *Journal of Energy Storage*, 134, 118131.

[16] Bahrami, H.-R. and Ghaedi, M., 2025b. Strategies for Passive Thermal Management of Lithium-Ion Batteries in Microgravity: Combining PCMs, Metal Foams, Fins, and Nanoparticles. *Case Studies in Thermal Engineering*, 106247.

[17] Bahrami, H.-R. and Ghaedi, M., 2025c. Enhancement of charging time in a shell-and-tube latent heat storage system using innovative inner-tube motion profiles: A numerical study. *Energy Conversion and Management:* X, 26, 100981.

[18] Bahrami, H.-R. and Ghaedi, M., 2025d. Recent advancements in the thermal management of lithium-ion batteries. In: *Advances in Heat Transfer.* Elsevier, 209–247.

[19] Behera, U.S., Sangwai, J.S. and Byun, H.-S., 2025. A comprehensive review on the recent advances in applications of nanofluids for effective utilization of renewable energy. *Renewable and Sustainable Energy Reviews,* 207, 114901.

[20] Beig, A.R. and Muyeen, S.M., 2016. 4 - Wind energy. In: M.H. Rashid, ed. *Electric Renewable Energy Systems.* Boston: Academic Press, 60–77.

[21] Bejger, A., Frank, E. and Bartoszko, P., 2021. Failure Analysis of Wind Turbine Planetary Gear. *Energies*, 14 (20), 6768.

[22] Bretado-de Los Rios, M.S., Rivera-Solorio, C.I., Gijón-Rivera, M. and Nigam, K.D.P., 2025. Enhancing the performance of a salt gradient solar pond using a coiled flow inverter heat exchanger with nanofluids. *Energy Reports*, 13, 4609–4623.

[23] Chappidi, S., Kumar, A. and Singh, J., 2023. Geothermal energy extraction from abandoned oil and gas wells using mono and hybrid nanofluids. *Geothermics*, 114, 102794.

[24] Chekifi, T., Boukraa, M. and Benmoussa, A., 2024. Artificial Intelligence for thermal energy storage enhancement: A Comprehensive Review. *Journal of Energy Resources Technology*, 146 (6), 060802.

[25] Chen, G.-R., Liao, T.-W., Hsieh, C.-C., Barman, J., Huang, C.-Y. and Jeffrey Kuo, C.-F., 2025. Using the Taguchi method and grey relational analysis to optimize the parameter design of flat-plate collectors with nanofluids, and phase change materials in an integrated solar water heating system. *Energy Conversion and Management:* X, 26, 100910.

[26] Çinici, O.K., Canlı, M.E., Çakıroğlu, R. and Acır, A., 2022. Optimization of melting time of solar thermal energy storage unit containing spring type heat transfer enhancer by Taguchi based grey relational analysis. *Journal of Energy Storage*, 47, 103671.

[27] Dashtbozorg, A., Safarianbana, B. and Shanbedi, M., 2025. Enhancing PV/T systems performance: A comprehensive study on MXene/methanol nanofluid in two-phase closed thermosyphons. *Solar Energy Materials and Solar Cells,* 279, 113274.

[28] Deshmukh, K., Kadam, S., Shirke, M., Nehe, S., Karmare, S. and Galande, S., 2025. Performance evaluation of TiN/water nanofluid charged U-pipe evacuated tube solar thermal collector with twisted tape inserts. *Heat and Mass Transfer*, 61 (1), 5.

[29] Feng, L., Turner, I.W., Anh, V.V. and Liu, F., 2025. Review of classical and nonlocal nanofluid models for solar collectors. *Renewable and Sustainable Energy Reviews*, 212, 115382.

[30] Garud, K.S. and Lee, M.-Y., 2022. Thermodynamic, environmental and economic analyses of photovoltaic/thermal-thermoelectric generator system using single and hybrid particle nanofluids. *Energy*, 255, 124515.

[31] Gholipour, S., Afrand, M. and Kalbasi, R., 2021. Introducing two scenarios to enhance the vacuum U-tube solar collector efficiency by considering economic criterion. *Journal of the Taiwan Institute of Chemical Engineers*, 124, 228–237.

[32] Hamidatou, S., Nadir, M., Togun, H., Abed, A.M., Deghoum, K., Hadjad, A. and Ahmadi, G., 2023. Experimental investigation of a novel heat exchanger for optimizing heat transfer performance using Al2O3-water nanofluids. *Heat and Mass Transfer*, 59 (9), 1635–1646.

[33] Hamzat, A.K., Omisanya, M.I., Sahin, A.Z., Ropo Oyetunji, O. and Abolade Olaitan, N., 2022. Application of nanofluid in solar energy harvesting devices: A comprehensive review. *Energy Conversion and Management*, 266, 115790.

[34] Hota, S.K., Mata-Torres, C., Cardemil, J.M. and Diaz, G., 2022. Techno-economic assessment of carbon-based nanofluid dispersions in solar stills for rural coastal locations in the Northern and Southern hemispheres. *Desalination and Water Treatment*, 245, 72–84.

[35] Huang, J., Han, X., Zhao, X., Khosa, A.A. and Meng, C., 2022. The stability, optical behavior optimization of Ag@SiO2 nanofluids and their application in spectral splitting photovoltaic/thermal receivers. *Renewable Energy*, 190, 865–878.

[36] Jakeer, S., Reddy, S.R.R., Thameem Basha, H., Cho, J. and Sathishkumar, V.E., 2025. Activation energy and Coriolis force impact on three-dimensional dusty nanofluid flow containing gyrotactic microorganisms: Machine learning and numerical approach. *Nanotechnology Reviews*, 14 (1), 20250179.

[37] Jayanthi, N., Venkatesh, M., Suresh Kumar, R. and Sekar, S., 2024. A review on performance of nanofluids in various heat pipe solar collector. Proceedings of the Institution of Mechanical Engineers, Part E: *Journal of Process Mechanical Engineering*, 238 (1), 5–14.

[38] Ji, J., Zhang, C., Zhang, X. and Chen, Y., 2025. Recent research advances in wind turbine thermal management technologies. *Renewable and Sustainable Energy Reviews*, 208, 114983.

[39] Jing, D., Liu, G., Li, Y., Xiong, T., Xing, M. and Yan, G., 2025. A systematic review of supercooling in the solidification for cold storage: theoretical analysis, behavioral characteristics and control technologies. *International Journal of Heat and Mass Transfer*, 250, 127336.

[40] Joseph, A., Sreekumar, S. and Thomas, S., 2020. Energy and exergy analysis of SiO2/Ag-CuO plasmonic nanofluid on direct absorption parabolic solar collector. *Renewable Energy*, 162, 1655–1664.

[41] Joseph, A. and Thomas, S., 2022. Energy, exergy and corrosion analysis of direct absorption solar collector employed with ultra-high stable carbon quantum dot nanofluid. *Renewable Energy*, 181, 725–737.

[42] Karaaslan, I. and Menlik, T., 2021. Numerical study of a photovoltaic thermal (PV/T) system using mono and hybrid nanofluid. *Solar Energy*, 224, 1260–1270.

[43] Kazem, H.A., Al-Waeli, A.H.A., Chaichan, M.T. and Sopian, K., 2021. Numerical and experimental evaluation of nanofluids based photovoltaic/thermal systems in Oman: Using silicone-carbide nanoparticles with water-ethylene glycol mixture. *Case Studies in Thermal Engineering*, 26, 101009.

[44] Khan, A.A., Danish, M., Rubaiee, S. and Yahya, S.M., 2022. Insight into the investigation of Fe3O4/SiO2 nanoparticles suspended aqueous nanofluids in hybrid photovoltaic/thermal system. *Cleaner Engineering and Technology*, 11, 100572.

[45] Kumar, R. and Hassan, M.A., 2024. Biosurfactant Augmented Characterization and Heat Transport Assessment of MWCNT-H2O Nanofluid in Solar Collector. *International Journal of Thermophysics*, 45 (5), 66.

[46] Kumar, R., Kharub, M., Sharma, R., Hrisheekesha, P.N., Goel, V., Bhattacharyya, S., Tyagi, V.V. and Varun, 2023. A novel design for solar collector used for water heating application having nanofluid as working medium: CFD modeling and simulation. *Environmental Science and Pollution Research*, 30 (2), 3942–3952.

[47] Lari, M.O. and Sahin, A.Z., 2017. Design, performance and economic analysis of a nanofluid-based photovoltaic/thermal system for residential applications. *Energy Conversion and Management*, 149, 467–484.

[48] Lari, M.O. and Sahin, A.Z., 2018. Effect of retrofitting a silver/water nanofluid-based photovoltaic/thermal (PV/T) system with a PCM-thermal battery for residential applications. *Renewable Energy*, 122, 98–107.

[49] Lee, D. and Jo, B., 2021. Thermal energy storage characteristics of binary molten salt nanofluids: Specific heat and latent heat. *International Journal of Energy Research*, 45 (2), 3231–3241.

[50] Liu, B., Zhang, X. and Ji, J., 2021. Review on solar collector systems integrated with phase-change material thermal storage technology and their residential applications. *International Journal of Energy Research*, 45 (6), 8347–8369.

[51] Liu, C. and Yang, H., 2022. Multi-objective optimization of a concrete thermal energy storage system based on response surface methodology. *Applied Thermal Engineering*, 202, 117847.

[52] Liu, Q., Lv, C., Wen, M. and Wang, Y., 2025a. Research on the heat transfer performance of a ground heat exchanger under the synergistic effect of nanofluid and phase change material. *Energy Conversion and Management*, 326, 119490.

[53] Liu, Q., Lv, C., Wen, M. and Wang, Y., 2025b. Numerical investigation on the nanofluid heat transfer enhancement in ground heat exchanger. *Renewable Energy*, 244, 122757.

[54] Liu, S., Zhang, L., Lu, J., Zhang, X., Wang, K., Gan, Z., Liu, X., Jing, Z., Cui, X. and Wang, H., 2025. Advances in urban wind resource development and wind energy harvesters. *Renewable and Sustainable Energy Reviews*, 207, 114943.

[55] López-Núñez, O.A., Lara, F., González-Angeles, A., Cardenas-Robles, A., Ramírez-Minguela, J.J. and Alfaro-Ayala, J.A., 2024. Assessment of thermohydraulic performance and entropy generation in an evacuated tube solar collector employing pure water and nanofluids as working fluids. *Heliyon*, 10 (8), e29309.

[56] Mahamude, A.S.F., Harun, W.S.W., Kadirgama, K., Farhana, K., Ramasamy, D., Samylingam, L. and Aslfattahi, N., 2021. Thermal performance of nanomaterial in solar collector: State-of-play for graphene. *Journal of Energy Storage*, 42, 103022.

[57] Manna, I., 2009. Synthesis, characterization and application of nanofluid—an overview. *Journal of the Indian Institute of Science*, 89 (1), 21–33.

[58] Meraje, W.C., Huang, C., Ujihara, M. and Kuo, C.J., 2022. Performance enhancement study of ag@ $SiO_2$ core-shell nanoplate plasmonic hybrid spectral splitting nanofluid based photovoltaic thermal system in a temperate region. *International Journal of Energy Research*, 46 (11), 15008–15026.

[59] Michael Joseph Stalin, P., Arjunan, T.V., Matheswaran, M.M., Dolli, H. and Sadanandam, N., 2020. Energy, economic and environmental investigation of a flat plate solar collector with CeO2/water nanofluid. *Journal of Thermal Analysis and Calorimetry*, 139 (5), 3219–3233.

[60] Mirshafiee, S.A., Kahani, M. and Ahmadi, M.H., 2025. Energy and exergy analysis of a multi-generation system for power, freshwater, and hydrogen production using geothermal water. *Journal of Thermal Analysis and Calorimetry*, 150 (8), 6539–6559.

[61] Pathak, S.K., Tyagi, V.V., Chopra, K., B., K., Pandey, A.K., Goel, V., Saxena, A. and Ma, Z., 2023. Energy, exergy, economic and environmental analyses of solar air heating systems with and without thermal energy storage for sustainable development: A systematic review. *Journal of Energy Storage*, 59, 106521.

[62] Pisarevsky, M.I., Struchalin, P.G., Balakin, B.V., Kutsenko, K.V. and Maslov, Y.A., 2024. Experimental study of nanofluid heat transfer for geothermal applications. *Renewable Energy*, 221, 119631.

[63] Rativa, D., De Araujo, Renato.E., Gómez-Malagón, L.A., De Araújo, R.P. and Vital, C.V.P., 2025. Lenticular Thermal Solar Collector. *Solar Energy*, 287, 113184.

[64] Ren, F., Li, Q. and Wang, P., 2025. Optimizing heat transfer in phase change thermal energy storage systems: A bionic method using alveolar vessel fins and nanofluids. *Applied Thermal Engineering*, 266, 125668.

[65] Rostami, S., Rostamzadeh, H. and Fatehi, R., 2021. A new wind turbine driven trigeneration system applicable for humid and windy areas, working with various nanofluids. *Journal of Cleaner Production*, 296, 126579.

[66] Said, Z., Hachicha, A.A., Aberoumand, S., Yousef, B.A.A., Sayed, E.T. and Bellos, E., 2021. Recent advances on nanofluids for low to medium temperature solar collectors: energy, exergy, economic analysis and environmental impact. *Progress in Energy and Combustion Science*, 84, 100898.

[67] Sainz-Mañas, M., Bataille, F., Caliot, C., Vossier, A. and Flamant, G., 2022. Direct absorption nanofluid-based solar collectors for low and medium temperatures. A review. *Energy*, 260, 124916.

[68] Shahzad, F., Jamshed, W., Sajid, T., Nisar, K.S. and Eid, M.R., 2021. Heat Transfer Simulation for 3D MHD Rotating Hybrid NanoFluid Flow Between Parallel Plates in Parabolic Trough Solar Collector: A Numerical Study. *Journal of Engineering Thermophysics*, 30 (4), 704–726.

[69] Sohani, A., Shahverdian, M.H., Sayyaadi, H., Samiezadeh, S., Doranehgard, M.H., Nizetic, S. and Karimi, N., 2021. Selecting the best nanofluid type for A photovoltaic thermal (PV/T) system based on reliability, efficiency, energy, economic, and environmental criteria. *Journal of the Taiwan Institute of Chemical Engineers*, 124, 351–358.

[70] Sreekumar, S., Shah, N., Mondol, J.D., Hewitt, N. and Chakrabarti, S., 2022. Numerical investigation and feasibility study on MXene/water nanofluid based photovoltaic/thermal system. *Cleaner Energy Systems*, 2, 100010.

[71] Srinivasan, N.K. and Ponnusamy, C., 2025. Impact of surfactant groups on the stability, thermophysical and phase change characteristics of water-based single and hybrid nanofluid

PCMs for cool energy storage applications. *Thermal Science and Engineering Progress*, 59, 103298.

[72] Tafavogh, M. and Zahedi, A., 2021. Design and production of a novel encapsulated nano phase change materials to improve thermal efficiency of a quintuple renewable geothermal/hydro/biomass/solar/wind hybrid system. *Renewable Energy*, 169, 358–378.

[73] Talesh Bahrami, H.R., Aminian, E. and Saffari, H., 2020. Energy Transfer Enhancement Inside an Annulus Using Gradient Porous Ribs and Nanofluids. *Journal of Energy Resources Technology*, 142 (12), 122102.

[74] Thakur, P.P., Khapane, T.S. and Sonawane, S.S., 2021. Comparative performance evaluation of fly ash-based hybrid nanofluids in microchannel-based direct absorption solar collector. *Journal of Thermal Analysis and Calorimetry*, 143 (2), 1713–1726.

[75] Venkatesh, R., Sharma, P., Subbarao, Ch.C., Mohanavel, V., Veerababu, K., Jadhav, A., Ravichandran, M., Soudagar, M.E.M. and Alotaibi, M.A., 2025. Effect of paraffin with salt hydrates PCM and hybrid Al2O3/Tio2 nanofluid on thermal and energy storage characteristics of solar thermal heat exchanger. *Journal of Thermal Analysis and Calorimetry*, 150 (10), 8029–8042.

[76] Venkatesh, R., Singh, R.P., Nagarajan, N., Prabagaran, S., Mohanavel, V., Rabadiya, D., Soudagar, M.E.M., Alotaibi, M.A. and Seikh, A.H., 2025. Thermal performance assessment of dish collector-integrated cooking application using TiO2/SiO2 hybrid nano-enhanced coated receiver. *Journal of the Brazilian Society of Mechanical Sciences and Engineering*, 47 (3), 148.

[77] Venkateswarlu, B., Chavan, S., Joo, S.W., Kim, S.C. and Nisar, K.S., 2025. A numerical investigation of heat transfer performance in a prismatic battery cooling system using hybrid nanofluids. *Case Studies in Thermal Engineering*, 66, 105719.

[78] Wang, S., Zhang, J. and Zhu, Z., 2022. Wind-Thermal-Energy Storage System Optimization: Evidence from Simulations of the Economical Consumption of Wind Energy. *Mathematical Problems in Engineering*, 2022, 1–13.

[79] Xiao, L., Gan, L.-N., Wu, S.-Y. and Chen, Z.-L., 2021. Temperature uniformity and performance of PV/T system featured by a nanofluid-based spectrum-splitting top channel and an S-shaped bottom channel. *Renewable Energy*, 167, 929–941.

[80] Xu, B., Zeng, R., Zheng, N. and Sun, Z., 2025. Energy flow control of nanofluid-based direct absorption solar collectors with functional optical coatings for efficient solar harvesting. *Journal of Central South University*.

[81] Xu, M., Yang, X., Dai, Z., Wang, R., Wang, J., Xie, J., Liu, M., Shi, Y., Chen, S., Xue, M., Cai, J. and Jiang, G., 2024. Experimental research of convective heat transfer between nanofluids and high-temperature dense granite in deep geothermal reservoirs. *Renewable Energy*, 235, 121245.

[82] You, T., Yan, T., Huang, Y., Huang, S. and Cui, H., 2025. Thermo-mechanical coupling performance analysis of phase change material and nanofluid synergized high-efficiency energy piles for shallow geothermal energy utilization. *Energy*, 334, 137691.

[83] Zamen, M., Kahani, M., Rostami, B. and Bargahi, M., 2022. Application of $Al_2$ $O_3$ /water nanofluid as the coolant in a new design of photovoltaic/thermal system: An experimental study. *Energy Science & Engineering*, 10 (11), 4273–4285.

[84] Zwalnan, S.J., Duvuna, G.A., Abakr, Y.A. and Banda, T., 2021. Design and Performance Evaluation of a Multi-Temperature Flat Plate Solar Collector. *International Journal of Renewable Energy Development-Ijred*, 10 (3), 537–549.

# 12 Boiling, Condensation, and Phase-Change Heat Transfer Mechanisms in Nanofluids

Santosh Kumar Singh[1], Alok Kumar Das[2], Amit Rai Dixit[3] and Arun Kumar Tiwari[4*]

This chapter deals with the growing importance of nanofluids, liquids containing nanoscale solid particles, in advancing phase-change heat transfer (PCHT) processes such as boiling, condensation, and evaporation. These processes are vital to energy conversion, electronics cooling, and solar thermal storage systems. Nanofluids enhance heat transfer rates by improving thermal conductivity, modifying viscosity, and influencing fluid–surface interactions. Studies report up to a 45% increase in critical heat flux (CHF) with alumina–water nanofluids. The chapter also examines key microscale parameters, including the onset of nucleate boiling (ONB), bubble elongation (OBE), and flow instability (OFI), which govern system performance. It highlights how nanoparticle size, shape, and concentration impact boiling dynamics, surface wettability, and stability. Applications extend to renewable energy systems such as solar collectors and desalination units, where nanofluids enable compact, efficient designs. Experimental insights into microstructured evaporators and condensation surfaces that demonstrate dropwise behavior are also discussed. Finally, emerging challenges such as nanoparticle aggregation, fouling, and long-term stability are addressed, alongside future directions in hybrid nanofluids, eco-friendly formulations, and AI-driven optimization for next-generation clean energy technologies.

## 1. Introduction

In recent years, nanofluids or liquids that have a stable suspension of nanoparticles have become an important breakthrough in thermal engineering. Nanofluids are developed by suspending very fine particles in common base liquids such as water, oil, or ethylene glycol, which have attracted increasing attention due to their ability to transfer heat much more effectively than other fluids. Their unique formulation places them between the characteristics of solid and liquid heat transfers, leading to increased efficiency in thermal systems (Kumar Singh *et*

[1-3] Mechanical Engineering Department, Indian Institute of Technology (Indian School of Mines), Dhanbad, 826004.

[4] Mechanical Engineering Department, Institute of Engineering & Technology, Dr. A.P.J. Abdul Kalam Technical University, Lucknow, 226021, India.

* Corresponding author: aruntiwari@ietlucknow.ac.in

DOI: 10.1201/9781003686002-12

*al.*, 2023). The development of nanofluids has created new opportunities in the evolution of high-performance compact heat transfer systems. Among the various methods of heat transfer, phase-change heat transfer (PCHT) is notable for its high efficiency because it makes use of latent heat. In practical applications, including power plants, electronic cooling systems, desalination plants, air conditioners, and thermal energy storage systems, PCHT involves boiling, condensation, melting, and solidification processes. These systems can also be improved by adding nanofluids to increase their energy efficiency, reduce the size of equipment, and improve their reliability in operation. Due to these advantages, nanofluids are quickly emerging as a potential contender for the design of the next-generation and sustainable thermal systems (Alsabery *et al.*, 2023).

Some of the nanoparticles used to form nanofluids (synthetic colloidal suspensions) are alumina oxide ($Al_2O_3$), copper oxide (CuO), titania dioxide ($TiO_2$), silica ($SiO_2$), graphene, and carbon nanotubes (CNTs) dispersed in a base fluid. These are nanoparticles that are usually one hundred and one nanometers in size (Kumar Singh *et al.*, 2023). Nanofluids are not similar to normal fluids due to their distinct thermal characteristics. These differences include higher thermal conductivity, Brownian motion of small particles, microscale convection, and improved fluid-surface interaction. Together, these effects increase the heat transport capacity of nanofluids by enabling them to conduct more energy, cool faster, and alter the properties of wetting or adhesion to hot surfaces. However, to make nanofluids work in a practical environment, the particles must be dispersed (they are in good dispersion) and avoid agglomeration of the particles. To overcome these issues, researchers are inclined to either use surfactants (compounds that ensure particles do not get mixed) or modify the surface of the nanoparticles to make them more stable in the fluid (Singh *et al.*, 2024; Singh *et al.*, 2025b).

Renewable energy can be useful for applying nanofluids in thermal systems. It has been demonstrated that nanofluids can be used to increase the heat transfer and solar absorption rates in solar thermal collectors, increasing the efficiency of the collector. Photovoltaic/thermal (PV/T) hybrid systems can also use nanofluids, since they are capable of cooling photovoltaic panels in a more efficient way, and at the same time, capture thermal energy (Kumar Singh *et al.*, 2024; Singh *et al.*, 2023, Singh *et al.*, 2026). Their role in the performance augmentation of solar power systems is highly appraised and a novel arena to explore (Singh *et al.*, 2025a).

Phase-change heat transfer (PCHT) enables a large amount of energy to be transferred with very low temperature change, with the latent heat of a substance. When a liquid boils, it becomes vapor, which quickly carries heat away from warm surfaces. This is particularly beneficial in boilers, heat pipes, and cooling electronics (Paul *et al.*, 2024). This chapter is carefully structured to offer a clear and complete understanding of how nanofluids improve phase-change heat transfer (PCHT). Key processes such as melting, solidification, condensation, and boiling are explained, along with the typical limitations encountered when traditional fluids are used in these systems.

## 2. Boiling Heat Transfer in Nanofluids

When the cold liquid comes in contact with a hot surface, the fluid experiences dual-stage heat interaction known as boiling, which leads it to become vapor. It is the latent heat whose effectiveness is responsible for transferring energy and extracting heat from the surface (Chu *et al.*, 2023). Boiling mainly occurs in two forms.

Heat applied to a stationary liquid over its saturation temperature on a free liquid-vapour interface on a submerged hot surface causes this effect, which is referred to as pool boiling or vessel boiling (Xu *et al.*, 2021). Flow boiling, in its turn, occurs within tubes or minor openings, heating the liquid on its way through them. This movement can be due to a pump or natural convection, depending on the design of the system (Zhang *et al.*, 2023).

There are multiple zones in flow boiling, which are indicated schematically in Figure 2. In single-phase convection, the heat is absorbed in the liquid passing through the heated part, yet it is not evaporated. In this phase, the sensible heat is taken up by the wall and slowly increases the temperature of the liquid. The fluid begins to boil or saturate, and upon further heating, tiny bubbles of vapour develop and slide down the blaze. The boiling point at this stage is known as the nucleate boiling point. The high rate of heat transfer in this region is very much enhanced by the fact that bubbles are rapidly and continuously formed in the region. Interestingly, even though the transfer of heat during this phase is extremely effective, the temperature of the wall is practically kept constant.

The further the fluid flows on the heated surface, the more vapor is produced and the thinner the layer of liquid that forms on the inner wall. This is the liquid-film evaporation zone or the forced convection boiling region, where evaporation is gradual owing to the constant heating and fluid flow. During this phase, the major part of the fluid is held near boiling or saturation temperature. After full evaporation of the liquid film, only vapor is left in the flow. It is at this stage that single-phase vapor convection replaces, and the temperatures of the vapor and the wall start to increase simultaneously since more heat is added (Jamialahmadi *et al.*, 2008). Overall, the flow boiling process illustrated in Figure 2 involves complex interactions between convection, bubble nucleation, film evaporation, and phase transition, making it one of the most efficient mechanisms for transferring thermal energy.

To understand pool boiling, it is important to analyze the corresponding boiling heat transfer curve, as shown in Figure 1. Pool boiling usually goes through four main stages:

- Natural convection
- Nucleate boiling
- Transition boiling, and
- Film boiling.

It starts at point A, referred to as the onset of nucleate boiling (ONB), when bubbles begin to develop on the heated surface. Before this stage, the liquid

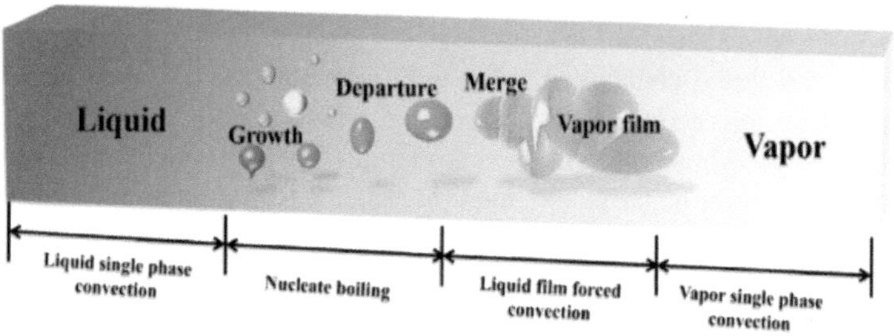


**Fig. 1** Bubble formation in flow boiling process (Chu *et al.*, 2023). License number, 6134650097501

remains inert (still) and most of the heat is conducted and transferred through natural convection; thus, the efficiency of heat transfer is low. As soon as the boiling process starts at ONB, bubbles start to form and detach themselves. This leads to a high flow of fluid, and this significantly increases the process of convective heat transfer. This causes a high heat flux at the expense of only a slight rise in wall temperature compared to a high heat flux at the expense of very high wall temperature. As the heat flux enhances further, it becomes maximum at point C, which is the critical heat flux (CHF) or the burnout point. Past this threshold, excessive amounts of vapor form and liquid are no longer able to moisturize the surface, and dry spots can appear locally. In real-world practice, as a safety limit to heat flux, the departure, or presence, of nucleate boiling (DNB) point is frequently employed. The bubbles coalesce to produce a continuous film of steam on the surface as the wall temperature increases past the region of CHF. Since vapor is far less thermally conducting than liquid, heat transfer resistance also rises, and the heat flux drops rapidly. This is the transition boiling zone. Lastly, a stable vapor layer forms and covers the heating surface, marking the transition of the process to the film boiling phase. In this case, the primary methods of transferring heat include radiation and vapor conduction. The further rise in the wall temperature makes radiation heat transfer more significant, which again raises the overall heat flux slightly (Pare and Ghosh, 2022).

Overall, the pool boiling curve in Figure 2 illustrates how different boiling regimes evolve with wall temperature, highlighting the balance between efficient heat removal during nucleate boiling and performance deterioration in film boiling.

Experiments conducted in the last 20 years have proven that nanofluids are capable of significantly enhancing the heat transfer in phase-change processes, in particular, boiling and condensation. Even low concentrations of nanoparticles can change surface properties, e.g., the wettability, roughness, or the counts of bubble nucleation locations of nanofluids. This will lead to better boiling action and critical heat flux (CHF). Nanoparticles often create a thin porous layer on the boiling surface. This layer lowers the contact angle and makes the surface more hydrophilic or water-attracting, and this makes the liquid able to rewet dry regions. This postpones the transition between nucleate boiling to film boiling, and the

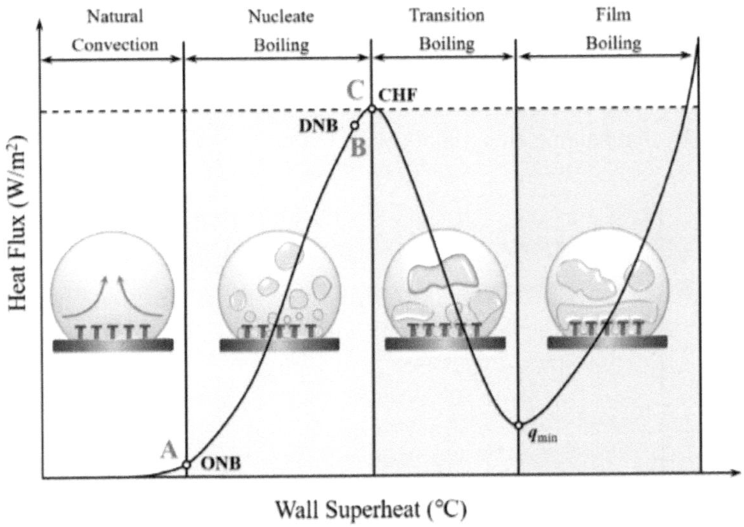


**Fig. 2** Pool boiling curve (Chu *et al.*, 2023). License number, 6134650097501

system can absorb more heat demands with higher safety. These advancements are particularly crucial in small, high-heat systems where thermal performance and safety are crucial, such as nuclear reactor heat exchangers, power electronics, and microchannel cooling devices.

Figure 3 makes a comparison of the critical heat flux (CHF) behaviour of pure water and alumina-water nanofluid on clean heater surfaces and surfaces with nanoparticles, at various flow speeds. The results show an increase of CHF by 50 percent in pool boiling (in 0 m/s, when the pool does not flow) and a noticeable increase even in forced convection when the fluid does flow. This can be mainly attributed to nanoparticle depositions on the heater surface, resulting in a reduction of the contact angle that was approximately 65 degrees down to only 12 degrees. Due to this change, the surface was made much more wettable, and the liquid-surface contact was enhanced. The result is more stable bubble behavior, enhanced rewetting of dry patches, and even distribution of heat. These effects are useful in high-performance thermal systems (like a nuclear heat exchanger, power electronics, and microchannel cooler) since they have a longer operating range, increase the safety of the system, and slow burnout (Ahn *et al.*, 2010). Figure 3 presents the influence of nanoparticle concentration on CHF enhancement as reported by You *et al.*, (You *et al.*, 2003). Experiments showed that the critical heat flux (CHF) is astonishingly 200 percent higher when 0.005 g/L of alumina nanoparticles (or some 0.0001 percent by volume) was introduced into water rather than pure water. This observation shows that even minute doses of nanoparticles can be very effective in increasing the efficiency of heat transfer. This is mainly because surface properties are modified by deposits of nanoparticles, including increased wettability and modified surface texture, and not merely the increased thermal conductivity. However, excessive concentration of nanoparticles might lead to a cumbersome coating that will serve as thermal insulation on the hot surface. This does not make boiling even more efficient. Thus, the best results are achieved by maintaining an optimal concentration of nanoparticles.

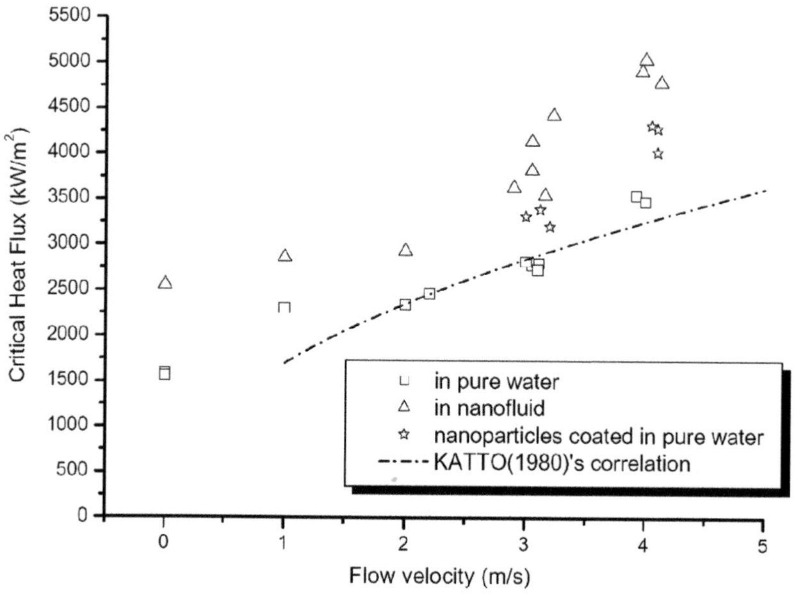


**Fig. 3** Critical heat flux against flow velocity of water and nanofluid (Ahn *et al.*, 2010). License number 6134650533728

These findings reflect the two mechanisms of nanofluid-induced improvement, namely (i) improvement in surface properties due to nanoparticle coating, and (ii) modification in bubble behavior due to an increase in nucleation and rewetting properties. The results of the experiment presented in this study are strong experimental evidence that nanofluids can be effectively used in the future in boiling systems. They can be used to achieve higher heat removal rates, smaller system designs, better thermal stability, and reliability in harsh applications.

**Table 1** Summary of the heat transfer studies with nanofluids.

| *Author* | *Nanofluid* | *Heat transfer phenomenon* | *findings* |
|---|---|---|---|
| He *et al.*, (2016) | ZnO–water and ZnO–ethylene glycol | Pool boiling in a cylindrical vessel | Increasing heat flux significantly improved the HTC of both nanofluids. |

*Contd.*

**Table 1** *Contd.*

| *Author* | *Nanofluid* | *Heat transfer phenomenon* | *findings* |
|---|---|---|---|
| Sharma and Unune (2022) | Ag/ZnO nanofluid (0.02–0.1 wt%) | Pool boiling on EDM-textured heater surfaces | At 0.1 wt%, CHF ↑ by 80.43% and HTC ↑ by 252.98% vs. DI water; EDM enhanced nucleation sites and lowered contact angle. |
| (Hegde *et al.*, 2012) | CuO–water (0.01–0.5 vol%) | Pool boiling | Optimum concentration 0.2 vol%; CHF improved by >60% vs. deionized water. |
| (Karimzadehkhouei *et al.*, 2017) | $TiO_2$–water and CuO–water | Pool boiling on a flat plate | $TiO_2$ showed decreasing enhancement with concentration, while CuO maintained stable improvement. |
| (Xing *et al.*, 2016) | Functionalized MWCNT–water | Pool boiling on a flat plate | Covalently functionalized nanofluid gave 53.4% higher HTC than the base fluid. |
| (Manetti *et al.*, 2017) | $Al_2O_3$–water (0.0007–0.007 vol%) | Pool boiling on copper surfaces (Ra = 0.05–0.23 μm) | HTC increased by 75% (smooth) and 15% (rough) vs. DI water for a 100–800 $kW/m^2$ heat flux range. |
| (Suriyawong and Wongwises, 2010) | $TiO_2$–water (0.00005–0.01 vol%) | Pool boiling on Cu and Al tubes | At 0.0001 vol%, HTC decreased on Al but increased on Cu tubes due to surface–nanoparticle interaction. |
| (Peng *et al.*, 2009) | CuO–R113 refrigerant | Flow boiling in a horizontal smooth tube | At 0.5 wt%, HTC was enhanced by 29.7% compared with pure refrigerant. |
| (Kim *et al.*, 2018) | $Al_2O_3$ and reduced graphene oxide (RGO)–water | Pool boiling at atmospheric pressure | CHF increased up to 54% ($Al_2O_3$) and 10–37% (RGO) with increasing concentration. |
| (Wang and Su, 2016) | $\gamma$-$Al_2O_3$–water (0.1–0.5 vol%) | Flow boiling in a vertical tube | Nusselt number and heat transfer rate increased with heat flux, nanoparticle concentration, and pressure; a maximum enhancement ≈ 86%. |

Despite the huge potential of nanofluids in boiling processes, issues exist that are of concern. The most difficult problem is to keep nanoparticles in a uniform suspension. They might cause system dependability problems by settling or depositing in large amounts, since they could clog small channels or cause permanent fouling in hot surfaces. Another problem is the inadequate preparation and testing of nanofluids. The inconsistency in results across studies is due to differences in surfactant use, mixing procedures, and methods of measurement. To be effective in real systems, nanofluids need to be optimized regarding the type, size, and quantity of nanoparticles, and the surface structures should be more directed. These measures are critical to extending the benefits of the laboratory testing to real thermal systems.

## 3. Condensation Heat Transfer in Nanofluids

Condensation, which is the conversion of vapor into a liquid, is a very important process in most heat transfer systems, including a refrigerator, condensers, and thermosyphons. Removal of latent heat in such systems is critical in ensuring that the design is small and towards ensuring high performance (Fieg and Roetzel, 1994). Another technique that can be used to increase condensation is the use of nanofluids, which are base fluids with tiny suspended nanoparticles. These fluids can enhance surface wettability, thermal conductivity, and enhance how the liquid film drains on the surface, leading to a better condensation process. Experiments with CuO water nanofluids in grooved heat pipes demonstrated that there was a substantial rise in the maximum heat flux and overall thermal efficiency Liu *et al.*, Sun and Zheng, 2006). They attributed this enhancement to the reduced film thickness within the grooves and to enhanced capillary action. On the same note, Dutta *et al.*, (2004) found that $Fe_3O_4$-$H_2O$ nanofluids decreased the thermal resistance of a thermosyphon by about 15% and showed better condensation rates because the film conductivity of the liquid was greatly enhanced. Analytical modelling by Dalkilic and Wongwises (2009) demonstrated that thermophoretic movement of nanoparticles towards a cold wall increases the local heat transfer by raising the concentration of nanoparticles close to the interface, but Brownian diffusion has the opposite effect. Generalizing on the same, Liu *et al.*, (2010) numerically studied the condensation process of the nanofluid in a square microchannel and observed up to 20 percent enhancement in the condensation heat-transfer coefficient (HTC) with an increase in nanoparticle concentration (0-5 percent). Additional study by Huminic and Huminic (2011) found that the condensate film thickness reduces as the nanoparticle loading increases, resulting in more intense hydrodynamic shear and quicker heat transfer as per multiphase modeling. Due to its thermal conductivity, Ag produced the greatest improvement among the tested nanoparticles, and $TiO_2$ the least. In addition to this, (Parise, 2012) were able to demonstrate that the migration of nanoparticles towards the condensing wall is positively correlated with effective film conductivity and reduced wall superheat. They also observed that condensation in ferrofluids is enhanced further by the application of a magnetic field, as this controls the movement of particles.

Figure 4 depicts the variation of nanoparticle volume fraction within the condensate film as a function of the dimensionless temperature (T*) and the Brownian-to-thermophoretic diffusivity ratio, as reported by Heysiattalab *et al.*, (Heysiattalab *et al.*, 2016). In the case of condensation, nanoparticles in nanofluids are not distributed uniformly. They are concentrated most closely around the cold wall (where the temperature is lowest, T = 0) and least around the vapor-liquid interface (where the temperature is highest, T = 1). This is because thermophoretic forces will cause the nanoparticles to move towards the colder wall, and away from the hot vapor side. It is then that nanoparticles gather on the cold surface, increasing the heat transfer and raising the conductivity of the local thermal conductivity. The findings prove the thesis that the intersection of thermophoresis and Brownian motion has a strong effect on the behavior of nanofluids when condensing a film, which directly affects the local heat transfer and the overall efficiency of the condensation process.

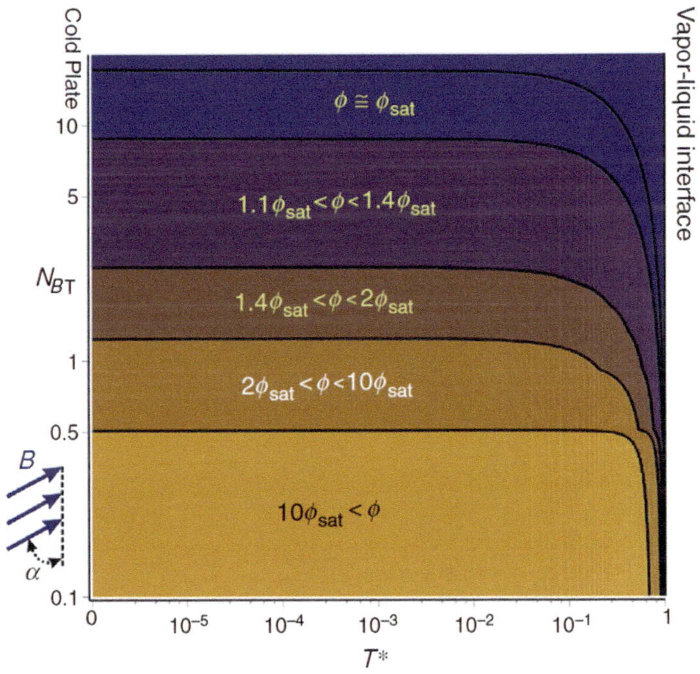


**Fig. 4** NBT Ratio Distribution as a Function of Scaled Temperature dimensionless temperature (T*). License number 6134650742828

More evidence that nanoparticle-modified surfaces with heterogeneous wettability promote dropwise condensation, higher droplet departure frequency, and slower film formation, further validating that heterogeneous wettability

**Table 2** Summary of the studies on nanoparticles.

| *Study* | *Method* | *Size of nanoparticle* | *Concentration* | *Findings* |
|---|---|---|---|---|
| (Turkyilmazoglu, 2015) | Analytical | Cu, Ag, CuO, $Al_2O_3$, $TiO_2$ / not mentioned | 0–10% | Demonstrated that nanoparticle type significantly alters condensation performance. Silver and copper nanoparticles showed superior enhancement due to higher thermal conductivity. Curved surface geometry further improved heat transfer by minimizing film thickness. |
| (El Mghari *et al.*, 2015) | Numerical | Cu, Al, CuO, $Al_2O_3$, diamond /not mentioned | 0–5% | Found up to 20% increase in HTC in square microchannels using nanofluids, particularly at higher Re and nanoparticle concentrations. The effect is pronounced in mini-channel regimes. |
| (Malvandi *et al.*, 2016) | Analytical | $Al_2O_3$, $TiO_2$, $Fe_3O_4$/ ~100 nm | Not specified | Studied nanoparticle migration during filmwise condensation over cylinders and in magnetic fields. Found that thermophoresis enhances local nanoparticle concentration at the cold wall, improving thermal conductivity and heat transfer. Magnetic fields further enhance migration in ferrofluids. |
| (Heysiattalab *et al.*, 2016) | Analytical | $Fe_3O_4$/ <100 nm | Not specified | Enhancement in heat transfer depends on the Brownian-to-thermophoretic diffusivity ratio (NBT). Higher NBT weakens the enhancement by promoting uniform distribution. |
| (Avramenko *et al.*, 2015) | Analytical | Parametric (generic) | 0–1% | Developed models showing how nanoparticle diffusion direction and type (Brownian vs. thermophoretic) influence film thickness and HTC. Showed that optimal concentration gradients enhance condensation. |
| (Huminic and Huminic, 2013) | Experimental / Numerical | CuO and $Fe_3O_4$ / 4–5 nm | 0, 2, 5.3% | Experimentally observed 15–20% reduction in thermal resistance in thermosyphons using CuO–water. Numerically showed improved heat flux and condensate removal with $Fe_3O_4$–water. |
| (Liu *et al.*, 2010) | Experimental | CuO/50 nm | 1% | Investigated heat pipes; found maximum heat flux increased significantly with CuO nanofluid due to reduced film resistance and improved capillary action. |
| (Parise, 2012) | Mathematical | Cu/not mentioned | 0–2% | Showed that using copper nanoparticles improves film condensation due to enhanced conductivity and reduced film thickness. |

with nanoparticles promotes dropwise condensation, was recently furnished by experiments that, in themselves, led to up to 25 percent enhancement in condensation heat transfer. On overall, these research works prove the idea that nanofluids are more effective in facilitating condensation because the liquid film is thinned by nanofluids, the surface is wetted more quickly, and the removal of condensate is accelerated. Nevertheless, the concentration of nanoparticles, the choice of material, and the coating design are still crucial, and excessive deposition may increase flow resistance and decrease the durability.

Evaporation, whereby heat is exchanged between a liquid and vapor, is one of the processes of heat transfer. This process is used in many modern thermal systems, including heat pipes, fuel injectors, desalination units, heat pumps with small-sized evaporators, and microchannel cooling systems of electronics. The ability of the fluid to absorb and release heat during evaporation is significant to the effectiveness of the systems. Regular fluids, however, are generally hampered by low thermal conductivity, incomplete surface tension and slower heat transfer. To resolve such problems, researchers have started using nanofluids, that is, fluids that are specifically formulated by means of the addition of tiny nanoparticles to standard base fluids. These nanofluids have already shown considerable potential in enhancing the efficiency of evaporation and enhancing the reliability and efficiency of thermal systems.

Nanofluids can significantly modify the surface-related and thermal properties of fluids to improve the mechanism of evaporation. Nanofluids affect not only the evaporation rate but also the internal fluid motion within droplets, the particle distribution on heated surfaces, and the morphology of the residues formed after evaporation.

Over time, researchers have found that nanofluid evaporation is a highly complex process, revealed through mathematical modeling, simulations, and experiments. Chon *et al.,* (Chon *et al.*, 2007) studied how nanoparticle size affects nanofluid droplet evaporation. They found that smaller nanoparticles increase fluid viscosity, which suppresses thermal convection and slows evaporation, forming a broader, faint ring-like deposit. This data proves that nanoparticle size is critical in decreasing the surface-tension-driven flows and Brownian motion. This balance directly influences the distribution of heat in the droplet as well as the evaporation pattern that will occur.

Sefiane and Bennacer (2009) dealt with the role of changes in wettability caused by nanoparticles by examining ethanol sessile droplets containing aluminum nanoparticles. It was found that nanofluid droplets possessed much higher initial contact angles compared to pure ethanol droplets with the help of atomic force microscopy (AFM) research technique. This change affects heat transfer. A larger contact angle could limit direct heat conduction, and this means that the droplet is not in contact with as much surface area. However, it also accelerates evaporation near the triple-phase contact line, or the junction of the three phases: liquid, solid, and vapor. The evaporating droplet starts to be covered with nanoparticles. This changes the surface energy of the material, affecting the manner in which the

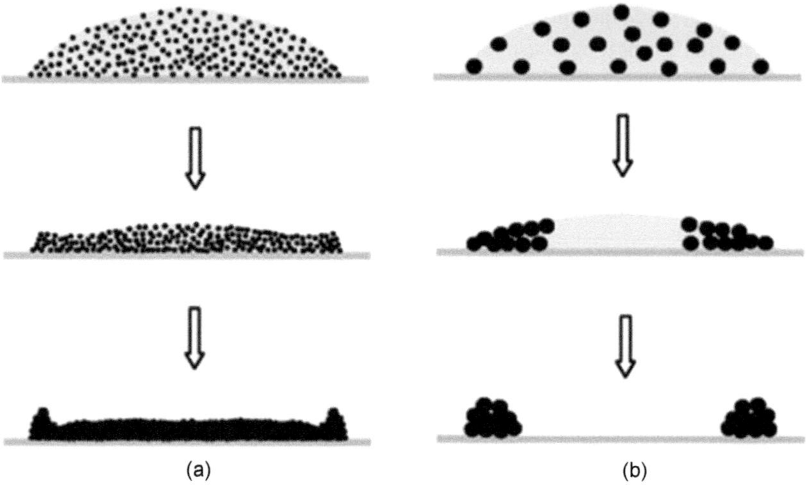


**Fig. 5** Droplet evaporation (a) smaller particles (Au of 2-nm), and (b) larger particles ($Al_2O_3$ of 47-nm) (Chon *et al.*, 2007). License number 6134650887169

subsequent droplets are dispersed and evaporated. Such surface changes prove that nanofluids contact and cover the surface with the lapse of time, which leads to a dynamic and changing evaporation process instead of the process of evaporation of regular liquids. In their experiment, Chen *et al.,* (2010) aimed at comparing the behavior of nanofluids prepared using various nanoparticles, namely Laponite, Fe 2 O 3, and silver (Ag). These nanoparticles were dispersed using deionized water and stabilized using polyvinylpyrrolidone (PVP). These results indicated that the rate of evaporation and the apparent heat of vaporization, which is the energy required to evaporate the droplet, were greatly dependent on the composition of the nanoparticles. The Laponite nanofluid can potentially hold more heat energy in the droplet as it exhibited a greater heat of vaporization but a small decline in the evaporation rate. The silver nanofluid, on the other hand, made the evaporation accelerate drastically. Chen *et al.,* (2011) further explored how nanoparticles affect surface tension during evaporation. This reduction enhanced evaporation in thin films and small droplets by strengthening capillary action and promoting easier liquid spreading on surfaces.

Conversely, a low surface tension may result in the liquid film being unstable, and hence, uneven deposit of particles, or dry-out could occur. These results prove that the use of nanoparticles to alter interfaces is a good method of controlling evaporation. Nevertheless, it also mentions the importance of possessing absolute control of the stability and quantity of particles to avoid the negative consequences. A Zhao *et al.,* (2011) provided a higher system-level view and an analytical examination of nanofluid film evaporation within a microchannel. It was demonstrated that upon evaporation, nanoparticles tend to settle on the channel walls to create a porous layer of foulant. Initially, this coating can be beneficial since it enhances the wettable surface and retains remnants of water, thereby preventing the surface from drying up and prolonging it. This renders the heat transfer smoother and more consistent. Thickening over time, however, causes the layer to begin acting as a thermal barrier, thereby decreasing the overall effectiveness of the evaporator and increasing resistance to heat flow. One recent dual effect is emphasized in this work, namely, nanoparticle fouling is capable of enhancing microscale capillary action and wetting, which, when uncontrolled, can also cause hindrance in heat transfer through augmented thermal resistance. Thus, once we strike the correct balance between useful surface rehydration and harmful thermal insulation, the long-term performance will be ensured. Finally, the behavior of nanoparticles in relation to surface architecture and heat conduction is also an important factor in deciding the efficiency of evaporation dynamics over time.

Gan and Qiao (2011) made a significant contribution to the behavior of nanoparticles in evaporating droplets, particularly in fuel droplets containing aluminum (Al) nanoparticles. Their experiment surveyed the influence of nanoparticle agglomeration, or the clumping of particles, on evaporation by integrating simulation and experiment. They tested with 80 nm uniformly dispersed aluminum particles to simulate a 1 mm fuel droplet and found that Brownian motion, internal fluid movement, and differences in the settling speeds of particles lead to the eventual clustering of the particles. The simulations showed that such clusters can significantly influence evaporation in case the drop lifetime is more than the agglomeration time. The agglomerates reduce the evaporation rate and modify the temperature ratio within the droplet. Nonetheless, evaporation is not affected much in short-lived droplets due to the lack of time in which the droplets can extensively cluster. This can be proved by showing that the evaporation of nanofluids depends greatly on time and is closely connected to the movement and contact of nanoparticles. The size of particles used and their ability to remain stable during dispersion are essential in the applications of fuel sprays, because high rates of evaporation are desired and sedimentation should be avoided, as it would affect the performance of the spray. Wei *et al.,* (2016) studied the impact of circulation patterns on the distribution of the particles and the evaporation rates. In addition, Fu *et al.,* (2017) performed an analysis on the use of $Al_2O_3$-$H_2O$ nanofluids evaporating on external surfaces.

These studies highlight the complex link between surface traits, fluid motion, and nanoparticles in enhancing evaporation heat transfer. Nanoparticles behave as the critical agent, improving heat interaction and flow regimes. The right balance of variables such as particle size, shape, material, concentration, stability, and the fluid capacity to interact with the surface it comes in contact with can be the factor that contributes to either assisted evaporation or hindered evaporation. A good understanding of these interconnected mechanisms is essential to exploit the potential of nanofluids. The next key challenge is to build predictive models capable of considering this complexity and help in selecting the most suitable nanofluid formulations to use in specific applications to achieve effective, stable, and high-performance evaporation in future thermal systems.

Mahian *et al.,* (2017) examined the impact of nanoparticles, such as SiO2 and Cu, in working fluids on the rate of evaporating water in a solar still that operates in combination with a heat exchanger. Their configuration, schematically illustrated in Figure 6, was to evaluate the evaporation with varying operating temperatures and nanoparticle concentrations (up to 4 percent). Through their studies, they have found that nanofluids can significantly enhance the rate of evaporation in solar stills, especially where the inlet temperature into the heat exchanger is low. In the same way, the addition of the nanoparticles enhanced evaporation by up to 10 percent when the inlet temperature was lower than 50°C. It was mainly due to enhanced thermal conductivity and the enhanced localized convection within the system. Only a 1 percent improvement with higher temperatures made the effect of nanofluids decrease significantly. This means that when thermal input is already high, then the benefit of nanofluids is reduced since there is less room to gain. Interestingly, however, they further found that when using a heat exchanger in the solar still as compared to a standard (conventional) still, overall productivity could indeed be less in low-temperature conditions. This unexpected finding was because the heat losses that the exchanger provided were more than the negligible gains that the nanofluid offered, particularly at a lower temperature difference. Based on these results, it is possible to emphasize that the most advantageous results with nanofluids in the field of solar thermal can be achieved through a careful combination of system design and operating conditions.

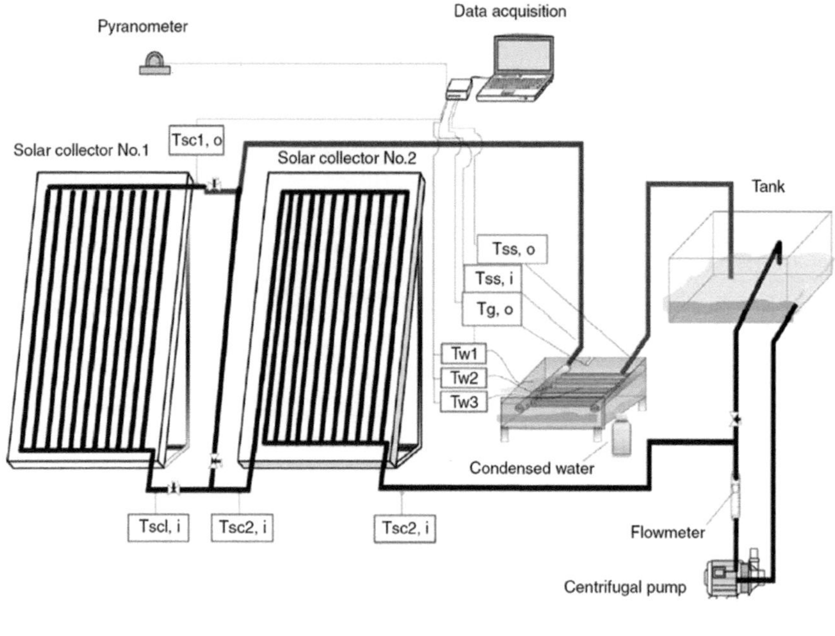


**Fig. 6** Schematic of the Nanofluid-Assisted Solar Still Experiment (Mahian *et al.*, 2017). License number 6134651287489.

In conclusion, although nanofluids have a lot of promise for enhancing evaporation in thermal systems, operating temperature, the kind of heat source, and system design all affect how effective they are in practice.

## 4. Evaluation Metrics for Nanofluid

To enhance and comprehend the flow boiling in microchannels with nanofluids, certain metrics of evaluation should be applied. The most significant ones are:

- Critical Heat Flux (CHF) – the maximum heat flux before the surface dries out.
- Onset of Nucleate Boiling (ONB) – the point where bubbles start to form.
- Onset of Bubble Elongation (OBE) – where bubbles stretch along the channel due to flow.
- Onset of Flow Instability (OFI) – when boiling becomes unstable and may cause performance issues.

These parameters are necessary in determining boiling stability, heat transfer efficiency, and safe operation of small-scale thermal systems such as the microchannel heat sink.

### *4.1 Critical Heat Flux (CHF) Behavior*

The surface wettability has a great effect on the nucleate boiling behavior and the CHF in boiling heat transfer, especially in high heat flux and microscale cases. This property of a liquid to remain in contact with a solid surface is called wettability and is dependent on the surface energy of the material and the interfacial forces that exist between the material and the liquid. When the contact angle is less than 90, the surface is usually well wetted and therefore allows the freezing of the liquid on it. Enhanced boiling system wettability results in easier rewetting of boiling spots of liquids, reduces dry-out, and enhances the separation of bubbles off the surface. This is the reason why enhancing the wettability of heating surfaces has become an important tactic in thermal management, particularly in microchannel boiling and high-flux cooling.

A significant change was introduced in the context of the ability to influence wettability with the advent of nanofluids, i.e., fluids with suspended nanoparticles. Compared to normal fluids, nanofluids are capable of behaving in a dynamic way where the solid–liquid interface is modified in the course of boiling. By settling on the hot surface, the nanoparticles alter the structure and surface energy of the nanoparticle at the nanoscale, which subsequently influences heat transfer and mass during boiling. Such a capacity to dynamically modify features of surfaces makes nanofluids have a unique benefit in enhancing boiling heat transfer.

- *Mechanism of Structural Disjoining Pressure:* The pioneering observation of the influence of nanofluid on the wettability can be attributed to the studies of Wasan and Nikolov (Wasan and Nikolov, 2003), who found that in nanofluid films on heated surfaces, there exists structural disjoining pressure. This pressure is caused by the ordering of nanoparticles in confined liquid films by electrostatic and van der Waals forces. Nanofluids possessing self-assembled particle layering, a part of which is the formation of a thin, stable liquid film, when confined between a solid wall and a vapor bubble, are not susceptible to rupture, unlike conventional fluids. This film suppresses the premature formation of insulating vapor blankets by inhibiting the dry out and increasing the period during which the liquid is in contact with the hot wall. Based on this, Wen (Wen, 2008) and Xu (Xu and Xu, 2012) explained how this disjoining pressure controls the dynamics of the three-phase contact line. They suggested that the presence of nanoparticles around the contact line decreases the local contact angle and alters the interfacial energy equilibrium. Along with the enhanced wetting efficiency of surfaces, this reconfiguration enhances a more uniform distribution of the liquid on the boiling surface.

- *Experimental Observations of Wettability Enhancement:* Previous studies reveal that the boiling nanoparticles lie at the surface to form the porous microstructure shell, which enhances the capillary wicking, surface area, and contact angle, and the boiling heat interaction and the CHF limit (Kim *et al.*, 2006).
- *CHF Enhancement with Nanoparticle Concentration and Mass Flux:* Kim *et al.*, (Kim *et al.*, 2009) studied the application of varying the mass flux and nanoparticle concentration on the base fluid. Nanofluids continued to perform at low fluxes, also underlining their two-fold benefit of both improving heat transfer at high loads and being able to stay in a steady flow at varying conditions. On the same note, Vafaei and Wen (2010) analyzed the effect of the concentration of nanoparticles on the improvement of CHF. Small additions recorded massive CHF increases, but the increments no longer increased proportionally at high concentrations, but instead turned down at an optimum concentration, as their results showed. This effect is usually ascribed to an increase in concentrations of the nanoparticles or high-viscosity and can slow down the fluid flow and lower the convectional heat transfer. However, nanofluids are highly productive in CHF because of the advantages of enhanced wettability, heat transfer, and bubble behavior at high concentrations.
- *Implications for Microchannel and High-Flux Applications:* Nanofluids have been put in perspective as an attractive solution to handling thermal loads in nuclear thermal hydraulics, compact heat exchangers, and microelectronic cooling systems, where the CHF is a rate-limiting factor because of the cumulative impact of enhanced surface wettability, which has been validated both theoretically and experimentally. Nanofluids are a passive, self-regulating system to enhance safe operating conditions without having a complex control system through the enhancement of rewetting capability of the surface and delaying a transition to film boiling.

In summary, nanoparticle-induced surface wettability modification is a key mechanism enhancing boiling performance in nanofluids. Rooted in nanoscale physics yet impacting macroscopic heat transfer, it bridges colloidal science and thermal engineering, advancing the design and safety of next-generation thermal systems.

## 4.2 Influence of Nanoparticle on CHF

The addition of nanoparticles to conventional base fluids extends the critical heat flux (CHF)—the point where rapid vapor formation disrupts liquid contact and severely reduces heat transfer efficiency. Nanofluids are used to enhance CHF via a combination of interfacial, thermal, and dynamic effects due to the physicochemical special interactions at the boiling interface. This part discusses the immense influence of nanoparticles on surface wettability, bubble dynamics, nucleation, and dryout properties, which eventually lead to improved and safer thermal systems.

### 4.2.1 Surface Wettability

One of the most established and notable methods through which nanofluids improve the CHF is their ability to increase surface wettability, or the ability of the fluid to remain in contact with the heated surface. The higher the wettability, the less likely that there will be dry patches or vapor blankets, which form due to the availability to keep hot spots constantly rewetted by the liquid. It was found in an innovative investigation that nanoparticles in nanofluids generate a structural disjoining force at the nanoscale, in reality, forming a stable and nanoparticle-rich thin film between the bubble of vapor and the heated surface (Wasan and Nikolov, 2003). This movie does not want to break; it serves as a barrier that postpones drying up. This was subsequently shown to be supported by Wen (Wen, 2008) and Xu and Xu ((2012), who showed that this thin film decreases the dry contact area by keeping a continuous liquid at the wall, leading to a higher CHF threshold. These findings were also confirmed through experimental studies. Kim *et al.,* (2018) found that nanoparticles are likely to settle on the heater surface, forming a porous layer that increases capillarity and decreases the contact angle. This change assists the surface in retaining the liquid in a better way and enhances the actions of the bubbles during boiling. Also, Vafaei and Wen (2010) observed that the increase in CHF is not a straightforward linear relationship where there is an optimal level of nanoparticles that may result in plateauation or decrease, since at that point, there is the problem of increasing viscosity or agglomeration of the particles. Collectively, these studies make surface wettability modification an underlying mechanism where nanofluids enhance boiling performance, especially at high thermal loads.

### 4.2.2 Bubble Dynamics

Also, the bubble behavior, i.e., the formation, growth, rupture, and contact of vapor bubbles with the fluid and the surface, is affected by nanoparticles indirectly but substantially. The size and frequency of bubble departure can have a great influence on the performance of heat transfer in boiling systems. The forces of drag, surface tension, and buoyancy are altered when nanoparticles are present in the system, and this influences the behavior of bubbles during boiling. According to the theory of bubble force balance suggested by Xu and Xu (2012), nanoparticles decrease the surface tension (F sigma) and base diameter of bubbles (DB). The result of this is smaller and more frequent bubbles that burst out of the surface at a faster rate. These fast bubble processes reduce the amount of local hot spots and promote enhanced heat transport by making sure that the surface remains in better contact with the liquid phase. A constant and steady boiling regime can immensely enhance system safety and performance in processes needing a high heat flux, such as microelectronics cooling and miniature heat exchangers.

### *4.2.3 Active Nucleation Sites*

Another important process by which nanofluids improve CHF is by altering nucleation sites, which are small cavities or surface defects where bubbles form. The density and activity of these sites are influenced by nanoparticles. Yu and Sur (Yu *et al.*, 2015) explain that nanoparticles accumulate into these surface cavities as the boiling process continues, altering their geometry and potentially stabilizing the boiling process by enhancing a more uniform bubble formation. Narayan and Anoop (Narayan *et al.*, 2007), however, provided a corroborating viewpoint that the ratio of the surface roughness to the size of the nanoparticles controls whether the cavities are filled or fresh nucleation sites are developed above the surfaces. When the roughness elements are significantly bigger, particles can fill in the voids and blunt nucleation when their size is considerably smaller. Conversely, the optimal ratios would enhance a nucleation process through semi-porous deposits that would act as new nucleation sites. The nucleation activity is also affected by the thermal boundary layer thickness. According to a study proposal, the superheat of the wall has reached a critical point at which nucleation takes place, and this hinges upon the boundary layer (Hsu, 1962). Other researchers report that nanofluids can modify the boundary layer profile by modifying thermal conductivity, viscosity, and specific heat (Liu and Yu, 2011). These changes alter the onset and intensity of boiling by influencing the ease with which heat is removed from the surface as well as the rate of bubble formation. In general, there are two benefits of change in the behavior of nucleation sites: it slows down the creation of premature vapors film and enhances the creation of more vigorous bubbles at smaller superheats.

### *4.2.4 Dryout Behavior*

Lastly, nanoparticle-enhanced CHF also varies with wetout behavior, especially under subcooled and saturated conditions of boiling. Dryout is the critical point at which the liquid will not be able to rewet the surface, and hence causes a significant rise in the surface temperature and possible burnout. Do Nascimento and Moreira (2019) have gained strong evidence that the behavior of nanofluids varies with the thermal regime. One of the significant aspects of subcooled boiling CHF improvement is the enhanced wettability and active nucleation that allow the liquid to maintain a good contact with the surface even during the rise in temperature. These factors inhibit the appearance of dry spots and expand the nucleate boiling area. This is not the case in saturated boiling. The process is limited by the liquid replenishment cycle, whereby thin liquid films are generated near the surface and evaporate rapidly. Dryout occurs, irrespective of the size and concentration of the nanoparticles, when replenishment by the bulk liquid is insufficient. Interestingly, at saturated nanofluids, it was reported (Jabardo and Bandarra Filho, 2000; Revellin and Thome, 2008) that CHF did not change much when the concentration of the particles or their diameter was altered, which means that in this regime, the particle-enhanced wetting was not the dominant mechanism but was the fluid dynamics of evaporation and replenishment.

## 4.3 Additional Evaluation Metrics

Beyond CHF, nanofluid performance in phase-change heat transfer depends on key parameters, such as the Onset of Nucleate Boiling (ONB), Onset of Bubble Elongation (OBE), and Onset of Flow Instability (OFI). These transitions govern boiling behavior, flow stability, and heat transfer efficiency, which are critical for microscale and high heat flux thermal systems.

### 4.3.1 Onset of Nucleation Boiling

An important turning point in boiling heat transfer is the Onset of Nucleate Boiling (ONB), which occurs when the system switches from single-phase convective heat transfer to two-phase boiling. In addition to improving the heat transfer coefficient, this transition adds new dynamics to the interaction between the surface, liquid, and vapor. The liquid can stay in a single-phase state at higher wall superheats because nanofluids are known to postpone the start of nucleate boiling. Improved surface wettability and decreased wall superheat fluctuations are primarily responsible for this seemingly paradoxical behavior. Nanofluids slow down the phase transition by suppressing premature vapor nucleation and preserving a superior liquid film on the surface. In a numerical investigation, Soleimani *et al.,* (2020) examined alumina-HFE 7100 nanofluids and reported that the increase in nanoparticle concentration resulted in a delay in ONB, which was evident from lower surface temperatures observed during boiling initiation.

### 4.3.2 Onset of Bubble Elongation

Geometric constraints cause bubble dynamics to take a different form in microchannel and mini-channel heat exchangers. The Onset of Bubble Elongation (OBE), or the moment when bubbles change from spherical to elongated slugs, is one crucial parameter in this context. Pressure drop, flow stability, and wall heat transfer are all significantly impacted by this transition. Edel and Mukherjee (2015) conducted experiments using nanofluids and observed a more uniform distribution of OBE locations compared to the highly scattered distribution seen with pure water. Consequently, OBE is more than just a morphological milestone; it also represents deeper fluid-structure interactions that have been altered by nanoparticles.

### *4.3.3 Onset of Flow Instability*

Another crucial performance criterion is flow instability during boiling, which is frequently measured as the OFI. It indicates when the flow starts to behave erratically or oscillatorily in terms of temperature, velocity, or pressure. Particularly in microchannel systems and high-flux electronic cooling applications, these instabilities can result in flow reversal, local dryout, or even thermal runaway. Although this effect is not yet entirely consistent across all studies, the addition of nanoparticles to the base fluid seems to suppress or delay OFI. There are two possible causes: First, by maintaining improved liquid-wall contact and reducing the contact angle and surface wettability, the local thermal gradients that frequently cause instabilities are lessened. In conclusion, ONB, OBE, and OFI clarify how nanofluids enhance boiling stabilizing bubbles, improving stability under heat flux changes, and expanding the thermal margin, showing that nanoparticles boost and regulate boiling dynamics.

## 5. Conclusion and Future Scope

Research on phase-change heat transfer in nanofluids has transformed modern thermal system design. Overall, the discussions in this chapter show that adding nanoparticles to base fluids alters both microscopic and macroscopic heat transfer mechanisms during phase change. At the microscale, nanoparticles influence bubble dynamics, enhance nucleation, and delay surface dry-out, while at the macroscale, they promote liquid rewetting, uniform flow, and reduced thermal instability. The nanofluids possess several advantages within a condensation system, such as enhanced dropwise condensation, wicked capillary, and thin film, which accelerate condensate removal and enhance the heat transfer coefficient. Nanofluids are optimal in solar-powered and evaporation systems because they can respond to changes in surface tension and contact angles, in addition to amplifying the rate of evaporation by improving the internal convection and thermal conductivity. These properties render nanofluids especially desirable to concentrated solar power (CdSP) units, photovoltaic/thermal hybrid systems, and solar desalination, whose performance is often constrained by heat transfer. The integration of nano-enhanced phase change materials (nano-PCMs) into thermal energy storage (TES) systems opens even more possibilities for nanotechnology, and by allowing rapid charging and discharging with minimal degradation, stable and dispatchable energy storage of intermittent renewable sources is possible.

However, there is still a list of practical and technical challenges despite the amazing improvements. Among the key issues is the stability of nanofluids during long-term use. Issues such as particle agglomeration, sedimentation, and surface fouling can be detrimental to the system performance, particularly in high-temperature or closed-loop applications. To ensure that nanoparticles can be dispersed uniformly and can be used with system materials on a long-term basis, surface functionalization, pH regulation, and appropriate surfactant usage are needed. More so, as variations in combinations have different impacts on thermal conductivity, viscosity, and interfacial heat transport, the research on optimization of nanoparticle type, size, concentration, and shape remains an open area. In the context of the prospects of the research, some of the most promising directions are revealed:

- Combining various nanoparticles (hybrid or multi-component nanofluid), it is anticipated that research on ternary or composite nanofluids will yield customized solutions for particular heat transfer applications.
- Bio-based and eco-friendly nanofluids – The development of biodegradable base fluids and environmentally friendly nanoparticles will enable sustainable thermal management systems aligned with circular economy goals.
- Surface interaction engineering – Future work should explore micro/nanoscale surface functionalization and additive manufacturing. Integrating engineered surfaces with modified nanofluids can greatly augment the boiling stability and critical heat flux (CHF).
- Integration with emerging technologies – Integrating nanofluid-based phase-change systems with hydrogen production, fuel cells, and advanced reactors holds great promise. Their application in microgravity could also revolutionize space thermal management and aerospace systems.
- Scaling up nanofluids needs long-term testing, low-cost synthesis, and standard protocols to ensure reliable industrial performance.

In conclusion, nanofluids as phase-change heat transfer media represent a novel advancement in thermal engineering, enabling compact, efficient systems by bridging nanoscale materials science with macroscale energy technologies. To maximize the potential of nanofluids in transforming the energy infrastructure in the world, future studies must focus on interdisciplinary research in materials science, thermofluid engineering, and system modeling.

## References

[1] Ahn, H.S., Kim, H., Jo, H., Kang, S., Chang, W. and Kim, M.H. 2010. Experimental study of critical heat flux enhancement during forced convective flow boiling of nanofluid on a short heated surface. *International Journal of Multiphase Flow,* 36, 375–384.

[2] Alsabery, A.I., Abosinnee, A.S., Al -Hadraawy, S.K., Ismael, M.A., Fteiti, M.A., Hashim, I., Sheremet, M., Ghalambaz, M. and Chamkha, A.J. 2023. Convection heat transfer in enclosures with inner bodies: A review on single and two-phase nanofluid models. *Renewable and Sustainable Energy Reviews,* 183, 113424.

[3] Avramenko, A.A., Shevchuk, I.V., Tyrinov, A.I. and Blinov, D.G. 2015. Heat transfer at film condensation of moving vapor with nanoparticles over a flat surface. *International Journal of Heat and Mass Transfer,* 82, 316–324.

[4] Chen, R.-H., Phuoc, T.X. and Martello, D. 2010. Effects of nanoparticles on nanofluid droplet evaporation. *International Journal of Heat and Mass Transfer,* 53, 3677–3682.
[5] Chen, R.-H., Phuoc, T.X. and Martello, D. 2011. Surface tension of evaporating nanofluid droplets. *International Journal of Heat and Mass Transfer,* 54, 2459–2466.
[6] Chon, C.H., Paik, S., Tipton, J.B. and Kihm, K.D. 2007. Effect of nanoparticle sizes and number densities on the evaporation and dryout characteristics for strongly pinned nanofluid droplets. *Langmuir,* 23, 2953–2960.
[7] Chu, H., Yu, X., Jiang, H., Wang, D. and Xu, N. 2023. Progress in enhanced pool boiling heat transfer on macro- and micro-structured surfaces. *International Journal of Heat and Mass Transfer,* 200, 123530.
[8] Dalkilic, A.S. and Wongwises, S. 2009. Intensive literature review of condensation inside smooth and enhanced tubes. *International Journal of Heat and Mass Transfer,* 52, 3409–3426.
[9] Do Nascimento, F.J., Moreira, T.A. and Ribatski, G. 2019. Flow boiling critical heat flux of DI-water and nanofluids inside smooth and nanoporous round microchannels. *International Journal of Heat and Mass Transfer,* 139, 240–253.
[10] Dutta, A., Som, S. and Das, P. 2004. Film condensation of saturated vapor over horizontal noncircular tubes with progressively increasing radius of curvature drawn in the direction of gravity. *J. Heat Transfer,* 126, 906–914.
[11] Edel, Z. and Mukherjee, A. 2015. Flow boiling dynamics of water and nanofluids in a single microchannel at different heat fluxes. *Journal of Heat Transfer,* 137, 011501.
[12] El Mghari, H., Louahlia-Gualous, H. and Lepinasse, E. 2015. Numerical study of nanofluid condensation heat transfer in a square microchannel. *Numerical Heat Transfer, Part A: Applications,* 68, 1242–1265.
[13] Fieg, G.P. and Roetzel, W. 1994. Calculation of laminar film condensation in/on inclined elliptical tubes. *International journal of heat and mass transfer,* 37, 619–624.
[14] Fu, S., Tso, C., Fong, Y. and Chao, C.Y. 2017. Evaporation of Al2O3-water nanofluids in an externally micro-grooved evaporator. *Science and Technology for the Built Environment,* 23, 345–354.
[15] Gan, Y. and Qiao, L. 2011. Evaporation characteristics of fuel droplets with the addition of nanoparticles under natural and forced convections. *International Journal of Heat and Mass Transfer,* 54, 4913–4922.
[16] He, Y., Li, H., Hu, Y., Wang, X. and Zhu, J. 2016. Boiling heat transfer characteristics of ethylene glycol and water mixture based ZnO nanofluids in a cylindrical vessel. *International Journal of Heat and Mass Transfer,* 98, 611–615.
[17] Hegde, R.N., Rao, S.S. and Reddy, R. 2012. Investigations on heat transfer enhancement in pool boiling with water-CuO nano-fluids. *Journal of thermal science,* 21, 179–183.
[18] Heysiattalab, S., Malvandi, A. and Ganji, D. 2016. Anisotropic behavior of magnetic nanofluids (MNFs) at filmwise condensation over a vertical plate in presence of a uniform variable-directional magnetic field. *Journal of Molecular Liquids,* 219, 875–882.
[19] Hsu, Y. 1962. On the size range of active nucleation cavities on a heating surface. *Journal of heat transfer,* 84, 207–213.
[20] Huminic, G. and Huminic, A. 2011. Heat transfer characteristics of a two-phase closed thermosyphons using nanofluids. *Experimental Thermal and Fluid Science,* 35, 550–557.
[21] Huminic, G. and Huminic, A. 2013. Numerical study on heat transfer characteristics of thermosyphon heat pipes using nanofluids. *Energy Conversion and Management,* 76, 393–399.
[22] Jabardo, J.M.S. and Bandarra Filho, E.P. 2000. Convective boiling of halocarbon refrigerants flowing in a horizontal copper tube–an experimental study. *Experimental thermal and fluid science,* 23, 93–104.
[23] Jamialahmadi, M., Müller -Steinhagen, H., Abdollahi, H. and Shariati, A. 2008. Experimental and theoretical studies on subcooled flow boiling of pure liquids and multicomponent mixtures. *International Journal of Heat and Mass Transfer,* 51, 2482–2493.

[24] Karimzadehkhouei, M., Shojaeian, M., Şendur, K., Mengüç, M.P. and Koşar, A. 2017. The effect of nanoparticle type and nanoparticle mass fraction on heat transfer enhancement in pool boiling. *International Journal of Heat and Mass Transfer,* 109, 157–166.

[25] Kim, J.H., Kim, J.M., Jerng, D.W., Kim, E.Y. and Ahn, H.S. 2018. Effect of aluminum oxide and reduced graphene oxide mixtures on critical heat flux enhancement. *International Journal of Heat and Mass Transfer,* 116, 858–870.

[26] Kim, S., Bang, I.C., Buongiorno, J. and Hu, L. 2006. Effects of nanoparticle deposition on surface wettability influencing boiling heat transfer in nanofluids. *Applied Physics Letters,* 89.

[27] Kim, S.J., Mckrell, T., Buongiorno, J. and Hu, L.-W. 2009. Experimental study of flow critical heat flux in alumina-water, zinc-oxide-water, and diamond-water nanofluids.

[28] Kumar Singh, S., Kumar Tiwari, A. and Paliwal, H.K. 2023. A holistic review of MXenes for solar device applications: Synthesis, characterization, properties and stability. *FlatChem,* 39, 100493.

[29] Kumar Singh, S., Kumar Tiwari, A. and Paliwal, H.K. 2024. Techno-economic assessment of retrofitted parabolic trough collector for Kalina power cycle. *Applied Thermal Engineering,* 236, 121550.

[30] Liu, D. and Yu, L. 2011. Single-phase thermal transport of nanofluids in a minichannel.

[31] Liu, Z.Liu, Z.-H., Li, Y.H., Li, Y.-Y. and Bao, R. 2010. Thermal performance of inclined grooved heat pipes using nanofluids. *International journal of thermal sciences,* 49, 1680–1687.

[32] Mahian, O., Kianifar, A., Heris, S.Z., Wen, D., Sahin, A.Z. and Wongwises, S. 2017. Nanofluids effects on the evaporation rate in a solar still equipped with a heat exchanger. *Nano energy,* 36, 134–155.

[33] Malvandi, A., Ghasemi, A., Ganji, D. and Pop, I. 2016. Effects of nanoparticles migration on heat transfer enhancement at film condensation of nanofluids over a vertical cylinder. *Advanced Powder Technology,* 27, 1941–1948.

[34] Manetti, L.L., Stephen, M.T., Beck, P.A. and Cardoso, E.M. 2017. Evaluation of the heat transfer enhancement during pool boiling using low concentrations of $Al_2O_3$-water based nanofluid. *Experimental Thermal and Fluid Science,* 87, 191–200.

[35] Narayan, G.P., Anoop, K. and Das, S.K. 2007. Mechanism of enhancement/deterioration of boiling heat transfer using stable nanoparticle suspensions over vertical tubes. *Journal of Applied Physics,* 102.

[36] Pare, A. and Ghosh, S.K. 2022. The chronological study on parametric evolution of pool boiling with nanofluids: An experimental review. *Thermal Science and Engineering Progress,* 34, 101420.

[37] Parise, J.A.R. 2012. A simulation model for the application of nanofluids as condenser coolants in vapor compression heat pumps.

[38] Paul, U.K., Mohtasim, M.S., Kibria, M.G. and Das, B.K. 2024. Nano-material based composite phase change materials and nanofluid for solar thermal energy storage applications: Featuring numerical and experimental approaches. *Journal of Energy Storage,* 98, 113032.

[39] Peng, H., Ding, G., Jiang, W., Hu, H. and Gao, Y. 2009. Heat transfer characteristics of refrigerant-based nanofluid flow boiling inside a horizontal smooth tube. *International journal of refrigeration,* 32, 1259–1270.

[40] Revellin, R. and Thome, J.R. 2008. A theoretical model for the prediction of the critical heat flux in heated microchannels. *International Journal of Heat and Mass Transfer,* 51, 1216–1225.

[41] Sefiane, K. and Bennacer, R. 2009. Nanofluids droplets evaporation kinetics and wetting dynamics on rough heated substrates. *Advances in Colloid and Interface Science,* 147, 263–271.

[42] Sharma, P.O. and Unune, D.R. 2022. Augmentation of pool boiling performance using Ag/ZnO hybrid nanofluid over EDM assisted robust heater surface modification. *Colloids and Surfaces A: Physicochemical and Engineering Aspects,* 655, 130150.

[43] Singh, S.K., Das, A.K. and Dixit, A.R. 2025a. Performance of proton exchange membrane in solar-integrated Kalina cycle systems for green hydrogen production. *Thermal Science and Engineering Progress,* 60, 103413.

[44] Singh, S.K., Dixit, A.R. and Das, A.K. 2026. Multiphysics modeling of solid oxide steam electrolysis with varying support structures for green hydrogen production. *Thermal Science and Engineering Progress,* 69, 104458.

[45] Singh, S.K., Tiwari, A.K. and Ajbar, W. 2025b. Optimization and performance enhancement of parabolic trough collectors using hybrid nanofluids and ANN modeling. *Journal of the Taiwan Institute of Chemical Engineers,* 169, 105984.

[46] Singh, S.K., Tiwari, A.K. and Paliwal, H.K. 2023. Thermo-economic assessment of hybrid Kalina cycle and organic Rankine cycle system using a parabolic trough collector solar field. *Thermal Science and Engineering Progress,* 46, 102132.

[47] Singh, S.K., Tiwari, A.K. and Said, Z. 2024. MXene nanofluid enhanced parabolic trough collectors: An integrated energy, exergy, environmental, and economic study for enhanced energy generation. *Solar Energy,* 276, 112658.

[48] Soleimani, A., Sattari, A. and Hanafizadeh, P. 2020. Thermal analysis of a microchannel heat sink cooled by two-phase flow boiling of $Al_2O_3$ HFE-7100 nanofluid. *Thermal Science and Engineering Progress,* 20, 100693.

[49] Sun, D.-W. and Zheng, L. 2006. Vacuum cooling technology for the agri-food industry: Past, present and future. *Journal of food engineering,* 77, 203–214.

[50] Suriyawong, A. and Wongwises, S. 2010. Nucleate pool boiling heat transfer characteristics of $TiO_2$–water nanofluids at very low concentrations. *Experimental Thermal and Fluid Science,* 34, 992-999.

[51] Turkyilmazoglu, M. 2015. Analytical solutions of single and multi-phase models for the condensation of nanofluid film flow and heat transfer. *European Journal of Mechanics-B/Fluids,* 53, 272–277.

[52] Vafaei, S. and Wen, D. 2010. Critical heat flux (CHF) of subcooled flow boiling of alumina nanofluids in a horizontal microchannel. *Journal of Heat Transfer,* 132.

[53] Wang, Y. and Su, G. 2016. Experimental investigation on nanofluid flow boiling heat transfer in a vertical tube under different pressure conditions. *Experimental Thermal and Fluid Science,* 77, 116–123.

[54] Wasan, D.T. and Nikolov, A.D. 2003. Spreading of nanofluids on solids. *Nature,* 423, 156–159.

[55] Wei, Y., Deng, W. and Chen, R.-H. 2016. Effects of internal circulation and particle mobility during nanofluid droplet evaporation. *International Journal of Heat and Mass Transfer,* 103, 1335–1347.

[56] Wen, D. 2008. Mechanisms of thermal nanofluids on enhanced critical heat flux (CHF). *International Journal of Heat and Mass Transfer,* 51, 4958–4965.

[57] Xing, M., Yu, J. and Wang, R. 2016. Effects of surface modification on the pool boiling heat transfer of MWNTs/water nanofluids. *International Journal of Heat and Mass Transfer,* 103, 914–919.

[58] Xu, L. and Xu, J. 2012. Nanofluid stabilizes and enhances convective boiling heat transfer in a single microchannel. *International Journal of Heat and Mass Transfer,* 55, 5673–5686.

[59] You, S., Kim, J. and Kim, K. 2003. Effect of nanoparticles on critical heat flux of water in pool boiling heat transfer. *Applied physics letters,* 83, 3374–3376.

[60] Yu, L., Sur, A. and Liu, D. 2015. Flow boiling heat transfer and two-phase flow instability of nanofluids in a minichannel. *Journal of Heat Transfer,* 137, 051502.

[61] Zhang, P., Wang, T., Jiang, Y. and Guo, C. 2023. Measurement of transient liquid film and its effect on flow boiling heat transfer in non-circular microchannels. *International Journal of Thermal Sciences,* 184, 108004.

[62] Zhao, J.-J., Duan, Y.-Y., Wang, X.-D. and Wang, B.-X. 2011. Effect of nanofluids on thin film evaporation in microchannels. *Journal of Nanoparticle Research,* 13, 5033.

# 13 Artificial Intelligence, Nanochemistry, and Ionanofluids for Sustainable Development

Aditya Shrivastav[1] and Susanta Das[1,2*]

Interdisciplinary convergence areas that hold great promise to develop sustainable technologies are artificial intelligence (AI), nanochemistry, and ionanofluids (INFs). With machine learning and materials modelling, AI is increasing the speed of nanomaterials and working fluids discovery and optimization by providing predictive structure-property- performance correlations to complement more traditional experimentation. The Nanochemistry processes enable the synthesis, control, and functionalization of nanoparticles (NPs) to enhance their thermophysical properties for energy, thermal management, and biomedical applications. Ionanofluids, which are nanomaterials dispersed in ionic liquids, exhibit high thermal conductivity and stability, along with multiple functionalities that support their use in solar thermal systems, electronic cooling, and industrial waste-heat recovery. The chapter constitutes a comprehensive platform that integrates AI-driven optimization, nanochemical design, and ionanofluid engineering within a sustainability-oriented framework. It discusses the principles of material design, considerations, and case studies and bibliometric analyses of emerging applications. Among the technical, environmental, and economic challenges, key issues are critically discussed to define the possible ways of deploying in scaling down to sustainable energy systems.

## 1. Introduction

Three overlapping areas where sustainable thermal technologies are making breakthroughs are artificial intelligence (AI), nanochemistry, and ionanofluids (INFs). The global energy systems constantly face problems with poor heat transfer, significant energy loss, and the environmental effects of traditional working fluids and material design methods. The substitution of AI in energy and materials science has transformed research and development trajectories by allowing the quick finding of novel materials, designing novel materials rationally,

1 School of Engineering, Ajeenkya D Y Patil University, Charholi Bk. via Lohegaon, Pune-412105, Maharashtra, India.
2 School of Science, Ajeenkya D Y Patil University, Charholi Bk. via Lohegaon, Pune-412105, Maharashtra, India.
* Corresponding author: susanta2024das@gmail.com

DOI: 10.1201/9781003686002-13

and optimizing new materials behind modern thermal management systems (Bai and Zhang, 2025; Peivaste *et al.*, 2025; Batool *et al.*, 2024; Chávez-Angel *et al.*, 2025). The machine learning algorithms based on AI have become powerful methods to accelerate the screening of materials, predict properties, and optimize the design by overcoming the time constraints, cost constraints, and scalability constraints of a traditional trial-and-error experimentation (Bai and Zhang, 2025; Batool *et al.*, 2024; Peivaste *et al.*, 2025; Wang, 2023). The use of data-driven methods has become standard to determine the best compositions and synthesis conditions of materials with high performance in fuel cells, thermoelectric devices, and heat exchangers in seconds. In addition, AI can also increase the precision of predictive models, the efficiency of experiments, and the sustainability of design strategies, thus minimizing the amount of waste of materials, their energy usage, and environmental footprint throughout the lifecycle of materials (Batool *et al.*, 2024; Chávez-Angel *et al.*, 2025).

The second pillar of this triad, nanochemistry, is the basis of synthesizing and functionalizing nanomaterials with both structural control and precision and physicochemical and surface capabilities. Top-down and bottom-up methods, such as chemical reduction, self-assembly, and similar techniques, are usually used to manufacture nanoparticles (NPs) with a certain size, morphology, and surface area to fulfil application-specific performance needs (Vinukonda *et al.*, 2025; Durgam and Oroszi, 2025). Despite their benefits, traditional nanofluids often lack agglomeration control, long-term stability, and consistency in thermophysical behaviour, which restrict their practical applications. Surface functionalization strategies (PEGylation, ligand exchange, and targeted chemical modification) are essential for enhancing colloidal stability, biocompatibility, and selective reactivity, and for expanding the range of applications of nanomaterials in energy conversion and biomedical systems. These are also supplemented by in situ characterization and real-time quality control methods that provide reproducibility, reliability, and scalability of nanomaterials for high-efficiency thermal systems (Vinukonda *et al.*, 2025; Durgam and Oroszi, 2025; Szczyglewska *et al.*, 2023; Rizvi *et al.*, 2022).

Meanwhile, ionanofluids, i.e., hybrid systems that combine ionic liquids (ILs) and dispersed nanomaterials, have also become the next generation of heat transfer fluids that can display a higher thermal conductivity, controllable viscosity, and increased colloidal stability (Chandra *et al.*, 2025; De Castro *et al.*, 2024; Lingala, 2023). These synthetic fluids take advantage of the inherent thermal stability, low vapour pressure, and tuneable physicochemical characteristics of ionic liquids coupled with the increased thermophysical performance of nanoparticles, giving a greatly improved thermophysical performance. For example, amino acid anion ionic liquid-based ionanofluids containing carbon nanotubes have demonstrated significant improvements in thermal conductivity, specific heat capacity, and performance, making them suitable for energy-efficient heat exchangers and thermal energy storage systems. Thousand percent thermal conductivity improvements and slashed viscosity in relation to conventional nanofluids have been reported, which result in less pumping power demand and better system-level thermal efficiency in solar thermal collectors and electronic cooling device applications, among others (Chandra *et al.*, 2025; Huminic and Huminic, 2024; De Castro *et al.*, 2024).

Nevertheless, the progress in these areas does not mean that the creation of ionanofluids is mostly where empirical data is concerned, and there is still not much integration of AI-based optimization, standard sustainability assessment, and economic viability testing. The combination of AI-based materials discovery, then precision nanochemical synthesis, and high-performance ionanofluid engineering provides unprecedented chances to overcome these drawbacks and revamp thermal management technologies to support the sustainable development agenda (Lingala, 2023; Peivaste *et al.*, 2025; Wang, 2023; Almutairi, 2023). There should be an integrated structure that ties up the material composition, synthesis parameters, thermophysical performance, and sustainability metrics (Wang, 2023; Almutairi, 2023).

Such mutually supportive advances can accelerate innovation cycles, expand synthesis frontiers, and enable mass production of next-generation heat transfer fluids that align with technological progress and global sustainability challenges. The synergistic innovation of intelligent computation systems and nanoscale material engineering not only provides greater energy efficiency and environmental performance but also resilient, adaptive, and resource-efficient thermal systems that can meet the multidimensional challenges of the new energy scenario. Therefore, the current study will contribute to the paradigm shift in the development of thermal systems by being more intelligent, environmentally friendly, and sustainable with the simultaneous use of AI, nanochemistry, and ionanofluids (Chandra *et al.*, 2025; Bai and Zhang, 2025; Wang, 2023; Lingala, 2023).

This chapter is cohesive and presents the central principles of nanochemistry in nanofluid development, composition, and benefits of ionanofluids, AI-controlled workflows in optimizing thermophysical properties, sustainability analysis and life-cycle concerns, a few representative case studies of new technological applications, and obstacles and future opportunities of large- scale implementation in the quest for sustainable development.

## 2. Fundamentals of Nanochemistry in Nanofluid Development

The creation of nanofluids has scientific foundations and relies on the concept of nanochemistry that provides the principles and methods to accurately control the properties of nanoparticles to enhance the heat transfer and the work of the energy systems (Cardoso *et al.*, 2025). Nanochemistry can regulate the pathways of synthesis, surface modification, and dispersion schemes that are highly relevant to optimizing nanofluid stability, interfacial compatibility, and active functioning (Razzaq *et al.*, 2025; Rahman *et al.*, 2024). Through these nanochemical controls, there are direct effects on the key performance measures used in thermal applications, which are thermal conductivity, viscosity, and the life span of operations.

The three main methods commonly used to create nanoparticles expected to be used in nanofluids are chemical, physical, and green methods. Chemical, sol-gel processing, co-precipitation methods, and hydrothermal methods allow regulating the size, composition, and morphology of the particles and adjusting the reaction variables such as reactant concentration, pH, temperature, and reaction duration (Altammar, 2023; Ijaz *et al.*, 2020). The products of such techniques tend to be nanoparticles of 5–100 nm size, where ratios of surface area to volume are amplified, and thermal conductivity rises. Structural and morphological characteristics are very easily controlled in the physical synthesis pathways, such as mechanical alloying and vapor deposition, but scaling, energy density, and environmental impact are of concern. In contrast, the green synthesis methods rely on plant extracts, microorganisms, or biogenic waste as reducing and capping agents, which offer a greener alternative to the classical chemical pathways and still have reasonable control over the morphology and surface chemistry of particles (Cardoso *et al.*, 2025; Altammar, 2023; Ijaz *et al.*, 2020). Such approaches are becoming more desirable in sustainable nanofluid design because they are less toxic and the general environmental footprint of the life-cycle is smaller.

Nanoparticle agglomeration and long-term colloidal stability inhibition are two of the most crucial problems of nanofluid formulation. Surface functionalization is the largest resolution to this issue in nanochemistry; either by covalent attachment to the surfaces of the nanoparticles (e.g., carboxyl, hydroxyl, and amine groups) or by adsorbing surfactants (i.e., sodium dodecylbenzene sulfonate, sodium dodecyl sulphate, cetyltrimethylammonium bromide, gum Arabic, etc.) to the surfaces of nanoparticles. It is necessary to ensure a stable thermophysical behaviour and minimize the sedimentation during long thermal cycles (Zapata-Hernandez *et al.*, 2021; Razzaq *et al.*, 2025; Cardoso *et al.*, 2025).

Surfactant selection, surface modification, and pH control can also be combined with nanochemistry to keep the particles apart and prevent sedimentation during an extended work time (Kalsi *et al.*, 2023; Manimaran *et al.*, 2025). To keep nanoparticle growth, oxidation, and surface degradation down, nanofluids are stabilized in a way that increased thermal conductivity can be retained and reproducible rheological performance can be attained in cyclic heating and cooling. The most popular technique is the two-step synthesis strategy, that is, nanoparticles are synthesized and then dispersed into the base fluid because it is simple to scale and attain high process efficacy, but this approach needs a high nanochemical control to avoid post-dispersion instability (Kalsi *et al.*, 2023; Razzaq *et al.*, 2025; Cardoso *et al.*, 2025).

With nanochemical principles, nanoparticle properties such as size, aspect ratio, porosity, and surface energy can be engineered at the nanoscale, which directly influence nanofluid thermal conductivity, viscosity, and interfacial heat transfer efficiency. High aspect ratio nanoparticles (carbon nanotubes and derivatives of graphene) may often raise thermal conductivity by over 20–40% even at low volume fractions (less than 1 vol%), but at the expense of potentially increased viscosity. The density and the chemical properties of the surface functional groups also further diversify dispersion stability and thermophysical performance, particularly in hybrid nanofluids (Manimaran *et al.*, 2025; Altammar, 2023). Moreover, base fluid characteristics and nanoparticle surface chemistry are rationally developed to construct nanofluids to suit a specific working environment, i.e., temperatures, flow regimes, and compatibility constraints of the materials used in the solar thermal systems, electronic cooling, and nuclear reactor heat exchangers (Cardoso *et al.*, 2025; Rahman *et al.*, 2024; Altammar, 2023).

In general, nanotechnology is the foundation of all nanofluid production to nanoparticle production and surface functionalization, formulation, dispersion, and use, which can be used to create versatile and viable high-performance working fluids in advanced thermal management systems (Manimaran *et al.*, 2025). Further improvements in green synthesis, atom-level surface engineering, and in-situ stability will enhance the reliability, scalability, and sustainability of nanofluids, accelerating their integration into next-generation green energy infrastructure (Cardoso *et al.*, 2025; Rahman *et al.*, 2024).

## 3. Ionanofluids (INFs): Composition, Properties, and Advantages

The latest generation of nanofluid is ionanofluids (INFs); the stable dispersions of engineered nanoparticles in ionic liquids (ILs), which are more multifunctional, exhibit better thermophysical performance, and show a higher degree of operational stability as compared to a classical nanofluid (Chandra *et al.*, 2025; Duarte *et al.*, 2024; Moulefera *et al.*, 2024). The nanofluid systems with ionic liquids may be used with base fluids, which are organic salts that melt at temperatures below 100 C, and can be used to catalogue the inherent advantages of ILs, such as negligible vapour pressure, high heat stability, and variable physicochemical properties (Moulefera *et al.*, 2024). The type of nanoparticles, concentration, and dispersion medium (pure ILs or mixtures of IL and water), as well as the choice of ionic liquid cation anion pair, have a close impact on the resulting thermophysical, chemical, and mechanical behaviour of ionanofluids (Chandra *et al.*, 2025; Duarte *et al.*, 2024; Moulefera *et al.*, 2024).

The ionanofluids are typically categorized by the type of ionic liquid (e.g. imidazolium, pyrrolidinium, ammonium, or phosphonium derivatives), the nature of dispersed nanoparticles (carbon-based nanomaterials, metal oxides or polymeric particles), and the nature of the base fluid (neat ILs or IL-water mixtures) because the ionic liquid is typically low volatility, flammable and has high decomposition temperatures. Vapor pressures of zero (or close to zero, less than 1076 Pa) are also advantageous to minimize the loss of evaporation, and at the same time, high ion-nanoparticle interaction precludes the possibility of sedimentation and agglomeration in long-term application (Chandra *et al.*, 2025; Moulefera *et al.*, 2024).

As the studies of the thermophysical characterization indicate, the thermal conductivity of ionanofluids (compared to the analogous base ionic liquids) can increase by 2070% and dramatically surpass the conventional nanofluids based on water and glycol (with an equal load of nanoparticles less than 1 vol) (Jóźwiak *et al.*, 2020; Moulefera *et al.*, 2024). Amino acid- based ionic liquids (AAILs) have been demonstrated to be 4-fold more effective at specific heat capacity than conventional heat transfer fluids and have a higher-energy storage density and reduced working fluid volume (Chandra *et al.*, 2025; Das *et al.*, 2021). Additionally, ionanofluids have wide thermal operating temperature ranges, typically from below ambient temperatures up to 250°C or higher, due to the high thermal breaking points and low vapor pressure of IL components. (Chandra *et al.*, 2025; Duarte *et al.*, 2024). These fluids are also highly electrochemically stable, oxidation-resistant, and non-flammable, thus they can be employed in safety-critical thermal systems (Faizan *et al.*, 2021).

Besides the application of enhancing thermal conductivity, the ionanofluids possess a useful synergy between nanoparticles and ionic liquids. Examples of carbon-based nanomaterials that can significantly enhance the optical absorption, transfer of charges, and photothermal conversion efficiency include graphene and carbon nanotubes, so ionanofluids are particularly promising in solar thermal collectors and hybrid photovoltaic-thermal (PV-T) systems (Moulefera *et al.*, 2024). The experimental study has revealed that optimized ionanofluid preparations can maintain a stable thermophysical performance of hundreds of heating-cooling cycles, featuring low viscosity drift (less than 10%) and bottom nanoparticle sedimentation (Faizan *et al.*, 2021; Moulefera *et al.*, 2024).

There are several advantages associated with the application of ionanofluids in terms of sustainability, including reduction of evaporative losses, longevity of the service, and ionic liquids may be recycled, ultimately leading to the reduction of negative environmental impact caused by frequently replacing a fluid (Chandra *et al.*, 2025; Faizan *et al.*, 2021). It is possible to develop less toxic and more biodegradable ionic liquids, and thus it is possible to have environmentally friendly ionanofluids that could be utilized at large-scale levels of industrial application (Duarte *et al.*, 2024). However, there are concerns of sustainability, such as the costliness of certain ionic liquids, the energy-intensive process of production, and the dearth of toxicological data on protracted exposure to the environment. Whereas IL recycling has already been developed through distillation and solvent extraction, the recovery and economic viability of using the method at large scale industrial level must be further substantiated.

The new case studies observe that the potential of ionanofluids is viable in renewable and industrial thermal processes. Ionanofluids composed of 1-ethyl-3-methylimidazolium acetate and graphene oxide are efficient by at least 37%. The dispersion of nanoparticles is stable and has a two-year history of continuous operation (Moulefera *et al.*, 2024). Similarly, the AAIL-based ionanofluids would never be inferior to the conventional nanofluids in terms of thermal stability, heat capacity, and working life (Chandra *et al.*, 2025). Even though such promising results have been obtained, extensive techno-economic assessments and conventional life-cycle analyses are still not performed to fully determine the benefits of sustainability in the long term (Sundar *et al.*, 2024; Jóźwiak *et al.*, 2020; Das *et al.*, 2021). In general, ionanofluids could represent a universal and high-performance system for sustainable thermal engineering, which combines the do-not-lintel thermal systems of ionic liquids with the do-not-lintel thermal systems of nanoparticles. They conduct thermal better, are stable over the long-term, more electrochemically strong, and less volatile, which makes them a good candidate for the new generation of uses in solar energy harvesting, electronic cooling, and waste recovery in industries. Nevertheless, the problems related to expenses, massive recyclability, and extensive assessment of environmental risks ought to be tackled as well to ensure their prevalence (Chandra *et al.*, 2025; Jóźwiak *et al.*, 2020).

## 4. Artificial Intelligence (AI) in Nanofluid Research and Development

AI is transforming the research and development field in nanofluids since it can predict and optimize the most critical thermophysical attributes, such as viscosity, thermal conductivity, specific heat capacity, and long-term stability. Rational design, stability analysis, and discovery of the next-generation nanofluid systems in high-performance thermal management and sustainable energy utilization have become highly developed with the assistance of machine learning (ML), deep learning (DL), data-driven modelling, and high-throughput analytics (Ajila *et al.*, 2024; Ganga *et al.*, 2023; Alfaleh *et al.*, 2023). Complex and nonlinear correlations between nanofluid composition, nanoparticle properties, operating conditions of synthesis, and operating conditions are correlations that are not easily solved using traditional empirical and semi-empirical correlations and can be learned by AI models using large volumes of experimental measurements, numerical modelling, and molecular dynamics simulations.

The support vector machines (SVM), random forests (RF), gradient boosting regression (GBR), decision trees, extra trees, and artificial neural networks (ANNs) are such algorithms that are superior predictors with large temperature, nanoparticle size, concentration, and base fluid composition. Some typical input parameters include nanoparticle diameter (5–100 nm), volume fraction (<1–5 vol%), base fluid properties, temperature (20–250°C), and surface functionalization descriptors, and some common output parameters include thermal conductivity, viscosity, and stability indices. In recent times, the decision tree-based and ensemble regression models have shown better predictive power of systems of viscosity and thermal conductivity of water-based nanofluids, depending on the properties of nanoparticles, such as their type, size, and concentration, as the most significant ones (Singh and Sharma, 2023; Ganga *et al.*, 2023; Alfaleh *et al.*, 2023).

Besides prediction, AI-based methodologies also can be used to optimize nanofluid preparations in a fully automated way, through the use of ML-oriented optimization schemes that are systematic in deciding what are the most valuable parameters, such as nanoparticle loading, base fluid composition, and operating temperature, to achieve the desired thermal performance using the minimal number of experiments (Ajila *et al.*, 2024; Imran *et al.*, 2025). Hyperparameter optimization techniques, including grid search and Bayesian optimization techniques, are applied to improve the model generalization and minimize overfitting, and improve interpretability through feature selection and dimensional reduction. Such quantitative metrics as coefficient of determination ($R^2$), root mean square error (RMSE), mean absolute error (MAE), and cross-validation accuracy are commonly used to find out whether the model is doing a good job and can be trusted to make predictions. This is because of the combination of supervised and unsupervised learning to be able to swiftly virtual prototype nanofluids to apply them in heat exchangers, microchannel cooling, and thermal energy storage systems (Singh and Sharma, 2023; Ganga *et al.*, 2023; Alfaleh *et al.*, 2023; Imran *et al.*, 2025).

It is also essential to employ AI-based analytical methods to understand and predict the stability of nanofluids in a dynamic operating regime. The time-dependent experimental data, morphological variables, and compositional variables are processed using the advanced ML models to categorize the dispersion stability and forecast the long-term sedimentation behaviour. The stability of silica nanoparticle-assisted smart water systems has been effectively simulated using ANN-based models that associate greater stability of dispersion with interfacial and microfluidic interactions. Similarly, the Stability and performance of hybrid and ternary nanofluids in relation to magnetic field, chemical reactivity, and input energy parameters have been evaluated through the use of ML techniques that enable one to design robust working fluids of high-efficiency cooling and energy conversion apparatus (Imran *et al.*, 2025; Khan *et al.*, 2025).

Recent applications of artificial intelligence to the study of nanofluid proceed broadly in the pipeline format: (i) collection of information about the experiment and simulations and literature; (ii) preprocessing and feature engineering of information; (iii) model selection and training; (iv) interpretation of statistical numbers; (v) explainable artificial intelligence analysis to ascertain the dominant parameters; and (vi) validation of the optimized formulations using experiment and simulation. This class of AI-based discovery pipeline offers adaptive discovery and virtual screening of large compositional spaces, which can quickly enable the discovery of thermally efficient, theologically useful, and economical nanofluids. The explainable ML models also improve trust and interpretability by demonstrating the relationship between the model's predictions and the physical phenomena, thereby making the implementation of photovoltaic thermal systems, electronic cooling, and industry heat transfer systems easier (Ganga *et al.*, 2023).

Recent studies exemplify the revolutionizing of AI in nanofluids. Ajila *et al.*, (2024) used a mixture of ML algorithms and molecular dynamics simulation to predict thermal conductivity and viscosity and discovered that the size of nanoparticles and volume fraction are the most crucial governing parameters. Alfaleh *et al.*, (2023) established that the SVM-based model is more effective in the prediction of thermophysical properties compared to classical empirical correlations. Explainable ML models enabled Kanti *et al.*, (2024) to optimise silicon dioxide, graphene oxide, and titanium dioxide nanofluids, and can now conduct heat with great efficiency with far greater interpretability. As claimed by Singh and Sharma (2023), the application of data-driven optimization of industrial cooling systems with the assistance of GBR models turned out to be successful, and Riyadi *et al.*, (2024) implemented ANN-based frameworks in the study of ternary and hybrid magnetohydrodynamic nanofluids and described the complex processes of heat and mass transfer. Overall, these advances represent a paradigm shift in nanofluid, nanochemistry, and ionanofluid research, where AI-based approaches help researchers find solutions more quickly, cheaply, and sustainably. By organizing facts, schemes, and physical insights, AI plays a key role in developing intelligent, high-performance, and environmentally friendly thermal management systems (Imran *et al.*, 2025).

## 5. Sustainability Assessment

The sustainability of the ionanofluids is typically quantified in a life cycle analysis (LCA) model that includes 4 macro steps, i.e., extraction of raw materials, production and synthesis, use, and end-of-life disposal. Each stage directly affects the overall environmental footprint and sustainability profile of ionanofluids. At the raw material stage, ionic liquids (ILs) and nanomaterials, such as multi-walled carbon nanotubes (MWCNTs) or graphene oxide (GO), are selected for their high thermal, structural, and chemical stability. Subsequently, functional nanomaterials are prepared and nano-dispersed in low ionic viscosity, thermal conductivity anions of acetate or amino acid functional groups that provide a wide operating temperature range (0–200°C). The ionic liquid matrix and nanostructures have good interfacial interaction in these systems, which leads to long-term colloidal stability and long lifetime of operation. Ionanofluids are emerging as next-generation heat transfer fluids (HTFs) for advanced heat exchangers and flat-plate solar thermal collectors, where their thermophysical performance is significantly enhanced. The parameters that significantly affect the environment at the disposal phase are toxicity, biodegradability, and recyclability, as well as recovery of ILs and nanoparticles (Chandra *et al.*, 2025; Moulefera *et al.*, 2024).

In quantitative terms, the current literature shows that the CO2-equivalent of the life-cycle emissions can be reduced by some 15–35% when ionanofluids are used in place of traditional heat transfer fluids, primarily due to an augmented thermal efficiency, and a decline in the amount of energy needed to operate it. The ionanofluids typically possess high specific heat capacity (up to –10 $Jg^{-1}{}^{\circ}C^{-1}$) in comparison with water (4.184 $Jg^{-1}{}^{\circ}C^{-1}$) and thermal conductivity additions of greater than 40 in comparison with the conventional IL-based nanofluids. All these improvements can be performed at a lower fluid volume, reduced pumping power, and an increase in the rate of heat transfer. It has been argued that a 20–30% reduction in energy consumption during the operation of thermal systems and a 10–25% reduction in water use in systems that depend on cooling are possible due to increased heat exchange efficiency and decreased circulation requirements. Enhanced colloidal stability, which commonly lasts between months and in certain instances, years, lowers the maintenance and replacement needs, which is one of the factors that minimizes the use of resources and generation of waste (Maciel *et al.*, 2019; Ordoñez *et al.*, 2025).

It has been experimentally proven that the collector efficiency of the solar thermal can increase by up to 37.4 when the temperature of the outlet is doubled, because of the utilization of the graphene oxide-based ionanofluids, and the greater the ability of the nanomaterials to lower energy consumption and carbon footprint in thermal management systems is also demonstrated (Chandra *et al.*, 2025; Moulefera *et al.*, 2024). However, LCA trade-offs reveal that synthesis stages of certain nanoparticles and ionic liquids, in which the energy-intensive reaction is still present and, therefore, contributes to the embodied energy, and localized emissions during the production process are present. This highlights the importance and the importance of green synthetic pathways and scalable production strategies to offset the environmental cost on the upstream (Maciel *et al.*, 2019; Ordoñez *et al.*, 2025).

Eco-friendliness and green chemistry are two core values of the sustainable development of the ionanofluid. The lack of halogenated anions and long alkyl chains in next-generation ionic liquids and, in particular, amino-acid-based ILs, has been shown to increase biodegradability and biocompatibility to a considerable extent, and this lowers the potential of ecotoxicity risk. The conventional ionic liquids, despite their thermal resistance, are persistent and harmful to the water and soil environment. On the other hand, the recently developed amino-acids- based IL preparations and engineered nanofluid preparations are much less ecotoxic and more environmentally friendly. In addition, ionic liquids do not have a high vapor pressure or thermal stability (up to approximately 300°C), therefore, recovery, reuse, and recyclability can be achieved, which significantly decreases the amount of waste generated during the life cycle (Goncalves *et al.*, 2021; Banerjee *et al.*, 2025; Khoo *et al.*, 2024).

In a long-term LCA analysis of ionanofluids, it has been revealed that the fluids can be compensated by savings in the cumulative energy of the fluids on the basis of the increased production energy of the fluids within the first 6–12 months of operations. Their classification as beneficial to the environment HTFs is, in the long run, supported by experimental E. coli uptake, which is necessary to guarantee both long-term stability and a low level of nanoparticle sedimentation. Chandra *et al.*, (2025) report that amino-acid anion-based ionanofluids achieved 21–40% of high thermal conductivity with low viscosity and high colloidal stability, which is why they are best applied in efficient and safe heat transfer processes. Similarly, Moulefera *et al.*, (2024) found that graphene oxide-based ionanofluids improve the functionality of the solar collector by 37.4% and can sustain its functionality for 2 years, which is credible and can be used in the operation of renewable energy. Halogen-free IL structures associated with reduced toxicity and enhanced sustainability are repeatedly mentioned in complementary reviews, which do not contradict the green chemistry guidelines (Chandra *et al.*, 2025; Moulefera *et al.*, 2024).

Despite these advantages, there are still sustainability issues, such as higher costs for initial materials, the energy-intensive process of nanoparticle production, and the lack of adequate recycling infrastructure scale. Overcoming these drawbacks with AI-controlled material optimization, low-energy synthesis, and recycling-oriented solutions of the circular economy will be required to support the mass adoption of it. Another field of green thermal fluid is the ionanofluids: improved energy efficiency and minimal carbon emission, improved recyclability, and consistent with the concept of life-cycle sustainability. When comparing the results based on quantitative LCA measures, such as the CO2 emissions, energy consumption, and water consumption, ionanofluids suggest a positive correlation between environmental benefits and production trade-offs. Their nanochemical engineering methods enable pronounced reductions of the energy expenditure of operation and overall environment, as the ionanofluids can be a solution to the sustainability of future energy and thermal management systems through the mechanism of ionic liquid systems (Chandra *et al.*, 2025; Moulefera *et al.*, 2024; Maciel *et al.*, 2019; Ordoñez *et al.*, 2025).

## 6. Case Studies and Emerging Applications

Recent research emphasises the viable and tested application of ionanofluids (INFs) in renewable energy, electronic cooling, and intensifying industrial processes, where they can enhance efficiency, stability, and sustainability. Nanochemistry-informed material design may be coupled with AI-based optimization to ensure that the conversion of the formulation of ionanofluids into pilot-scale and operational thermal systems could be done with greater speed, and the performance, scalability, and environmental benefits of nanochemical systems could be measured quantitatively (Hamida *et al.*, 2025; Moulefera *et al.*, 2025). Combining tuning physicochemical properties of ionic liquids with AI-based control strategies is now shown to offer predictable and reproducible thermal behaviour in realistic operating conditions, which explains their utility in the next generation energy technologies.

The ionanofluids can be of great potential in solar thermal collections and in combined photovoltaic-thermal (PV-T) systems, in which experimental and simulation-based analyses consistently demonstrate improved heat transfer and exergy. Graphene nanoplatelet (GNP)-based ionanofluids that have been prepared with the 1-ethyl-3-methylimidazolium acetate ([Emim][Ac]) as their ionomer have proven to have collector temperature factors exceeding 20–30°C and improvement in the exergy efficiency of the order of 5% when used in the outdoor conditions (Hamida *et al.*, 2025). Verified CFD simulations using experimental results also support a uniform temperature of the absorber plate and reduced losses of heat to the surroundings in the case of ionanofluids working fluid.

The utilization of MXene-based ionanofluids should be optimized with the help of artificial intelligence, and ANN and Response Surface Methodology (RSM) have been successfully optimized using multi-objective optimization to obtain nanoparticle loading, inlet temperature, and flow rate. Such AI-controlled models were found to reduce the number of experiment trials by more than 50% and are potentially found to be 15–18% more thermally efficient in PV-T systems, a good prospect for scalable deployment (Hamida *et al.*, 2025; Moulefera *et al.*, 2025; Das *et al.*, 2021).

Ionanofluids and AI-based thermal control systems have been demonstrated to be non-toxic and advantageous in either performance or energy storage in both electric vehicles (EVs) and energy storage. Optimized coolant compositions and copper-based heat exchangers have been shown to achieve a 20% lower peak battery temperature of coolant systems controlled by AI when high-rate charge/discharge cycles are used (Zhou *et al.*, 2025; Kumar *et al.*, 2025). Such reductions are guaranteed through radical hardware-in-loop simulation and pack-level experimental validation, which guarantee reductions of the threat of thermal runaway and an increase in battery life by 10–15% (Kumar *et al.*, 2025).

Besides cooling, AI-assisted predictive diagnostics (estimation of state of charge, fault prediction, adaptive balancing) have extended system-level performance and operation safety, which justifies the viability of the ionanofluid-assisted EV thermal platforms. Ionanofluid applications in chemical engineering have been particularly seen in electronics cooling, which must be precise in thermal conductivity, viscosity, and other stability requirements in the long term. It has already been demonstrated that Amino-acid-based ionic liquid ionanofluids can be characterized by low viscosity (<30 mPa.s) and high thermal conductivity (>25% improvement) and can therefore be utilized to immerse and convectively cool power electronics and data-centre components (Chandra *et al.*, 2025). CFD simulations and experimental heat-sink measurements showed that the temperature of the hotspots can be reduced by 12–18% compared to traditional coolants, and adaptive coolants, rather than conventional ones, may be regulated in real time with the help of AI (Kumaran and Hemavathi, 2024; Shi *et al.*, 2022).

Increases in heat transfer coefficients and energy recovery efficacy have been demonstrated in ionanofluids in industrial heat exchangers, systems with process intensification. Nanofluid preparations incorporating either Al2O3 or SiC superficial nanoparticles have recorded up to 28% thermal conductivity optimisation, which has been verified in pilot-scale heat-exchangers (Abushammala *et al.*, 2025; Maghrabie *et al.*, 2021). Diffusion and deposition CFDML models have been established to simulate the fouling behaviour and service periods to reduce unplanned downtimes by about 30%. Foul and scaling can be predicted using AI-based predictive maintenance, which has been effective in machine learning and near-infrared spectroscopy, thus reducing operational downtimes and the number of resources used. System-based energy studies indicate net energy savings of 10–18% when ionanofluids are optimized through AI and put into the heat-exchanger systems, which go directly to the decarbonization in industries (Ramalingam *et al.*, 2020).

Overall, AI, nanochemistry, and ionanofluids integration is a model of sustainable energy and industrial systems that has been demonstrated to work and is scalable. Empirical experiments, CFD modelling, and artificial intelligence-enhanced optimization studies all point to the fact that ionanofluids can deliver steady gains in performance and resist the constraints on long- term stability, safety, and efficiency. This integrated solution provides a robust platform to implement high-performance thermal systems to balance the needs of performance with the sustainability challenge (Hamida *et al.*, 2025; Moulefera *et al.*, 2025).

## 7. Bibliometric Analysis

This bibliometric work presents a critical and graphic survey of the fields of AI, nanochemistry, and ionanofluids, insofar as sustainable development is involved. The given paper will describe the spread of scientific life in this quickly developing interdisciplinary issue in the world, trace the intellectual history, and present the hottest research centres with the help of the enormous dataset and the conducted comparison of the data results in Lens.org and PubMed. The analysis will be based on using two tools: bibliometric network construction will be mapped with the assistance of the VOSviewer, and the subsequent tools will be developed using Python scripts: a global research output map, a publication trend chart, and a word cloud.

Figure 1 and Figure 2 were created with the help of VOSviewer, and their network clustering visualizations are based on two different binary databases of bibliometrics, Lens and PubMed, respectively. They both centre their attention on the problems of AI, Nanochemistry, and Ionanofluids for Sustainable Development. Even though the map provided by PubMed (Figure 2) can be considered a more biomedical-oriented approach with a less extensive scope of themes, the Lens-based map (Figure 1) presents a broad interdisciplinary methodology with the emphasis on technical, environmental, and biological issues. Figure 1 displays four significant clusters with the help of the network, and each cluster is coloured according to a specific theme dimension. The red cluster in which such phrases as machine learning, renewable energy, climate modelling, and optimization of AI are interconnected demonstrates how AI is used to control the environment and sustainable systems. The green cluster deals with nanochemistry and materials research by such terms as nanoparticles, ionanofluids, nano catalysts, and green synthesis. The blue cluster includes such phrases as drug delivery, biosensors, biocompatibility, and nanomaterials that are to be used in health, which is already the case in chemical and biological applications. Lastly, the yellow cluster reflects cross-cutting domains, and this includes but is not limited to environmental protection, industrial applications, policy frameworks, and sustainability. This ecosystem of networks illustrates how Lens can visualize the integrated research ecosystem by elucidating how AI and nanochemistry might be integrated in environmental, technological, and life science sectors, so that sustainable development may be achieved. In its turn, a narrower biomedical approach is shown in Figure 2, which was retrieved from PubMed. The terms that dominate in the network clustering in this case are those associated with molecular biology, biotechnology, and nanomedicine. The primary groups are focused on such terms as nanoparticles synthesis, drug nanocarriers, biosensors, gene regulation, and toxicological evaluation. The biological concentration of PubMed adds to the reduced concentration around the terms related to health, specifically, as far as molecular diagnostics and therapeutic design are concerned. Although the term artificial intelligence (AI) is still discussed, its main value in the network is to provide computational assistance to biological modelling, and not to initiate research on sustainability.

When looking at the two images, one can conclude that even though PubMed has split this topic into the biomedical and life-science aspects, Lens offers a multidisciplinary approach and maps diverse applications of AI, nanochemistry, and sustainability. Although PubMed is more molecular and clinical in its application and is particularly applied to nanomedicine, Lens introduces the implementation of AI methods to environmental sustainability, materials optimization, and energy efficiency. The combined effect of the choice of data sources on the thematic

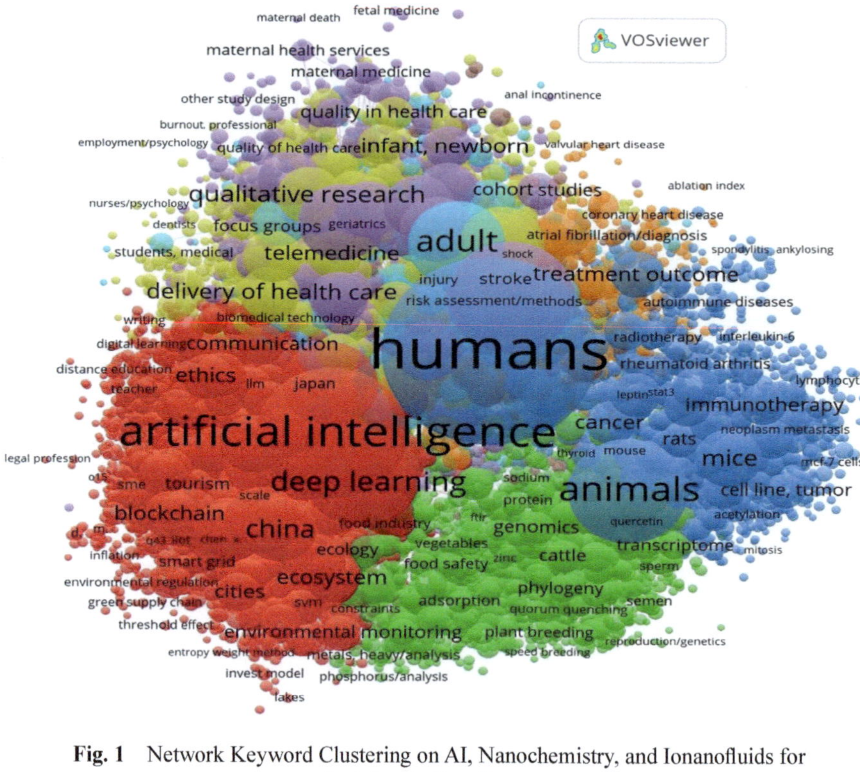


**Fig. 1** Network Keyword Clustering on AI, Nanochemistry, and Ionanofluids for Sustainable Development

(*Source:* Lens Database; created using VOSviewer) (Lens, 2025; Van Eck and Waltman, 2010).

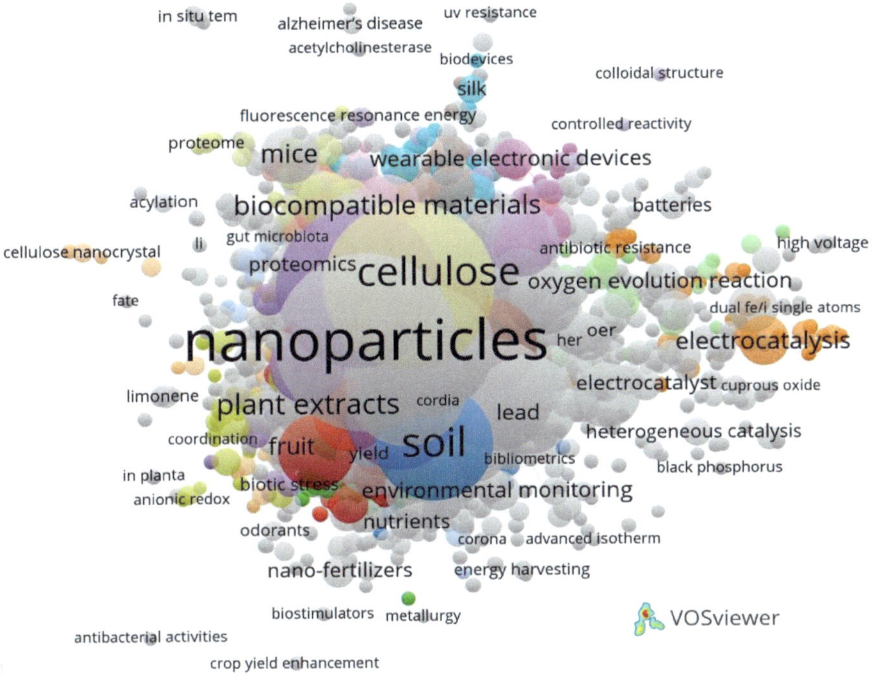


**Fig. 2** Network Keyword Clustering on AI, Nanochemistry, and Ionanofluids for Sustainable Development
(*Source:* PubMed Database; created using VOSviewer) (National Centre for Biotechnology Information, 2025; Van Eck and Waltman, 2010).

organization of bibliometric outputs can be seen by combining these two network maps, which show that both databases are complementary to each other; PubMed provides the depth of discipline, whereas Lens provides the breadth of concepts.

The visualizations of temporal overlays achieved in VOSviewer are presented in Figures 3 and 4, which reveal the progress of the research in the directions of artificial intelligence, nanochemistry, and ionanofluids development to guarantee sustainable development through time. Such maps illustrate how, in both sets of data, priorities in research have changed over the years. To reflect the history of activity of keywords, the overlay map in Figure 3 (Lens) is coloured by a colour gradient, that is, between blue (old publications) and yellow (recent publications). The blue colour is taken to denote the first phase that envelops papers published during the 1990s and early 2000s that focused on the ion-based fluids and basic computer modelling, and fundamental nanotechnology. The field developed middle-stage research 2010–2018 (marked in green), which also involved the investigation of energy-efficient fluid systems, AI-assisted process optimization, and the use of nanostructured materials to attain sustainability. Recent trends in research also revolve around areas such as artificial intelligence, machine learning, green nanochemistry, sustainable energy systems, and eco-efficient manufacturing,

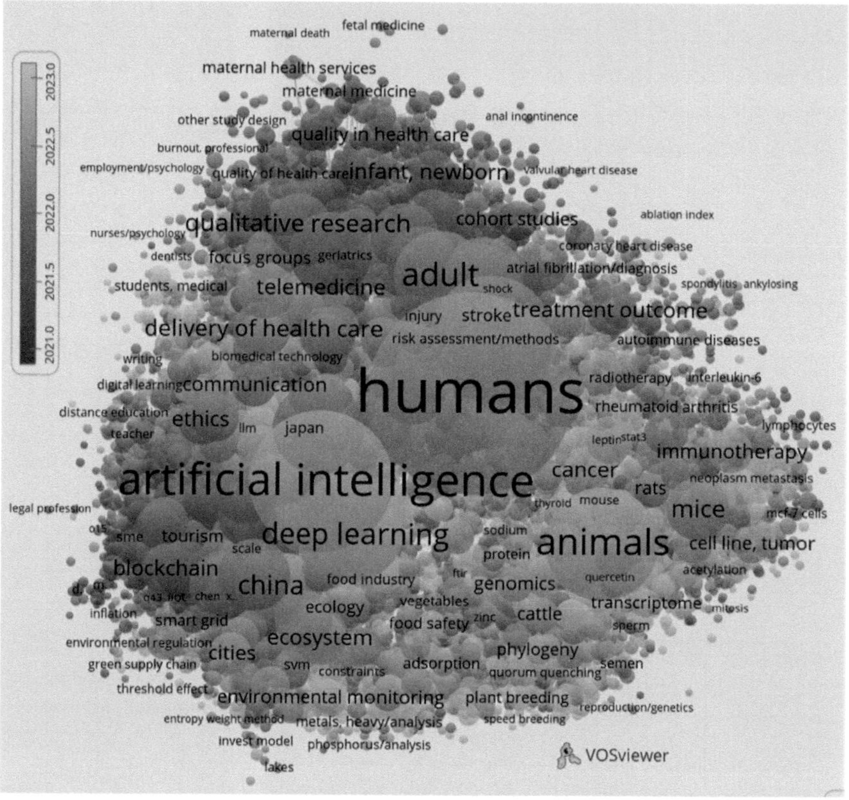


**Fig. 3** Overlay Visualization on AI, Nanochemistry, and Ionanofluids for Sustainable Development
(*Source:* Lens Database; created using VOSviewer) (Lens, 2025; Van Eck and Waltman, 2010).

and are depicted by yellow colours. It indicates the progressive adoption of AI in the nanochemical and environmental research paradigm, not only in terms of the characterization of materials and experimental synthesis but in the description of data and the creation of intelligent systems.

The PubMed Figure 4, however, illustrates a more limited time track that puts greater emphasis on the aspect of clinical translation and biomedical research and less emphasis on broader aspects of sustainability. The main areas in mid-phase research (green) during the period of 2010 to 2018 are drug delivery mechanisms, biocompatibility, and cellular imaging, and early articles (blue) are on nanoparticle production, toxicological tests, and biosensors. It is possible to note that PubMed focuses on the biomedical and pharmaceutical sphere (more recent yellow clusters (since 2018)), the trend shifts towards AI-driven molecular modelling, deep learning-based drug development, and precision nanomedicine. PubMed visualization is a specialized development that is limited to health sciences, as opposed to

Lens overlay, which considers environmental and industrial sustainability. In conclusion, comparing Figures 3 and 4, it is possible to say that whereas Lens illustrates the multidisciplinary extrapolation of AI and nanochemistry to the sustainability, energy, and policy domain, PubMed represents a more specific direction of biomedicine, beginning with the nanomaterials and biosensing and continuing with the AI-assisted clinical research. Even though it is stressed that the history of research in Lens seems to be more dispersed and geographically, whereas in PubMed, it seems to be more clinical and linear, the difference leads to the fact that the sphere develops in many dimensions.

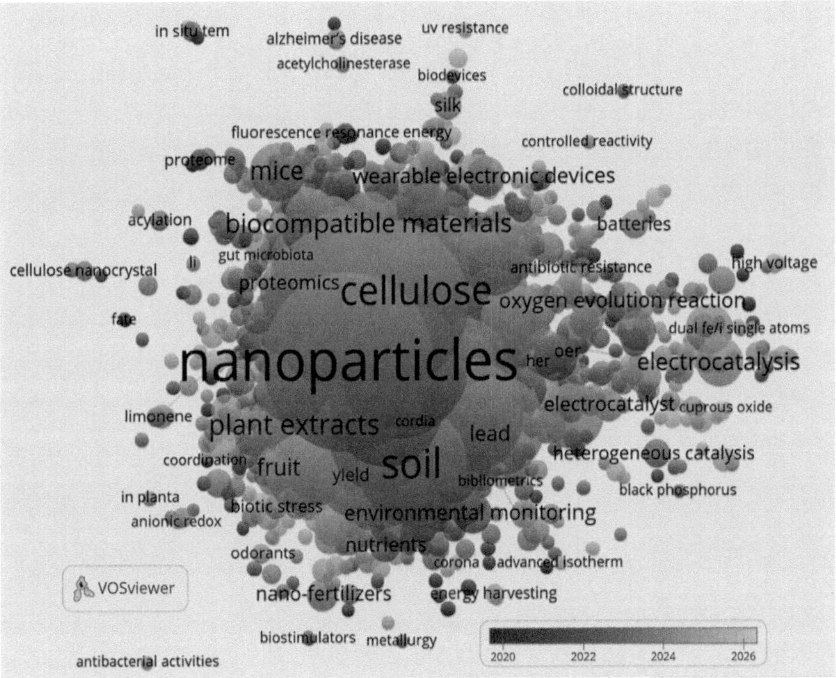


**Fig. 4** Overlay Visualization on AI, Nanochemistry, and Ionanofluids for Sustainable Development

(*Source:* PubMed Database; created using VOSviewer) (National Centre for Biotechnology Information, 2025; Van Eck and Waltman, 2010).

Figures 5 and 6 illustrate the density plot of the Lens and PubMed respectively that map the intensity and focus of the study in color-coded matrixes of the frequency of the keywords. These values shed light on the intellectual hotspots in both datasets for research. Figure 5 visualizes the density of the results that we got from Lens Database, which depicts an interdisciplinary and multilayered research framework with high-density areas (identified by yellow and red colour) around the terms, such as artificial intelligence, deep learning, nanochemistry, sustainability, green synthesis, and ionanofluids. These hotspots introduce the main spheres of scholarly concern, and the complex of AI algorithms and nanochemical composition of eco-friendly materials, environmentally friendly production, and energy conservation is the subject matter. New literature on the subject of smart materials, environmental sensors, and data-driven sustainability can be regarded as the peripheral, but significant, components of green and blue. The convergence of the innovation of research in materials science, computational information and sustainability goals is represented by Lens as a heterogeneous ecosystem, according to the density gradient.

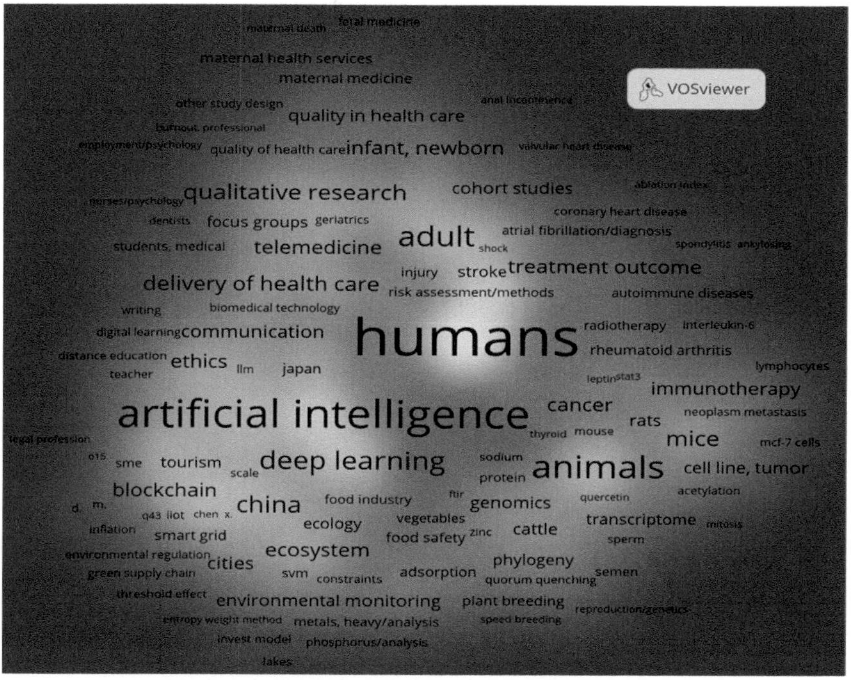


**Fig. 5** Density visualization on AI, Nanochemistry, and Ionanofluids for Sustainable Development (*Source:* Lens Database; created using VOSviewer) (Lens, 2025; Van Eck and Waltman, 2010).

Figure 6 shows a lower biomedical density profile obtained from the Lens Database, with the most significant clusters represented by keywords such as cancer, biosensors, nanoparticles, drug delivery, and gene expression. These are the primary aspects that demonstrate the abundance of therapeutic systems, biomedical diagnostics, and clinical nanotechnology in the data set of PubMed. The use of PubMed as a biomedical research depository is justified by the appearance of AI-related words, including machine learning and deep learning, occupying their supportive or peripheral positions. The PubMed density map does not depict the entire transdisciplinary landscape (as illustrated in Lens) because there are no crucial clusters on the map regarding industrial and environmental sustainability.

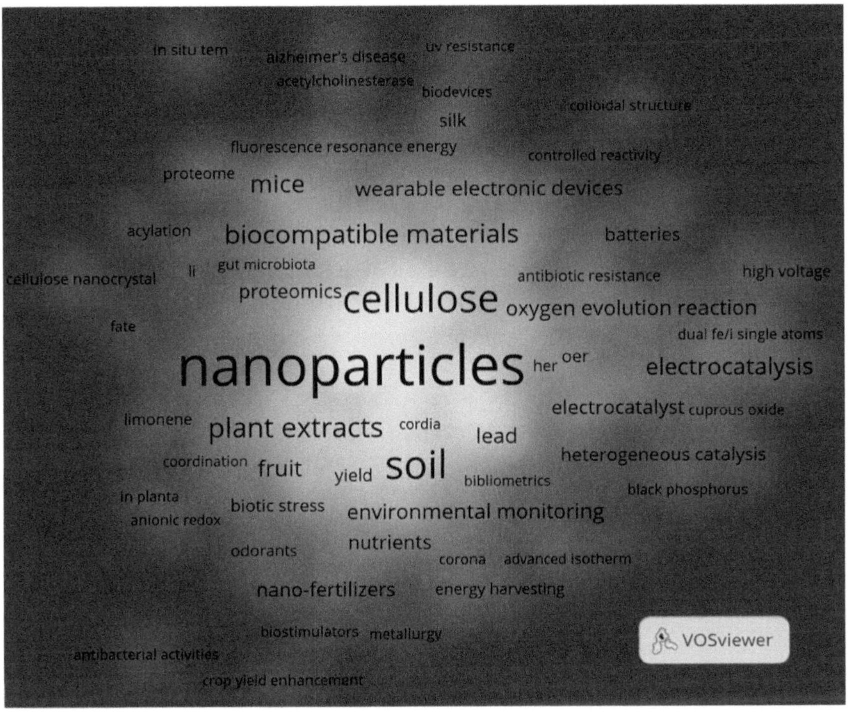


**Fig. 6** Bibliometric Analysis on AI, Nanochemistry, and Ionanofluids for Sustainable Development using VOS Viewer and data from PubMed: Density Chart (National Centre for Biotechnology Information, 2025; Van Eck and Waltman, 2010).

Bringing Figures 5 and 6 together, one can see that, whilst PubMed is a very narrow-focused density concentration on the fields of biological and pharmaceutical sciences, in Lens, the distribution of research density is multidisciplinary and includes the scope of the health, environment, materials, and energy domains. In this comparison, one can notice that Lens is the most effective one in exposing breadth on related topics, whereas PubMed is the most effective one in mapping depth on a particular topic. Even though the definition of the world research work in AI and nanochemistry varies depending upon the area of the source of data, the juxtaposition of these two density maps demonstrates that these two spheres are vast and filled with applications.

The cloud shown in Figure 7 provides an additional, frequency-driven outlook on the most essential issues that shape the scholarly discussion of the considered area. This mode of visualization allows giving a succinct and clear picture of the most prominent words by simply increasing the font size of the most common words in the literature corpus and correspondingly the frequency of the words in the literature corpus. The resultant graphic highlights the words forming the vocabulary foundation of the discipline instantaneously, and this creates a concept replica of the field under investigation. The strongest words in the cloud include: health, artificial, intelligence, machine learning, and development. Health is the primary sphere of operation, and the ultimate objective is sustainable development, which is why the list of such phrases is overwhelming, as the research is the basis of applying data-oriented technologies, grounded in artificial intelligence and machine learning. The combination of words is the most important goal of the field. These are finer grains and more details provided by an auxiliary layer of quite big yet very important keywords. This layer contains technological and methodological terms that characterize the tools and frameworks in use that consist of model, system, network, digital, analysis, and blockchain. It also contains such words as disease, cancer, medical, patient, care, environmental, sustainable, agriculture, and energy that define the field of application and goals. Such a stratification of keywords provides an idea about the scope of applications that are both medical and environmental conservation. Lastly, the third level of the terms, which are ethics, education, social, management, and policy, reveals the major issues of the non-technical aspects of the research. The presence of the above words means that there is more and more academic discourse and the realization of the ethical implications, social effects, and political mechanisms that can help sustain the use of these potent technologies in a responsible manner. In conclusion, the word cloud, however, can divide the field into its basic ingredients: its core element of AI technology, an enormous variety of applications to the environment and health, and a critical stratum of socio-ethical issues.

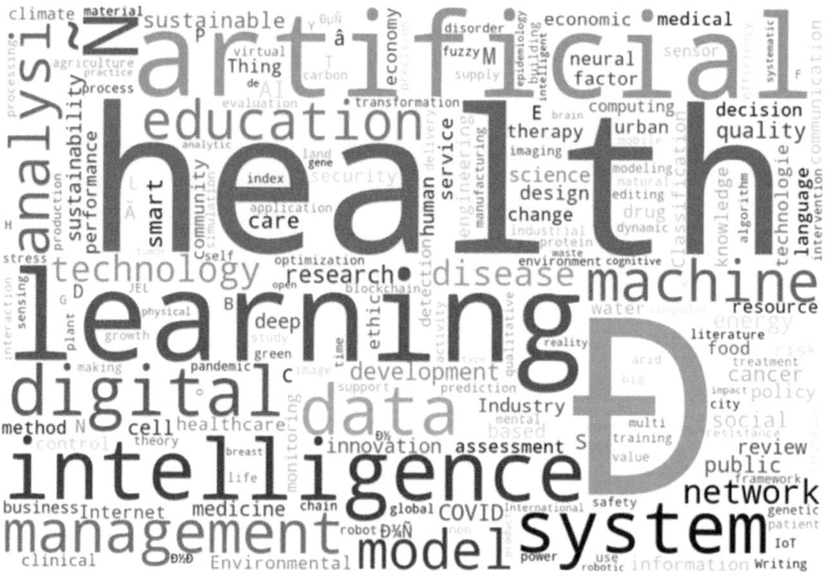


**Fig. 7** Word Cloud on AI, Nanochemistry, and Ionanofluids for Sustainable Development (*Source:* Lens Database) (Lens, 2025; Van Eck and Waltman, 2010).

Figure 13.8, an area chart, following the number of publications per year since 2000 and 2025, gives a good visual clue on the temporal dynamics of the quantity of research generated in this area. Such an example of a research topic that has gained popularity within a remarkably short amount of time is, perhaps, the best example of the galactic emergence of the field. There are two radically different periods that can be distinguished by the timeline. The former one is signified by an even-as-you-like strip of insignificant to no output regarding publication, the so-called period of incubation, which took place between 2000 and approximately 2018. During these eighteen years, there were a few pioneer scientists who were engaged in the nexus of artificial intelligence, nanochemistry, and sustainable development. The graph, however, shows a sharp turnaround point in 2019. The field was reaching an exponential development point starting this year. The fact that the number of publications was virtually vertical is a sign of a sudden and extensive inflow of funding, intellectual activity, and research interest. All in all, the number of publications will exceed 18,000 compared to nearly zero in 2018. There were strong forces in the world that were combined to achieve such a dynamic growth. The main reason is the fact that the technologies of deep learning and artificial intelligence are being created and democratized, and it is now accessible to all academics across all fields. This was accompanied by the revolution of big data, which became the foundation of AI-based discovery, especially in genomics, medical imaging, and satellite environmental sensing. The end of 2019 outbreak of the COVID-19 pandemic is likely to have made a significant contribution to this trend. Research based on AI to analyse health data has increased exponentially as a consequence of the pandemic because of the unprecedented global demands of fast, data-driven research in drug discovery, diagnostics, and public health. At the same time, technical solutions to the environmental problem also gained more research as the world became more concerned about the changes in climate and the UN Sustainable Development Goals. At that, the trend chart of the publication is a good illustration of how the technical preparedness and the pressing issues of the world can come together and make a new, active sphere of study.

The geographical perspective of the research world is revealed in Figure 9, a choropleth globe map that shows the number of documents per country. This representation will be required to determine the most appropriate countries and possible knowledge gaps, and to learn about the world dynamics of the creation of knowledge in this business. The scale of colour, passing along with dark blue (low production) and the bright yellow (high output), demonstrates a study setting very articulately when a small segment of the very developed countries is predominant. As the map shows, Germany is the most significant contributor to this area of study in terms of the volume of national funding allocated to it: perhaps, the reason why Germany is shown on the map in bright yellow is the outstanding national funding programs in biotechnology, artificial intelligence, and green technologies, as well as by the outstanding research facilities of its universities and research institutes, including the Max Planck and Fraunhofer societies. It is a close-up of two world scientific giants of China and the United States, which are projected in high-output

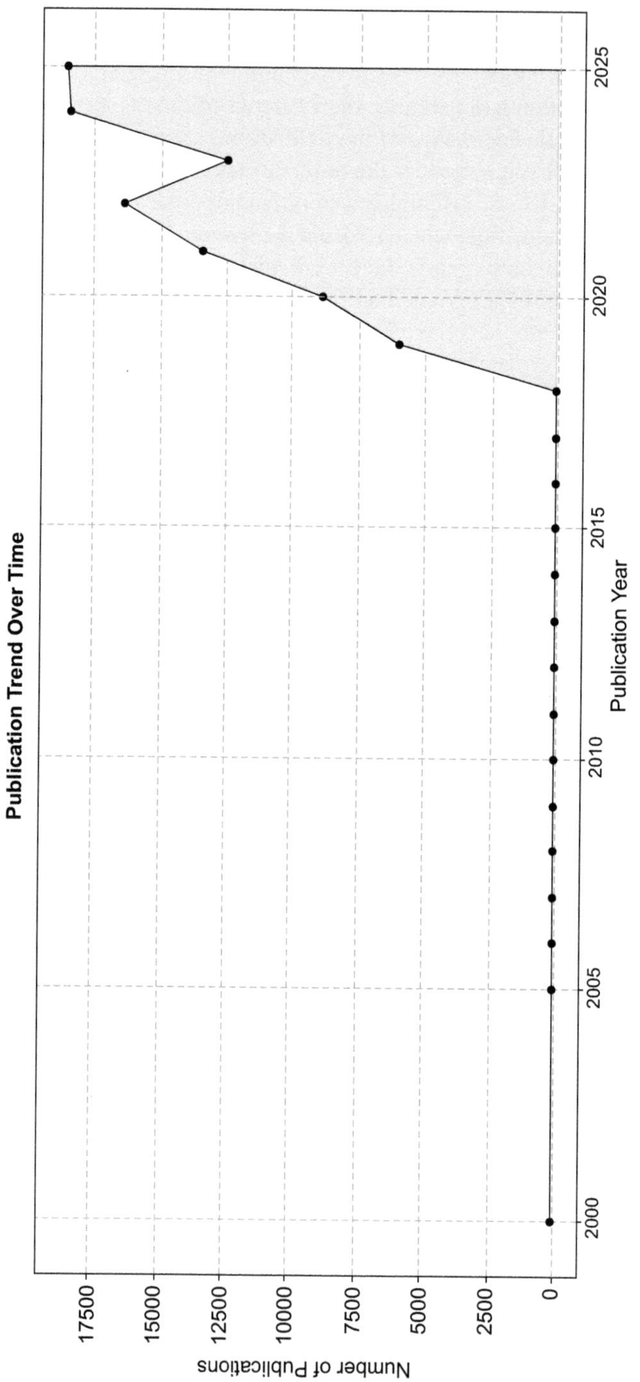


**Fig. 8** Global Trend on research of AI, Nanochemistry, and Ionanofluids for Sustainable Development (*Source:* Lens Database) (Lens, 2025; Van Eck and Waltman, 2010).

reddish-orange colour. Their contributions should be remembered, bearing in mind that they have made significant investments in the sphere of AI studies and have huge scientific teams. Others are also given in purple colours and include a list of industrialized countries in the form of the UK, Canada, Australia, and Italy, and the emerging scientific superpowers, with India being included. These nations form a second group of very prolific research centres. However, the photo shows the extreme inequalities of research abilities of the world population as well. Dark blue represents very low and insignificant production of research in large areas of the world, as is the case with most of Africa, South America, as well as parts of Central and Southeast Asia. The existence of such sharp inequality proves that the gap in the research exists in the entire world. The countries and regions that are most vulnerable to the effects of climate change and the onset of diseases, which are the same issues that this paper attempts to discuss, are the ones least able to be included in the scientific community and the formulation of a local solution. The implications of this conclusion are far-reaching regarding the aspect of sustainable development of the research problem, as the actual sustainability must be inclusive and global.

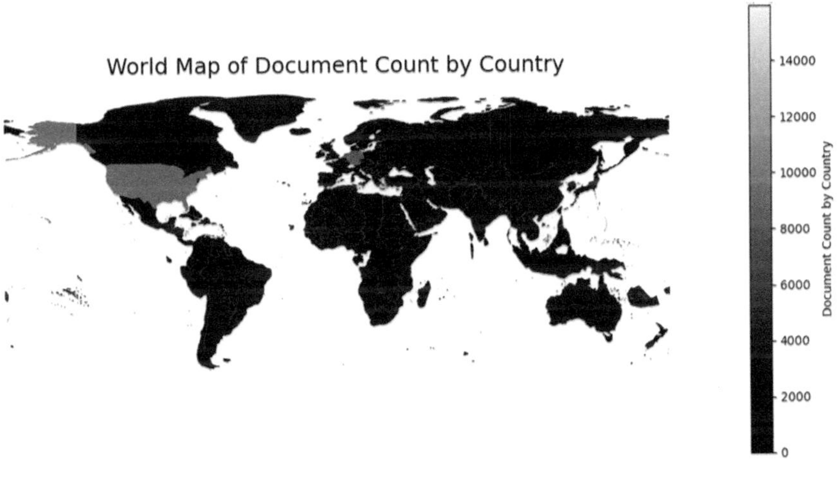


**Fig. 9** Global research on AI, Nanochemistry, and Ionanofluids for Sustainable Development (*Source:* Lens Database) (Lens, 2025; Van Eck and Waltman, 2010).

## 8. Challenges, Limitations, and Future Perspectives

Despite the high potential that ionanofluids have shown in the field of energy and environmental usage, the commercialization of these fluids on a large-scale level is limited because of a set of technical, economic, and regulatory factors. One of the most severe technical bottlenecks among them is nanoparticle agglomeration, which deteriorates the colloidal stability, increases viscosity, and eventually reduces heat transfer efficiency (Iravani *et al.*, 2024; Józiak and Boncel, 2020).

The non-uniform dispersion of nanoparticles may also cause blockage of channels in micro- and nano-scale flow systems, negatively impacting the stability of a system and its safety of operation. The stability of nanoparticles in the long term is especially problematic, since each physicochemical interaction between nanoparticles and ionic liquid matrices changes with time and temperature, shear rate, and exposure to the environment. Moreover, the process of nano chemical scale-up (i.e., going from laboratory to industrial-scale) creates challenges in terms of reproducibility, control over quality, and cost-stability, considering that the demands of stoichiometry, temperature regulation, and purification of ionic liquids and nanomaterials are very high (Razzaq *et al.*, 2025; Iravani *et al.*, 2024).

Some technical solutions that can be taken into action are the adoption of surface-grafted nanoparticles and covalently functionalized ionic liquids that can greatly reduce agglomeration without excess surfactants that raise viscosity and cost. From a feasibility standpoint, grafting techniques using silanes, amino acids, or polymer brushes can be applied in the current chemical reactors and hence can be used with current industrial plants. Moreover, closed-loop production systems can be improved with inline ultrasonication and high shear mixing units that can guarantee homogeneous dispersion during scale-up, decreasing the inter-batch variability of the product but at relatively low power consumption levels (Razzaq *et al.*, 2025; Hai *et al.*, 2024).

AI has provided an effective opportunity to curb most of these technical difficulties. The synthesis pathways are the most preferable, the aggregation trends are the most preferential, and the stabilizing formulations can be suggested at different operation regimes with the help of the predictive and optimization models that are trained on a large scale of experimental and simulation data (Razzaq *et al.*, 2025; Hai *et al.*, 2024). Nevertheless, ionanofluids are nonequilibrium multiscale and nonlinear systems, in which atomic-scale interactions are extrapolated to macroscopic thermophysical behaviour, making it difficult to generalize models and interpret them. Hybrid AI schemes that combine machine learning with molecular dynamics, CALPHAD-based thermodynamics, and reduced-order physical models are a viable solution and allow for predicting accuracy and physical consistency. These hybrid schemes decrease the requirement of massive experimentation campaigns, which decreases the cost of the development and expedites the time-to-market.

The use of ionanofluids in sustainable technologies is also restricted by economic and regulatory limitations. A significant issue is the high costs of production, which are mainly caused by the high cost of ionic liquid precursors, nanoparticle synthesis, and highly regulated processing conditions during production. There are no standardized manufacturing protocols on large scales, hence the variability of costs and the lack of incentive to invest in industry. Another major obstacle is regulatory uncertainty because detailed guidelines to regulate the environmental and health effects of ionanofluids are yet to be developed (Lourenco *et al.*, 2021; Mukherjee *et al.*, 2023). Issues surrounding nanoparticle toxicity, persistence, and bioaccumulation, especially when it comes to metallic and carbon-based nanomaterials, require a lot of toxicological and life-cycle analysis. Furthermore, cross-study comparability and certification cannot be facilitated by the unstandardized reporting measures of zeta potential, particle size distribution, viscosity, and thermal performance (Goyal *et al.*, 2025).

Cost-consciously, it can be recommended that the next generation of research should focus on low-cost, bio-derived ionic liquids (e.g., amino acid- or choline-based ILs), and on the low-cost nanoparticles that can be found on a large scale (e.g., metal oxides), which have good performance-to-cost ratios. Early techno-economic studies indicate that the substitution of fluorinated or long-chain imidazolium ILs with bio-based alternatives would help cut the material costs by 30-50 and retain reasonable thermal performance. Simultaneously, life-cycle costing and environmental risk assessment introduced at the initial stages of the design phase can help avoid expensive redesign and regulatory delays at later stages (Goyal *et al.*, 2025). In the future, it will be necessary to defeat these challenges in a combined and interdisciplinary approach that includes AI, nano chemistry, green engineering, and policy development. Future studies should focus on AI-based materials informatics tools that can remove thousands of ionanofluid compositions to find those with a balance between performance and stability, cost, and environmental safety (Das *et al.*, 2021). Creation of open-access databases of experimentally verified ionanofluid properties would also increase model reliability and decrease unnecessary experimentation. At the same time, the green synthesis with the use of renewable feedstocks, solvent-free reactions, and energy-efficient reactors can enhance the economic and environmental viability of the large-scale production (Mohammed *et al.*, 2025; Hussain *et al.*, 2025).

It is also important that standardized testing procedures and certification systems should be developed, preferably under the umbrella of international regulatory and standards bodies, to allow standardized benchmarking of performances and regulatory acceptance. This kind of standardization would reduce the barriers to entry into the industry and increase the speed of technology transfer. To sum up, to overcome the technical, economic, and regulatory obstacles related to ionanofluids, a multidisciplinary approach should be taken, involving the efforts of materials scientists, AI experts, toxicologists, economists, and policymakers. Ionanofluids can become a sustainable thermal management, energy conversion, and industrial process optimization cornerstone technology due to the presence of actionable and cost-conscious design principles and responsible governance systems (Das *et al.*, 2021; Mohammed *et al.*, 2025; Hussain *et al.*, 2025).

## 9. Conclusion

The convergence of AI, nanochemistry, and ionanofluids offers a period of revolution in sustainable thermal technologies, which have unprecedented prospects of handling worldwide energy issues through novel designs of materials and optimization. This multidisciplinary approach to a combination of AI-based predictive modelling, nanochemical production, and ionanofluid synthesis has enabled the rapid fabrication of high-performance fluids in heat transfer, reduced energy consumption, and green operations. Significant advances have been achieved, including the development of amino acid anion-based ionanofluids that are highly efficient and stable by as much as 40%, which empirical studies have demonstrated increase the thermal performance and are used in solar thermal collector systems, electronic cooling implementation, and renewable energy systems. Bibliometric analysis proves that the topic of research is developing exponentially, and the quantity of publications is increasing several times compared to the situation in 2019, which indicates that it aligns with the Sustainable Development Targets of energy efficiency and climate resilience. However, certain challenges, such as nanoparticle agglomeration, high cost of production, ambiguity in regulation, and standardization, still persist and need an interdisciplinary intervention of materials scientists, information gurus, and policymakers. To handle these challenges and enable scalable innovations that are environmentally friendly, including AI-aided material informatics, green synthesis processes, and life-cycle assessment, must fall under the priorities of future research. Nevertheless, eventually, not only does this three-way technology change thermal management, but also opens the way towards more intelligent, more environmentally conscious technology, which will result in a more stable and sustainable energy future and the union between scientific breakthrough and practical use.

## References

[1] Abushammala, O. *et al.*, (2025) 'Smart predictive maintenance of heat exchangers using AI and Near-Infrared (NIR) spectroscopy,' *Proceedings of the... International Conference on Fluid Flow, Heat and Mass Transfer*. https://doi.org/10.11159/ffhmt25.173.

[2] Ajila, F. *et al.*, (2024) 'Prediction of nanofluid thermal conductivity and viscosity with machine learning and molecular dynamics,' *Thermal Science*, 28(1 Part B), pp. 717–729. https://doi.org/10.2298/tsci230312005a.

[3] Alfaleh, A. *et al.*, (2023) 'Predicting thermal conductivity and dynamic viscosity of nanofluid by employment of Support Vector Machines: A review,' *Energy Reports*, 10, pp. 1259–1267. https://doi.org/10.1016/j.egyr.2023.08.001.

[4] Almutairi, R. (2023) 'Ionanofluids: A review on its properties and thermal applications,' *Materials Research Proceedings*, 36, pp. 16–37. https://doi.org/10.21741/9781644902790-3.

[5] Altammar, K.A. (2023) 'A review on nanoparticles: characteristics, synthesis, applications, and challenges,' *Frontiers in Microbiology*, 14. https://doi.org/10.3389/fmicb.2023.1155622.

[6] Bai, X. and Zhang, X. (2025) 'Artificial Intelligence-Powered Materials Science,' *Nano-Micro Letters*, 17(1). https://doi.org/10.1007/s40820-024-01634-8.

[7] Banerjee, S. *et al.*, (2025). Revolutionizing organic synthesis through green chemistry: metal-free, bio-based, and microwave-assisted methods,' *Frontiers in Chemistry*, 13. https://doi.org/10.3389/fchem.2025.1656935.

[8] Batool, M. *et al.*, (2024) 'Application of artificial intelligence in the materials science, with a special focus on fuel cells and electrolyzers,' *Energy and AI*, 18, p. 100424. https://doi.org/10.1016/j.egyai.2024.100424.

[9] Cardoso, B.D. *et al.*, (2025) 'Progress in nanofluid technology: from conventional to green nanofluids for biomedical, heat transfer, and machining applications,' *Nanomaterials*, 15(16), p. 1242. https://doi.org/10.3390/nano15161242.

[10] Chandra, A. *et al.*, (2025) 'Heat transfer fluids: amino acid anion ionic liquid based IoNanofluids with remarkable thermal conductivity and low viscosity,' *RSC Advances*, 15(29), pp. 23146–23164. https://doi.org/10.1039/d5ra03752b.

[11] Chávez-Angel, E. *et al.*, (2025) 'Applied artificial intelligence in materials science and material design,' *Advanced Intelligent Systems*. https://doi.org/10.1002/aisy.202400986.

[12] Das, L. *et al.*, (2021) 'State-of-the-art ionic liquid & ionanofluids incorporated with advanced nanomaterials for solar energy applications,' *Journal of Molecular Liquids*, 336, p. 116563. https://doi.org/10.1016/j.molliq.2021.116563.

[13] De Castro, C. a. N. *et al.*, (2024) 'Possible new heat transfer fluid: the IONanofluid of 1-Ethyl-3-methylimidazolium dicyanamide + Nano-Titanium Oxide—Studying its thermal conductivity and viscosity,' *Journal of Chemical & Engineering Data*, 69(6), pp. 2227–2235. https://doi.org/10.1021/acs.jced.4c00167.

[14] Duarte, T. a. G. *et al.*, (2024) 'A glance at novel ionanofluids incorporating Silk-Derived carbon dots,' *Chemistry of Materials*, 36(3), pp. 1136–1152. https://doi.org/10.1021/acs.chemmater.3c01370.

[15] Durgam, L.K. and Oroszi, T.L. (2025) 'Revolutionizing healthcare: the transformative potential of nanotechnology in medicine,' *Frontiers in Drug Delivery*, 5. https://doi.org/10.3389/fddev.2025.1556426.

[16] Faizan, M. *et al.*, (2021) 'A critical review on thermophysical and electrochemical properties of Ionanofluids (nanoparticles dispersed in ionic liquids) and their applications,' *Journal of the Taiwan Institute of Chemical Engineers*, 124, pp. 391–423. https://doi.org/10.1016/j.jtice.2021.02.004.

[17] Ganga, S. *et al.*, (2023) 'Modelling of Viscosity and Thermal Conductivity of Water-Based Nanofluids using Machine-Learning Techniques,' *International Journal of Mathematical Engineering and Management Sciences*, 8(5), pp. 817–840. https://doi.org/10.33889/ijmems.2023.8.5.047.

[18] Gonçalves, A.R.P. *et al.*, (2021) 'Ionic Liquids-A review of their toxicity to living organisms,' *International Journal of Molecular Sciences*, 22(11), p. 5612. https://doi.org/10.3390/ijms22115612.

[19] Goyal, A. *et al.*, (2025) 'Advancing biofuel production through nanomaterials: a review of efficiency improvements, challenges, and future directions,' *Nanocomposites*, 11(1), pp. 151–172. https://doi.org/10.1080/20550324.2025.2518755.

[20] Hai, T. *et al.*, (2024) 'Integrating artificial neural networks, multi-objective metaheuristic optimization, and multi-criteria decision-making for improving MXene-based ionanofluids applicable in PV/T solar systems,' *Scientific Reports*, 14(1). https://doi.org/10.1038/s41598-024-81044-3.

[21] Hamida, M.B.B. *et al.*, (2025) 'Optimization of MXene-based aqueous ionic liquids for solar systems using conventional and AI-based techniques.,' *PubMed*, 15(1), p. 20565. https://doi.org/10.1038/s41598-025-06702-6.

[22] Huminic, G. and Huminic, A. (2024) 'Capabilities of advanced heat transfer fluids on the performance of flat plate solar collector,' *Energy Reports*, 11, pp. 1945–1958. https://doi.org/10.1016/j.egyr.2024.01.044.

[23] Hussain, M. *et al.*, (2025) 'Integrating artificial intelligence and machine learning with numerical simulation for enhanced thermal performance of ternary nano-fluid,' *Journal of Computational Design and Engineering*. https://doi.org/10.1093/jcde/qwaf041.

[24] Ijaz, I. *et al.*, (2020) 'Detail review on chemical, physical and green synthesis, classification, characterizations and applications of nanoparticles,' *Green Chemistry Letters and Reviews*, 13(3), pp. 223–245. https://doi.org/10.1080/17518253.2020.1802517.

[25] Imran, M. *et al.*, (2025) 'ANN-based thermal analysis of 3D MHD hybrid nanofluid flow over a shrinking sheet via LMA,' *Scientific Reports*, 15(1). https://doi.org/10.1038/s41598-025-18363-6.

[26] Iravani, S. *et al.*, (2024) 'MXenes and artificial intelligence: fostering advancements in synthesis techniques and breakthroughs in applications,' *RSC Advances*, 14(49), pp. 36835–36851. https://doi.org/10.1039/d4ra06384h.

[27] Jóźwiak, B. and Boncel, S. (2020) 'Rheology of ionanofluids –A review,' *Journal of Molecular Liquids*, 302, p. 112568. https://doi.org/10.1016/j.molliq.2020.112568.

[28] Jóźwiak, B. *et al.*, (2020) 'Remarkable thermal conductivity enhancement in Carbon-Based ionanofluids: Effect of Nanoparticle morphology,' *ACS Applied Materials & Interfaces*, 12(34), pp. 38113–38123. https://doi.org/10.1021/acsami.0c09752.

[29] Kalsi, S. *et al.*, (2023) 'Thermophysical properties of nanofluids and their potential applications in heat transfer enhancement: A review,' *Arabian Journal of Chemistry*, 16(11), p. 105272. https://doi.org/10.1016/j.arabjc.2023.105272.

[30] Kanti, P.K. *et al.*, (2024) 'Experimental and explainable machine learning approach on thermal conductivity and viscosity of water based graphene oxide based mono and hybrid nanofluids,' *Scientific Reports*, 14(1). https://doi.org/10.1038/s41598-024-81955-1.

[31] Khan, R. *et al.*, (2025) 'AI-Integrated Design of Nanofluid-Based PVT Systems Containing Nano-Enhanced PCM: Combining Optimized GPR Models with Grasshopper Optimization and the CoCoSo Algorithm,' *Case Studies in Thermal Engineering*, p. 107081. https://doi.org/10.1016/j.csite.2025.107081.

[32] Khoo, Y.S. *et al.*, (2024) 'Techniques for recovery and recycling of ionic liquids: A review,' *The Science of the Total Environment*, 922, p. 171238. https://doi.org/10.1016/j.scitotenv.2024.171238.

[33] Kumar, P.H. *et al.*, (2025) 'Integrated thermal and battery management for electric vehicles: Experimental validation and simulation-based optimization of lithium-ion batteries,' *Energy Exploration & Exploitation* [Preprint]. https://doi.org/10.1177/01445987251337094.

[34] Kumaran, A.T. and Hemavathi, S. (2024) 'Optimization of Lithium-ion battery thermal performance using dielectric fluid immersion cooling technique,' *Process Safety and Environmental Protection*, 189, pp. 768–781. https://doi.org/10.1016/j.psep.2024.06.117.

[35] Lens (2025) *"("Artificial Intelligence" OR "Machine Learning" OR "Deep Learning") AND ("Nanochemistry" OR "Nanomaterials" OR "Nanotechnology") AND ("Ionanofluid" OR "Ionanofluids" OR "Ionic Liquid Nanofluid" OR "Nanofluid") AND ("Sustainable Development" OR "Green Chemistry" OR "Environmental Sustainability" OR "Clean Energy" OR "Renewable Energy")"*. Available at: https://www.lens.org/lens/search/scholar/list?q=(%22Artificial%20Intelligence%22%20O R%20%22Machine%20Learning%22%20OR%20%22Deep%20Learning%22)%20%20A ND%20(%22Nanochemistry%22%20OR%20%22Nanomaterials%22%20OR%20%22Na notechnology%22)%20%20AND%20(%22Ionanofluid%22%20OR%20%22Ionanofluids %22%20OR%20%22Ionic%20Liquid%20Nanofluid%22%20OR%20%22Nanofluid%22) %20%20AND%20(%22Sustainable%20Development%22%20OR%20%22Green%20Che mistry%22%20OR%20%22Environmental%20Sustainability%22%20OR%20%22Clean %20Energy%22%20OR%20%22Renewable%20Energy%22)&p=0&n=10&s=_score&d= %2B&f=false&e=false&l=en&authorField=author&dateFilterField=publishedYear&order By=%2B_score&presentation=false&preview=true&stemmed=true&useAuthorId=false (Accessed: 27 September 2025).

[36] Lingala, S.S. (2023) 'Ionic-Liquid-Based Nanofluids and their Heat-Transfer Applications: A Comprehensive review,' *ChemPhysChem*, 24(22). https://doi.org/10.1002/cphc.202300191.

[37] Lourenço, M.J. *et al.*, (2021) 'The Balance between Energy, Environmental Security, and Technical Performance: The Regulatory Challenge of Nanofluids,' *Nanomaterials*, 11(8), p. 1871. https://doi.org/10.3390/nano11081871.

[38] Maciel, V.G. *et al.*, (2019) 'State-of-the-art and limitations in the life cycle assessment of ionic liquids,' *Journal of Cleaner Production*, 217, pp. 844–858. https://doi.org/10.1016/j.jclepro.2019.01.133.

[39] Maghrabie, H.M. *et al.*, (2021) 'Intensification of heat exchanger performance utilizing nanofluids,' *International Journal of Thermofluids*, 10, p. 100071. https://doi.org/10.1016/j.ijft.2021.100071.

[40] Manimaran, M. *et al.*, (2025) 'Critical review on the stability and thermal conductivity of water-based hybrid nanofluids for heat transfer applications,' *RSC Advances*, 15(18), pp. 14088–14125. https://doi.org/10.1039/d5ra00844a.

[41] Mohammed, H.I. *et al.*, (2025) 'The role of nanotechnology and artificial intelligence in optimizing thermal energy systems,' *Applied Energy*, 400, p. 126576. https://doi.org/10.1016/j.apenergy.2025.126576.

[42] Moulefera, I. *et al.*, (2024) 'Novel application for graphene oxide-based ionanofluids in flat plate solar thermal collectors,' *Scientific Reports*, 14(1). https://doi.org/10.1038/s41598- 024-67874-1.

[43] Moulefera, I. *et al.*, (2025) 'Innovative application of graphene nanoplatelet-based ionanofluids as heat transfer fluid in hybrid photovoltaic-thermal solar collectors,' *Scientific Reports*, 15(1). https://doi.org/10.1038/s41598-025-91040-w.

[44] Mukherjee, S. *et al.*, (2023) 'Nanofluids: Critical issues, economics and sustainability perspectives,' *Particuology*, 87, pp. 147–172. https://doi.org/10.1016/j.partic.2023.06.021.

[45] National Center for Biotechnology Information (2025) *AI, nano chemistry*. Available at: https://pubmed.ncbi.nlm.nih.gov/?term=AI%2C+nano+chemistry (Accessed: 27 September 2025).

[46] Ordoñez, D.F. *et al.*, (2025) 'Unveiling the environmental costs of lignocellulosic film production with ionic liquids: the case of 1-ethyl-3-methylimidazolium acetate,' *RSC Sustainability*, 3(7), pp. 3002–3008. https://doi.org/10.1039/d5su00212e.

[47] Peivaste, I. *et al.*, (2025) 'Artificial intelligence in materials science and engineering: Current landscape, key challenges, and future trajectories,' *Composite Structures*, 372, p. 119419. https://doi.org/10.1016/j.compstruct.2025.119419.

[48] Rahman, M.A. *et al.*, (2024) 'Review on Nanofluids: Preparation, properties, stability, and thermal performance augmentation in heat transfer applications,' *ACS Omega* [Preprint]. https://doi.org/10.1021/acsomega.4c03279.

[49] Ramalingam, S. *et al.*, (2020) 'Assessment of heat transfer characteristics and system physiognomies using hybrid nanofluids in an automotive radiator,' *Chemical Engineering and Processing - Process Intensification*, 150, p. 107886. https://doi.org/10.1016/j.cep.2020.107886.

[50] Razzaq, I. *et al.*, (2025) 'Nanofluids for Advanced Applications: A comprehensive review on preparation methods, properties, and environmental impact,' *ACS Omega*, 10(6), pp. 5251–5282. https://doi.org/10.1021/acsomega.4c10143.

[51] Riyadi, T.W.B. *et al.*, (2024) 'Nanofluid heat transfer and machine learning: Insightful review of machine learning for nanofluid heat transfer enhancement in porous media and heat exchangers as sustainable and renewable energy solutions,' *Results in Engineering*, p. 103002. https://doi.org/10.1016/j.rineng.2024.103002.

[52] Rizvi, M. *et al.*, (2022) 'Functionalization of nanomaterials: synthesis and characterization,' in *ACS symposium series*, pp. 1–26. https://doi.org/10.1021/bk-2022-1418.ch001.

[53] Shi, H. *et al.*, (2022) 'Optimization on thermal management of Lithium-Ion batteries using computational fluid dynamics and air-cooling methods,' *International Journal of Electrochemical Science*, 17(5), p. 220550. https://doi.org/10.20964/2022.05.46.

[54] Singh, H.M. and Sharma, D.P. (2023) 'Machine learning model to predict the efficiency of thermal conductivity of oxide nanofluids,' *Waves in Random and Complex Media*, pp. 1–20. https://doi.org/10.1080/17455030.2023.2172233.

[55] Sundar, L.S. *et al.*, (2024) 'Thermophysical Properties, and Figures-of-Merit Analyses of nanodiamond/CuO Ionanofluids: Experimental and Artificial Neural Network Predictions,' *Journal of Ionic Liquids*, p. 100113. https://doi.org/10.1016/j.jil.2024.100113.

[56] Szczyglewska, P. *et al.*, (2023) 'Nanotechnology–General aspects: A chemical reduction approach to the synthesis of nanoparticles,' *Molecules*, 28(13), p. 4932. https://doi.org/10.3390/molecules28134932.

[57] Van Eck, N.J. and Waltman, L. (2010) 'Software survey: VOSviewer, a computer program for bibliometric mapping', *Scientometrics*, 84(2), pp. 523–538. https://doi.org/10.1007/s11192-009-0146-3.

[58] Vinukonda, A. *et al.*, (2025) 'Synthesis of nanoparticles using advanced techniques,' *Next Nanotechnology*, 8, p. 100169. https://doi.org/10.1016/j.nxnano.2025.100169.

[59] Wang, Z. (2023) 'Application of Artificial Intelligence in New Energy Materials,' *ACM*, pp. 282–288. https://doi.org/10.1145/3653081.3653128.

[60] Zapata-Hernandez, C. *et al.*, (2021) 'Surfactants versus surface functionalization to improve the stability of graphene nanofluids,' *Journal of Dispersion Science and Technology*, 43(11), pp. 1717–1724. https://doi.org/10.1080/01932691.2021.1880429.

[61] Zhou, S. *et al.*, (2025) 'Thermal management for EV power batteries based on INFO-ADRC algorithm,' *International Journal of Green Energy*, pp. 1–18. https://doi.org/10.1080/15435075.2025.2486194.

# 14 Adaptive Heat Transfer Enhancement through Magnetically Tunable Hybrid Nanofluids

Ihsan Ur Rahman[1*], Oronzio Manca[2], Bernardo Buonomo[3] and Sergio Nardini[4]

Effective thermal management is a critical challenge across various industrial and engineering applications, where conventional heat transfer fluids often fall short of higher efficiency demands. The development of nanofluids represented a significant leap forward, but the emergence of magnetic hybrid nanofluids offers excellent potential for adaptive and precise control over heat transfer processes. However, an inclusive understanding of their thermophysical behavior and heat transfer performance remains a key research gap. This chapter delves into the advanced field of adaptive heat transfer enhancement by investigating the unique capabilities of magnetically tunable hybrid nanofluids. These fluids, composed of multiple nanoparticle types suspended in a base fluid, enable their thermal properties to be dynamically controlled through the treatment of an external magnetic field. This magnetic excitation provides a superior degree of control, influencing nanoparticle alignment, thermal conductivity, and viscosity in real-time. The capability to tune these properties adaptively allows for the optimization of heat transfer rates, vital for enhancing energy efficiency and performance in systems extending from electronics cooling to solar thermal collectors. The chapter discusses the core concepts, characterization, and experimental outcomes, presenting insights into rheological behavior and improved thermal performance, opening the way for future innovations in thermal management.

## 1. Introduction

Over the past period, nanotechnology has fundamentally transformed heat transfer methodologies. This initiated advanced strategies that greatly enhance thermal efficiency across a wide range of engineering applications. (Adogbeji *et al.*, 2024). At the forefront of these innovations are nanofluids, which are developed colloidal fluids comprising nanoparticles (NPs) (1 to 100 nanometers). These nanoparticles are suspended in base fluids such as water, ethylene glycol (EG), or oil (Mohd Rijal *et al.*, 2023). This revolutionary concept, first introduced by Choi and Eastman

[1-4] Dipartimento di Ingegneria, Università degli Studi della Campania "Luigi Vanvitelli", Via Roma 29, 81031 Aversa, Caserta, Italy.

* Corresponding author: ihsanur.rahman@unicampania.it

DOI: 10.1201/9781003686002-14

(1995), addresses the inherent inefficiency of traditional heat transfer fluids (HTF) due to their minimal thermal conductivity. By incorporating metallic, oxide, or carbon-based nanoparticles, nanofluids achieve superior thermal conductivity and better convective heat transfer characteristics. Directing to significant improvements in thermal efficiency and overall system performance (Tsegaye Moltot *et al.*, 2025). Their utility spans numerous modern industrial systems, such as solar collectors, electronics cooling, heat exchangers, automotive applications, power plants, and biomedical fields, where efficient thermal management is paramount (Ullah *et al.*, 2024).

Building upon the success of conventional nanofluids, an advanced class known as hybrid nanofluids (HNFs) has emerged. It is composed of two or more distinct categories of NPs dissolved in a base fluid (Sajid and Ali, 2018). This synergistic combination aims to optimize and enhance thermal and physical properties more than what mono-component nanofluids can accomplish (Kalsi *et al.*, 2025). HNFs offer several notable advantages, including significantly increased thermal conductivity, a broader concentration range, lowered pressure drops, and reduced frictional damage. This makes them promising replacements for conventional nanofluids (Jat *et al.*, 2024; Alqawasmi *et al.*, 2023; Choudhary *et al.*, 2024). Their improved thermal performance makes them increasingly essential in many fields where demand for enhanced energy efficiency and compact system designs is high, including renewable energy, advanced solar thermal collectors, electronics cooling, energy storage, drug delivery, and polymer processing (Ullah *et al.*, 2024; Kalsi *et al.*, 2025). For example, studies have shown that specific HNFs can show superior thermal performance related to mono-component nanofluids and even greater enhancements than conventional nanofluids (Anwar *et al.*, 2025).

Despite the substantial advancements offered by nanofluids and HNFs, conventional (passive) approaches present inherent limitations. These approaches deter their extensive adoption and optimal execution in dynamic thermal systems. A primary challenge is the stability of the nanoparticle dispersion, as issues such as particle aggregation and sedimentation can occur over time (Adogbeji *et al.*, 2025). This instability can degrade thermal performance, increase fluid viscosity, and potentially lead to fouling or damage within heat transfer systems. Furthermore, passive nanofluid systems generally possess fixed thermophysical properties once formulated. Which means they cannot adapt to fluctuating thermal loads or varying operating conditions in real-time (Adogbeji *et al.*, 2024). While higher nanoparticle concentrations typically correlate with improved heat transfer. They can also lead to expanded fluid viscosity and pressure drops. This necessitates higher pumping power and consequently reduces overall thermal efficiency (Li *et al.*, 2022). Moreover, a notable scarcity of research exists for certain complex flow regimes. Particularly for HNFs in turbulent and transitional flows. A comprehensive understanding of the effect of magnetic waveform alterations and pulsating factors remains underexplored. These limitations underscore the need for more adaptable and controllable thermal management solutions.

The aforementioned challenges highlight a critical motivation for tunable and active thermal management systems. The competence to dynamically control and optimize heat transfer performance in response to changing operating conditions is crucial for next-generation engineering applications. This necessity has spurred interest in magnetic nanofluids. These are suspensions of magnetic NPs that can be manipulated by external magnetic fields (Adogbeji *et al.*, 2024; Alsangur *et al.*, 2024). The usage of a magnetic field allows for real-time control over several fluid characteristics. Including nanoparticle alignment, dispersion, and the fluid's thermophysical properties like thermal conductivity and viscosity. This magnetic tunability offers a novel and accurate method to optimize heat transfer processes. Magnetic fields can induce mixing, disturb boundary layers, and prevent particle aggregation. Consequently, continuing or increasing heat transfer efficiency and stability (Li *et al.*, 2022). Among these innovations, magnetic hybrid nanofluids (MHNFs) have emerged as highly promising candidates. It combines the superior thermal properties of HNFs with the adaptive control offered by magnetic fields (Adogbeji *et al.*, 2025). Such advanced fluids, when exposed to varied magnetic field strengths and pulsing parameters. By means of offering a novel and accurate methodology for managing and optimizing heat transfer processes in many applications (Alsangur *et al.*, 2024). This adaptive capability promises superior thermal efficiency, enhanced system performance, and significant energy savings across a wide scale of engineering applications, making MHNFs pivotal for future thermal management systems.

This chapter aims to comprehensively investigate and elucidate the adaptive heat transfer enhancement mechanisms of MHNFs across various applications. This study aims to deliver a detailed experimental analysis of how external magnetic fields, applied through varying strengths and pulsating parameters, influence the thermophysical properties and thermal-hydraulic performance of MHNF systems. The primary focus will be on evaluating their heat transfer performance, quantified by thermal conductivity and heat transfer coefficient, as well as their pressure drop and viscosity characteristics. By systematically examining these parameters for multiple nanofluids across various volume fractions, this research intends to offer novel insights into the real-time control and optimization of heat transfer progressions for advanced thermal management applications.

This chapter is divided into various sections, starting with a detailed description (Section 2) of the fundamental concepts of MHNFs, including the types of nanoparticles utilized, their working mechanism, and the impact on thermophysical properties. Further, sections 3 & 4 discuss how MHNFs are synthesized and characterized, respectively. While, section-5 explains the applicability of MHNFs in various sectors, such as electronic cooling, battery thermal management, and solar thermal collectors. Moreover, section 6 elaborates on the computational and modelling approaches. Finally, the chapter concludes with section 7, which explains current challenges faced by MHNFs and future directions for further research developments.

# 2. Core Concepts of Magnetically Tunable Hybrid Nanofluids

## 2.1 Thermal Nanoparticles and Magnetic Nanoparticles

The fundamental distinction between thermal and magnetic nanoparticles lies in their primary contributions to a nanofluid's properties and their responsiveness to external stimuli, as shown in Figure 1. Thermal nanoparticles, including aluminum oxide ($Al_2O_3$), titanium dioxide ($TiO_2$), copper (Cu), silver (Ag), carbon nanotubes (CNTs), and graphene, are primarily utilized to boost the intrinsic thermal conductivity of base fluids (Bakthavatchalam *et al.*, 2020; Tsegaye Moltot *et al.*, 2025). Their high thermal conductivity and expansive surface areas enable more efficient thermal transport. Consequently, making them valuable additions to nanofluids for improved heat transfer capabilities. In contrast, magnetic nanoparticles, notably iron oxides like magnetite ($Fe_3O_4$) and hematite ($Fe_2O_3$), nickel (Ni), and cobalt (Co), possess inherent magnetic properties that allow for their manipulation by external magnetic fields (Alsangur *et al.*, 2024). This magnetic responsiveness provides an additional dimension of control over the nanofluid's thermophysical properties and flow behavior. HNFs and specifically MHNFs leverage the synergistic effects of combining these distinct nanoparticle types to achieve enhanced thermal performance, improved stability, and dynamic tunability (Alsangur *et al.*, 2024).

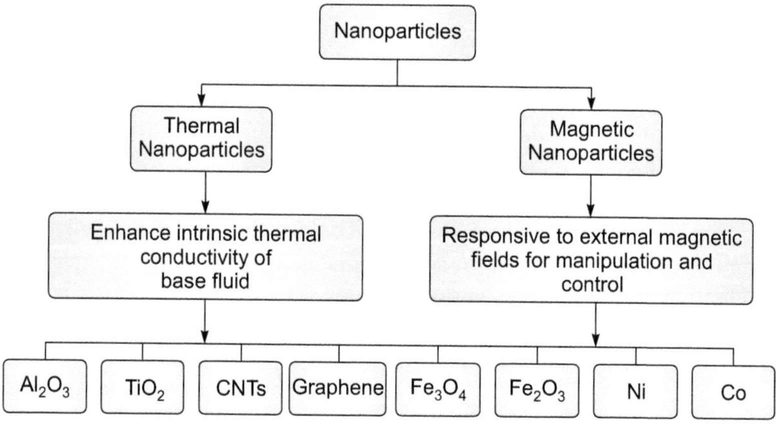


**Fig. 1** Nanoparticles and their respective types based on thermal and magnetic properties.

For instance, studies frequently investigate combinations such as $Fe_3O_4/TiO_2$, $Fe_3O_4/MgO$, and $Fe_3O_4/ZnO$ to optimize heat transfer characteristics (Adogbeji, 2015). Other hybrid systems like $Fe_3O_4$/CNT, $Fe_3O_4$/graphene, $Cu/Al_2O_3$, and $Ag/TiO_2$ have also been explored, demonstrating the versatility of this approach (Tekir *et al.*, 2022; Shahsavar *et al.*, 2016). This combination aims to integrate the corresponding characteristics of several NPs to overcome the limits of conventional mono-component nanofluids. The advantages include stability, viscosity, and fixed thermal properties, offering a more optimized and adaptable solution for advanced thermal management.

## 2.2 Working Principle

The interaction mechanisms of magnetic nanoparticles within a fluid under external magnetic fields are complex and pivotal for the tunable properties of MHNFs. External magnetic fields actively control the position and orientation of magnetic NPs. This provides a precise method for controlling the fluid's characteristics in real cases. A primary mechanism involves the alignment of magnetic moments of NPs with the local magnetic field, which can initiate the formation of chain-like clusters or structures. These aggregations are critical as they significantly impact the fluid's thermophysical properties, creating routes of smaller thermal resistance within the fluid (Karimi *et al.*, 2015; Wright *et al.*, 2007; Philip *et al.*, 2007). The principles of magnetohydrodynamics (MHD) are fundamental to understanding these interactions, as electrically conductive nanoparticles within the nanofluid respond to magnetic fields. Thereby altering the fluid flow dynamics (Hussain *et al.*, 2023). This includes inducing nanoparticle motion and clustering, disturbing the boundary layers, and exerting a resistive Lorentz force that can reduce fluid velocity and increase temperature through Joule heating (Abbas *et al.*, 2025; Tsegaye Moltot *et al.*, 2025; Ullah *et al.*, 2024). Magnetic fields can also influence the level of turbulence in the flow and prevent particle aggregation, thereby maintaining or boosting heat transfer efficiency and stability. The extent of this control is influenced by several magnetic field parameters, including the strength, waveform type (e.g., sine, square, triangular waves), frequency, and voltage. Optimal conditions, such as 4V and 60Hz across all waveform types for $Fe_3O_4/TiO_2$ nanofluids, have been identified to maximize heat transfer enhancement (Adogbeji *et al.*, 2024). The direction of the magnetic field relative to the temperature gradient or fluid flow also performs a fundamental role in shaping the resulting thermal operation (Alsangur *et al.*, 2024).

## 2.3 Influence on Thermophysical Properties

The dynamic contact of magnetic fields with hybrid nanofluids profoundly influences their convective heat transfer (CHT), thermal conductivity, and viscosity characteristics. These characteristics are discussed in the following sections;

### 2.3.1 Thermal Conductivity

Thermal conductivity is significantly enhanced by the application of magnetic fields, often due to the formation of aligned nanoparticle chains that create efficient heat transfer pathways. This enhancement generally increases with higher nanoparticle volume concentration and temperature (Alsangur *et al.*, 2024). Though strong magnetic fields can sometimes lead to a reduction in thermal conductivity at higher temperatures due to increased hydraulic resistance and particle agglomeration (Kiani *et al.*, 2021). Specific magnetic field parameters, such as voltage and frequency, can be optimized to realise peak thermal conductivity enhancement. The choice of nanoparticles also critically affects thermal conductivity, with materials like graphene and CNTs c ontributing substantially due to their inherently notable thermal conductivity and higher aspect ratios (Mohd Rijal *et al.*, 2023).

### 2.3.2 Viscosity

Viscosity is a critical thermophysical parameter that is highly sensitive to magnetic fields, nanoparticle concentration, and temperature. Typically, viscosity increases with higher NP volume fractions due to increased particle-particle and particle-fluid connections. Conversely, an increase in temperature generally leads to a decrease in viscosity (Khaleduzzaman *et al.*, 2014). When a magnetic field is employed, it can cause nanoparticles to align or cluster. It causes the formation of structures that increase the fluid's resistance to flow and consequently elevate its viscosity (Shima *et al.*, 2009). This increase in viscosity can lead to higher pressure drops across the system, necessitating greater pumping power and potentially affecting overall thermal efficiency (Adogbeji, 2015).

### 2.3.3 Convective Heat Transfer

Regarding CHT, magnetic fields offer a unique mechanism for enhancement by enabling precise manipulation of fluid flow, induction of mixing, disturbance of thermal boundary layers, and controlled migration of nanoparticles (Li *et al.*, 2022). MHNFs generally exhibit superior CHT coefficients compared to conventional base fluids or mono-component nanofluids, benefiting from the synergistic combination of their thermal and magnetic properties (Tsegaye Moltot *et al.*, 2025). Optimal CHT performance is often observed at smaller NP concentrations (e.g., 0.0125% vol and 0.00625% vol for $Fe_3O_4/TiO_2$ and $Fe_3O_4$/MgO MHNFs). Where nanoparticle dispersion is improved, and detrimental aggregation effects and hydraulic resistance are minimized (Adogbeji *et al.*, 2024). Higher fluid flow rates, characterized by increased Reynolds numbers, also correlate positively with enhanced CHT performance (Heidari *et al.*, 2021). The dynamic tunability offered by magnetic fields, through variations in strength, waveform, frequency, and voltage, provides an adaptive strategy to optimize heat transfer rates in practical applications. This responds effectively to fluctuating thermal demands in various engineering applications.

## 3. Preparation Methods

In general, magnetic hybrid nano fluids are prepared with a two-step method as shown in Figure 2. This approach is also known as the top-down approach, which is a widely employed and cost-effective method. This mechanism involves two stages, where in 1st step, nanoparticles are synthesized by various physical or chemical approaches. While in the 2nd step, these pre-prepared NPs are then added to the base fluid, and thereafter, the steps shown in Figure 2 are followed.

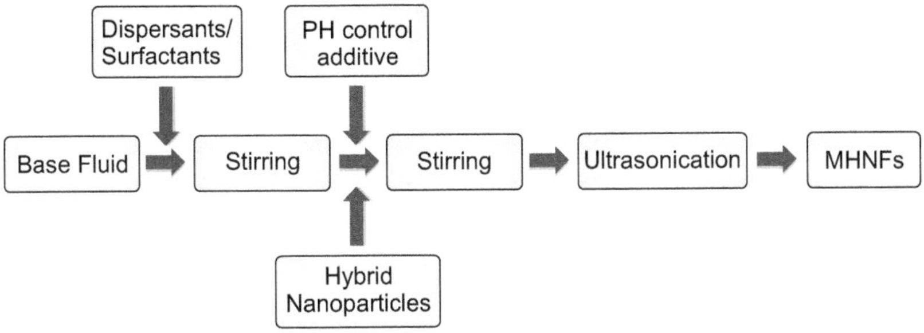


**Fig. 2** Preparation of magnetic hybrid nano fluids.

### *3.1 Selection of Base Fluid*

To pick up an appropriate base fluid is a foundational step in the synthesis of both nanofluids and hybrid nanofluids. Because it profoundly influences their thermophysical properties and ultimate performance. Common base fluids include water, particularly often deionized water (DIW), ethylene glycol (EG), and various types of oils. Deionized water is a frequently utilized base fluid (Adogbeji *et al.*, 2025). Particularly for MHNFs comprising $Fe_3O_4/TiO_2$, or $Fe_3O_4/MgO$ nanoparticles. For instance, a study prepared $Fe_3O_4/TiO_2$ and $Fe_3O_4/MgO$ MHNFs by dispersing nanoparticles in DIW at concentrations of 0.00625-0.3 vol%. The thermal properties of DIW typically include a density of 997 $kg/m^3$, thermal conductivity of 0.613 W/m.K, and a specific heat capacity (Cp) of 4179 J/kg.K. Ethylene glycol ($C_2H_6O_2$) is another prevalent base fluid, often chosen for its different thermal properties and suitability for specific applications. $Al_2O_3$–$Cu/C_2H_6O_2$ hybrid nanofluids and $Al_2O_3/C_2H_6O_2$ nanofluids have been examined using ethylene glycol. In ternary hybrid nanofluids, combinations like $Al_2O_3$, Cu, and $TiO_2$ nanoparticles are frequently distributed in ethylene glycol (Qamar *et al.*, 2025). The thermophysical properties of ethylene glycol are reported as a density of 1115 $kg/m^3$, Cp of 2430 J/kg.K, and thermal conductivity of 0.253 W/m.K (Tsegaye Moltot *et al.*, 2025). Various oils also serve as base fluids, offering different viscosities and temperature range capabilities. For example, SAE 10W-30 synthetic blend motor oil has been utilized as a base fluid for $Fe_3O_4$–GO/EO hybrid nanofluids. The properties reported include a density of 884 $kg/m^3$, thermal conductivity of 0.1440 W/m.K, Cp of 1910 J/kg.K, and a Prandtl number of 6450 (Aruna and Niranjan, 2025). Moreover, kerosene-based $Fe_3O_4$ magnetic nanofluids have shown thermal conductivity increases of up to 34.0% for a 1.0 vol% concentration (Yu *et al.*, 2010).

## 3.2 Stability of Hybrid Nanofluids

Maintaining the stability of dual-component or hybrid nanofluid systems is crucial for their effective and prolonged use. Where particle aggregation and sedimentation can severely degrade thermal performance and even damage heat transfer systems (Kalsi *et al.*, 2025). This instability is primarily driven by van der Waals forces and surface energy, which promote particle clustering. Several methods are employed to assess and enhance stability. Visual inspection is a common qualitative method, as can be observed in Figure 3. For example, no sedimentation was observed by (Adogbeji *et al.*, 2025) for $Fe_3O_4$/MgO MHNFs with volumetric fractions of 0.00625%, 0.0125%, and 0.025% over a 30-day monitoring period. While slight settling occurred at 0.05% and, significant settling occurred at 0.1%. Viscosity measurements can also indicate stability, with consistent viscosity over time suggesting good dispersion (Osman *et al.*, 2019; Giwa *et al.*, 2021). UV-Vis spectroscopy provides quantitative insights by relating absorption rates to particle volumetric fraction. Where smaller absorption rates correlate with reduced particle volumetric fractions and reduced sedimentation (Suseel Jai Krishnan *et al.*, 2022). For $Fe_3O_4/TiO_2$ (18 nm) MHNFs at an 80:20 hybridization ratio, a sedimentation percentage of 31.97% was observed, while for $Fe_3O_4$/MgO, it was 15.59% (Adogbeji *et al.*, 2024). Zeta potential measurements, which quantify the surface charge of nanoparticles, are critical, as values greater than ±30 mV generally indicate excellent stability (Li *et al.*, 2022). However, Adogbeji *et al.* (2024) conducted a study on $Fe_3O_4/TiO_2$ at an 80:20 ratio, which reported a zeta potential of ±20.3 mV and conductivity of 0.1010 mS/cm. It suggests the need for further optimization to reach higher stability thresholds. Furthermore, pH control is vital. Adjusting the pH to an optimal value, such as 8 for $Fe_3O_4$-$H_2O$ nanofluids, can significantly improve stability by preventing agglomeration and reducing transmittance (Mei *et al.*, 2019). Studies show that pH levels in MHNFs can decrease with increasing temperature. This indicates heightened acidity, which

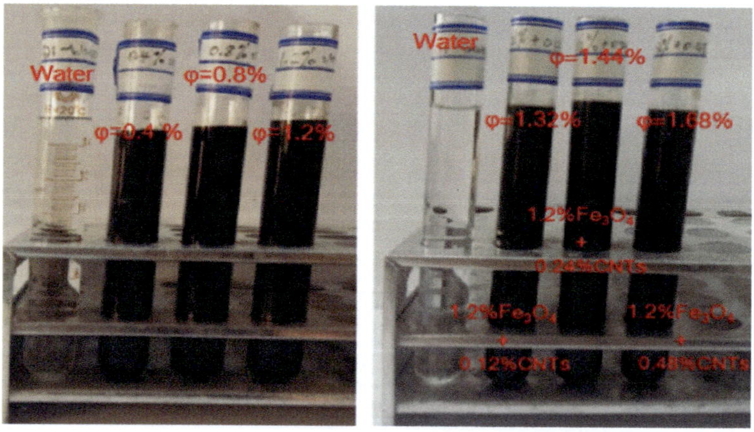


**Fig. 3** Real photographs of HNFs (Li et al., 2022).

is an important consideration for maintaining long-term stability (Chakraborty and Panigrahi, 2020).

## 3.3 Dispersion Techniques

Several advanced techniques are employed to counteract nanoparticle aggregation and ensure uniform dispersion during the preparation of magnetically tunable hybrid nanofluids.

### 3.3.1 Ultrasonication

Ultrasonication is a widely used method that applies high-frequency sound waves. These break apart nanoparticle clusters and create a stable, homogeneous dispersion (Kalsi *et al.*, 2023). In the preparation of $Fe_3O_4/TiO_2$ and $Fe_3O_4$/MgO MHNFs, nanoparticles were ultrasonically dispersed in deionized water for 4 hours using a Qsonica Q-700 ultrasonicator. An automated sequence of 5-second pulses with 2-second intervals was employed to keep stability (Giwa *et al.*, 2020). CNTs were mixed into distilled water with $Fe_3O_4$ using an ultrasonic cleaner for 60 minutes (frequency=40 kHz and a power=380 W) (Li *et al.*, 2022).

### 3.3.2 Addition of Surfactants

On the other hand, surfactants are chemical agents added to nanofluids to enhance stability by creating repulsive forces between nanoparticles, thereby inhibiting aggregation (Chakraborty and Panigrahi, 2020). Gum Arabic (GA) is a commonly employed surfactant, utilized at a 0.75 wt% ratio to safeguard the stability of MHNFs composed of $Fe_3O_4$ and $TiO_2$. With a purity of higher than 98.5%, it plays a critical role in enhancing the homogeneous distribution of MHNFs (Adogbeji *et al.*, 2024). Another example is oleic acid, which was coated onto $Fe_3O_4$ nanoparticles to improve their stability when dispersed in deionized water or kerosene, consequently preventing agglomeration (Chien and Weng, 2018).

### 3.3.3 Field-assisted Dispersion Techniques

Another technique is the field-assisted dispersion technique. The application of magnetic fields offers dynamic control over nanoparticle distribution. Magnetic fields can induce mixing within the fluid, disturb boundary layers, and actively prevent particle aggregation. Thus, maintaining or even enhancing heat transfer efficiency and overall stability (Stuyven *et al.*, 2009). By manipulating the position and positioning of magnetic nanoparticles, external magnetic fields can assist in the creation of chain-like structures. These are crucial for creating pathways of lower thermal resistance and preventing uncontrolled clustering (Nkurikiyimfura *et al.*, 2011). Maintaining a uniform magnetic field, such as one generated by a solenoid, is important for stable dispersion. where it minimizes magnetic forces that could otherwise lead to particle realignment or aggregation (Kalsi *et al.*, 2025). This adaptive capability of magnetic fields provides a precise mechanism for controlling and optimizing heat transfer processes in practical uses.

## 4. Characterization of Hybrid Nanofluids

The effective utilization of MHNFs in various thermal systems relies heavily on their thorough characterization and comprehensive testing under diverse conditions. This includes evaluating their fundamental thermophysical properties, understanding their flow behavior, and assessing their heat transfer potentials in relevant model systems. The following sections highlight these characterization techniques.

### *4.1 Measurement of Thermal Conductivity*

Usually, thermal conductivity rises with the growth in temperature and concentration of NPs. It is a key property for assessing the heat transfer potential of MHNFs. Generally, its measurement often involves specialized instruments and controlled environments. The KD2-pro instrument is a commonly employed tool. This instrument is capable of evaluating thermal conductivity in the range of 0.02–2 W/m-K with a producer-specified 5% uncertainty. It has been used to conduct continuous measurements over 24 hours, spanning temperatures with a range of 10°C to 50°C at 5°C intervals. (Adun *et al.*, 2021) utilized the KD2 analyzer by following the ASTM D7896-19 standard; the results are shown in Figure 4. Moreover, studies on $Fe_3O_4/TiO_2$ nanofluids utilized this setup, with Gum Arabic (GA) acting as a surfactant to boost stability. Similarly, investigations into $Fe_3O_4$/CNT hybrid nanofluids also employed the KD2-Pro (Decagon devices Inc., USA) to find out thermal conductivity (Alsarraf *et al.*, 2019). (Amani *et al.*, 2017) revealed that the thermal conductivity of $MnFe_2O_4$/water nanofluids exhibited the greatest improvement of 26.4% at 60°C with a nanoparticle concentration of 3vol% without a magnetic field. While a significant 36.5% enhancement was observed under a 400 Gauss (G) magnetic field at 20°C, with the same 3 vol%.

Another technique, the 3ω AC hot wire method, coupled with a 3D Helmholtz coil method to produce a consistent magnetic field, has been used for $Fe_3O_4$/CNT–water and $Fe_3O_4$/Graphene–water MHNFs. Measurements at room temperature such as 24°C showed that $Fe_3O_4$/Graphene–water samples achieved maximum thermal conductivity enhancements of approximately 51% in a parallel magnetic field and 21% in a perpendicular magnetic field, both at 5 weight percent (wt%) under 250 G. Without a magnetic field, the highest enhancements were 6.7% for $Fe_3O_4$/CNT–water and 14.6% for $Fe_3O_4$/Graphene–water at 5 wt% (Alsangur *et al.*, 2024).

Moreover, a transient hot wire method with ±5 error in measurement indicated that a 0.078 vol% magnetite nanofluid could achieve a thermal conductivity enrichment of up to 300% at 82 G (Shima *et al.*, 2009). For $Fe_3O_4$/water magnetic nanofluids, experimental studies demonstrated a maximum thermal conductivity

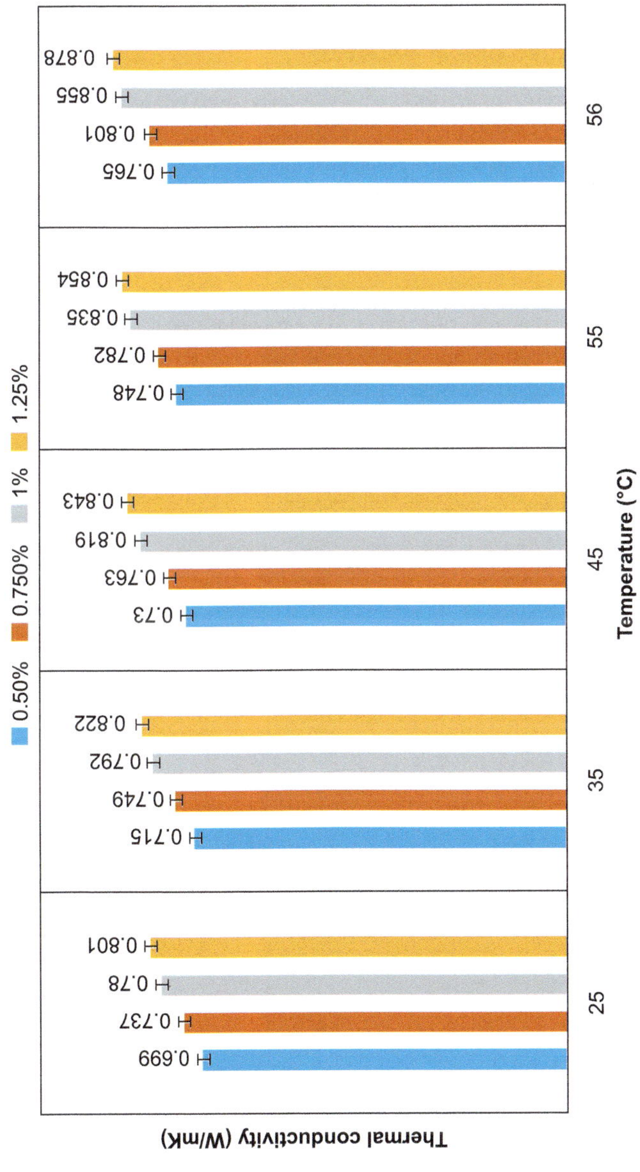


**Fig. 4** Thermal conductivity vs temperature values for 1:1:1 ratios of $Al_2O_3$–ZnO-$Fe_3O_4$ (Adun et al., 2021). Reprinted with permission from Elsevier.

increment of about 90% at a 3.0% solid volume fraction, with measurements performed across temperatures from 20°C to 55°C (Afrand *et al.*, 2016). The thermal conductivity of rGO-$Fe_3O_4$ hybrid nanofluid was also found to be enhanced by up to 11% (Mehrali *et al.*, 2017).

### 4.2 Analysis of Rheological Behavior

The rheological behavior of MHNFs, particularly their viscosity and stability, is crucial for practical applications and is significantly influenced by static and dynamic magnetic fields. Viscosity meters integrated with data loggers are used to record data at short intervals, such as over 5 hours at 30°C, to assess nanofluid stability. $Fe_3O_4/TiO_2$ nanofluids demonstrated remarkable stability with no observed sedimentation over a 30-day monitoring period. While 0.05% showed slight settling, and 0.1% exhibited substantial settling with a sedimentation factor (SF) of 344.44%. Viscosity measurements of $Fe_3O_4$-MgO/water ferrofluid samples revealed a consistent trend of decreasing fluid viscosity as temperature increased across all samples. Even in the case of DIW. The trend of viscosity against time for waste-based hybrid nanofluid containing $Fe_3O_4$ and CNTs nanoparticles with various concentrations is shown in Figure 5. However, certain ferrofluid samples demonstrated slightly higher viscosity related to DIW. On the contrary, upper concentrations like 0.2% and 0.3% vol led to the highest viscosity. This indicates an increased pressure drop within the system, requiring greater pump power (Adogbeji *et al.*, 2025).

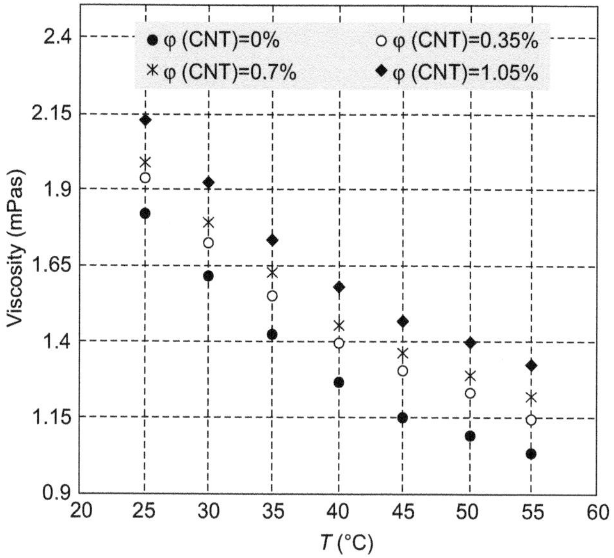


**Fig. 5** Viscosity vs temperature cure at various concentrations of CNTs while containing 0.7% of $Fe_3O_4$ (Shahsavar and Bahiraei, 2017). Reprinted with permission from Elsevier.

Whereas the application of magnetic fields can further influence viscosity. Studies show that the magnetite–CNT/oil hybrid nanofluids experience magnetorheological properties and experience viscosity enhancement due to the development of heterogeneous assemblies like chains and columns in the existence of a magnetic field (Felicia and Philip, 2014). The dynamic viscosity for various nanofluids generally drops with rising temperature. Also, higher nanoparticle volume fractions can lead to higher overall viscosity (Li *et al.*, 2022). Interestingly, research on magnetite nanofluids found that at low volume concentrations up to 0.0057 vol%, a significant enhancement was observed for thermal conductivity enhancement around 50% without a notable increase in viscosity. This makes them ideal for cooling applications (Shima *et al.*, 2009). The rGO-$Fe_3O_4$ hybrid nanofluid was observed to exhibit Newtonian behavior and showed 5.2% sedimentation after 60 days, indicating high stability (Mehrali *et al.*, 2017).

### *4.3 Heat Transfer Performance*

Testing the heat transfer performance of MHNFs in model systems like flow loops, mini/microchannels, and heated tubes provides direct evidence of their efficacy. Experimental setups typically include a heated copper tube, a flow meter, a pressure transducer, thermocouples, and a data collection system for forced convection heat transfer studies. One of the experimental setups is shown in Figure 6. For $Fe_3O_4/TiO_2$ nanofluids in a heated copper tube, 4 V at 60 Hz throughout all waveforms (sine, square, triangular), which were optimal magnetic field conditions. Greater frequencies alongside high voltages, however, led to enhanced magnetic field intensity and a drop in the heat transfer rates. At a higher nanoparticle concentration of 0.1 vol%, 16.5% increase in the heat transfer was observed in the absence of a magnetic field. Whereas this was reduced with

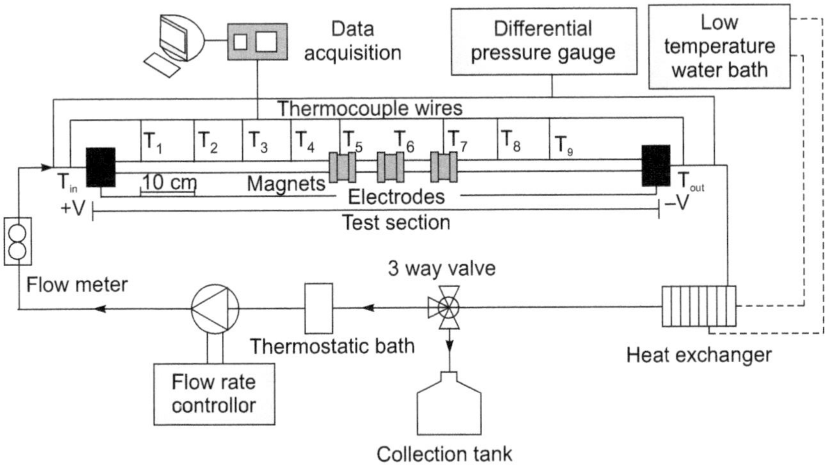


**Fig. 6** Schematic diagram of experimental setup (Li *et al.*, 2022).

the presence of a magnetic field, resulting in enhancements of 15.3%, 13.26%, and 12.59% for sine, square, and triangular waveforms. Contrarywise, at lesser volume fractions, for example, 0.0125% vol, heat transfer improvements with magnetic fields significantly surpassed, reaching 27.87%, 28.21%, and 26.74% for sine, square, and triangular waveforms (Adogbeji *et al.*, 2025). Moreover, Lee *et al.* (2021) explored the impact of the magnetic field on the CHT coefficient for an MHNF (base fluid water and EG, $Fe_3O_4$, and multiwall CNTs (MWCNTs) nanoparticles). An increase in the concentration of NPs and the magnetic field caused a rise in the CHT to some extent. Alternatively, a further expansion in concentration caused a drop in the CHT coefficient, as shown in Figure 7.

In systems using $Fe_3O_4$–CNTs hybrid nanofluids in a horizontal rotary tube, annular magnets were employed. This resulted in a 61.54% increase in Nu at Re of 996 when a magnetic field was engaged in association with conditions without a field. The maximum average Nu enhancements were 67.9% in the existence of a magnetic field and 20.89% without it, in reference to deionized water. This enhancement is attributed to nanoparticle aggregation. Triggered by an increase in local thermal conductivity near the magnets, which disrupts the wall boundary layer. The 1.44 vol% $Fe_3O_4$–CNTs–water nanofluid showed optimal heat transfer enhancement in the magnetic field (Li *et al.*, 2022). For $Fe_3O_4$-MgO/DIW MHNFs, the transition flow regime occurred at significantly higher Reynolds numbers than DIW. It showed maximum heat transfer enhancements of 31.6% for a specific vol% of MHNFs. This indicates optimal performance at these lower concentrations (Adogbeji *et al.*, 2025).

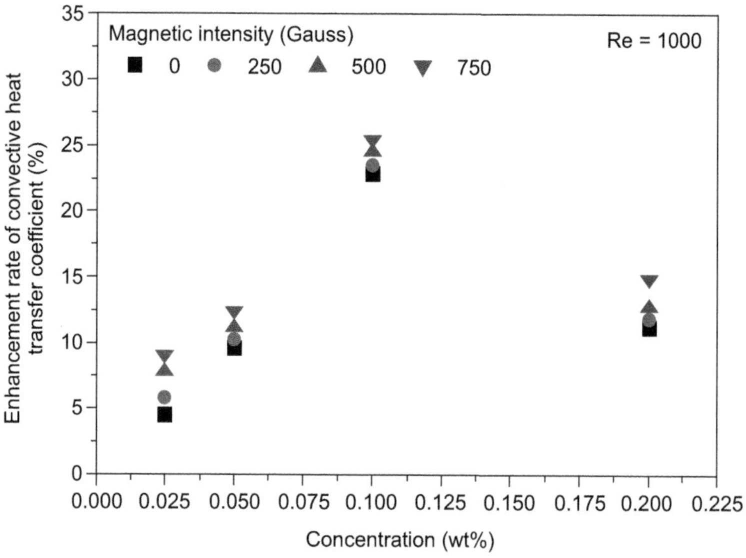


**Fig. 7** $Fe_3O_4$-MWCNT hybrid nanofluid CHT coefficient under varying magnetic field at Re = 1000 (Lee *et al.*, 2021).

Double pipe heat exchangers utilizing a non-even magnetic field produced by a wire demonstrated that the magnetic source intensified fluid circulation, leading to enhanced heat transfer. Notably, a triangular tube geometry proved 15% more efficient for heat transfer than a smooth tube (Azizi *et al.*, 2023). Moreover, numerical simulations of water–$Fe_3O_4$ ferrofluid in a square passage with permanent magnets showed that the magnetic field created mixing and boundary layer disturbance. Consequently resulted in a 40.79% higher average convective heat transfer coefficient with four magnets associated with no magnet with ferrofluid. This was 58.19% higher than pure water. Stronger magnets and a higher concentration of nanoparticles also increased heat transfer rates. Conversely, increasing the Re diminished the magnetic field's effectiveness on heat transfer (Bahiraei *et al.*, 2019). For $Fe_3O_4$-water nanofluids in a rotary tube with a parallel magnetic field, increasing NP mass fraction was favorable for heat transfer. Where the Nu was increased by up to 14.6% in laminar flow and 22.1% in turbulent flow in the lack of a magnetic field. Though the magnetic field decreased the exergy efficiency due to higher flow resistance (Mei *et al.*, 2019). Overall, MHNFs like the tetra hybrid nanofluid (Cu, $Al_2O_3$, $Fe_3O_4$, $H_2O$) can improve the heat transfer rate up to 16.783% compared to conventional fluids (Ullah *et al.*, 2024).

## 5. Applications of Magnetically Tuned Hybrid Nanofluids

MHNFs denote a substantial progress in thermal management. It offers enhanced performance and dynamic control across a broad spectrum of industrial and engineering applications. Their unique capacity to combine the superior thermal properties of HNFs with the controllability of magnetic fields makes them ideal for optimizing heat transfer. Further, this mechanism can improve system efficiency and enable adaptive responses to changing thermal demands. The following sections explore specific applications where these advanced fluids are demonstrating promising results and real-world impact. Whereas, Table 1 summarizes the research conducted on MHNFs.

### 5.1 Cooling of Electronic Devices

The ever-increasing power density of electronic devices and data centers necessitates highly efficient and adaptive cooling solutions (Rahman *et al.*, 2025a). MHNFs provide a promising solution for meeting these demands because of their advanced heat transfer capabilities and the potential for active thermal management (Qamar *et al.*, 2025; Sun *et al.*, 2024). For instance, $Fe_3O_4$-$H_2O_4$ magnetic nanofluid has been investigated for microchannel heat sink cooling applications. By demonstrating its ability to reduce temperature significantly compared to pure water. Studies showed that using this nanofluid with an NP volume fraction ($\phi$) of 0.4% decreased the temperature by 0.04%, while raising the volume fraction to 0.6% and 0.8% led to further temperature reductions of 0.07% and 0.08%, respectively (Hatami *et al.*, 2018). Moreover, Selvakumar and Suresh (2012) studied the utilization

**Table 1** Summary of the case studies conducted on magnetically tuned hybrid nanofluids.

| *Reference* | *Type of MHNFs* | *Magnetic field* | *Application* | *Outcomes* |
|---|---|---|---|---|
| (Alsarraf *et al.*, 2019) | MWCNT–$Fe_3O_4$/water Hybrid Nanofluid (e.g., 0.9% $Fe_3O_4$ + 1.35% CNT) | A constant non-uniform magnetic field is generated by three permanent magnet pairs. Maximum strength peaked at 1000 Gauss along the tube centerline | Cooling Systems/Heat Exchangers-Circular Tube | Maximum average Nusselt number increment of 109.31% related to the absent magnetic field case (at Reynolds number (Re) 500). |
| (Li *et al.*, 2022) | Fe3O4–Carbon Nanotubes (CNTs)–Water Nanofluid (1.44 vol%) | Non-uniform field generated by Annular magnets (Case III, using three cannulas) | Cooling Systems/Heat Exchangers-Circular Tube | The Nu number at Re = 996 grew by 61.54% compared to that without a magnetic field. |
| (Alsangur *et al.*, 2024) | $Fe_3O_4$/Graphene–water Magnetic Hybrid Nanofluid (5 wt%) | External uniform magnetic field of 250 G applied in the parallel and particular direction | Solar Thermal Collectors/ General Heat Transfer -Thermal Conductivity Study | Thermal conductivity enhancement of ~51% and ~21%. |
| (Alsangur *et al.*, 2024) | $Fe_3O_4$/CNT–water Magnetic Hybrid Nanofluid (5 wt%) | An external uniform magnetic field of 250 G is utilized in the parallel and vertical directions | Solar Thermal Collectors / Heat Transfer-Thermal Conductivity Study | The thermal conductivity enhancements observed were ~12% and 9%. |
| (Shima *et al.*, 2009) | Magnetite ($Fe_3O_4$) Nanofluid (Kerosene-based) | Uniform magnetic field applied parallel to the temperature slope, specifically tested up to 82 G | Thermal Management / Magneto-Rheology-Miniature Cooling/Damping | The ratio ($k/\eta$) of thermal conductivity ($k$) to viscosity ($\eta$) is controllable from 0.725 to 2.35 by tuning the employed field. Thermal conductivity enhancement reached ~300% at 82 G. |

*Contd.*

**Table 1** *Contd.*

| *Reference* | *Type of MHNFs* | *Magnetic field* | *Application* | *Outcomes* |
|---|---|---|---|---|
| (Bahiraei *et al.*, 2019) | Water–$Fe_3O_4$ Ferrofluid (Magnetic Nanofluid) | Magnetic field produced by four permanent magnets placed at different positions | Heat Transfer-Square Channel | The average CHT coefficient with four magnets was about 40.79% and 58.19% superior to zero magnetic field. |
| (Amani *et al.*, 2017) | $MnFe_2O_4$/water Nanofluid (3 vol%) | Regular magnetic field of 400 G intensity | Heat Transfer-Thermal Conductivity Study | The greatest thermal conductivity enhancement accomplished was 36.5% (at 20°C). |
| (Aruna and Niranjan, 2025) | $Fe_3O_4$–Graphene Oxide (GO)/SAE 10W-30 Engine Oil (Maxwell Hybrid Nanofluid) | Magnetic dipole effects (Magnetic parameter (M) is varied) | Automotive/Engine Lubrication and Cooling-Engine Oil | Magnetic dipole properties enhance heat transfer, minimize drag, and offer superior lubrication. Increasing the M enhances the temperature of the fluid. |
| (Akindele *et al.*, 2024) | $Al_2O_3 + Fe_3O_4 + CuO$ / Blood Tri-hybrid nanofluid | Inclined magnetic field | Biomedical Engineering-Targeted Thermal Therapies | Incorporation of $Fe_3O_4$ NPs shows a boost in the thermal conductivity of the medium and sensitivity to magnetic fields. |
| (Ahmad *et al.*, 2023) | Tri-hybrid nanofluids containing Ag | Localized magnetic field | Heat Transfer-Thermal Fluid Coupling | Achieved up to a 52% surge in Nu number |

of $Al_2O_3$-copper hybrid nanoparticles dispersed in water for electronic heat sink applications. In the same study, a 0.9°C reduction was observed in the temperature alteration between the fluid inlet and interface temperatures, accompanied by an average increase of 24.35% in the CHT coefficient.

Precise temperature control is vital for maintaining optimal operational conditions and extending the lifetime of sensitive electronic parts. The use of magnetically controllable nanofluids, which can tune their thermal conductivity and viscosity, holds exciting potential for advanced microfluidic devices. Especially micro/nanoelectromechanical systems (MEMS/NEMS), and other miniaturized technologies where dynamic thermal management is critical (Shima *et al.*, 2009).

## 5.2 Battery Thermal Management

Efficient thermal management is paramount for the safety, performance, and endurance of batteries, especially in electric vehicles (EVs) (Rahman *et al.*, 2025b). MHNFs present a compelling solution for this application by offering superior heat dissipation and lubrication properties (Ullah *et al.*, 2024). For example, SAE 10W-30 synthetic blended engine oil enhanced with $Fe_3O_4$ and graphene oxide (GO) nanoparticles under magnetic dipole effects has shown significant improvements in cooling and lubricating engines. This formulation acts as a superior lubricant and coolant. It stabilizes engine performance in demanding conditions like cold start-ups and high RPM use by facilitating urgent heat dissipation in critical areas such as the cylinder liner and the combustion chamber (Aruna and Niranjan, 2025). In automotive cooling applications, Dhairiyasamy *et al.* (2022) used $SiC/Al_2O_3$ hybrid nanofluids at 0.1–0.2% volume concentration, demonstrating an increase of 20% in the CHT coefficient without a substantial drop in the pressure. These advancements contribute to more proficient radiator geometries, quicker heat dissipation, and lowered risks of engine burning. Consequently, this results in improving vehicle safety and performance within the automotive industry.

## 5.3 Solar Thermal Collectors

Solar thermal collectors benefit immensely from advanced fluid absorption and heat transfer characteristics. This can be provided by utilizing MHNFs. By leveraging their enhanced thermal conductivity and the ability to be directed by external magnetic fields. These fluids can substantially boost the efficiency of solar energy systems (Qamar *et al.*, 2025; Samal and Ontela, 2025). A schematic view is shown in Figure 8. Various nanofluid compositions, such as $Fe_3O_4$/water, $Fe_3O_4$/TiN, MWCNT-$SiO_2$/Ag, CuO/Ag, photonic, and plasmonic nanoshell-based nanofluids, have been explored for solar thermal conversion and steam generation (Mei *et al.*, 2019).

Specifically, $Fe_3O_4$/Graphene–water MHNFs demonstrated a maximum thermal conductivity enhancement of approximately 51% (0.91 W/mK) at 5 wt% under a 250 G parallel magnetic field (Alsangur *et al.*, 2024). This substantial boost

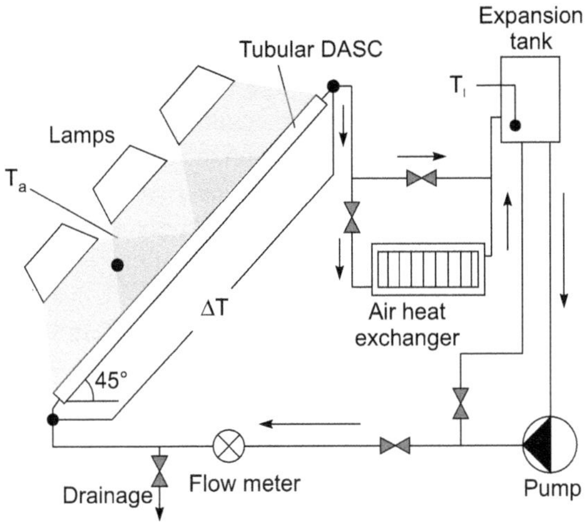


**Fig. 8** Schematic for experimental setup of direct absorption solar collector (DASC), with a circulation of MHNFs (Struchalin *et al.*, 2022).

in thermal conductivity, especially in the existence of a magnetic field, is crucial for improving thermal absorption. Moreover, it assists in overall energy capture in DASC and photovoltaic/thermal (PV/T) systems by advocating sustainable energy techniques (Tsegaye Moltot *et al.*, 2025).

## 6. Computational and Modeling Approaches

Computational and modeling approaches are indispensable tools for predicting and optimizing the complex thermal and hydrodynamic behaviors of MHNFs. Principally, under the impact of magnetic fields. These numerical studies convert complex governing partial differential equations into manageable ordinary differential equations. This can be achieved by using techniques like similarity transformations, allowing researchers to accurately simulate fluid characteristics across various industrial and biomedical contexts. Such approaches offer crucial insights where experimental validation may be difficult or resource-intensive.

### *6.1 CFD and Multiphase Modeling*

Computational Fluid Dynamics (CFD) is widely adopted for simulating the intricate flow structures and heat transfer mechanisms in MHNFs. It often employs commercial software packages such as COMSOL Multiphysics and various numerical methods, like the control volume method (Bahiraei *et al.*, 2019). Many numerical studies simplify the fluid as a homogeneous single-phase fluid (Alsarraf *et al.*, 2019), but more advanced approaches utilize multiphase models to capture the intricate interaction between nanoparticles and the base fluid. This is particularly important when particles are subjected to magnetic forces. For instance, one numerical study employed the two-phase fusion simulation to explore the hydrothermal features of water–$Fe_3O_4$ ferrofluid streaming in a square channel in the presence of permanent magnets. The simulation determined that using a configuration with four permanent magnets resulted in the average CHT coefficient being approximately 40.79%. It is higher than the case with the ferrofluid but no magnet, and 58.19% higher than pure water (Bahiraei *et al.*, 2019). Furthermore, COMSOL Multiphysics was utilized to simulate the distribution of a constant varying magnetic field generated by three permanent magnet pairs for MWCNT–$Fe_3O_4$/water HNFs flow. Revealing a maximum magnetic flux density peaking at the tube centerline under a 1000 Gauss magnetic field (Alsarraf *et al.*, 2019). CFD is also fundamental for investigating localized effects, with simulations showing that magnetic sources intensify fluid circulation and enhance heat transfer within double pipe heat exchangers (Azizi *et al.*, 2023).

### 6.2 Coupled Thermal-Magnetic Field Simulations

Simulating MHNFs requires a comprehensive coupling of thermal dynamics, fluid mechanics, and electromagnetic effects, often categorized under Magnetohydrodynamics (MHD). These models incorporate key thermal parameters such as Joule heating, viscous dissipation, and temperature-dependent properties. The presence of a magnetic field, particularly when employed vertically to the flow, is modeled via the Lorentz force, which acts as a resisting force, decreasing fluid velocity (Qamar *et al.*, 2025; Abbas *et al.*, 2025). For example, in the numerical analysis of Maxwell ternary nanofluid (Cu-$Al_2O_3$-$TiO_2$/Water) flow over a progressing needle, it is found that an increase in the magnetic field parameter (M) resulted in a reduction in fluid velocity. Whereas, a corresponding increase in temperature and concentration profiles demonstrates the coupled effects (Abbas *et al.*, 2025). Furthermore, when studying the unsteady flow of tangent hyperbolic ternary hybrid nanofluid, the magnetic field is often assumed to be time-dependent, such as $B_0 = B/(1\text{-}ct)^{1/2}$. The mathematical formulation for this model explicitly observes that this unsteady magnetic field, based on Maxwell's equations, can bring an electric field. This further influences charge transport and alters the nanofluid dynamics in MHD applications (Tsegaye Moltot *et al.*, 2025). Numerical results on hybrid nanofluid flow further confirm that the magnetic field strength ($\gamma$) and ferrohydrodynamic parameter ($\beta$) exhibit a positive outcome on the temperature turf but a negative effect on the velocity distribution (Gul *et al.*, 2020).

### 6.3 Case Studies

Computational studies consistently utilize extensive parametric analysis to determine the optimal conditions and material combinations for MHNFs. A common numerical technique employed is the BVP4C or BVP5C solver built into MATLAB, used to solve the ODEs derived from the governing flow equations (Abbas *et al.*, 2025; Qamar *et al.*, 2025; Tsegaye Moltot *et al.*, 2025; Aruna and

Niranjan, 2025). One significant comparative case study involved the unstable MHD flow of tangent hyperbolic ternary hybrid nanofluid ($Al_2O_3$-Cu-$TiO_2$/$C_2H_6O_2$) over a stretching sheet. The BVP5C method was used to demonstrate that increasing the volume percentage of the combined NPs (to $\Phi 1 = \Phi 2 = \Phi 3 = 0.15$) results in a significant increase in the Nu number (1.404), local skin friction coefficient (4.042), and Sherwood number (1.044). This confirms the superior heat transfer capabilities of the ternary hybrid nanofluid (Tsegaye Moltot *et al.*, 2025). Another powerful modeling approach combines numerical methods with Artificial Neural Networks (ANN). For a Maxwell hybrid nanofluid ($Fe_3O_4$-GO/SAE 10W-30 oil) flow in a sloped cylinder, data generated by the bvp4c solver was used to train an ANN model (Levenberg-Marquardt algorithm). Which accurately predicted performance metrics, achieving a high coefficient of determination, $R = 0.9986$ (Aruna and Niranjan, 2025). Furthermore, simulations employing the Keller Box method for Casson hybrid nanofluids (Ag-$TiO_2$/Water) discovered that increasing "M" results in a growth in the velocity profile (Mohd Rijal *et al.*, 2023). Comparative modeling also established that the thermal efficiency of the HNFs ($Al_2O_3$–Cu/$C_2H_6O_2$) is better than that of the single nanofluid ($Al_2O_3$/$C_2H_6O_2$) (Qamar *et al.*, 2025).

## 7. Challenges and Future Perspectives

The journey toward widespread adoption of MHNFs necessitates addressing several technical and economic hurdles while capitalizing on their remarkable potential for dynamic thermal control. Although computational models confirm the superior thermal performance of these advanced fluids. Whereas, practical realization requires rigorous focus on stability, energy economics, scalable manufacturing, and clear commercialization pathways. Furthermore, future research directions are crucial for transforming theoretical efficacy into reliable industrial utility.

### 7.1 *Stability*

A critical barrier facing the application of MHNFs is ensuring their long-term stability and preventing particle aggregation and subsequent sedimentation under actual operating conditions. The complex hybrid nature of these fluids means they may contain NPs with both positively and negatively charged surface characteristics, complicating the equilibrium process (Kalsi *et al.*, 2025). Numerical models often rely on the hypothesis of homogeneous nanoparticle distribution, but this examines potential clusters and sedimentation drifts that can occur in reality. Thereby limiting the precision of the findings (Tsegaye Moltot *et al.*, 2025). Experimental evidence confirms these issues, particularly at higher concentrations; for instance, $Fe_3O_4$-$TiO_2$ nanofluids at 0.1% volume concentration exhibited substantial settling with a sedimentation factor (SF) of 344.44% over 30 days (Adogbeji *et al.*, 2025). Conversely, lower concentrations of $Fe_3O_4$-MgO MHNFs showed greater stability (SF around 8.0–9.0%). Even highly stable fluids exhibit some settling; rGO-$Fe_3O_4$

hybrid nanofluid showed a sedimentation of 5.2% after 60 days (Mehrali *et al.*, 2017). Therefore, future studies must focus on optimizing nanofluid preparations to address viscosity and stability. Moreover, it is important to explore long-term stability in varied ecological conditions, containing extreme temperatures and exposure to several chemicals, to ensure commercial reliability.

### 7.2 Energy Utilization

The utilization of a magnetic field can significantly enhance the heat transfer efficiency of the MHNFs system. This regularly comes with an energy trade-off in terms of increased flow resistance, which elevates the required pumping power. The magnetic field, especially when parallel to the flow movement, causes nanoparticles to aggregate into chain-like structures that augment friction between the aggregated particles and the base fluid, increasing flow resistance. Experimental data on $Fe_3O_4$/water nanofluids in a rotary tube explained that increasing the parallel magnetic induction intensity (up to 300 G) could enhance the resistance coefficient by up to 11.5% in turbulent flow (at 5.0% mass fraction). This resulted in a pressure drop increase of 12.5% (Mei *et al.*, 2019). Numerical investigations of MWCNT-$Fe_3O_4$/water HNFs subjected to a 1000 Gauss magnetic field revealed that pressure loss increased by up to 25.02% compared to the no-field case (Alsarraf *et al.*, 2019). This higher flow resistance ultimately leads to a decrease in exergy efficiency and a decline in the comprehensive evaluation index of the system. However, this trade-off can sometimes be minimized. Research on magnetite nanofluids found that at low volume concentrations (up to 0.0057 vol%), a 50% thermal conductivity enhancement was observed without a notable increase in viscosity (Shima *et al.*, 2009). This suggests optimization pathways for highly efficient cooling applications.

### 7.3 Adaptable Preparation Methods

The transition of MHNFs from lab-scale experimentation to large-scale industrial production remains a significant hurdle. Recent manufacturing methods often require complex equipment and high-energy processes. which makes them costly and challenging to implement commercially. Furthermore, the lack of standardized synthesis protocols results in inconsistent findings across different production methods, even when the nanofluid composition and concentration are the same. Two primary preparation techniques are utilized, such as physical and chemical synthesis. The chemical synthesis technique is often preferred due to its simplicity and cost-effectiveness (Kalsi *et al.*, 2025). For the future, fostering new and accessible synthesis methods, such as one-pot and green synthesis approaches, will be decisive for making MHNF manufacture more cost-profitable and realistic for large-scale adoption. Achieving precise control over nanoparticle shape is also essential, as the shape affects sedimentation, agglomeration, and overall thermal properties (Mohd Rijal *et al.*, 2023).

### 7.4 Potential Commercialization

MHNFs possess tremendous potential for adoption in various high-demand thermal management sectors, including solar energy systems, heat exchangers, electronic components, biomedical fields, and automotive systems. Their unique ability to be dynamically controlled via magnetic fields makes them promising for adaptive cooling in micro/nanoelectromechanical systems (MEMS/NEMS) and targeted biomedical applications (Tsegaye Moltot *et al.*, 2025; Qamar *et al.*, 2025; Abbas *et al.*, 2025). Some MHNF types show high commercial viability, such as the Maxwell hybrid nano coolant ($Fe_3O_4$-GO/SAE 10W-30 oil), which is deemed a valuable alternative for the automotive industry as it enhances lubrication and cooling, prolonging engine life (Aruna and Niranjan, 2025). The magnetic tunability of the key performance ratio of magnetite nanofluids, ranging from 0.725 to 2.35, provides a pathway for manufacturing smart fluids capable. This can be advantageous for switching roles between an efficient coolant and a damper (Shima *et al.*, 2009). Despite these benefits, commercialization faces significant challenges. These include the high cost of nanoparticle synthesis, potential environmental and health concerns. Moreover, there is an absence of regulatory frameworks and standardized testing methods. While challenges still exist for integrating fluids with new properties into existing systems (Kalsi *et al.*, 2025). To facilitate widespread adoption, comprehensive economic analyses and robust cost models for large-scale production are necessary to confirm that the performance gains outweigh the production costs.

## 8. Conclusion

Magnetically modulated hybrid nanofluids (MHNFs) demonstrate superior thermal performance and increased heat transfer efficiency in association with traditional nanofluids. This recent advance makes them particularly relevant for a large range of applications across several sectors. In addition to enhancing thermal enhancement strategies, MHNFs also strengthen the adaptability of thermal management systems.

This chapter gives an in-depth overview of the latest innovations in MHNFs, offering insights for further investigation. The case studies analyzed show that the thermal conductivity and CHT coefficient of MHNFs are influenced by both the nanoparticle concentration and the applied magnetic field strength. However, these effects can also impact the pressure drop and fluid viscosity, with consequent impacts on energy requirements and pumping efficiency.

The use of MHNFs thus proves advantageous for improving thermal performance, although it still presents challenges related to stability, energy balance, and potential commercialization. These critical issues represent both an opportunity and a stimulus for future research developments.

## Acknowledgement

This research is funded under the National Recovery and Resilience Plan (NRRP), Mission 4 Component 2 Investment 1.3, grant No. PE0000021 – "Network 4 Energy Sustainable Transition-NEST", Spoke 6, ENERTERM – UniCampania – Funded by European Union –Next Generation EU, CUP E13C22001890001.

## References

[1] Abbas, A., Kiran, L., Ghachem, K., Abdennaji, T.S., Alshammari, B.M., Kolsi, L., Khan, I. and Khan, M.S. 2025. Effects of variable heat rise/fall on MHD Maxwell ternary nanofluid (Copper-Alumina-Titanium Dioxide/Water) flow over a moving needle. *Scientific Reports,* 15, 25209.

[2] Adogbeji, V.O. 2015. Experimental Investigation of Magnetic Hybrid Nanofluids for enhancing heat transfer in forced convection internal flow through transition and turbulent regimes. *Transition,* 62, 5000.

[3] Adogbeji, V.O., Sharifpur, M. and Meyer, J.P. 2024. Experimental investigation of heat transfer enhancement, thermal efficiency, and pressure drop in forced convection of magnetic hybrid nanofluid ($Fe_3O_4/TiO_2$) under varied magnetic field strengths and waveforms. *Case Studies in Thermal Engineering,* 63, 105313.

[4] Adogbeji, V.O., Sharifpur, M. and Meyer, J.P. 2025. Experimental investigation of heat transfer, thermal efficiency, pressure drop, and flow characteristics of Fe3O4-MgO magnetic hybrid nanofluid in transitional flow regimes. *International Journal of Thermal Sciences,* 209, 109515.

[5] Adun, H., Kavaz, D., Dagbasi, M., Umar, H. and Wole-Osho, I. 2021. An experimental investigation of thermal conductivity and dynamic viscosity of Al2O3-ZnO-Fe3O4 ternary hybrid nanofluid and development of machine learning model. *Powder Technology,* 394, 1121–1140.

[6] Afrand, M., Toghraie, D. and Sina, N. 2016. Experimental study on thermal conductivity of water-based Fe3O4 nanofluid: Development of a new correlation and modeled by artificial neural network. *International Communications in Heat and Mass Transfer,* 75, 262–269.

[7] Ahmad, S., Ali, K., Castellanos, H.G., Aryanfar, Y., Rashid, F.L., Hendy, A.S., Deifalla, A., Ragab, A.E., Khan, M. and Gomaa, H.G. 2023. Complex dynamics of induced vortex formation and thermal-fluid coupling in tri-hybrid nanofluid under localized magnetic field: a novel study. *Scientific Reports,* 13, 21140.

[8] Akindele, A., Obalalu, A., Oladapo, O., Ohaegbue, A., Prasad, B. and Bajaj, M. Thermal Performance of Ternary Hybrid Nanofluid Flow over an Inclined Magnetic Surface with Solar Radiative Heat Flux. E3S Web of Conferences, 2024. EDP Sciences, 05007.

[9] Alqawasmi, K., Alharbi, K.A.M., Farooq, U., Noreen, S., Imran, M., Akgül, A., Kanan, M. and Asad, J. 2023. Numerical approach toward ternary hybrid nanofluid flow with nonlinear heat source-sink and fourier heat flux model passing through a disk. *International Journal of Thermofluids,* 18, 100367.

[10] Alsangur, R., Doganay, S., Ates, İ., Turgut, A., Cetin, L. and Rebay, M. 2024. Magnetic field dependent thermal conductivity investigation of water based Fe3O4/CNT and Fe3O4/graphene magnetic hybrid nanofluids using a Helmholtz coil system setup. *Diamond and Related Materials,* 141, 110716.

[11] Alsarraf, J., Rahmani, R., Shahsavar, A., Afrand, M., Wongwises, S. and Tran, M.D. 2019. Effect of magnetic field on laminar forced convective heat transfer of MWCNT–Fe3O4/water hybrid nanofluid in a heated tube. *Journal of Thermal Analysis and Calorimetry,* 137, 1809–1825.

[12] Amani, M., Amani, P., Kasaeian, A., Mahian, O. and Wongwises, S. 2017. Thermal conductivity measurement of spinel-type ferrite MnFe2O4 nanofluids in the presence of a uniform magnetic field. *Journal of Molecular Liquids,* 230, 121-128.

[13] Anwar, M.S., Khan, M., Hussain, Z., Muhammad, T. and Puneeth, V. 2025. Investigation of heat transfer characteristics in MHD hybrid nanofluids with variable viscosity and thermal radiations. *Journal of Radiation Research and Applied Sciences,* 18, 101240.

[14] Aruna, J. and Niranjan, H. 2025. An artificial neural network approach to comparative aspects: A predictive analysis of magnetic dipole on the heat transfer of maxwell hybrid nano coolants flow in an inclined cylinder. *Case Studies in Thermal Engineering,* 68, 105825.

[15] Azizi, Y., Bahramkhoo, M. and Kazemi, A. 2023. Influence of non-uniform magnetic field on the thermal efficiency hydrodynamic characteristics of nanofluid in double pipe heat exchanger. *Scientific Reports,* 13, 407.

[16] Bahiraei, M., Hangi, M. and Rahbari, A. 2019. A two-phase simulation of convective heat transfer characteristics of water–Fe3O4 ferrofluid in a square channel under the effect of permanent magnet. *Applied Thermal Engineering,* 147, 991–997.

[17] Bakthavatchalam, B., Habib, K., Saidur, R., Saha, B.B. and Irshad, K. 2020. Comprehensive study on nanofluid and ionanofluid for heat transfer enhancement: A review on current and future perspective. *Journal of Molecular Liquids,* 305, 112787.

[18] Chakraborty, S. and Panigrahi, P.K. 2020. Stability of nanofluid: A review. *Applied Thermal Engineering,* 174, 115259.

[19] Chien, Y.-C. and Weng, H.C. 2018. Magnetic Nanofluid Droplet Impact on an AAO Surface with a Magnetic Field. *Applied Sciences,* 8, 1059.

[20] Choi, S.U. and Eastman, J.A. 1995. Enhancing thermal conductivity of fluids with nanoparticles. Argonne National Lab.(ANL), Argonne, IL (United States).

[21] Choudhary, P., Choudhary, S., Jat, K., Loganathan, K. and Eswaramoorthi, S. 2024. Impacts of unsteady MHD hybrid nanofluid over a non-linear stretchable porous sheet with thermal radiation and gyrotatic microorganisms. *International Journal of Thermofluids,* 23, 100788.

[22] Dhairiyasamy, R., Ahmed, M.H., Abdelrhman, Y., Afzal, A., Essa, F.A., Alsehli, M., Aly, A.A. and Saleh, B. 2022. Experimental investigation of thermophysical properties and heat transfer characteristics of hybrid nanofluids based on particle size. *Thermal Science,* 26, 4193-4206.

[23] Felicia, L. J. and Philip, J. 2014. Magnetorheological properties of a magnetic nanofluid with dispersed carbon nanotubes. *Physical Review E,* 89, 022310.

[24] Giwa, S.O., Sharifpur, M., Ahmadi, M.H., Sohel Murshed, S.M. and Meyer, J.P. 2021. Experimental Investigation on Stability, Viscosity, and Electrical Conductivity of Water-Based Hybrid Nanofluid of MWCNT-Fe2O3. *Nanomaterials,* 11, 136.

[25] Giwa, S.O., Sharifpur, M. and Meyer, J.P. 2020. Experimental study of thermo-convection performance of hybrid nanofluids of Al2O3-MWCNT/water in a differentially heated square cavity. *International Journal of Heat and Mass Transfer,* 148, 119072.

[26] Gul, T., Khan, A., Bilal, M., Alreshidi, N.A., Mukhtar, S., Shah, Z. and Kumam, P. 2020. Magnetic Dipole Impact on the Hybrid Nanofluid Flow over an Extending Surface. *Scientific Reports,* 10, 8474.

[27] Hatami, M., Mohammadi-Rezaei, S., Tahari, M. and Jing, D. 2018. Recent developments in magneto-hydrodynamic Fe3O4 nanofluids for different molecular applications: A review study. *Journal of Molecular Liquids,* 250, 244-258.

[28] Heidari, H., Mohebbi, R. and Kazemi, A. 2021. Forced convection heat transfer of Ag–MgO/water micropolar hybrid nanofluid inside a stairway channel. *International Journal of Modern Physics C,* 32, 2150147.

[29] Hussain, S.M., Imtiaz, M., Bibi, K., Rehman, S., Jamshed, W., Eid, M.R. and El Din, S.M. 2023. Error analysis of zirconium and zinc oxides/kerosene oil-based hybrid nanofluid flow between rotating disks: An innovative case study. *Case Studies in Thermal Engineering,* 51, 103549.

[30] Jat, K., Sharma, K. and Makinde, O.D. 2024. Novel features of radiating hybrid nanofluid flow past a nonlinear stretchable porous sheet with different nanoparticles shape. *Numerical Heat Transfer, Part B: Fundamentals,* 1-25.

[31] Kalsi, S., Kumar, S., Kumar, A., Alam, T. and Dobrotă, D. 2023. Thermophysical properties of nanofluids and their potential applications in heat transfer enhancement: A review. *Arabian Journal of Chemistry,* 16, 105272.

[32] Kalsi, S., Kumar, S., Kumar, A., Alam, T., Sharma, A. and Yadav, A.S. 2025. A review on hybrid nanofluids for heat transfer: advancements, synthesis, challenges and applications. *Discover Applied Sciences,* 7, 698.

[33] Karimi, A., Goharkhah, M., Ashjaee, M. and Shafii, M.B. 2015. Thermal Conductivity of Fe2O3 and Fe3O4 Magnetic Nanofluids Under the Influence of Magnetic Field. *International Journal of Thermophysics,* 36, 2720-2739.

[34] Khaleduzzaman, S.S., Rahman, S., Selvaraj, J., Mahbubul, I.M., Sohel, M.R. and Shahrul, I.M. 2014. Nanofluids for Thermal Performance Improvement in Cooling of Electronic Device. *Advanced Materials Research,* 832, 218-223.

[35] Kiani, M., Omiddezyani, S., Nejad, A.M., Ashjaee, M. and Houshfar, E. 2021. Novel hybrid thermal management for Li-ion batteries with nanofluid cooling in the presence of alternating magnetic field: An experimental study. *Case Studies in Thermal Engineering,* 28, 101539.

[36] Lee, A., Veerakumar, C. and Cho, H. 2021. Effect of Magnetic Field on the Forced Convective Heat Transfer of Water–Ethylene Glycol-Based Fe3O4 and Fe3O4–MWCNT Nanofluids. *Applied Sciences,* 11, 4683.

[37] Li, G., Wang, J., Zheng, H., Xie, G. and Sundén, B. 2022. Improvement of cooling performance of hybrid nanofluids in a heated pipe applying annular magnets. *Journal of Thermal Analysis and Calorimetry,* 147, 4731-4749.

[38] Mehrali, M., Sadeghinezhad, E., Akhiani, A.R., Tahan Latibari, S., Metselaar, H.S.C., Kherbeet, A.S. and Mehrali, M. 2017. Heat transfer and entropy generation analysis of hybrid graphene/Fe3O4 ferro-nanofluid flow under the influence of a magnetic field. *Powder Technology,* 308, 149-157.

[39] Mei, S., Qi, C., Liu, M., Fan, F. and Liang, L. 2019. Effects of paralleled magnetic field on thermo-hydraulic performances of Fe3O4-water nanofluids in a circular tube. *International Journal of Heat and Mass Transfer,* 134, 707-721.

[40] Mohd Rijal, I., Siti Shuhada, I., Noorehan, A., Norsyasya Zahirah Mohd, Z., Roselah, O. and Fazillah, B. 2023. Impact of Nanoparticles Shapes on Magnetohydrodynamic Flow and Heat Transfer of Casson Hybrid Nanofluids over a Moving Inclined Plate. *Journal of Advanced Research in Fluid Mechanics and Thermal Sciences,* 110, 79-96.

[41] Nkurikiyimfura, I., Yanmin, W., Zhidong, P. and Dawei, H. Enhancement of thermal conductivity of magnetic nanofluids in magnetic field. 2011 International Conference on Materials for Renewable Energy & Environment, 20-22 May 2011 2011. 1333-1337.

[42] Osman, S., Sharifpur, M. and Meyer, J.P. 2019. Experimental investigation of convection heat transfer in the transition flow regime of aluminium oxide-water nanofluids in a rectangular channel. *International Journal of Heat and Mass Transfer,* 133, 895-902.

[43] Philip, J., Shima, P.D. and Raj, B. 2007. Enhancement of thermal conductivity in magnetite based nanofluid due to chainlike structures. *Applied Physics Letters,* 91.

[44] Qamar, M., Khan, M., Yasir, M., Alqahtani, A.S. and Malik, M.Y. 2025. Heat transport features of MHD hybrid nanofluids beyond variable fluid characteristics. *Results in Engineering,* 26, 104987.

[45] Rahman, I.U., Nardini, S., Buonomo, B., Manca, O., Khan, H. and Siviero, B. 2025a. Thermal interface materials: A promising solution for passive heat dissipation in electronic appliances. *Thermal Science and Engineering Progress,* 62, 103673.

[46] Rahman, I.U., Nardini, S., Buonomo, B., Mohammed, H.J., Khan, H., Rehman, S.U. and Manca, O. 2025b. Graphene-Enhanced Composite Phase Change Materials for Battery Thermal Management.

[47] Sajid, M.U. and Ali, H.M. 2018. Thermal conductivity of hybrid nanofluids: A critical review. *International Journal of Heat and Mass Transfer,* 126, 211–234.

[48] Samal, S. and Ontela, S. 2025. Shape-driven heat transfer analysis in non-linear radiative casson bioconvective hybrid nanofluid flow in squeezing channel with inclined MHD. *Discover Molecules,* 2, 15.

[49] Selvakumar, P. and Suresh, S. 2012. Use of Al2O3–Cu /Water Hybrid Nanofluid in an Electronic Heat Sink. *IEEE Transactions on Components, Packaging and Manufacturing Technology,* 2, 1600–1607.

[50] Shahsavar, A. and Bahiraei, M. 2017. Experimental investigation and modeling of thermal conductivity and viscosity for non-Newtonian hybrid nanofluid containing coated CNT/Fe3O4 nanoparticles. *Powder Technology,* 318, 441–450.

[51] Shahsavar, A., Saghafian, M., Salimpour, M.R. and Shafii, M.B. 2016. Experimental investigation on laminar forced convective heat transfer of ferrofluid loaded with carbon nanotubes under constant and alternating magnetic fields. *Experimental Thermal and Fluid Science,* 76, 1–11.

[52] Shima, P.D., Philip, J. and Raj, B. 2009. Magnetically controllable nanofluid with tunable thermal conductivity and viscosity. *Applied Physics Letters,* 95.

[53] Struchalin, P.G., Kuzmenkov, D.M., Yunin, V.S., Wang, X., He, Y. and Balakin, B.V. 2022. Hybrid Nanofluid in a Direct Absorption Solar Collector: Magnetite vs. Carbon Nanotubes Compete for Thermal Performance. *Energies,* 15, 1604.

[54] Stuyven, B., Chen, Q., Moortel, W.V.D., Lipkens, H., Caerts, B., Aerts, A., Giebeler, L., Eerdenbrugh, B.V., Augustijns, P., Mooter, G.V.D., Humbeeck, J.V., Vanacken, J., Moshchalkov, V.V., Vermant, J. and Martens, J.A. 2009. Magnetic field assisted nanoparticle dispersion. *Chemical Communications,* 9, 47–49.

[55] Sun, L., Geng, J., Dong, K. and Sun, Q. 2024. The Applications and Challenges of Nanofluids as Coolants in Data Centers: A Review. *Energies,* 17, 3151.

[56] Suseel Jai Krishnan, S., Momin, M., Nwaokocha, C., Sharifpur, M. and Meyer, J.P. 2022. An empirical study on the persuasive particle size effects over the multi-physical properties of monophase MWCNT-Al2O3 hybridized nanofluids. *Journal of Molecular Liquids,* 361, 119668.

[57] Tekir, M., Taskesen, E., Gedik, E., Arslan, K. and Aksu, B. 2022. Effect of constant magnetic field on Fe3O4-Cu/water hybrid nanofluid flow in a circular pipe. *Heat and Mass Transfer,* 58, 707–717.

[58] Tsegaye Moltot, A., Haile Gorfie, E., Awgichew Zergaw, G. and Dessie, H. 2025. Unsteady MHD flow of tangent hyperbolic ternary hybrid nanofluid in a darcy-forchheimer porous medium over a permeable stretching sheet with variable thermal conductivity [version 2; peer review: 3 approved]. *F1000Research,* 14.

[59] Ullah, S., J. Algahtani, O. and Ali, A. 2024. Thermal Analysis of MHD-Modified Hybrid Nanofluid Flow Inside Convergent/Divergent Channel With Heat Generation/Absorption and Viscous-Ohmic Dissipation. *International Journal of Energy Research,* 2024, 3912044.

[60] Wright, B., Thomas, D., Hong, H., Groven, L., Puszynski, J., Duke, E., Ye, X. and Jin, S. 2007. Magnetic field enhanced thermal conductivity in heat transfer nanofluids containing Ni coated single wall carbon nanotubes. *Applied Physics Letters,* 91.

[61] Yu, W., Xie, H., Chen, L. and Li, Y. 2010. Enhancement of thermal conductivity of kerosene-based Fe3O4 nanofluids prepared via phase-transfer method. *Colloids and Surfaces A: Physicochemical and Engineering Aspects,* 355, 109–113.

# Index

# About the Editors

**Dr. Nikhil S. Mane** is an Assistant Professor at ADYPU, Pune, with a PhD in Mechanical Engineering from BITS Pilani, specializing in nanofluids, thermal management and energy systems. His research spans nanofluid synthesis, characterization studies for thermal applications, and the exploration of advanced cooling techniques. His work focuses on heat transfer enhancement in heat pipes using nanofluids and the role of nano additives in Phase Change Materials for electric vehicle battery cooling applications. With a strong track record of publications in prestigious peer-reviewed journals and international conferences, he continues to contribute to cutting-edge developments in thermal engineering.

**Dr. Zafar Said** is a Distinguished Associate Professor in the Department of Mechanical and Aerospace Engineering at United Arab Emirates University. His research focuses on renewable and solar energy systems, artificial intelligence in energy, and advanced thermal sciences, contributing to SDGs 7, 9, 11, and 13. (clean energy, innovation, sustainable cities, and climate action). He has published more than 350 peer-reviewed publications in top journals such as Progress in Energy and Combustion Science, Physics Reports, Advanced Energy Materials, Journal of Materials Chemistry A, Nano Energy, and Renewable and Sustainable Energy Reviews, alongside books, book chapters, conference papers, and patents, including five granted US patents and two Indian patents, with several works among the most highly cited. Dr. Said is ranked consistently among the World's Top 2% Scientists by Stanford/Elsevier in energy and has received numerous international awards, including the Khalifa Award for Education – Distinguished University Professor (2025), Best Academic Research at the Dubai Award for Sustainable Transport (2024), and multiple Research & Innovation Awards from the UAE Ministry of Energy and Infrastructure. He is a sought-after plenary and keynote speaker at leading global conferences.

**Dr. Sayantan Mukherjee** is an Associate Professor the Department of Mechanical Engineering, Gandhi Academy of Technology and Engineering, Orissa, India. He was a Post-Doctoral Research Fellow at the Institute for Plasma Research, a research institute under the Department of Atomic Energy, India, and has a PhD in Thermal Engineering from KIIT Deemed to be University. His research expertise encompasses nanofluids, heat transfer enhancement, boiling phenomena, energy analysis, and microchannel cooling. He has authored 55 publications in prestigious journals, book chapters, and conference proceedings with an H-index of 14. Dr. Mukherjee is actively engaged in collaborative research focused on applying nanofluids in microchannel cooling, boiling heat transfer, droplet evaporation, and direct solar collectors.

**Dr. Naser Ali** is a Research Scientist at the Kuwait Institute for Scientific Research (KISR) and Head of the Association of Scientific Researchers. His research spans nanofluids, nanocoatings, heat transfer, dispersion stability, thermophysical characterization, and surface technology. He has authored 66 publications in prestigious journals, book chapters, and conference proceedings with an H-index of 21. He holds 25 US granted patents and has received several distinguished awards, including the Jaber Al-Ahmad 2024 Prize for Young Researchers in the Scientific Subfields of Applied Sciences (Engineering Sciences) and the Kuwait Science Club Grand Award at the 2024 International Invention Fair in the Middle East (IIFME). He currently leads several funded research projects as Principal Investigator and is actively involved in various national and international collaborations.

For Product Safety Concerns and Information please contact our EU representative GPSR@taylorandfrancis.com Taylor & Francis Verlag GmbH, Kaufingerstraße 24, 80331 München, Germany